UNDERSTANDING
SEA-LEVEL RISE
AND VARIABILITY

In Memoriam: M.B. Dyurgerov

The Editors and Authors of this volume wish to honor the memory of Dr Mark B. Dyurgerov and acknowledge his valuable contributions to it. He will be missed by the glaciological and sea-level communities as an honest broker and an excellent scientist.

UNDERSTANDING SEA-LEVEL RISE AND VARIABILITY

EDITED BY
JOHN A. CHURCH
CENTRE FOR AUSTRALIAN WEATHER AND CLIMATE RESEARCH, A PARTNERSHIP BETWEEN CSIRO AND THE BUREAU OF METEOROLOGY, HOBART, AUSTRALIA

PHILIP L. WOODWORTH
PROUDMAN OCEANOGRAPHIC LABORATORY, LIVERPOOL, UK

THORKILD AARUP
INTERGOVERNMENTAL OCEANOGRAPHIC COMMISSION, UNESCO, PARIS, FRANCE

AND

W. STANLEY WILSON
NOAA SATELLITE & INFORMATION SERVICE, SILVER SPRING, MARYLAND, USA

WILEY-BLACKWELL
A John Wiley & Sons, Ltd., Publication

Blackwell Publishing was acquired by John Wiley & Sons in February 2007. Blackwell's publishing program has been merged with Wiley's global Scientific, Technical and Medical business to form Wiley-Blackwell.

Registered office: John Wiley & Sons Ltd, The Atrium, Southern Gate, Chichester, West Sussex, PO19 8SQ, UK

Editorial offices: 9600 Garsington Road, Oxford, OX4 2DQ, UK
The Atrium, Southern Gate, Chichester, West Sussex, PO19 8SQ, UK
111 River Street, Hoboken, NJ 07030-5774, USA

For details of our global editorial offices, for customer services and for information about how to apply for permission to reuse the copyright material in this book please see our website at www.wiley.com/wiley-blackwell

Library of Congress Cataloguing-in-Publication Data

Understanding sea-level rise and variability / edited by John A. Church ... [et al.].
p. cm.
Includes bibliographical references and index.
ISBN 978-1-4443-3451-7 (hardcover : alk. paper) – ISBN 978-1-4443-3452-4 (pbk. : alk. paper)
1. Sea level. I. Church, John, 1951-

GC89.U53 2010
551.45′8–dc22

2010012130

ISBN: 978-1-4443-3452-4 (paperback); 978-1-4443-3451-7 (hardback)

A catalogue record for this book is available from the British Library.

Set in 10 on 12.5 pt Minion by Toppan Best-set Premedia Limited
Printed in Singapore by Markono Print Media Pte Ltd

1 2010

Contents

Editor Biographies

John A. Church, FTSE

John Church is an oceanographer with the Centre for Australian Weather and Climate Research and the Antarctic Climate and Ecosystems Cooperative Research Centre. He was co-convening lead author for the chapter on sea level in the IPCC Third Assessment Report. He was awarded the 2006 Roger Revelle Medal by the Intergovernmental Oceanographic Commission, a CSIRO Medal for Research Achievement in 2006, and the 2007 Eureka Prize for Scientific Research.

Philip L. Woodworth

Philip Woodworth works at the Proudman Oceanographic Laboratory in Liverpool. He is a former Director of the Permanent Service for Mean Sea Level (PSMSL) and Chairman of Global Sea Level Observing System (GLOSS). He has been a lead or contributing author for each of the IPCC Research Assessments. He was awarded the Denny Medal of IMAREST in 2009 for innovation in sea-level technology and the Vening Meinesz Medal of the European Geosciences Union in 2010 for work in geodesy.

Thorkild Aarup

Thorkild Aarup is Senior Program Specialist with the Intergovernmental Oceanographic Commission of UNESCO and serves as technical secretary for the Global Sea Level Observing System (GLOSS) program. He has a PhD in oceanography from the University of Copenhagen.

W. Stanley Wilson

Stan Wilson has managed programs during his career, first at the Office of Naval Research where he led the Navy's basic research program in physical oceanography, then at NASA Headquarters where he established the Oceanography from Space program, and finally at NOAA where he helped organize the 20-country coalition in support of the Argo Program of profiling floats. Currently the Senior Scientist for NOAA's Satellite & Information Service, he is helping transition Jason satellite altimetry from research into a capability to be sustained by the operational agencies NOAA and EUMETSAT.

Contributors

T. Aarup, Intergovernmental Oceanographic Commission, UNESCO, Paris, France (t.aarup@unesco.org)

W. Abdalati, Earth Science & Observation Center, CIRES and Department of Geography, University of Colorado, Boulder, CO, USA (waleed.abdalati@colorado.edu)

D. Alsdorf, School of Earth Sciences, The Ohio State University, Columbus, OH, USA (alsdorf@geology.ohio-state.edu)

Z. Altamimi, Institut Géographique National, Champs-sur-Marne, France (altamimi@ensg.ign.fr)

F. Antonioli, Department of Environment, Global Change and Sustainable Development, Ente per le Nuove Tecnologie, l'Energia e l'Ambiente, Rome, Italy (fabrizio.antonioli@enea.it)

M. Anzidei, Istituto Nazionale di Geofisica e Vulcanologia, Rome, Italy (marco.anzidei@ingv.it)

J. Benveniste, ESRIN, European Space Agency, Frascatti, Italy (Jerome.Benveniste@esa.int)

N.B. Bernier, Department of Oceanography, Dalhousie University, Halifax, Canada (natacha.bernier@phys.ocean.dal.ca)

G. Blewitt, Nevada Bureau of Mines and Geology, University of Nevada, Reno, NV, USA (gblewitt@unr.edu)

H. Bonekamp, European Organisation for the Exploitation of Meteorological Satellites, Darmstadt, Germany (Hans.Bonekamp@eumetsat.int)

A. Cazenave, Laboratoire d'Etudes en Géophysique et Océanographie, Toulouse, France (anny.cazenave@cnes.fr)

D.P. Chambers, College of Marine Science, University of South Florida, St. Petersburg, FL, USA (chambers@marine.usf.edu)

J.A. Church, Centre for Australian Weather and Climate Research, A Partnership between CSIRO and BoM, and the Antarctic Climate and Ecosystems Cooperative Research Centre, Hobart, Australia (John.Church@csiro.au)

J.G. Cogley, Department of Geography, Trent University, Peterborough, Ontario, Canada (gcogley@trentu.ca)

J. Davis, Harvard-Smithsonian Center for Astrophysics, Cambridge, MA, USA (jdavis@cfa.harvard.edu)

C.M. Domingues, Centre for Australian Weather and Climate Research, A Partnership between CSIRO and BoM, Melbourne, Australia (Catia.Domingues@csiro.au)

M.R. Drinkwater, European Space Agency, ESTEC, The Netherlands (mark.drinkwater@esa.int)

M.B. Dyurgerov, INSTAAR, University of Colorado, Boulder, CO, USA (deceased)

J.S. Famiglietti, University of California, Irvine, CA, USA (jfamigli@uci.edu)

L.-L. Fu, Jet Propulsion Laboratory, Pasadena, CA, USA (llf@jpl.nasa.gov)

W.R. Gehrels, School of Geography, University of Plymouth, Plymouth, UK (w.r.gehrels@plymouth.ac.uk)

J.E. Gilson, Scripps Institution of Oceanography, La Jolla, CA, USA (jgilson@ucsd.edu)

V. Gornitz, NASA/GISS and Columbia University, New York, NY, USA (vgornitz@giss.nasa.gov)

J.M. Gregory, NCAS-Climate, Department of Meteorology, University of Reading, UK and Met Office, Hadley Centre, UK (j.m.gregory@reading.ac.uk)

R. Gross, Jet Propulsion Laboratory, California Institute of Technology, Pasadena, CA, USA (richard.gross@jpl.nasa.gov)

S. Gulev, P.P. Shirshov Institute of Oceanology, Moscow, Russia (gul@sail.msk.ru)

B.J. Haines, Jet Propulsion Laboratory, California Institute of Technology, Pasadena, CA, USA (bruce.j.haines@jpl.nasa.gov)

E. Hanna, Department of Geography, University of Sheffield, Sheffield, UK (e.hanna@sheffield.ac.uk)

D.E. Harrison, Pacific Marine Environmental Laboratory, NOAA, Seattle, WA, USA (d.e.harrison@noaa.gov)

K.J. Horsburgh, Proudman Oceanographic Laboratory, Liverpool, UK (kevinh@pol.ac.uk)

J.R. Hunter, Antarctic Climate and Ecosystems Cooperative Research Centre, Hobart, Tasmania, Australia (john.hunter@utas.edu.au)

P. Huybrechts, Earth System Sciences and Department of Geography, Vrije Universiteit Brussel, Brussel, Belgium (phuybrec@vub.ac.be)

E.R. Ivins, Jet Propulsion Laboratory, California Institute of Technology, Pasadena, CA, USA (eri@fryxell.jpl.nasa.gov)

G.C. Johnson, Pacific Marine Environmental Laboratory, NOAA, Seattle, WA, USA (gregory.c.johnson@noaa.gov)

M. Johnson, formerly Climate Program Office, NOAA, Silver Spring, MD, USA (now retired; mjohnson.pe@gmail.com)

T. Knutson, Geophysical Fluid Dynamics Laboratory, NOAA, Princeton, NJ, USA (tom.knutson@noaa.gov)

A. Köhl, Institut für Meereskunde, University of Hamburg, Hamburg, Germany (armin.koehl@zmaw.de)

C.-Y. Kuo, National Cheng Kung University, Taiwan (kuo70@mail.ncku.edu.tw)

J. Laborel, Université de la Méditerranée Aix-Marseille II, Marseille, France (rutabaga1@wanadoo.fr)

J.L. LaBrecque, Earth Science Division, NASA, Washington DC, USA (john.labrecque@nasa.gov)

K. Lambeck, Research School of Earth Sciences, Australian National University, Canberra, Australia and Antarctic Climate and Ecosystems Cooperative Research Centre, Australia (kurt.lambeck@anu.edu.au)

F.W. Landerer, Max Planck Institute for Meteorology, Hamburg, Germany (now at Jet Propulsion Laboratory, Pasadena, CA, USA) (felix.w.landerer@jpl.nasa.gov)

K. Laval, Laboratoire de Météorologie Dynamique, Paris, France (laval@lmd.jussieu.fr)

F.G. Lemoine, NASA Goddard Space Flight Center, Greenbelt, MD, USA (frank.g.lemoine@nasa.gov)

P.-Y. Le Traon, Operational Oceanography, IFREMER, Centre de Brest, Brest, France (Pierre.yves.le.traon@ifremer.fr)

D.P. Lettenmaier, University of Washington, Seattle, WA, USA (dennisl@u.washington.edu)

E.J. Lindstrom, Earth Science Division, NASA, Washington DC, USA (eric.j.lindstrom@nasa.gov)

J.A. Lowe, The Hadley Centre, Met Office, UK (jason.lowe@metoffice.gov.uk)

B. MacKenzie, Institute of Marine Engineering, Science and Technology, London, UK (bev.mackenzie@imarest.org)

J. Marotzke, Max Planck Institute for Meteorology, Hamburg, Germany (jochem.marotzke@zmaw.de)

R.E. McDonald, The Hadley Centre, Met Office, UK (ruth.mcdonald@metoffice.gov.uk)

K.L. McInnes, CSIRO, Aspendale, Australia (kathleen.mcinnes@csiro.au)

M.A. Merrifield, Department of Oceanography, University of Hawai'i, Honolulu, Hawai'i, HI, USA (markm@soest.hawaii.edu)

L. Miller, NOAA Laboratory for Satellite Altimetry, Silver Spring, MD, USA (laury.miller@noaa.gov)

P.C.D. Milly, US Geological Survey, Princeton, NJ, USA (cmilly@usgs.gov)

G.A. Milne, Department of Earth Sciences, University of Ottawa, Ontario, Canada (gamilne@uottawa.ca)

G.T. Mitchum, College of Marine Sciences, University of South Florida, St. Petersburg, FL, USA (mitchum@marine.usf.edu)

J.X. Mitrovica, Department of Earth and Planetary Sciences, Harvard University, Cambridge, MA, USA (jxm@eps.harvard.edu)

A.W. Moore, Jet Propulsion Laboratory, California Institute of Technology, Pasadena, CA, USA (angelyn.moore@jpl.nasa.gov)

R.E. Neilan, Jet Propulsion Laboratory, California Institute of Technology, Pasadena, CA, USA (ruth.neilan@jpl.nasa.gov)

R.S. Nerem, Department of Aerospace Engineering Sciences, University of Colorado, Boulder, CO, USA (nerem@colorado.edu)

R.J. Nicholls, School of Civil Engineering and the Environment, and the Tyndall Centre for Climate Change Research, University of Southampton, Southampton, UK (r.j.nicholls@soton.ac.uk)

E.C. Pavlis, University of Maryland and Space Geodesy Laboratory, NASA Goddard Space Flight Center, Greenbelt, MD, USA (epavlis@umbc.edu)

S. Piotrowicz, Climate Program Office, NOAA, Silver Spring, MD, USA (steve.piotrowicz@noaa.gov)

H.P. Plag, Nevada Bureau of Mines and Geology,University of Nevada, Reno, NV, USA (hpplag@unr.edu)

S.C.B. Raper, Department for Air Transport and the Environment, Manchester Metropolitan University, Manchester, UK (s.raper@mmu.ac.uk)

R. Rayner, Institute of Marine Engineering, Science and Technology, London, UK (ralph@ralphrayner.org)

E. Rignot, Centro de Estudios Cientificos, Valdivia, Chile; Jet Propulsion Laboratory, California Institute of Technology, Pasadena, CA, USA and University of California, Department of Earth System Science, Irvine, CA, USA (eric.rignot@jpl.nasa.gov)

D. Roemmich, Scripps Institution of Oceanography, La Jolla, CA, USA (droemmich@ucsd.edu)

M. Rothacher, GeoForschungsZentrum, Potsdam, Germany (markus.rothacher@ethz.ch)

D.L. Sahagian, Environmental Initiative, Lehigh University, Bethlehem, PA, USA (dork.sahagian@lehigh.edu)

T. Schöne, GeoForschungsZentrum, Potsdam, Germany (tschoene@gfz-potsdam.de)

C.K. Shum, School of Earth Sciences, The Ohio State University, Columbus, OH, USA (ckshum@osu.edu)

M.G. Sideris, Department of Geomatics Engineering, University of Calgary, Alberta, Canada (sideris@ucalgary.ca)

D. Stammer, University of Hamburg, Hamburg, Germany (detlef.stammer@zmaw.de)

K. Steffen, CIRES (Cooperative Institute for Research in Environmental Sciences), University of Colorado, Boulder, CO, USA (konrad.steffen@colorado.edu)

W. Sturges, Department of Oceanography, Florida State University, Tallahassee, FL, USA (sturges@ocean.fsu.edu)

T. Suzuki, Japan Agency for Marine-Earth Science and Technology, Yokohama, Japan (tsuzuki@jamstec.go.jp)

V. Swail, Environment Canada, Downsview, Canada (val.swail@ec.gc.ca)

M.E. Tamisiea, Proudman Oceanographic Laboratory, Liverpool, UK (mtam@pol.ac.uk)

R.H. Thomas, EG&G Services, NASA/GSFC/Wallops Flight Facility, Wallops Island, VA, USA (robert_thomas@hotmail.com)

E. Thouvenot, Strategy & Programmes Directorate, CNES, Toulouse, France (eric.thouvenot@cnes.fr)

P. Tregoning, The Australian National University, Canberra, Australia (paul.tregoning@anu.edu.au)

A.S. Unnikrishnan, National Institute of Oceanography, Goa, India (unni@nio.org)

L.L.A. Vermeersen, Delft Institute of Earth Observation & Space Systems (DEOS), Delft University of Technology, The Netherlands (l.l.a.vermeersen@tu.delft.nl)

H. von Storch, GKSS, Geesthacht, Germany (hvonstorch@web.de)

J.M. Wahr, University of Colorado, Boulder, CO, USA (john.wahr@colorado.edu)

R. Weisse, GKSS, Geesthacht, Germany (weisse@gkss.de)

N.J. White, Centre for Australian Weather and Climate Research, A Partnership between CSIRO and BoM, and the Antarctic Climate and Ecosystems Cooperative Research Centre, Hobart, Australia (Neil.White@csiro.au)

J.K. Willis, Jet Propulsion Laboratory, California Institute of Technology, Pasadena, CA, USA (joshua.k.willis@jpl.nasa.gov)

C.R. Wilson, University of Texas, Austin, TX, USA (crwilson@mail.utexas.edu)

W.S. Wilson, NOAA Satellite & Information Service, Silver Spring, MD, USA (stan.wilson@noaa.gov)

J. Wolf, Proudman Oceanographic Laboratory, Liverpool, UK (jaw@pol.ac.uk)

C.D. Woodroffe, School of Earth and Environmental Sciences, University of Wollongong, NSW, Australia (colin@uow.edu.au)

P.L. Woodworth, Proudman Oceanographic Laboratory, Liverpool, UK (plw@pol.ac.uk)

K. Woth, GKSS, Geesthacht, Germany (woth@gkss.de)

A.J. Wright, Faculty of Earth and Life Sciences, Department of Marine Biogeology, Vrije Universiteit, Amsterdam, The Netherlands (alex.wright@falw.vu.nl)

S. Zerbini, Department of Physics, University of Bologna, Italy (susanna.zerbini@unibo.it)

Foreword

Sea-level variability and change are manifestations of climate variability and change. The 20th-century rise and the recently observed increase in the rate of rise were important results highlighted in the Intergovernmental Panel on Climate Change (IPCC) Fourth Assessment Report completed in 2007.

In the last few years, there have been a number of major coastal flooding events in association with major storms such as Hurricane Katrina in 2005 and the Cyclones Sidr and Nargis in 2007 and 2008 respectively. The loss of life has been measured in hundreds of thousands and the damage to coastal infrastructure in billions of dollars. Such major coastal flooding events are likely to continue as sea level rises and have a greater impact as the population of the coastal zone increases.

The rate of coastal sea-level rise in the 21st century and its impacts on coasts and islands as expressed in the 2007 IPCC report contained major uncertainties. Incomplete understanding of the ocean thermal expansion, especially that of the deeper parts of the ocean, and uncertainties in the estimates of glacier mass balance and the stability of ice sheets are among the many factors which limit our ability to narrow projections of future sea-level rise. In particular, the instability of ice sheets requires special attention because it could lead potentially to a significant increase in the rate of sea-level rise over and above that of the 2007 IPCC report.

The World Climate Research Programme has led the development of the physical scientific basis that underpins the IPCC Assessments. On 6–9 June 2006 it organized a workshop in Paris, France, that brought together the world's specialists on the many aspects of the science of sea-level change to provide a robust assessment of our current understanding as well as the requirements for narrowing projections of future sea-level rise. The present book is based on the deliberations at the workshop and provides a comprehensive overview of present knowledge on the science of sea-level change.

The findings in this book will help set priorities for research and for observational activities over the next decade that will contribute to future assessments of the IPCC. In turn, the improvements in these assessments will better inform governments, industry, and society in their efforts to formulate sound mitigation and adaptation responses to rising greenhouse gas concentrations and sea level, and their economic and social consequences. In that respect, information on global and regional sea-level comprises an important product of a climate service. Its generation cuts across many disciplines and observation systems and requires effective coordination among many organizations.

Michel Jarraud
Secretary-General, World Meteorological Organization

Wendy Watson-Wright
Assistant Director-General, UNESCO
Executive Secretary, Intergovernmental Oceanographic
Commission of UNESCO

Deliang Chen
Executive Director, International Council for Science

Acknowledgments

The World Climate Research Programme, with the support of the Intergovernmental Oceanographic Commission of UNESCO, initiated the Sea-Level Workshop that led to this book. The completion of this book would not have been possible without the participation of attendees in the original workshop and their contributions to the various chapters, and of course without the help of the many sponsors and participating organizations listed below. We thank all of these people and organizations for their support. We would particularly like to express our appreciation to Emily Wallace (GRS Solutions) for her administrative and logistical support to the organizing committee prior to, during, and immediately following this workshop. We also thank Catherine Michaut (WCRP/COPES Support Unit, Université Pierre et Marie Curie) for administrative support and website development; as well as Pam Coghlan, Laurence Ferry, and Adrien Vannier (Intergovernmental Oceanographic Commission of UNESCO) for administrative logistical assistance prior to and during the workshop. We also thank Neil White, Lea Crosswell, Craig Macauley, Louise Bell, and Robert Smith for their efforts in the preparation of a number of the figures.

JAC acknowledges the support of the Australian Climate Change Science Program, the Wealth from Oceans Flagship, and the Australian Government's Cooperative Research Centres Program through the Antarctic Climate and Ecosystems Cooperative Research Centre. WSW acknowledges the financial support provided by the Research-to-Operations Congressional Earmark to NOAA.

John A. Church, Philip L. Woodworth,
Thorkild Aarup, and W. Stanley Wilson

Cosponsors

ACE CRC: Antarctic Climate and Ecosystems Cooperative Research Centre (Australia)
AGO: Australian Greenhouse Office (Australia)
BoM: Bureau of Meteorology (Australia)
CNES: Centre National d'Etudes Spatiales (France)
CNRS: Centre National de la Recherche Scientifique (France)

CSIRO: Commonwealth Scientific and Industrial Research Organization (Australia)
DFO: Department of Fisheries & Oceans (Canada)
EEA: European Environment Agency
ESA: European Space Agency
ESF-Marine Board: Marine Board of the European Science Foundation
EUMETSAT: European Organization for the Exploitation of Meteorological Satellites
EU: European Union
GEO: Group on Earth Observations
GKSS: GKSS Forschungszentrum (Germany)
IASC: International Arctic Science Committee
IAG: International Association of Geodesy
IAPSO: International Association for the Physical Sciences of the Oceans
IACMST: Interagency Committee on Marine Science and Technology (UK)
ICSU: International Council for Science
IFREMER: Institut Français de Recherche pour l'Exploitation de la Mer (France)
IGN: Institut Geographique National (France)
IOC of UNESCO: Intergovernmental Oceanographic Commission
IPY: International Polar Year
IRD: Institut de Recherche pour le Développement (France)
NASA: National Aeronautics and Space Administration (USA)
NSF: National Science Foundation (USA)
NOAA: National Oceanic and Atmospheric Administration (USA)
NERC: Natural Environment Research Council (UK)
Rijkswaterstaat (The Netherlands)
SCAR: Scientific Committee for Antarctic Research
TU Delft: Delft University of Technology (The Netherlands)
UKMO: The Met Office (UK)
UNESCO: United Nations Educational, Scientific and Cultural Organization
WCRP: World Climate Research Programme
WMO: World Meteorological Organization

Participating Organizations and Programs

Argo: International Argo Project
CryoSat: ESA's Ice Mission (ESA)
ENVISAT: Environmental Satellite (ESA)
ERS: European Remote Sensing satellite (ESA)
GCOS: Global Climate Observing System
GGOS: Global Geodetic Observing System
GLOSS: Global Sea-Level Observing System
GOCE: Gravity Field and Steady-State Ocean Circulation Explorer (ESA)

GOOS: Global Ocean Observing System
GRACE: Gravity Recovery and Climate Experiment (NASA)
ICESat: Ice, Cloud, and Land Elevation Satellite (NASA)
IGS: International GNSS Service
Jason: Ocean Surface Topography from Space (NASA/CNES)
SMOS: Soil Moisture and Ocean Salinity (ESA)

Abbreviations and Acronyms

AES40	North Atlantic wind and wave climatology developed at Oceanweather with support from Climate Research Branch of Environment Canada
ANU	Australian National University
AOGCM	atmosphere–ocean general circulation model
AR4	IPCC Fourth Assessment Report
BP	before present
CCM2	NCAR Community Climate Model version 2
cGPS	continuous GPS
CLASIC	Climate and Sea Level in parts of the Indian Subcontinent
CLIMBER	Climate and Biosphere model (of the Potsdam Institute for Climate)
CLIVAR	Climate Variability and Predictability project
CLM	Climate Version of the Local Model developed from the LM by the CLM Community (clm.gkss.de)
CNES	Centre National d'Etudes Spatiales (France)
CRF	celestial reference frame
CS3	POL barotropic model for the European Continental Shelf (1/9°×1/6° latitude by longitude or approximately 12 km resolution)
CSIRO	Commonwealth Scientific and Industrial Research Organisation (CSIRO); also to refer to the climate model developed by CSIRO
CSX	POL barotropic model for the European Continental Shelf (1/3°×1/2° latitude by longitude or approximately 35 km resolution)
CZMS	Coastal Zone Management Subgroup
DIVA model	Dynamic Interactive Vulnerability Assessment model

DORIS	Doppler Orbitography and Radiopositioning Integrated by Satellite
ECHAM3, ECHAM4, ECHAM5	atmosphere-only versions of the European Centre Hamburg climate model
ECHAM5-OM, ECHAM4/OPYC3, ECHAM5/MPI-OM1	alternative coupled models (atmosphere and ocean) versions of the European Centre Hamburg climate model
ECMWF	European Centre for Medium-Range Weather Forecasts
ENSO	El Niño Southern Oscillation
ENVISAT	Environmental Satellite (ESA)
EOF	empirical orthogonal function
EOP	Earth Orientation Parameters
ERA-40	reanalysis product provided by ECMWF (http://www.ecmwf.int/research/era/)
ERS-1, -2	European Remote Sensing satellites 1 and 2
ESA	European Space Agency
EUMETSAT	European Organisation for the Exploitation of Meteorological Satellites
GCM	general circulation model
GCN	GLOSS Core Network
GCOM2D	Global Coastal Ocean Model, depth-average version
GCOS	Global Climate Observing System
GEOSS	Global Earth Observation System of Systems
GFDL	Geophysical Fluid Dynamics Laboratory (of the National Oceanic and Atmospheric Administration)
GFO	GeoSat Follow-on Satellite
GGOS	Global Geodetic Observing System
GIA	glacial isostatic adjustment
GLIMS	Global Land Ice Measurements from Space
GLONASS	Global Orbiting Navigation Satellite System
GLOSS	Global Sea Level Observing System
GNSS	Global Navigation Satellite System
GOCE	Gravity Field and Steady-State Ocean Circulation Explorer
GODAE	Global Ocean Data Assimilation Experiment
GOOS	Global Ocean Observing System
GPS	Global Positioning System
GRACE	Gravity Recovery and Climate Experiment
HadAM3, HadAM3P, HadAM3H	variants of the Hadley Centre atmospheric climate model, version 3

HadCM2, HadCM3	versions of the Hadley Centre coupled climate model
HadRM2, HadRM3	versions of the Hadley Centre regional atmospheric climate model
IAG	International Association of Geodesy
ICESat	Ice, Cloud, and Land Elevation Satellite
IDS	International DORIS Service
IERS	International Earth Rotation and Reference Systems Service
IGFS	International Gravity Field Service
IGOS-P	Integrated Global Observing Strategy-Partnership
IGS	International GNSS Service
ILRS	International Laser Ranging Service
InSAR	interferometric synthetic aperture radar
IOC	Intergovernmental Oceanographic Commission
IPCC	Intergovernmental Panel on Climate Change
ISMASS	Ice Sheet Mass Balance and Sea Level project
ITRF	International Terrestrial Reference Frame
ITRS	International Terrestrial Reference System
IVS	International VLBI Service
JCOMM	WMO/IOC Joint Technical Commission for Oceanography and Marine Meteorology
JMA	Japan Meteorological Agency
JMA T106	JMA GCM with T106 spatial resolution (1.1°×1.1°)
ka	thousand years ago
KNMI	Royal Netherlands Meteorological Institute
LGM	Last Glacial Maximum
LSM	land-surface model
MEO	Medium Earth Orbit(er)
MIROC	Model for Interdisciplinary Research on Climate series of models
MIS	marine oxygen isotope stage
MLWS	mean low water springs
MWP	melt water pulse
NAO	North Atlantic Oscillation
NASA	National Aeronautics and Space Administration (USA)
NCAR	National Center for Atmospheric Research (USA)
NCEP	National Centers for Environmental Prediction (NOAA)

NOAA	National Oceanic and Atmospheric Administration (USA)
ODINAfrica	Ocean Data and Information Network for Africa
ORCHIDEE	French global land surface model
OSTM	Ocean Surface Topography Mission (radar altimeter mission)
PDI	power dissipation index
POL	Proudman Oceangraphic Laboratory (UK)
POLCOMS	POL Coastal-Ocean Modelling System (a three-dimensional model for shelf regions)
POM	Princeton Ocean Model
PRUDENCE	Prediction of Regional Scenarios and Uncertainties for Defining European Climate Change Risks and Effects (European Union-funded project)
PSMSL	Permanent Service for Mean Sea Level
RACMO	Regional Atmospheric Climate Model (KNMI)
RCAO	Rossby Centre Regional Atmosphere-Ocean model
REMO	Hamburg regional climate model
RLR	Revised Local Reference data set of the PSMSL
RSLR	relative sea-level rise
SAR	synthetic aperture radar
SLR	satellite laser ranging
SRALT	satellite radar altimetry
SRES	Special Report on Emissions Scenarios, and the scenarios therein
SST	sea-surface temperature
STOWASUS	Regional Storm, Wave and Surge Scenarios for the 2100 century
SWH	significant wave height
SWOT	Surface Water Ocean Topography (NASA)
TAR	IPCC Third Assessment Report
TE2100	Thames Estuary in 2100 project (of the UK Environment Agency)
TIGA-PP	Tide Gauge Benchmark Monitoring Pilot Project of the IGS
T/P	TOPEX/Poseidon radar altimeter satellite
TPW	true polar wander
TRF	terrestrial reference frame
TRIMGEO	Tidal Residual and Intertidal Mudflat Model

TRS	Terrestrial Reference System
UNESCO	United Nations Educational, Scientific and Cultural Organization
VLBI	very-long-baseline interferometry
WASA	Waves and Storms in the North Atlantic (European Union-funded project)
WCRP	World Climate Research Programme
WMO	World Meteorological Organization
WOCE	World Ocean Circulation Experiment
XBT	expendable bathythermograph

Introduction

Philip L. Woodworth, John A. Church, Thorkild Aarup, and W. Stanley Wilson

Millions of people are crowded along the coastal fringes of continents, attracted by rich fertile land, transport connections, port access, coastal and deep-sea fishing, and recreational opportunities. In addition, significant populations live on oceanic islands with elevations of only a few meters (Figure 1.1). Many of the world's megacities, cities with populations of many millions, are situated at the coast, and new coastal infrastructure developments worth billions of dollars are being undertaken in many countries. This coastal development has accelerated over the past 50 years (e.g. Figure 1.2), but it has taken place with an assumption that the stable sea levels of the past several millennia will continue; there has been little consideration of global sea-level rise.

Global sea-level rise and its resultant impact on the coastal zone, one of the consequences of global climate change, has been identified as one of the major challenges facing humankind in the 21st century. Impacts on the environment, the economy, and societies in the coastal zone will likely be large (e.g. Chapters 2 and 3 of this volume; Intergovernmental Panel on Climate Change (IPCC) Working Group 2 Report[1]; Stern Review of the Economics of Climate Change[2]; Millennium Ecosystem Assessment[3]). However, estimates of the timescales, magnitudes, and rates of future sea-level rise vary considerably, partly as a consequence of uncertainties in future emissions and the associated climate response, but also because of the lack of detailed understanding of the processes by which the many contributions to sea-level change will evolve in a future climate. The study of historical records of sea level and their proxies offers a means for understanding and quantifying the many uncertainties, as well as determining how a global monitoring system suitable for improved understanding of sea level change in the future might be established.

Minimizing future coastal impacts will require mitigation of greenhouse gas emissions, to avoid the most extreme scenarios of sea-level rise, and adaption to the rise that actually takes place. Optimal planning and policy decisions by

[1] http://www.ipcc-wg2.org/
[2] http://www.hm-treasury.gov.uk/independent_reviews/stern_review_economics_climate_change/sternreview_summary.cfm
[3] http://www.maweb.org

Understanding Sea-Level Rise and Variability, 1st edition. Edited by John A. Church, Philip L. Woodworth, Thorkild Aarup & W. Stanley Wilson.

Figure 1.1 Malé, the capital of the Maldive Islands. In common with most coral islands, the Maldives have elevations of only several meters (photo credit: Yann Arthus Bertrand/Earth from Above/UNESCO).

governments around the world as well as local-adaption decisions are currently constrained by our inadequate understanding of the response of sea level to increasing greenhouse gas concentrations. It is therefore imperative to identify and quantify the causes contributing to the presently observed sea-level change, in order that better models can be developed and more reliable predictions can be provided. Planners and decision-makers will need long-term forecasts of global sea-level rise, and also information on how short-term variability and long-term change of sea level will be expressed at regional and local scales.

In June 2006, a workshop was organized under the auspices of the World Climate Research Programme (WCRP) at the Intergovernmental Oceanographic Commission of United Nations Educational, Scientific and Cultural Organization (UNESCO) in Paris, with the aim of identifying the major uncertainties associated with sea-level rise and variability, as well as the research and observational activities needed for narrowing those uncertainties, thus laying the basis for improved projections of sea-level rise during the 21st century and beyond. It was sponsored by 34 organizations, and was attended by 163 scientists (from 29 countries) representing a wide range of expertise. The workshop also had the aim of obtaining consensus on sea-level observational requirements for the Global Earth Observation System of Systems (GEOSS) 10-Year Implementation Plan. Progress in major areas of research, their associated uncertainties and recommendations for future work were summarized in position papers circulated prior to and then discussed during the workshop. An interim report on research and observational priorities was prepared and is available[4] (see also Church et al. 2007). Subsequently, the position papers were revised, expanded, and peer-reviewed to constitute the chapters of this book. The chapters were then edited and assembled to provide a

[4] http://wcrp.ipsl.jussieu.fr/Workshops/SeaLevel/index.html

Figure 1.2 Increased coastal development on the Gold Coast (Queensland, Australia) from 1958 (a) to 2007 (b). Over this period the permanent population of the region increased by more than an order of magnitude from less than 40 000 in 1958 to over 480 000 in 2007 and with about 3.8 million visitors per year in 2008–9 (photo credit: Gold Coast City Council State Library).

coherent overview of the field of sea-level rise. Additional chapters were included to provide a review of future observational requirements and a synthesis of scientific findings.

This book is intended to complement the IPCC scientific assessments, by starting with the uncertainties the IPCC identified in sea-level rise and variability, and then focusing on the scientific and observational requirements needed to reduce those uncertainties. While the book provides consensus estimates of the present rate of global mean-sea-level rise, it does not provide new sea-level projections. In contrast, the IPCC Assessment does include projections, but does not provide the research and observational requirements needed to reduce uncertainties in those projections. Also, there are additional research and observational requirements relating to the impacts of sea-level rise which are beyond the scope of this book. These include information on the impact of waves on the coastline, including coastal inundation and erosion issues, information on local land motion and sediment budgets, information on the natural coastal environment, and the societal response to sea-level variability and rise.

Chapters 2 and 3 contain contributions on the topic of the impacts of sea-level rise. In many parts of the world, coastal developments and infrastructure are becoming increasingly vulnerable to sea-level rise and extremes (for example Figure 1.3), as Hurricanes Katrina, Sidr, and Nargis have demonstrated so clearly. The purpose of including these two chapters, which discuss representative major impacts, is to put the research described in subsequent chapters into context and to make it clear that it has great value, not only in a scientific context, but also to society in general.

The impacts of sea-level change occur at the local level and are a result of changes in *relative* sea level. These *relative* changes occur as a result of large-scale, basin-wide, and global-scale changes in sea level and land levels, as well as regional and local changes. These changes in sea and land levels compound the impacts from coastal storm surges and high wave conditions. Chapter 2 explains that local geological factors including those associated with seismic events, compaction of coastal sediments, and loss of coastal sediment supplies are important and need to be considered together with regional sea-level rise. Consideration of these effects requires impact studies on a variety of spatial scales.

Chapter 4 and the following chapters contain a detailed and systematic discussion of aspects of sea-level change. They have the aim of improving our understanding of the uncertainties associated with the various contributions to observed sea-level change, so that ultimately those uncertainties can be reduced and the observed sea-level rise be adequately explained, thereby removing what the eminent oceanographer Walter Munk (2002) has called the "enigma of 20th century sea level rise" (the inability to account for the observations).

It should be no surprise to anyone that global mean sea level is changing, and has always changed on a range of timescales. Chapter 4 discusses the evidence for changes on millennial and multicentury timescales with the use of geological and archaeological data, thereby providing a context for the modern-day observations summarized in Chapter 5.

Figure 1.3 Examples of coastal erosion. (a) Near Akpakpa/Cotonou (Benin) (photo credit: Adoté Blivi). (b) A beach house on the south shore of Nantucket Island off the northeast coast of the USA. This photograph was taken in 1995. While a number of recent storms had occurred, there had been no major events, and the collapse of the house was a result of long-term erosion (photo credit: Professor Stephen Leatherman, Florida International University). (c, d) Two views of Happisburgh on the east coast of England in 2007 (c) and 1998 (d). The dashed yellow lines are the locations of the top of the cliff in 1998 and 2007, respectively. This coast has soft low cliffs which are constantly eroded by waves off the North Sea (photo credit: http://www.Mike-Page.co.uk).

Figure 1.4 Taking a small core from a microatoll for radiocarbon dating. A height of 1.5 m above lowest astronomical tide was measured using the Global Positioning System (GPS), suggesting a higher sea level during the mid-Holocene in central Torres Strait, Australia (photo credit: Javier Leon, University of Wollongong).

These two chapters explain that a number of more precise observational techniques have been developed in recent years. Examples include new techniques for dating geological sea-level indicators (Figure 1.4) and methods for exploiting salt-marsh information to determine sea-level changes of the relatively recent past (within a few hundred years). Improvements in observations of sea level by tide gauges (Figure 1.5) include the use of advanced geodetic techniques (Global Positioning System (GPS), absolute gravimetry, and Doppler Orbitography and Radiopositioning Integrated by Satellite (DORIS)) to monitor vertical movements of the land on which they are located and thereby remove the effects of land-level changes in their records. Since 1993, tide-gauge data have been complemented by high-quality measurements from space by satellite altimetry (Figure 1.5) and space gravity (see Figure 1.10, below). Satellite altimetry has revolutionized our understanding of oceanography and sea-level rise. High-quality satellite-altimeter missions provide direct, near-global observations of the rate of sea-level rise and its temporal and spatial variability. Continuity between missions and careful and rigorous intercomparison of different missions and of satellite data with *in situ* observations is critical to gaining maximum benefit from this investment. Each observational technique, whether *in situ* or space-based, contains its temporal and spatial inadequacies which result in uncertainties in estimates of the rates of global sea-level change.

The main processes responsible for sea-level change, each of which are associated with climate variability and change, are internal changes in the ocean due to changes in the density of sea water, inputs to the ocean of additional water due to losses of ice from glaciers, ice caps, and ice sheets, modifications in the exchanges of water between ocean and land storage, and smaller modifications in the exchanges between ocean and atmosphere. The internal ocean changes are called steric changes. They result from modifications in the density of sea water throughout the water column and can be considered as a combination of thermosteric and halosteric change, due to changes in temperature and salinity respectively,

Figure 1.5 Tide gauges for measuring sea level from around the world. (a) A float and stilling well gauge at the Punta della Salute, Venice (photo credit: P.A. Pirazzoli). (b) An acoustic gauge at Kiribati, South Pacific (photo credit: National Tidal Centre, Australia). (c) The float gauge at Vernadsky, the site of the longest sea-level record in Antarctica (photo credit: British Antarctic Survey). (d) A radar tide-gauge installation at Liverpool, UK (photo credit: Proudman Oceanographic Laboratory). (e) The TOPEX/POSEIDON radar-altimeter satellite.

with the latter an order of magnitude smaller than the former in its importance to long-term, global sea-level rise. Steric changes are not globally uniform but have a spatial distribution which at many tropical and mid-latitude locations reflects changes in heat content.

The ocean temperature and salinity data sets from which steric changes are computed have their own temporal and spatial biases, with increasingly large gaps in coverage as one goes back in time, with relatively few measurements until recent years in the Southern Hemisphere. A different mix of observational methods (fixed moorings, research ships, ships of opportunity, profiling floats; Figure 1.6) have been used at different times, with consequent changes in sampling both in geographical position and vertically through the water column,

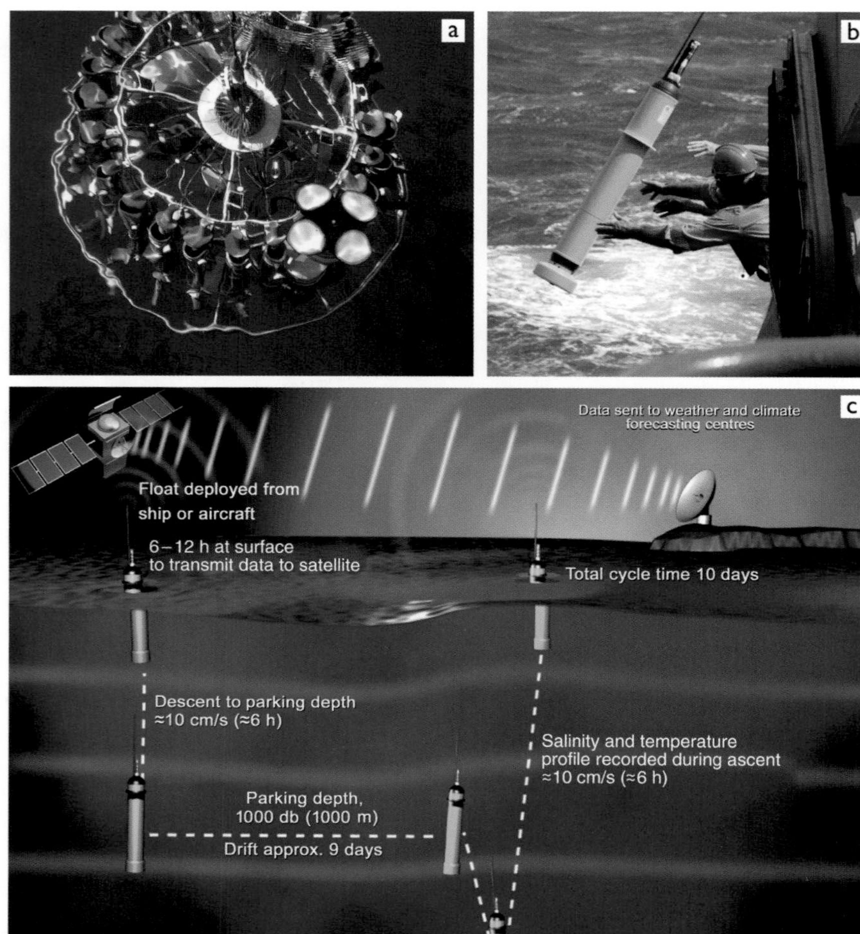

Figure 1.6 Techniques for measuring changes in ocean temperatures and salinities (and hence density). (a) Research ships can collect highly accurate temperatures and salinities using instruments lowered from ships at widely distributed locations; however, the measurements are sparse in space and time with the first comprehensive global observations completed as part of the World Ocean Circulation Experiment (WOCE; Siedler et al. 2001). (b) Deployment of an autonomous Argo float in the Southern Ocean. (c) These floats drift at depth for 10 days before profiling through the upper 2000 m of the water column, transmitting their data via satellite and then returning to drift at depth.

and with varying accuracies depending on the techniques employed. Chapter 6 discusses the uncertainties in the historical hydrographic data sets and in the numerical modeling of steric sea-level change, the progress made in recent years with the deployments of the Argo profiling float system (Figure 1.6), and requirements for monitoring of the ocean and for reliably determining steric sea-level change in the future.

Steric changes result in a change in the volume, but not the mass, of water in the ocean. On the other hand, melting (or negative mass-balance) of mountain glaciers and ice caps, and a melting or sliding of the great ice sheets of Antarctica and Greenland into the ocean, results in an increase in the mass of the ocean. It has been known for some time that mountain glaciers and ice caps contributed

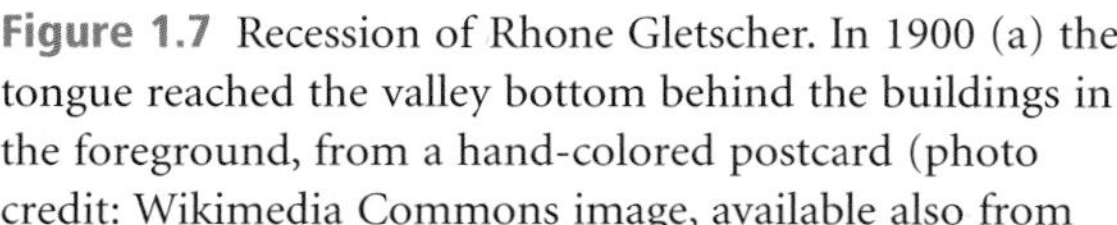

Figure 1.7 Recession of Rhone Gletscher. In 1900 (a) the tongue reached the valley bottom behind the buildings in the foreground, from a hand-colored postcard (photo credit: Wikimedia Commons image, available also from the United States Library of Congress Prints and Photographs Division). By 2008 (b) recession had already been so great that hardly any ice could be seen from this location (photo credit: http://www.swisseduc.ch/glaciers/).

to 20th century sea-level rise (Figure 1.7). However, as a number of papers have indicated, there has been an accelerating contribution over recent decades. The melting of glaciers and ice caps and ocean thermal expansion are responsible for the majority of the observed sea-level rise over recent decades. In spite of the enormous amount of ice stored in the Greenland and Antarctic Ice Sheets, equivalent to over 60 m sea-level change, the contributions of Greenland and Antarctica to 20th century sea-level change appear to have been smaller than that of glaciers and ice caps. However, recent observations indicate an enhanced, and possibly rapidly accelerating, contribution since the early 1990s. This is particularly true for the Greenland Ice Sheet (Figure 1.8), but there are also indications of an enhanced contribution from the West Antarctic Ice Sheet.

Chapter 7 discusses our current knowledge of changes in the cryosphere and indicates the basis of the main uncertainties. Measurements of the cryosphere are not straightforward. Only a small subset of glaciers worldwide are monitored regularly and with adequate precision, largely by the same *in situ* techniques that have been used since the 19th century. Altimetry and space gravity appear to offer ideal monitoring systems for the ice sheets. However, their time series are very short so far. An additional limitation is the need for adequate sampling in the coastal margins where narrow and swiftly flowing outlet glaciers transport ice to the ocean. These outlet glaciers are showing significant changes, most likely in response to warming in the adjacent oceans. Chapter 7 also reviews the many detailed requirements for measurements and for improved understanding via modeling of the dynamics of ice flows.

In addition to glaciers, ice caps, and ice sheets, water is stored on land in snow pack, surface water (lakes, dams, and rivers), and subsurface water (soil moisture,

Figure 1.8 Summer surface melting on the Greenland Ice Sheet and drainage into a crevasse called a moulin. Picture was taken north of Ilulissat on the Sermeq Avangnardleq outlet glacier (70°N, 500 m elevation) on the western slope of the Greenland Ice Sheet in August 2007 (photo credit: Koni Steffen).

ground water, and frozen ground or permafrost). Climate variability and change and the direct human intervention in regional hydrology, for example by dam building, irrigation schemes, and the mining of ground water, result in fluctuations in terrestrial water storage (Chapter 8). There has been significant scientific insight and progress in gathering the necessary hydrological information to attempt at least a partial understanding of the changes in the terrestrial water storages as a result of climate variations over the past few decades, particularly the last decade. However, estimates of the sea-level change as a consequence of direct human intervention contain many uncertainties, not only in magnitude, but even in sign. The two largest are the mining of ground water and the building of dams (Figure 1.9). These two terms are likely to at least partially offset each other over the 20th century, but probably have very different time histories.

Uncertainties in terrestrial water storage are among the largest of all the possible contributors to 20th century sea-level rise. Chapter 8 explains in detail why they occur and suggests that future monitoring systems, based especially on space gravity and a special space "water mission," will provide a real reduction in uncertainties.

Figure 1.9 Flooding behind the Three Gorges Dam, China (photo credit: International Space Station Earth Observations Experiment and Image Science & Analysis Laboratory, Johnson Space Center, National Aeronautics and Space Administration).

Geodetic techniques underpin much of our recent progress in *in situ* and space-based observations of sea-level change and factors contributing to that change. In fact, the new techniques have revolutionized the Earth sciences and are implicit in most of the discussion of measuring techniques referred to in every chapter of this book. The use of GPS in measuring vertical land movements at tide gauges referred to above provides just one example of the impact of new geodetic techniques on sea-level research. The progress is even more spectacular for space-based observations: now changes in ocean and ice-sheet volume are routinely made using satellite altimetry (e.g. Jason-1 for the ocean and the Ice, Cloud, and Land Elevation Satellite (ICESat) for the ice sheets and ice caps) and changes in their mass and the mass of water stored on land using time-varying space gravimetry measurements (e.g. Gravity Recovery and Climate Experiment (GRACE) satellite; Figure 1.10). The highest-resolution information on the Earth's gravitational field will also come from space-based observations (Gravity Field and Steady-State Ocean Circulation Explorer (GOCE) satellite; Figure 1.10).

Several of the techniques provide the basis for the fundamental reference frame, the International Terrestrial Reference Frame (ITRF), that is coordinated through the Global Geodetic Observing System (GGOS). Chapter 9 discusses the progress in development of a stable ITRF and the uncertainties in the measurements which depend upon it. Important recommendations from the workshop, also discussed in the chapter, are concerned with how one can enhance and sustain support for a robust and stable ITRF.

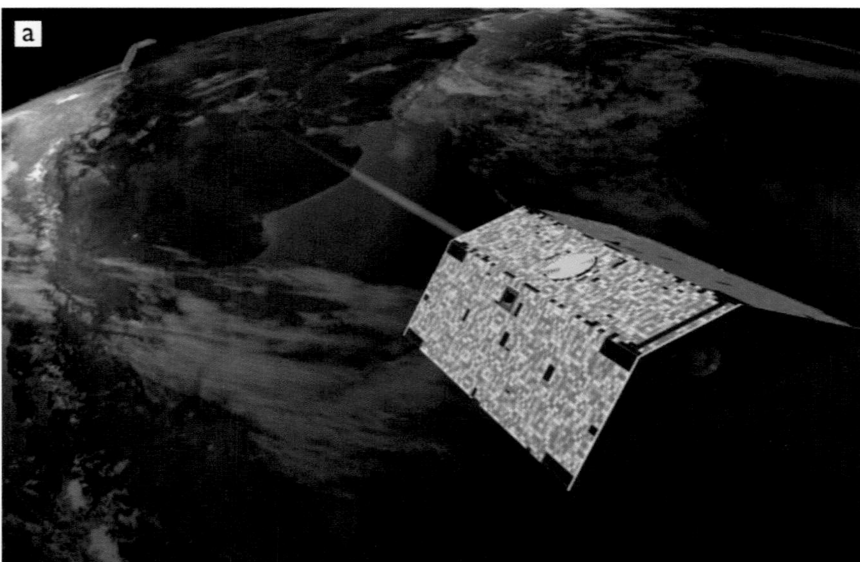

Figure 1.10 Space gravity missions provide precise measurements of the mass changes in the ocean, cryosphere, and hydrosphere (a; Gravity Recovery and Climate Experiment (GRACE) satellite) as well as accurate measurements of the Earth's gravity field (b; Gravity Field and Steady-State Ocean Circulation Explorer (GOCE) satellite). (GRACE image obtained from the mission home page http://www.csr.utexas.edu/grace/; GOCE image courtesy of European Space Agency/AOES-Medialab.)

Chapter 10 is concerned with how the solid Earth responds to changes in surface mass load. The Earth is still responding visco-elastically to removal of the great ice sheets of the last ice age, and the vertical land movements associated with such glacial isostatic adjustment (GIA) clearly manifest themselves in geological measurements (e.g. as raised beaches; Figure 1.11), tide-gauge records, and even modern satellite altimeter and space-gravity data. In addition, the changes in the present-day loads on the solid Earth (changes in ice sheets, glaciers, and terrestrial storage of water) have modified, and continue to modify, the shape of the gravitational field, leading to regional changes in sea level. The chapter discusses the uncertainties in models of past ice sheets and in the physics of GIA. These uncertainties limit the accuracy of the models to provide suitable corrections to tide-

Figure 1.11 Nineteenth-century boathouses on the island of Brämön, just south of Sundsvall on the Gulf of Bothnia coast of northern Sweden. The houses are now above present-day sea level as a result of vertical land motion due to GIA (photo credit: H-G. Scherneck, reprinted with permission from Lantmäteriet).

gauge and altimeter sea-level data required to remove GIA signals from those records. The chapter also investigates "fingerprint" analysis, the use of spatial variations in present-day rates of sea-level and gravity change to identify the various changing loads on the Earth. Fingerprints will also manifest themselves in the future, with any significant changes in the mass of the ice sheets during the 21st century resulting in spatially varying sea-level-rise contributions, with some vulnerable regions experiencing larger than global averaged sea-level rise and other regions experiencing less than the average rise.

This book is primarily concerned with the uncertainties in determining how global sea level has changed in the past and will change in the future. The uncertainties involved in a study of extreme sea levels are somewhat larger. However, knowledge of the character of extreme events is of great practical importance and provides the link between scientific insight, impacts, and policy and planning at the coast.

Impacts of rising sea levels will be felt most acutely through changes in the intensity and frequency of extreme events from the combined effects of high spring tides, storm surges, surface waves, and flooding rivers (Figure 1.12). Even if there are no changes in the climatology of surges and waves, an increase in mean sea level will result in more frequent flooding at a given level. Indeed, this change

Figure 1.12 Fire destroys homes on the beach as the storm surge from Hurricane Ike floods Galveston Island, Texas on September 12, 2008 (photo credit: David J. Phillip/AP/SIPA Press).

in frequency of flooding at a given level can be dramatic. For example, an analysis[5] of a century-long tide-gauge record in San Francisco Bay has indicated that flooding events (i.e. flooding to a given level) in the second half of the 20th century are 10 times more frequent than those same flooding events in the first half of the century. Thus, a "100-year" flooding event in the first half of the 20th century has become a "10-year" event in the second half of the 20th century. If uncertainties in predictions of one or more of these contributions to extreme sea levels are large, then design criteria for new coastal structures will be correspondingly inadequate, and those for existing structures will become increasingly out of date. Chapter 11 discusses the evidence for changes in extreme sea levels during the 20th century, especially how changes in extremes differ from those in mean sea level. It also covers past changes in storm surges, waves, and mid-latitude and tropical storms. The chapter discusses the limitations involved in modeling changes in storminess, and thereby the changes in storm surges, and points to the importance of considering interactions between the several contributions to an extreme sea level. The chapter also discusses results from detailed regional case studies.

An important message from the WCRP 2006 workshop was the concern in the community for continuity of the *in situ* and space-based observing systems which are relevant to studies of sea-level variability and rise. Continuous, long-term, high-quality, and stable measurements are absolutely critical to improving our understanding of past and present sea-level changes and to improving projections of future change. It was stressed that all systems must adhere to the Global Climate

[5] John Hunter, Antarctic Climate & Ecosystems Cooperative Research Centre (personal communication) performed this analysis applying the techniques described in Church et al. (2008). For a more complete analysis of this tide-gauge record in San Francisco, see Bromirski et al. (2003).

Observing System (GCOS) observing principles, which specify an open data policy and timely, unrestricted access to data products. Chapter 12 discusses both existing and new observing system technologies which are required to reduce sea-level uncertainties. It is important to note that all of the recommended systems are consistent with and complement requirements for monitoring as identified by study groups in related fields of research. While space systems are significantly more expensive than corresponding observations *in situ*, the costs are relatively small when compared with the costs associated with the impacts of sea-level rise and variability.

Each of the above chapters provides a review of its sub-sections of sea-level research. The final chapter, Chapter 13, provides a synthesis of findings: an overview of the progress and recommendations drawn from each chapter as well as a general guide to future research. While this book does not focus on new projections of sea-level rise, a role carried out by the IPCC, Chapter 13 does consider and compare various projections of sea-level rise, including a discussion of where the main uncertainties in such projections lie.

This book provides a current snapshot of sea-level research and the uncertainties associated with understanding of sea-level rise. While the IPCC conducts regular assessment and updates its projections of global sea-level rise, it does not lay out the necessary research agenda to improve understanding and reduce uncertainties. One anticipates that it will be necessary to repeat the WCRP 2006 workshop in some form, and produce a volume such as this together with its synthesis, at regular intervals in order to judge progress. This will also help ensure that the best possible and most recent information on sea-level rise is made available to the scientific community and, via processes such as the IPCC Scientific Assessments, to nations and the general public.

It is obvious that there are clear policy and planning implications from present and future rising sea levels. Much of society's past development has occurred in a period of relatively stable sea level. The world is now moving out of this period and future coastal planning and development must consider the inevitable increase of regional sea level. This will require a corresponding commitment to understanding sea-level rise and its implications through ongoing observations, research, modeling, and communication. This is a commitment only achievable through strengthened partnerships between the scientific, business, community, and government sectors.

References

Bromirski P.D., Flick R.E. and Cayan D.R. (2003) Storminess variability along the California coast: 1858–2000. *Journal of Climate* **16**, 982–93.

Church J., Wilson S., Woodworth P. and Aarup T. (2007) Understanding sea level rise and variability. Meeting report. *EOS Transactions of the American Geophysical Union* **88**(4), 43.

Church J.A., White N.J., Hunter J.R., McInnes K.L., Cowell P.J. and O'Farrell S.P. (2008) Sea-level rise. In: *Transitions, Pathways Towards Sustainable Urban Development in Australia* (P.W. Newton, ed.), pp. 191–209. CSIRO Publishing, Melbourne.

Munk W. (2002) Twentieth century sea level: an enigma. *Proceedings of the National Academy of Sciences USA* **99**, 6550–5.

Siedler G., Church J. and Gould J. (eds) (2001) *Ocean Circulation and Climate: Observing and Modelling the Global Ocean*. Academic Press, London.

2 Impacts of and Responses to Sea-Level Rise

Robert J. Nicholls

2.1 Introduction

Sea-level rise has been seen as a major threat to low-lying coastal areas around the globe since the issue of human-induced global warming emerged in the 1980s (e.g. Barth and Titus 1984; Milliman et al. 1989; Warrick et al. 1993). What is often less appreciated is that more than 200 million people are already vulnerable to flooding by extreme sea levels around the globe (Hoozemans et al. 1993; Mimura 2000). This population could grow to 800 million by the 2080s just due to rising population, including coastward migration (Nicholls 2004). These people generally depend on natural and/or artificial flood defenses and drainage to manage the risks, with the most developed and extensive artificial systems in Europe (especially around the southern North Sea) and East Asia. Most threatened are the significant populations (at least 20 million people today) already living below normal high tides in a range of countries such as Belgium, Canada, China, Germany, Italy, Japan, the Netherlands, Poland, Thailand, the UK, and the USA. Hurricane Katrina's impacts on New Orleans in 2005 remind us of what happens if those defenses fail. Increasing mean sea level and potentially more intense storms will exacerbate these risks. Despite these threats, the actual consequences of sea-level rise remain uncertain and contested. This reflects far more than just uncertainty in the magnitude of sea-level rise and climate change, with the success or failure of our ability to adapt to these challenges being a major uncertainty (Nicholls and Tol 2006).

This chapter focuses on understanding the threat of sea-level rise and its implications for climate science and policy, as well as coastal management. This includes consideration of the impacts of rising sea level on coastal areas, as well as the responses that can be implemented:

- mitigation: reducing greenhouse gas emissions and increasing sinks, and hence minimizing climate change, including sea-level rise, via climate policy; and/or
- adaptation: reducing the impacts of sea-level rise via behavioral changes including individual actions through to collective coastal management policy.

Understanding Sea-Level Rise and Variability, 1st edition. Edited by John A. Church, Philip L. Woodworth, Thorkild Aarup & W. Stanley Wilson.

The chapter is structured as follows. First climate change and sea-level rise are considered, including the distinction between global-mean and relative sea-level rise. Then the impacts of sea-level rise are considered from a physical and a socio-economic perspective. This leads to a discussion of methods and frameworks for considering the impacts of sea-level rise. Observed impacts from the 20th century and projected impacts from the 21st century are then considered, followed by a discussion of responses to the challenges of sea-level rise. The chapter then considers next steps in terms of research/application and concludes.

2.2 Climate Change and Global/Relative Sea-Level Rise

Human-induced climate change is expected to cause a profound series of changes including rising sea level, rising sea-surface temperatures, and changing storm, wave, and runoff characteristics (Figure 2.1). Although higher sea level only directly impacts coastal areas, these are the most densely populated and economically active land areas on Earth, which concentrates infrastructure such as ports and harbors, industry (e.g. oil refineries), and power stations, as well as an extensive built environment (Sachs et al. 2001; McGranahan et al. 2007). They also support important and productive ecosystems that are sensitive to sea level and other change (Kremer et al. 2004; Crossland et al. 2005). Rising global sea level is likely to accelerate through the 21st century. From 1990 to the last decade of the 21st century, a total rise in the range 18–59 cm has been projected by the Intergovernmental Panel on Climate Change (IPCC) Fourth Assessment Report (AR4) (Meehl et al. 2007). Including an allowance for an increased ice-sheet discharge, the projected range is 18–76 cm. However, even this scenario excludes uncertainties due to collapse of the large ice sheets, and as noted in the IPCC *Synthesis Report* (IPCC 2007), the quantitative AR4 scenarios do *not* provide an upper bound on sea-level rise during the 21st century. Thus, a global rise of sea level exceeding 1 m remains a low probability but physically plausible scenario for the 21st century, particularly because of uncertainties concerning ice-sheet dynamics and their response to global warming (Chapter 7). While these high-end scenarios may be relatively unlikely, their large potential impacts make them

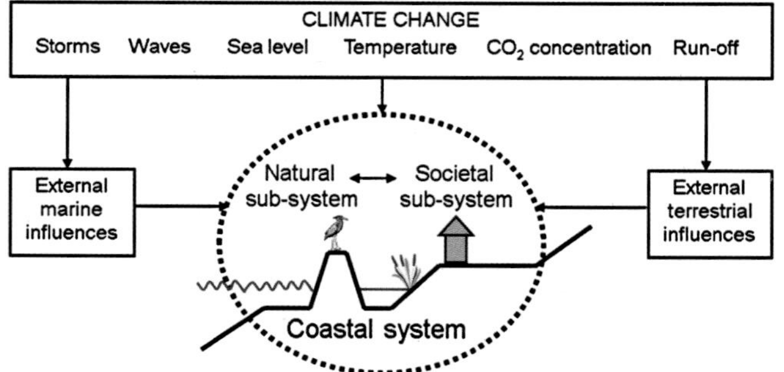

Figure 2.1 Climate change and the coastal system showing the major climate change factors, including external marine and terrestrial influences. The natural environment and coastal inhabitants interact directly, and are affected by external terrestrial and marine issues. Climate change and sea-level rise can directly or indirectly affect the coastal system (taken from Nicholls et al. 2007a).

highly significant in terms of climate risk (Stern 2006; Keller et al. 2008). It is worth noting that the current sea-level observations are at the high end of the projected range (Rahmstorf et al. 2007). There is also increasing concern about higher extreme sea levels due to more intense storms superimposed on these mean rises, especially for areas affected by tropical storms (Chapter 11). This would exacerbate the impacts of global sea-level rise, particularly the risk of more damaging floods and storms.

When analyzing sea-level rise impacts and responses, it is fundamental that impacts are a product of *relative* (or local) sea-level rise rather than global changes alone. Relative sea-level change takes into account the sum of global, regional, and local components of sea-level change: the underlying drivers of these components are (1) climate change such as melting of land-based ice (Chapter 7), thermal expansion of ocean waters and changing ocean dynamics (Chapter 6), and (2) non-climate uplift/subsidence processes such as tectonics, glacial isostatic adjustment (GIA; Chapters 9 and 10), and natural and anthropogenic-induced subsidence (Emery and Aubrey 1991). Hence relative sea-level rise (RSLR) is a response to both climate change and other factors and varies from place to place, with a few places experiencing relative sea-level fall, as illustrated by the measurements in Figure 2.2. Much of the world's coasts are experiencing a slow RSLR (see Sydney; Figure 2.2). Abrupt changes due to earthquakes occur at some sites (see Nezugaseki; Figure 2.2). Relative sea level is presently falling due to ongoing GIA (rebound) in some high-latitude locations that were formerly sites of large (kilometer-thick) glaciers, such as the northern Baltic and Hudson Bay (Chapter 10) (see Helsinki; Figure 2.2). In contrast, RSLR is more rapid than global-mean trends on subsiding coasts, including deltas such as the Mississippi Delta (see Grand Isle; Figure 2.2), the Nile Delta, and the large deltas of South and East Asia (Ericson et al. 2006; Woodroffe et al. 2006; Syvitski 2008) (Figure 2.3). Most dramatically, human-induced subsidence of susceptible areas due to drainage and

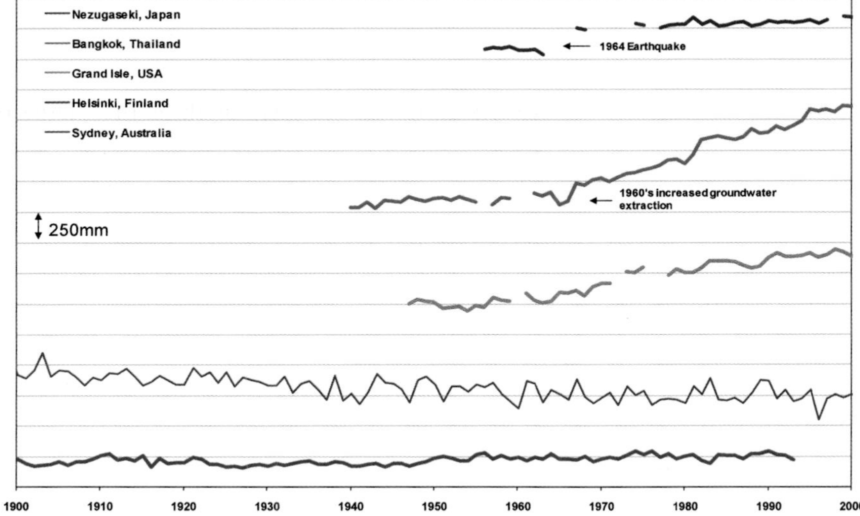

Figure 2.2 Selected relative sea-level records for the 20th century, illustrating different regions (offset for display purposes). Helsinki shows a falling trend (−2.5 mm/year), Sydney shows a gradual rise (0.5 mm/year), Grand Isle is on a subsiding delta (6.9 mm/year), Bangkok includes the effects of human-induced subsidence (17.5 mm/year since 1960), and Nezugaseki shows an abrupt rise due to an earthquake. Source: http://www.pol.ac.uk/psmsl/.

(a)

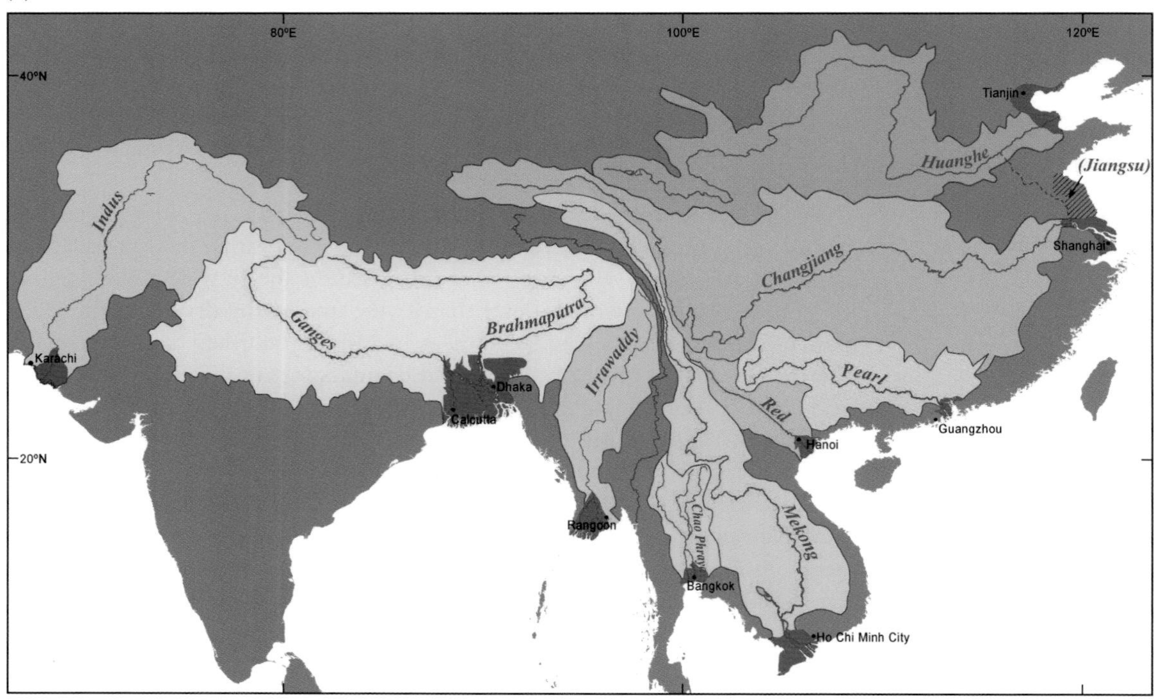

(b)

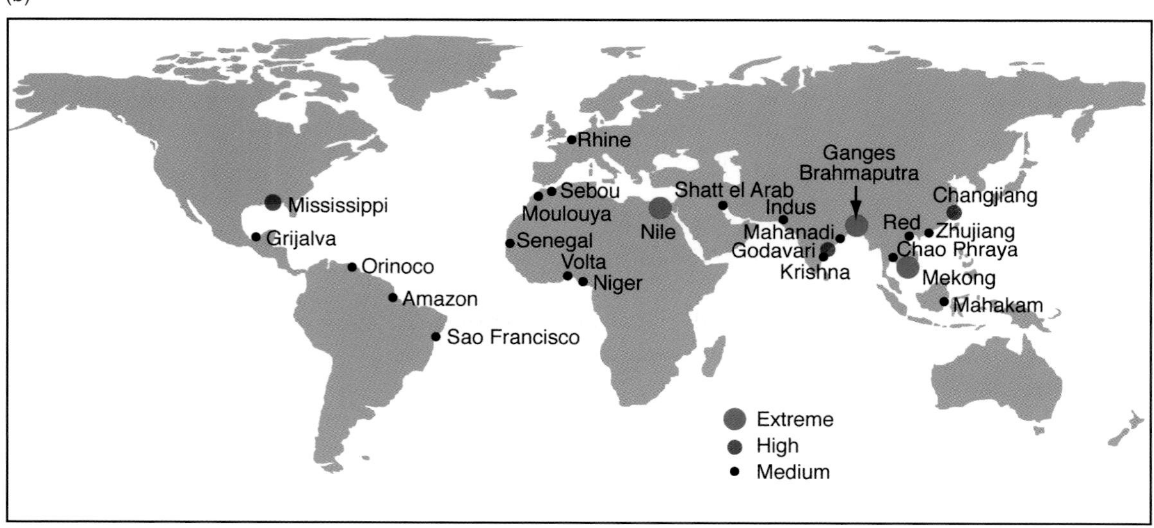

Figure 2.3 Vulnerable deltas. (a) The nine low-lying Asian megadeltas are all subsiding to some degree, so RSLR is already a problem that can only get worse. The present population of these deltas (the dark green areas) is about 250 million (adapted from Woodroffe et al. 2006, with kind permission of Springer Science and Business Media). (b) Relative vulnerability of deltas (in terms of displaced people) to present rates of RSLR to 2050 (extreme, ≥1 million; high, 1 million–50 000; medium, 50 000–5000). (Reproduced from Nicholls et al. 2007a, using data from Ericson et al. 2006.)

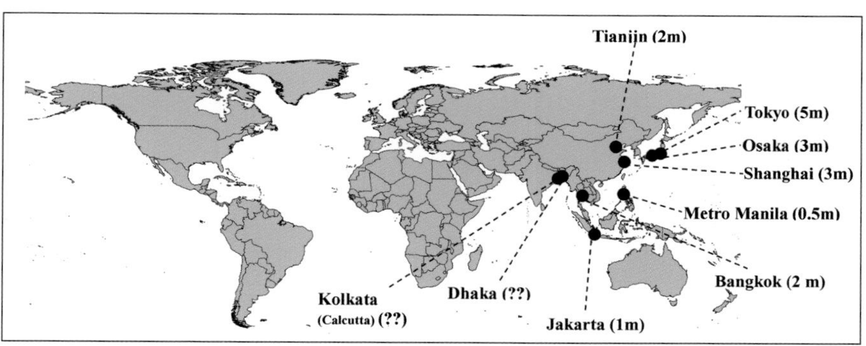

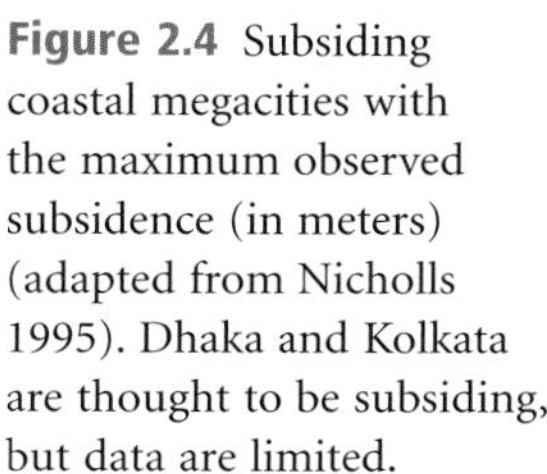
Figure 2.4 Subsiding coastal megacities with the maximum observed subsidence (in meters) (adapted from Nicholls 1995). Dhaka and Kolkata are thought to be subsiding, but data are limited.

withdrawal of groundwater can produce dramatic RSLR, especially in susceptible areas of cities (Milliman and Haq 1996; IGES 2007). Over the 20th century, parts of Tokyo and Osaka subsided up to 5 and 3 m, respectively, parts of Shanghai subsided up to 3 m, and parts of Bangkok subsided up to 2 m (Figure 2.4) (see Bangkok; Figure 2.2). Note that the maximum subsidence is reported because data on average subsidence are not available. Stopping shallow sub-surface fluid withdrawals greatly reduces subsidence, but natural "background" rates of subsidence will continue. These four example cities have all seen a combination of such policies combined with the provision of sophisticated flood defenses and pumped drainage to avoid submergence and/or frequent flooding. (In Bangkok, while subsidence declined in the center of the city from 1981 to 2002, it accelerated to a lesser degree over a wider surrounding area (Figure 2.5), so the problem is evolving, but not yet solved). In contrast, Jakarta and Metro Manila are subsiding cities where little systematic actions to manage and reduce the subsidence are in place as yet (e.g. Rodolfo and Siringan 2006). Without a wider approach which shares experience, enhanced subsidence will be repeated in expanding cities in similar geological settings over the 21st century: there are many candidates in deltaic settings, especially in Asia (see Nicholls et al. 2007b). More widely, most populated deltaic areas are experiencing enhanced human-induced subsidence to varying degrees (Ericson et al. 2006), and under current trends this will continue through the 21st century.

It is clear that global sea-level rise will continue far beyond the 21st century irrespective of future greenhouse emissions (Nicholls and Lowe 2006). This is because it takes centuries to millennia for the full ocean depth to adjust to a surface warming, resulting in ongoing thermal expansion and for ice sheets to respond to changes in climate. This inevitable rise has been termed the "commitment to sea-level rise". If global warming passes key and still somewhat uncertain thresholds for the irreversible breakdown of the Greenland or West Antarctic Ice Sheets, the committed rise could be 13–15 m, albeit over long timescales (many centuries or longer) (Chapter 7). Mitigation of climate change can reduce but not avoid some commitment to future sea-level rise. Hence, it appears that sea-level rise will remain a challenge for many generations to come. In essence, the commitment to sea-level rise leads to a commitment to adaptation

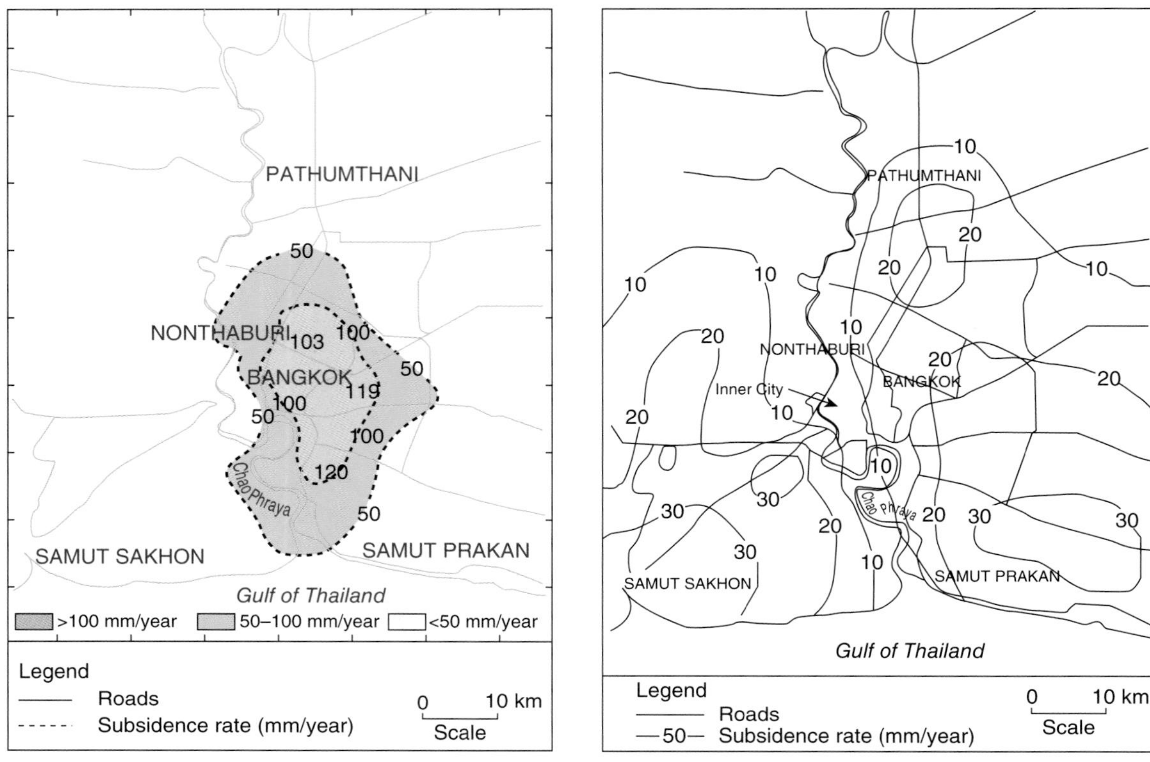

Figure 2.5 Contour lines of the rate of land subsidence for Bangkok, Thailand. Left: 1981; right: 2002. (Reproduced from Phien-Wej et al. 2006, with permission from Elsevier.)

to sea-level rise with fundamental implications for long-term human use of the coastal zone (Nicholls et al. 2007a).

2.3 Sea-Level Rise and Resulting Impacts

RSLR has a wide range of effects on the natural system, with the five main ones being summarized in Table 2.1. Flooding/submergence, ecosystem change, and erosion have received significantly more attention than salinization and rising water tables. Along with rising sea level, there are changes to all the processes that operate around the coast. The immediate effect is submergence and increased flooding of coastal land, as well as saltwater intrusion into surface waters. Longer-term effects also occur as the coast adjusts to the new environmental conditions, including wetland loss and change in response to higher water tables and increasing salinity, erosion of beaches and soft cliffs, and saltwater intrusion into groundwater. These lagged changes interact with the immediate effects of RSLR

Table 2.1 The main natural system effects of RSLR, including climate and non-climate interacting factors and examples of adaptation to these effects. Some interacting factors (e.g. sediment supply) appear twice as they can be influenced both by climate and non-climate factors. Adaptations are coded: P, protection; A, accommodation; R, retreat. (Adapted from Nicholls and Tol 2006.)

Natural system effect		Interacting factors		Adaptation responses	Example studies
		Climate	Non-climate		
1. Inundation, flood, and storm damage	a. Surge (flooding from the sea)	Wave/storm climate, erosion, sediment supply	Sediment supply, flood management, erosion, land reclamation	Dikes/surge barriers [P], building codes/ floodwise buildings [A], land-use planning/ hazard delineation [A/R]	Millman et al. (1989), Pernetta (1992), Hoozemans et al. (1993), Mimura (2000), Gornitz et al. (2002), Nicholls (2004), Dawson et al. (2005, 2007), Ericson et al. (2006), Thorne et al. (2007), Jacob et al. (2007)
	b. Backwater effect (flooding from rivers)	Runoff	Catchment management and land use		
2. Wetland loss (and change)		CO_2 fertilization of biomass production, sediment supply, migration space	Sediment supply, migration space, land reclamation (i.e. direct destruction)	Land-use planning [A/R], managed realignment/forbid hard defenses [R], nourishment/sediment management [P]	Titus (1988), Hoozemans et al. (1993), Lee (2001), Hartig et al. (2002), Scavia et al. (2002), Nicholls (2004), Gardiner et al. (2007), McFadden et al. (2007)
3. Erosion (of "soft" morphology)		Sediment supply, wave/storm climate	Sediment supply	Coast defenses [P], nourishment [P], building setbacks [R]	Stive et al. (1990, 2009), Mimura and Nobuoka (1996), Cowell et al. (2003a, 2003b, 2006), Van Goor et al. (2003) Stive (2004), Zhang et al. (2004), Dickson et al. (2007), Dawson et al. (2007), Walkden and Dickson (2008)
4. Saltwater Intrusion	a. Surface waters	Runoff	Catchment management (over extraction), land use	Saltwater intrusion barriers [P], change water abstraction [A/R]	Barth and Titus (1984), Sorensen et al. (1984), Hull and Titus (1986), Hay and Mimura (2005)
	b. Groundwater	Rainfall	Land use, aquifer use (over pumping)	Freshwater injection [P], change water abstraction [A/R]	Hull and Titus (1986), Sherif and Singh (1999), Essink (1996, 2001a, 2001b), Hay and Mimura (2005), Ranjan et al. (2006)
5. Rising water tables/impeded drainage		Rainfall, runoff	Land use, aquifer use, catchment management	Upgrade drainage systems [P], polders [P], change land use [A], land-use planning/ hazard delineation [A/R]	Barth and Titus (1984), National Research Council (1987), Titus et al. (1987)

and generally exacerbate them. For instance, coastal erosion will tend to degrade or remove natural protective features (e.g. salt marshes, mangroves, and sand dunes) so increasing the impact of extreme water levels and hence the risk of coastal flooding.

A mean rise in sea level also raises extreme water levels, as shown by Zhang et al. (2000) on the US East Coast and Aráujo and Pugh (2008) on the English Channel, and this is widely applied in impact studies for future conditions (Gornitz et al. 2002). Changes in storm characteristics could also influence extreme water levels both positively and negatively (Chapter 11) (von Storch and Woth 2008). For example, the widely debated increase in the intensity of tropical cyclones would increase in general terms extreme water levels in the areas affected, and changes in the tracks, intensity, and frequency of extra-tropical storms could change the occurrence of extreme water levels in mid latitudes (Meehl et al. 2007). Extra-tropical storms may also intensify in some regions. Changes in mean sea level will also influence the propagation of tides and surges and this could have additional significant effects (positive or negative) on extreme sea levels. While some case studies are available (Chapter 11), for most coastal regions, changes in extreme sea levels under climate change have not been systematically investigated to date due in large part to the uncertainties and low resolution of climate models, which do not resolve storms. This deficiency will need to be addressed to support adaptation, as discussed below.

Changes in natural systems as a result of RSLR have many important direct socioeconomic impacts on a range of sectors with the effect being overwhelmingly negative (Table 2.2). For instance, flooding can damage the extensive coastal infrastructure, towns and cities, and agricultural areas, and in the worst case lead to significant mortality as shown recently in Cyclone Nargis, Myanmar, in 2008. Erosion can similarly lead to losses of infrastructure and urban areas and have adverse consequences for sectors such as tourism and recreation. In addition to

Table 2.2 Summary of RSLR impacts on socioeconomic sectors in coastal zones. (Adapted from Nicholls et al. 2007a.)

Coastal socioeconomic sector	Sea-level rise natural system effect (Table 2.1)				
	Inundation, flood and storm damage	Wetland loss	Erosion	Saltwater intrusion	Rising water tables/ impeded drainage
Freshwater resources	X	x	–	X	X
Agriculture and forestry	X	x	–	X	X
Fisheries and aquaculture	X	X	x	X	–
Health	X	X	–	X	x
Recreation and tourism	X	X	X	–	–
Biodiversity	X	X	X	X	X
Settlements/infrastructure	X	–	X	X	X

X, strong; x, weak; –, negligible or not established.

these direct impacts there are indirect impacts such as adverse effects on human health: for example, mental health problems increase after a flood (Few et al. 2004), or the release of toxins from eroded landfills and waste sites which are commonly located in low-lying coastal areas, especially around major cities (e.g. Flynn et al. 1984). Economically, RSLR will also have direct and indirect effects (Sugiyama et al. 2008). Thus, RSLR has the potential to trigger a cascade of direct and indirect human impacts.

2.4 Framework and Methods for the Analysis of Sea-Level-Rise Impacts

RSLR does not happen in isolation and coasts are changing significantly due to non-climate-induced drivers (Crossland et al. 2005; Valiela 2006). Potential interactions with RSLR indicated in Table 2.1 need to be considered when assessing RSLR impacts and responses. For instance, a coast with a positive sediment budget may not erode given RSLR and vice versa (Cowell et al. 2003a, 2003b). These kind of problems require an integrated assessment approach to analyze the full range of interacting drivers, including the feedback of policy interventions (i.e. adaptation). Figure 2.6 presents a systems model of the impacts of sea-level

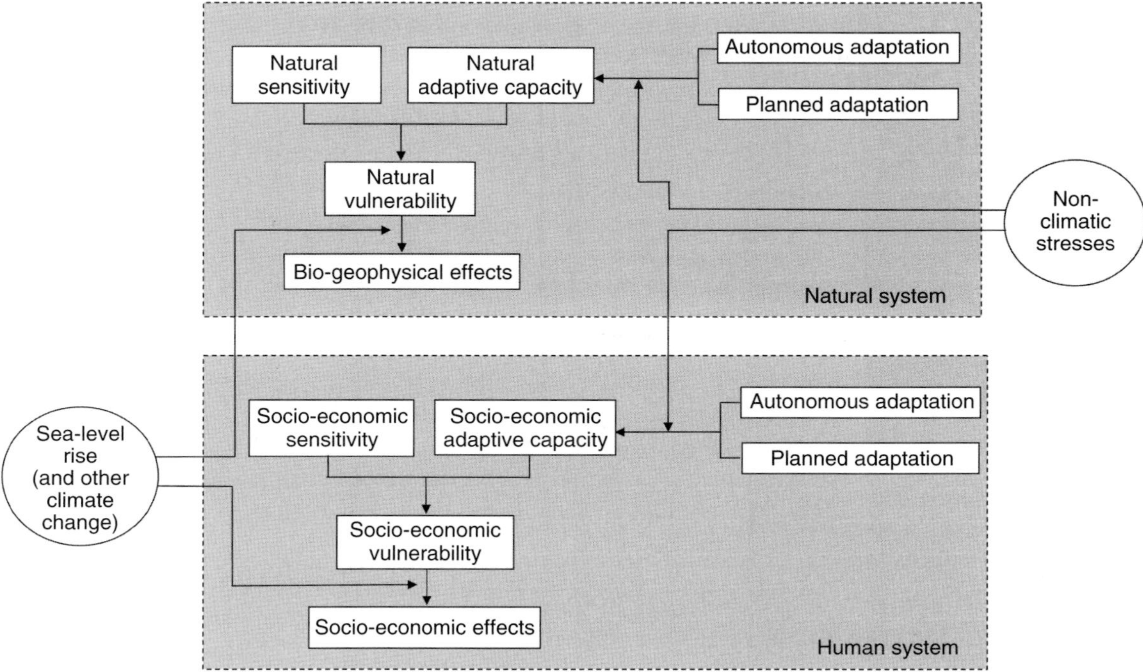

Figure 2.6 A conceptual framework for coastal impact and vulnerability assessment of sea-level rise. (Adapted from Klein and Nicholls 1999.)

rise on the coastal zone which builds on Figure 2.1 and captures the key interactions. This model highlights the varying implicit and explicit assumptions and simplifications that are made within all the available assessments, including their limitations. It characterizes the overall coastal system as interacting natural and socioeconomic systems, which have the potential to constrain each other's evolution. Sea-level rise is only one aspect of climate change for coastal areas, and all climate change drivers interact with other non-climate stresses, often exacerbating impacts. Lastly, the socioeconomic system is not passive as it influences the natural system through deliberate changes such as construction of sea dykes, destruction of wetlands, and building of port and harbor works, as well as unintended changes such as reductions of sediment and water fluxes due to the building of dams.

Both systems may be characterized by their *exposure*, *sensitivity*, and *adaptive capacity*[1] to change, both from sea-level rise and related climate change, and this may be modified by other *non-climate stresses*. Collectively, sensitivity and adaptive capacity, combined with exposure, determine each system's *vulnerability* to RSLR and other changes. Both systems are dynamic and interact in a coupled manner, including the effects of adaptation. *Autonomous adaptation* represent the spontaneous adaptive response to RSLR (e.g. increased vertical accretion of coastal wetlands within the natural system, or market price adjustments within the socioeconomic system). *Planned adaptation* (by the socioeconomic system) can serve to reduce vulnerability by a range of (anticipatory or reactive) measures. Adaptation normally reduces the magnitude of the impacts that would occur in their absence. Hence, impact assessments that do not take autonomous adaptation and/or (proactive or reactive) planned adaptation into account will generally estimate larger impacts (determining *worst-case* or *potential impacts* rather than *actual* or *residual impacts*) (Figure 2.7).

Using this conceptual framework, we can assess the numerous impact assessments of RSLR that are available. Given the complexity of impacts and especially adaptation to RSLR, these assessments have posed a wide range of questions and used diverse methods, so few studies are directly comparable. Hence, it is important to understand the scope and limitations of any study. While it is difficult to categorize the studies, different threads are apparent. Science-oriented studies have examined the natural responses to sea-level rise (e.g. Cahoon et al. 1999, 2006; Zhang et al. 2004) and transferred natural scientific knowledge into tools relevant to coastal management analysis (e.g. Stive et al. 1990; Capobianco et al. 1999; Cowell et al. 2006). There are also a range of policy-driven sub-national, national, and regional/global case studies (e.g. Nicholls and Mimura 1998; Mimura 2000; De La Vega-Leinert and Nicholls 2001; Nicholls 2004) which

[1] Exposure is the nature and degree to which a system is exposed to significant climatic variations (not indicated in Figure 2.6). Sensitivity is the degree to which a system is affected, either adversely or beneficially by climate-related stimuli such as sea-level rise. Adaptive capacity is the ability of a system to adjust to RSLR and climate change (including climate variability and extremes) to moderate potential damages, to take advantage of opportunities, or to cope with the consequences (McCarthy et al. 2001).

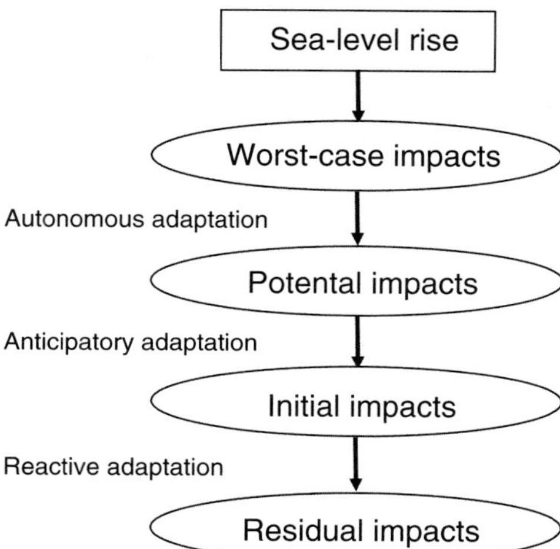

Figure 2.7 The different stages of adaptation and their influence on impacts.

combine information on physical changes to varying degrees with socioeconomic implications such as land values and people flooded. Some stress estimates of what is exposed to RSLR, although sometimes this exposure is erroneously presented as inevitable losses. Other analyses look at exposure and impacts based on arbitrary adaptation assumptions. Lastly, a range of integrated assessments have been undertaken which stress economics, costs, and estimates of optimum adaptation, usually based on cost-benefit analysis (e.g. Fankhauser 1995; Tol 2002a, 2002b, 2007) with a few examining more indirect economic effects (e.g. Darwin and Tol 2001; Fankhauser and Tol 2005; Bosello et al. 2007). These studies emphasize that protection is often cost-effective and, while there is a large impact potential, residual impacts could be orders of magnitude smaller. The new Dynamic Interactive Vulnerability Assessment (DIVA) tool combines several of these methodologies within an integrated assessment framework (e.g. Hinkel and Klein 2007; Nicholls et al. 2007c). Hence, there is a diversity of information from these studies which must be integrated carefully as discussed below.

2.5 Recent Impacts of Sea-Level Rise

Over the 20th century global sea level rose about 18 cm, which is faster than during the 19th century (Chapter 5). While this change may seem small, it has had many significant effects, most particularly in terms of the return periods of extreme water levels (Zhang et al. 2000; Woodworth and Blackman 2004; Chapter 11), and promoting an erosive tendency for coasts as widely observed (Bird 1985, 2000; Vellinga and Leatherman 1989). However, linking global sea-level rise quantitatively to impacts is quite difficult as the coastal zone has been subjected to multiple drivers of change over the 20th century (Rosenzweig et al. 2007). Good data on rising sea levels have only been measured in a few locations, and flood defenses have often been upgraded substantially through the 20th century, especially in those (wealthy) places where there are sea-level measurements. Most of this defense upgrade reflects expanding populations and wealth in the coastal flood plain and changing attitudes to risk, and RSLR may not have even been considered in the design. Equally, erosion can be promoted by processes other than sea-level rise (Table 2.1), and human reduction in sediment supply to the coast must contribute to the observed changes as noted by Bird (1985) and Syvitski et al. (2005). Decline in intertidal habitats such as salt marshes, mudflats, and mangroves is often linked to sea-level rise, but these systems are also subject to multiple drivers of change, including direct destruction (Hoozemans et al. 1993; Coleman et al. 2008). Hence, while global sea-level rise was a pervasive process, it is difficult to unambiguously link it to impacts, except in some special cases: most coastal change in the 20th century was a response to multiple drivers of change.

On the East Coast of the USA, RSLR occurred at variable rates between 2 and 4 mm/year over the 20th century. Both sea level and coastal change were measured during the 20th century, providing a laboratory for exploring shoreline response to sea-level rise. Comparing the rate of shoreline retreat and the long-term rate of RSLR away from inlets and engineered shores supports the concept of the Bruun Rule, where the shoreline retreat rate is 50–100 times the rate of RSLR (Zhang et al. 2004) (Mimura and Nobuoka 1996 found a similar relationship on a subsiding coastal plain in Japan). However, near inlets, the indirect effects of sea-level rise which cause the associated estuary/lagoon to trap beach-sized sediment can have much larger erosional effects on the neighboring open coasts than predicted by the Bruun Rule (Stive 2004). Hence, more general relationships are required to understand coastal change taking account of relative sea-level change, sediment supply, and coastal physiography (Cowell et al. 2006). Human responses to sea-level rise are even more difficult to document. Human abandonment of low-lying islands in Chesapeake Bay, USA, during the late 19th/early 20th century does seem to have been triggered by a small acceleration of RSLR (partly linked to acceleration in global sea-level rise) and resulting land loss (Kearney and Stevenson 1991; Leatherman 1992; Gibbons and Nicholls 2006).

There have certainly been impacts from RSLR resulting from large rates of subsidence, such as the Mississippi Delta where RSLR is 5 to 10 mm/year. Between 1978 and 2000 1565 km^2 of intertidal coastal marshes and adjacent lands were converted to open water, due to sediment starvation and increases in the salinity and water levels of coastal marshes due to human development and wider changes (Barras et al. 2003). By 2050, about 1300 km^2 of additional coastal land loss is projected if current global, regional, and local processes continue at the same rate. There have also been significant impacts of RSLR in deltas (Figure 2.3) and in and around subsiding coastal cities (e.g. Figure 2.4), resulting in increased waterlogging, flooding, and submergence, and the resulting need for management responses (Nicholls 1995; Rodolfo and Siringan 2006). The flooding in New Orleans during Hurricane Katrina in 2005 was significantly exacerbated by subsidence compared to earlier flood events such as Hurricane Betsy in 1965 (Dixon

Figure 2.8 Telegraph poles south of Bangkok, Thailand: built on subsiding land, they are now up to 1 km out to sea.

et al. 2006; Grossi and Muir-Wood 2006). The iconic medieval city of Venice has also been subject to subsidence, which has greatly increased the risk of flooding (Fletcher and Spencer 2005). In terms of response, all the major developed areas that were impacted by RSLR have been defended, even when RSLR was several meters over several decades. The future of New Orleans post-Katrina will be illuminating: the pre-existing dikes have been largely rebuilt, but now a longer-term plan is being developed which raises questions of upgraded protection versus retreat, and the relationship of the city to the surrounding delta. In less-developed areas coastal retreat has occurred, such as south of Bangkok where subsidence has led to a shoreline retreat of more than a kilometer (Figure 2.8). Similarly, around Galveston Bay in Texas subsidence of up to 3 m has occurred, and the San Jacinto Battleground State Historical Park, the site of the battle that won Texas its independence, is now partially submerged.

Hence observations through the 20th century reinforce the importance of understanding the impacts of sea-level rise in the context of multiple drivers of change: this will remain true under more rapid rises in sea level. Of these multiple drivers of change, human-induced subsidence is of particular interest, but this remains relatively unstudied in a systematic sense. Observations also emphasize the ability to protect against RSLR, especially for the most densely populated areas such as the subsiding Asian cities already discussed. Defenses have also been upgraded around the southern North Sea, including London and Hamburg: here subsidence is less important, whereas significant rises in extreme water levels have been observed as reclamation and dredging increase surge propagation.

2.6 Future Impacts of Sea-Level Rise

The future impacts of sea-level rise will depend on a range of factors, including the degree to which sea-level rise accelerates, the level and manner of coastal development, and the success (or failure) of adaptation. Assessments of the future impacts of sea-level rise have taken place on a range of scales from local to global. They all confirm potentially large impacts following Table 2.1, especially increases in inundation flooding and storm damage. Taking account of population exposure, sensitivity, and adaptive capacity, South and Southeast Asia and Africa appear to be most vulnerable in absolute terms due to storm-induced flooding combined with sea-level rise (Figure 2.9). Nicholls et al. (1999) estimated that for a roughly 40-cm global rise in sea level by the 2080s and the IS92a socioeconomic scenarios (the business-as-usual emissions (and socioeconomic) scenarios used prior to the Special Report on Emission Scenarios (SRES) scenarios), more than 70 million people might be impacted in Asia, and more than 10 million people in Africa. Significant residual risk will remain in other coastal areas of the world such as around the southern North Sea, and major flood disasters remain possible in any coastal region. Under the SRES climate and socioeconomic scenarios the

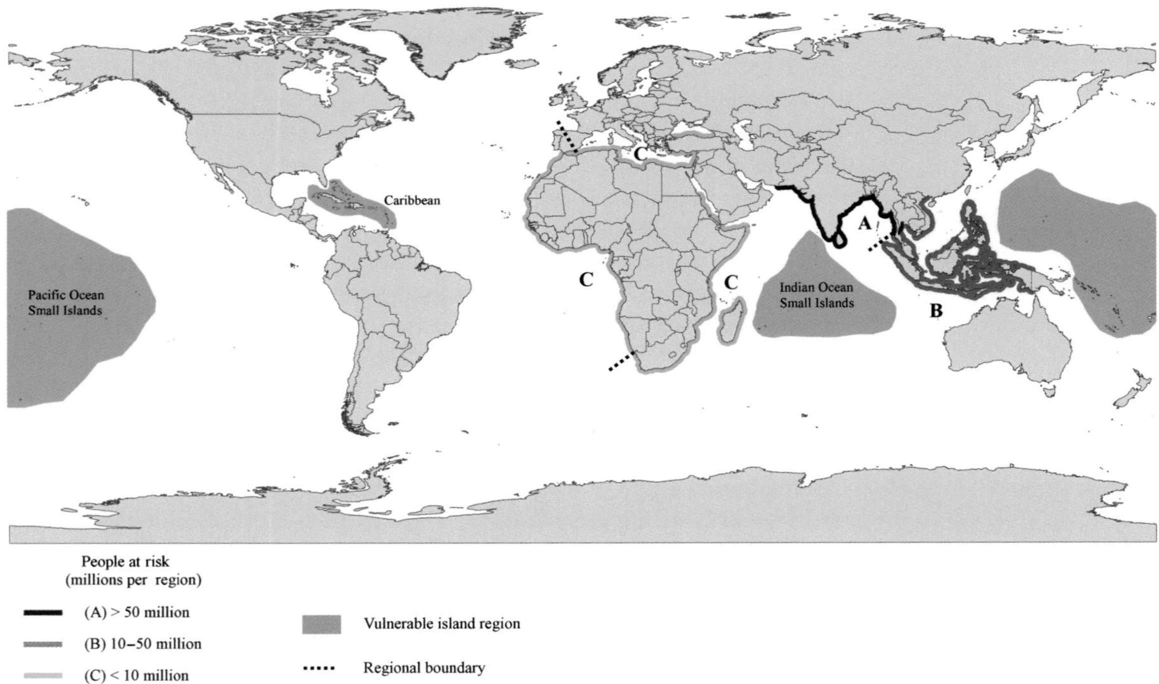

Figure 2.9 The regions most vulnerable to coastal flooding, based on an illustrative scenario for the 2080s, assuming a middle estimate of global sea-level rise by that time of 45 cm (note: the African coastline comprises three "regions"). (Adapted from Nicholls et al. 1999.)

absolute impacts are reduced compared to IS92a, as the rise in sea level is reduced, and more importantly gross domestic product/per-capita growth is larger, which gives a higher adaptive capacity (Nicholls 2004). However, the pattern of impacts remains the same. Small island regions in the Pacific Ocean, Indian Ocean, and Caribbean stand out as being especially vulnerable to flooding (Nurse et al. 2001), even though relatively few people are affected in global terms. The populations of low-lying islands such as the Maldives or Tuvalu face the real prospect of increased flooding, submergence, and forced abandonment. This perception may trigger a collapse in investment and general confidence blighting these areas and triggering abandonment long before it is physically inevitable (Barnett and Adger 2003; Gibbons and Nicholls 2006). The Alliance of Small Island States (AOSIS) – which represents 38 countries that are members of the United Nations – has raised considerable awareness about these issues and the vulnerability that some of the small island states face.

Adaptation can greatly reduce the impacts. Cost-benefit models that compare protection with retreat generally suggest that it is worth investing in widespread protection as populated coastal areas are often of high economic value due to the extensive infrastructure and urban areas (Fankhauser 1995; Tol 2007; Nicholls and Tol 2006; Sugiyama et al. 2008). (It is worth noting that if no economic growth is assumed, protection is much harder to justify and hence the impacts of sea-level rise depend on both climate and socioeconomic scenarios; Nicholls 2004; Anthoff et al. 2006.) With or without protection, small island and deltaic areas stand out as relatively more vulnerable in most analyses and the impacts fall disproportionately on poorer countries (Anthoff et al. 2006; Sugiyama et al. 2008).

Impacts on coastal ecosystems could also be severe, including the influence of coastal squeeze[2] if coasts are defended and armored, which exacerbates the direct losses of sea-level rise (Nicholls et al. 2007a; Sugiyama et al. 2008).

2.6.1 National-Scale Assessments

The available national-scale assessments generally comprise inventories of the potential impacts to a 1-m rise in sea level, with limited consideration of adaptation (Nicholls and Mimura 1998). These impacts are consistent with the global estimates of exposure already discussed. Considering just 18 countries, 180 million people are exposed to the full range of impacts shown in Table 2.1, assuming a 1-m global rise in sea level (by 2100) and no protection (Table 2.3). Low-lying coastal areas are most sensitive to sea-level rise; again particularly deltaic and small island settings. Coastal wetlands also appear highly threatened, although this may partly reflect the simple impact assumptions made in the studies, rather than actual impacts which could be lower if the wetlands are able to respond to sea-level rise.

[2] Coastal squeeze occurs where coastal habitats are retreating landward due to sea-level rise and this migration is limited by hard natural or defended coasts, leading to a squeeze of the habitats and ultimately their complete loss.

Table 2.3 Aggregated results of country studies, assuming existing development and a 1-m global rise in sea level. All impacts assume no adaptation, while adaptation assumes protection, except in areas of low population density. Costs are 1990 US dollars. (Adapted from Bijlsma et al. 1996.)

Country	Net people exposed		Capital value at loss		Land at loss		Wetland at loss	Incremental protection costs	
	No. of people (1000s)	% Total	Million US$	% GNP	km²	% Total	km²	Million US$	% GNP
Antigua	38	50	–	–	5	1.0	3	71	0.32
Argentina	–	–	5000	>5	3400	0.1	1100	>1800	>0.02
Bangladesh	71 000	60	–	–	25 000	17.5	5800	>1000	>0.06
Belize	70	35	–	–	1900	8.4	–	–	–
Benin	1350	25	118	12	230	0.2	85	>400	>0.41
China	72 000	7	–	–	35 000	–	–	–	–
Egypt	4700	9	59 000	204	5800	1.0	–	13 100	0.45
Guyana	600	80	4000	1115	2400	1.1	500	200	0.26
Japan	15 400	15	849 000	72	2300	2.4	–	>156 000	>0.12
Kiribati	9	100	2	8	4	12.5	–	3	0.10
Malaysia	–	–	–	–	7000	2.1	6000	–	–
Marshall Islands	20	100	160	324	9	80	–	>360	>7.04
Mauritius	3	<1	–	–	5	0.3	–	–	–
Netherlands	10 000	67	186 000	69	2165	5.9	642	12300	0.05
Nigeria	3200	4	17 000	52	18 600	2.0	16 000	>1400	>0.04
Poland	235	1	24 000	24	1700	0.5	36	1400	0.02
Senegal	110	>1	>500	>12	6100	3.1	6000	>1000	>0.21
St Kitts	–	–	–	–	1	1.4	1	50	2.65
Tonga	30	47	–	–	7	2.9	–	–	–
Uruguay	13	<1	1700	26	96	0.1	23	>1000	>0.12
USA	–	–	–	–	31 600	0.3	17 000	>156 000[a]	>0.03
Venezuela	56	<1	330	1	5700	0.6	5600	>1600	>0.03
Total	178 834		1 146 310		149 022		58 790	27124	

GNP, gross national product.
[a]See also Table 2.4 and discussion in main text.

In terms of adaptation, these studies have usually made very simple assumptions that are consistent with an inventory approach, such as costing protection for all areas, except those with a low population (a common threshold is fewer than 10 people/km²). These results suggest that the incremental costs of adaptation (only considering the capital costs of global sea-level rise) will pose a varying burden in relation to the present size of the national economies, particularly for many small island nations (Table 2.3). However, the issues of the adaptation process and the capacity of the coastal communities to adapt have not usually been considered in any detail.

Table 2.4 Potential cost of a 1-m global sea-level rise along the developed coastline of the USA (billions of 1990 US dollars). (Adapted from Neumann et al. 2001.)

Source	Measurement	Cost (billion US$)		
		Annualized estimate	Cumulative estimate	Annual estimate in 2065
Smith and Tirpak (1989)	Protection	NA	73–111	NA
Yohe (1990)	Property at risk of inundation	NA	321	1.37
Titus et al. (1991)	Protection	N/A	156	NA
Nordhaus (1991)	Protection	4.9	NA	NA
Fankhauser (1995)	Protection	1.0	62.6	NA
Yohe et al. (1996)	Protection and abandonment	0.16	36.1	0.33
Yohe and Schlesinger (1998)	Expected protection and abandonment	0.38	NA	0.4

NA, not available.

One important result concerns the importance of the scale of assessment. Sterr (2008) has investigated the vulnerability of Germany to global sea-level rise, at national, state (Schleswig–Holstein), and case studies (within Schleswig–Holstein) levels. As the scale of study increases, so the size of the hazard zones declined due to the use of higher-resolution elevation data. However, the potential impacts do not change significantly as the human values remain concentrated in the (smaller) hazard zones. Turner et al. (1995) examine the optimum response to sea-level rise in East Anglia, UK, using cost-benefit analysis. At the regional scale, it was worth protecting the entire coastal length against accelerated global sea-level rise. In contrast, at the scale of individual flood compartments, 20% of flood compartments should be abandoned even for present rates of RSLR as valuable assets are concentrated in certain locations, rather than distributed uniformly along the coast. This conclusion is consistent with current trends in coastal management policy for this region. Studies that do not take these distributions of assets into account will tend to overstate protection costs (Sugiyama et al. 2008).

Building on this conclusion, the potential cost of sea-level rise on the USA has received continuing analysis as summarized in Table 2.4 for a 1-m global sea-level rise scenario. In general, estimated costs have fallen as the early studies either ignored adaptation, or considered widespread protection. If lower-value areas are abandoned based on benefit-cost analysis[3], net costs are greatly reduced.

[3] Note that benefit-cost analysis does not reduce impacts to zero and it may not capture all the relevant values (e.g. ecosystems) and poorer people with less assets often live in the most vulnerable areas, both in the developed and developing world.

2.6.2 Regional- and Global-Scale Assessments

Compared to national assessments, regional and global assessments provide a more consistent basis to assess the broad-scale impacts of sea-level rise.

Coastal Flooding

There have been a number of studies of coastal flooding under sea-level rise. Globally, it was estimated that about 200 million people lived in the coastal flood plain (below the 1-in-1000-year surge-flood elevation) in 1990, or about 4% of the world's population (Hoozemans et al. 1993). Based on estimates of defense standards, on average 10 million people/year experienced coastal flooding in 1990 (Nicholls et al. 1999), while the DIVA model (version 2.03) has lower estimates of about 3 million people/year. The numbers of people flooded will change due to the competing influences of RSLR (due to local subsidence and global changes), changes in coastal population, and (possible) improving defense standards as people become more wealthy (see Nicholls et al. 1999). Figure 2.10 shows the numbers of people flooded per year for three SRES scenarios using DIVA, assuming two adaptation scenarios: (1) no dike upgrade and (2) an economically optimum "benefit-cost" upgrade. Coastal population follows the SRES scenario and increase to 2050, after which they decline in all cases. The global sea-level-rise scenarios are derived from CLIMBER-2 (the Climate and Biosphere model of the Potsdam Institute for Climate; see Figure 2.10 caption). These numbers are then downscaled to RSLR. The impact calculations assume that people do not move despite the flooding, although in reality many of these people may be forced to migrate (see environmental refugees, below).

Under a no-adaptation scenario, the number of people flooded grows with rising sea levels and growing coastal population (to 2050) in all cases: the greater the rise in global sea level, the greater the increase in the incidence of flooding. In the worst case, under the A1FI high scenario with a 1.07 m rise from 1990 to 2100, the number of people flooded increases roughly 70 times by 2100 to about 200 million people/year. Optimum dike upgrade is quite different in behavior and independent of the magnitude of the sea-level-rise scenario: the number of people flooded is projected to decline through the 21st century. This reflects that the dikes are raised more than the magnitude of sea-level rise as people adapt and become more risk adverse as they become more wealthy. This illustrates that the

Figure 2.10 *(Opposite)* A comparison of the number of people flooded assuming no dike upgrade and optimum dike upgrade under the B1, A1B, and A1FI SRES scenarios using the DIVA model (version 2.0.3). No dike upgrade assumes that the dikes (in 1995) are maintained at a constant elevation, while optimum adaptation assumes upgrade based on benefit-cost analysis. The global sea-level-rise scenarios are derived from CLIMBER-2 and include low, medium, and high estimates (from 1990 to 2100): B1, 0.16, 0.35, and 0.72 m rise; A1B, 0.21, 0.44, and 0.89 m rise; A1FI, 0.25, 0.52, and 1.07 m rise (Ganopolski et al. 2001). SLR, sea-level rise.

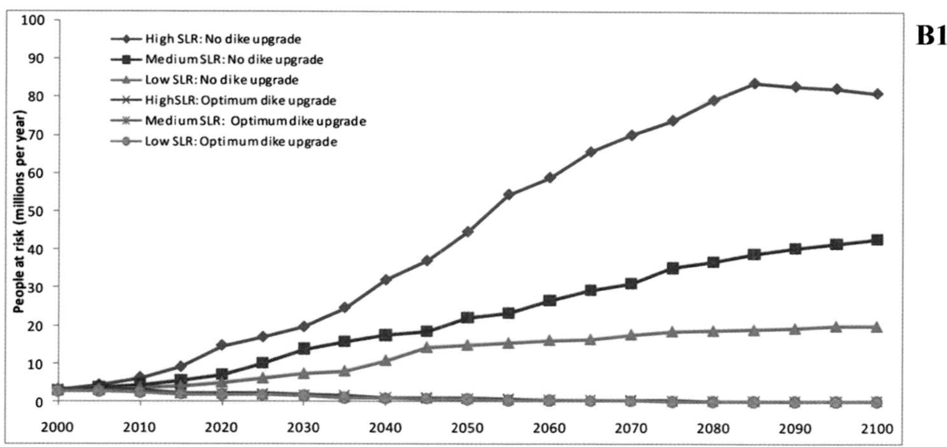
B1
People at risk (millions per year)
High SLR: No dike upgrade
Medium SLR: No dike upgrade
Low SLR: No dike upgrade
High SLR: Optimum dike upgrade
Medium SLR: Optimum dike upgrade
Low SLR: Optimum dike upgrade
0
10
20
30
40
50
60
70
80
90
100
2000
2010
2020
2030
2040
2050
2060
2070
2080
2090
2100

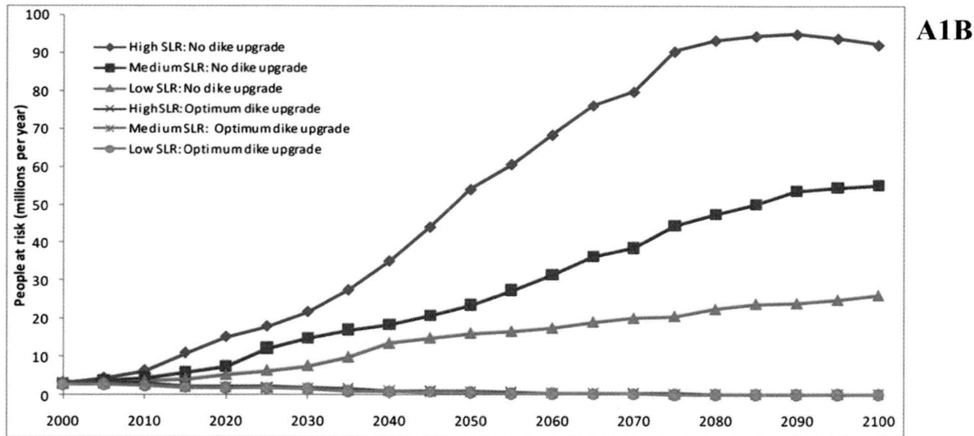
A1B
People at risk (millions per year)
High SLR: No dike upgrade
Medium SLR: No dike upgrade
Low SLR: No dike upgrade
High SLR: Optimum dike upgrade
Medium SLR: Optimum dike upgrade
Low SLR: Optimum dike upgrade
0
10
20
30
40
50
60
70
80
90
100
2000
2010
2020
2030
2040
2050
2060
2070
2080
2090
2100

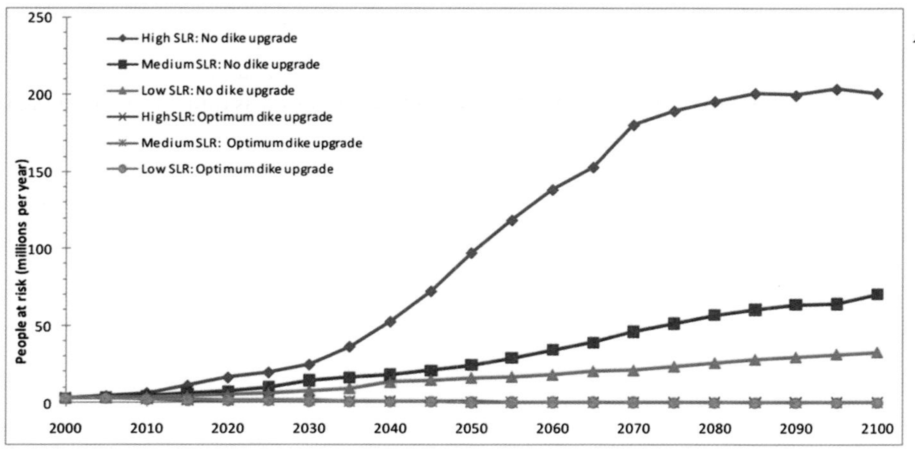
A1FI
People at risk (millions per year)
High SLR: No dike upgrade
Medium SLR: No dike upgrade
Low SLR: No dike upgrade
High SLR: Optimum dike upgrade
Medium SLR: Optimum dike upgrade
Low SLR: Optimum dike upgrade
0
50
100
150
200
250
2000
2010
2020
2030
2040
2050
2060
2070
2080
2090
2100

success or failure of adaptation is fundamental to understanding impacts, as discussed below.

Nicholls et al. (1999) and Nicholls (2004) found that the most vulnerable regions in relative terms are the small island regions of the Caribbean, Indian Ocean, and Pacific Ocean (Figure 2.9). However, absolute increases in the incidence of flooding are largest in the southern Mediterranean (largely in the Nile Delta), West Africa, East Africa, South Asia, and Southeast Asia: these five regions contain about 90% of the people flooded in all cases for the 2080s (Figure 2.9). This reflects the large populations of low-lying deltas in parts of Asia, and projections of rapid population growth around Africa's coastal areas. While developed country regions have relatively low impacts, sea-level rise still produces a significant increase in the number of people who would be flooded assuming no adaptation for sea-level rise. These results show that sea-level rise could have a profound impact on the incidence of flooding: the higher the total rise, the greater the increase in flood risk, all other factors being equal. Any increase in storminess would further exacerbate the predicted increase in coastal flooding.

Environmental Refugees

Sea-level rise is often associated with a large potential for environmental refugees forcibly displaced from their homes (Myers 2002). Potentially, many tens or even hundreds of millions of peoples could be so displaced, especially given that coastal populations are growing significantly worldwide. However, as illustrated in Figure 2.10, if we can successfully adapt to these challenges, this is a much smaller problem than is often assumed. Adaptation could include flood defenses for urban areas, and land-use planning for new developments to avoid the more risky areas. As a reference, Tol (2002a, 2002b) suggests that most coastal areas are worth protecting in a benefit-cost sense (protection costs are less than damage costs). This formulation suggests that fewer than or equal to 75 000 people/year will be displaced by a 1-m sea level through the 21st century, after allowing for protection: incrementally this is of order 1% of the potentially displaced population. This result has a large uncertainty, but it illustrates again that the success or failure of adaptation is a key element to understanding the scale of the problem.

Coastal Wetlands

Wetland losses are driven more by the rate of sea-level rise, rather than the total rise, as they have capacity to respond to inundation (Cahoon et al. 1999). All the studies suggest large additional losses due to global sea-level rise. Given a 1-m global rise in sea level by 2100, wetland losses could approach half of the present stock (Hoozemans et al. 1993; Nicholls et al. 1999; McFadden et al. 2007). Smaller rises in sea level still lead to substantial losses. Therefore, global sea-level rise is a significant *additional* stress which worsens the already poor prognosis for coastal wetlands worldwide (Nicholls 2004). Looking at 20 world regions, losses would be most severe on the Atlantic coast of North and Central America, the Caribbean, the Mediterranean, the Baltic, and all small island regions. It is noteworthy that coastal wetlands in many developed countries appear threatened by global sea-level rise. In these cases, direct destruction is less of an issue due to regulation and protection/compensation mechanisms, and accelerated global sea-level rise could be the major threat to these systems through the 21st century (e.g. Lee 2001; Gardiner et al. 2007).

Global Costs of Sea-Level Rise

Global estimates of the incremental costs of upgrading defense infrastructure[4] suggest the costs are much lower than the expected damage (Tol 2007). IPCC Coastal Zone Management Subgroup (CZMS) (1990) estimated the one-off costs of defending against a 1-m sea-level rise at US$500 billion. Hoozemans et al. (1993) doubled these one-off costs to $1000 billion. Looking at the total costs of sea-level rise including dryland and wetland loss and incremental defense investment, Fankhauser (1995) estimated annual global costs of $47 billion based on benefit-cost using the IPCC CZMS (1990) data. Tol (2002a, 2002b) made similar estimates using the Hoozemans et al. (1993) data, supplemented by other data sources and adding the costs of forced migration. For the IS92a scenario with a 66-cm rise by 2100, benefit-cost estimates of annual protection costs are below $10 billion, while the annual loss of land exceeds $100 billion by 2100 (Tol 2007). However, any failure in protection will lead to much higher costs. Sugiyama et al. (2008) noted that that the spatial distribution of infrastructure and wealth along the coast influences costs: the more wealth is concentrated, the smaller the necessary protection costs.

2.7 Responding to Sea-Level Rise

The two potential responses to sea-level rise are mitigation and adaptation. They operate at two very different scales with mitigation being by necessity a global-scale activity, while adaptation is sub-global (often a local to national activity) and a range of adaptation measures are available at any site. Hence, our understanding and assessment of sea-level rise also need to operate at multiple scales.

Given the high-impact potential already discussed and the commitment to sea-level rise (and hence adaptation) independent of future emissions, a rational response to global sea-level rise and climate change in coastal areas would be to identify the most appropriate mixture of mitigation and adaptation (Nicholls et al. 2007a). Mitigation can slow the global rise in sea level and reduce its impacts, and given its strong inertia, mitigation has an important additional effect of stabilizing the *rate* of global sea-level rise (rather than stabilizing sea level itself)

[4] These incremental costs should not be compared directly with projects such as the post-Katrina defense of New Orleans as they only reflect the sea-level component of these needs.

(Nicholls and Lowe 2006). Given that the rate of sea-level rise controls wetland losses, this reduces impacts. In the case of flooding, the absolute rise is of more concern, and impacts tend to be delayed rather than avoided due to the commitment to sea-level rise (but importantly gives more time to adapt and hence lowers annual costs). Hence, adaptation and mitigation need to be thought of as two reinforcing policies in coastal areas (Nicholls et al. 2007a). The fundamental goal of mitigation in the context of coastal areas is to reduce the risk of passing irreversible thresholds concerning the breakdown of the major ice sheets, and constrain the commitment to sea-level rise to a rate and ultimate rise which can be adapted to at a reasonable economic and social cost. The rest of the discussion of responses focuses on adaptation.

Adaptation involves responding to both mean and extreme RSLR. Given the large and rapidly growing concentration of people and activity in the coastal zone, autonomous (or spontaneous) adaptation processes alone will not be able to cope with RSLR. Further, adaptation in the coastal context is widely seen as a public responsibility (Klein et al. 2000). Therefore, all levels of government have a key role in developing and facilitating appropriate adaptation measures (Tribbia and Moser 2008). The required adaptation costs remain uncertain, but as large amounts are already invested in managing coastal floods, erosion, and other coastal hazards, the incremental costs of including global sea-level rise does not appear infeasible at a global scale over the coming decades (Nicholls 2007). However, in certain settings such as small islands, these costs could overwhelm local economies without external help (Fankhauser and Tol 2005; Nicholls and Tol 2006).

Planned adaptation options to RSLR are usually presented as one of three generic approaches (IPCC CZMS 1990; Bijlsma et al. 1996; Klein et al. 2001) (Figure 2.11):

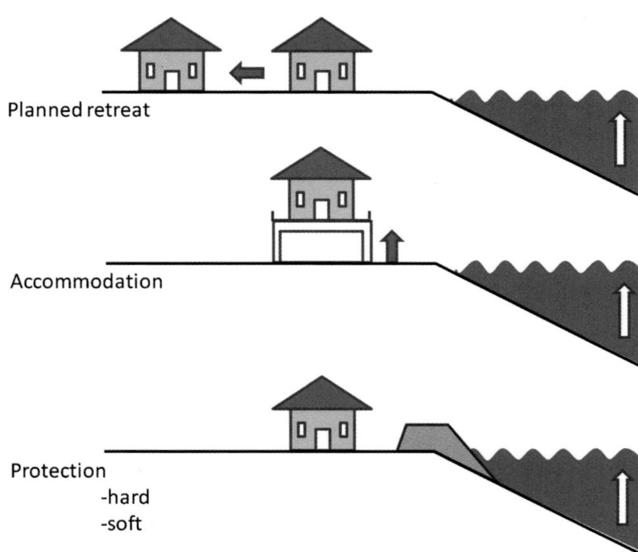

Figure 2.11 Generic approaches to adapt to RSLR as defined by IPCC CZMS (1990).

- *(planned) retreat*: all natural system effects are allowed to occur and human impacts are minimized by pulling back from the coast via land-use planning, development control, etc.;
- *accommodation*: all natural system effects are allowed to occur and human impacts are minimized by adjusting human use of the coastal zone via flood resilience, warning systems, insurance, etc.;
- *protection*: natural system effects are controlled by soft or hard engineering (e.g. nourished beaches and dunes or seawalls), reducing human impacts in the zone that would be impacted without protection.

Examples are given in Table 2.1.

Through human history, improving technology has increased the range of adaptation options in the face of coastal hazards, and there has been a move from retreat and accommodation to hard protection and active seaward advance via land claim (Van Koningsveld et al. 2008). Rising global sea level is one factor calling automatic reliance on hard protection into question, and the appropriate mixture of protection, accommodation, and retreat are now being more seriously evaluated. In practice, many responses may be hybrid and combine elements of more than one approach. Adaptation for one sector may also exacerbate impacts elsewhere: a good example is coastal squeeze of coastal ecosystems due to hard defenses. Coastal management needs to consider the balance between protecting socioeconomic activity/human safety and the habitats and ecological functioning of the coastal zone under rising sea levels (Nicholls and Klein 2005). This implies a need for more integrated responses, consistent with the ideals of integrated coastal management (see below).

Adaption to RSLR requires consideration of both global and local contributions to RSLR. For example, implementation of flood protection should not lead to complacency as a dynamic residual risk always remains for all protected areas. This risk will steadily increase with climate change and land subsidence. It is noteworthy that some authors report a tendency to ignore subsidence as an issue and to only focus on global climate change (Rodolfo and Siringan 2006). In addition risk grows as a result of socioeconomic growth in coastal areas (Evans et al. 2004; Ten Brinke et al. 2008).

The most appropriate timing for an adaptation response needs to be considered in terms of anticipatory versus reactive planned adaptation (or in practical terms, what should we do today, versus wait and see until tomorrow?). Present anticipatory decisions are made with more uncertainty than future reactive decisions, which will have the benefit of future knowledge. However, wait and see may lock in an adverse direction of development, increasing the potential impact of sea-level rise (see Figure 2.7). The coastal zone is an area where there is significant scope for anticipatory adaptation (Tompkins et al. 2005) as many decisions have long-term implications (e.g. Fankhauser et al. 1999). Examples of anticipatory adaptation in coastal zones include upgraded flood defenses and waste-water discharges, higher levels for reclamations and new bridges, and building setbacks to prevent development in areas threatened by erosion and flooding.

With a few exceptions, global sea-level rise will exacerbate existing pressures and problems, so there are important synergies in considering adaptation to climate change in the context of existing problems (Nicholls and Klein 2005). In some cases, the focus of climate change may help identify "win-win" situations where adaptation measures for sea-level rise are worthy of implementation just based on solving today's problems (Turner et al. 1995; Dawson et al. 2007). Also, adaptation measures are more likely to be implemented if they offer immediate benefits in reducing impacts of short-term climate variability as well as long-term climate change.

While there is limited experience of adaptation to climate change, there is considerable experience of adapting to climate variability and we can draw on this experience to inform decision-making under a changing climate. Importantly, adaptation to coastal problems is a *process*. Klein et al. (1999) suggested four stages operating within multiple policy cycles: (1) information and awareness-building, (2) planning and design, (3) evaluation, and (4) monitoring and evaluation. The constraints on approaches to adaptation due to broader policy and development goals should also be noted: this is one reason that adaptation fails to be optimum. Monitoring and evaluation is critical and yet easily ignored, and is essential to a "learning-by-doing approach" which is appropriate to the large uncertainties associated with adaptation and global and RSLR and coastal management in general (see National Research Council 1995; Willows and Cornell 2003).

Lastly, in many countries there is limited capacity to address today's coastal problems, let alone consider tomorrow's, including sea-level rise. Therefore, capacity-building should include developing coastal management, as already widely recommended (Ehler et al. 1997; Nicholls and Klein 2005; Adger et al. 2007). In the context of small islands, Tompkins et al. (2005) identified seven elements for an adaptation plan in addition to good information and science: (1) finance, (2) support networks, (3) legislation and enforcement, (4) links to other planning processes, (5) risk-management plans, (6) responsibility and ownership of development, and (7) education and communication.

2.8 Next Steps

In the following chapters of this book, the different aspects of the science of sea-level rise are examined. If we are to successfully manage global (and relative) sea-level rise, it is important that we progress all aspects of this science agenda and reduce the uncertainty on global sea-level rise and its regional distribution. Monitoring of sea-level rise and comparison with predictions is particularly important to identify and communicate any changes larger than expected (e.g. Rahmstorf et al. 2007). Three factors deserve particular attention as they could produce large rises in mean and extreme relative sea level and are associated with large uncertainty:

- a much better understanding of the physical basis of the potential collapse of the Greenland and West Antarctic Ice Sheet and the resulting low-probability/high-consequence impacts (see Chapter 7);

- an improved understanding of the regional distribution of sea-level rise and extreme sea-level events and how they will evolve in a warming world, including the analysis of their different components (mean sea level, surges, etc.) and their interaction (see Chapter 11);
- more emphasis on the non-climate components of sea-level rise, as these may exceed the climate component of sea-level rise in many densely populated deltaic areas during the 21st century. While this may be perceived as a local phenomenon compared to the global processes considered in this book, there are important commonalities between these sites both in terms of understanding and responding to these problems, and in impact terms they are a phenomenon of global importance (e.g. Nicholls 2004; Ericson et al. 2006). This suggests the need for more systematic monitoring of the areas that are most susceptible to human-induced subsidence, as these areas are easily identifiable based on geological criteria.

In terms of other aspects of understanding impacts and responses to sea-level rise, the following issues are important.

- Assessment of sea-level-rise impacts and responses within a coastal management context (including extreme sea levels) will quantify the relative importance of climate change versus non-climate drivers and will identify responses that address all the drivers of change in the coastal zone.
- There must be a focus on vulnerable coastal areas and types, such as small islands, deltas, coastal ecosystems, and low-lying cities that are a priority for impact and adaptation assessment.
- Understanding the process and hence the success (or failure) of adaptation is probably the largest uncertainty in understanding the residual impacts of sea-level rise. It deserves much more attention at local to global scales, including assessments of a wider range of adaptations across the spectrum of retreat, accommodate, and protect.
- Given the commitment to sea-level rise beyond 2100 and the consequent commitment to adaptation, long-term adaptation decisions need to be carefully considered, particularly the trade-offs between protection and retreat. This is more than a technical question as it will concern society's attitudes to residual risk which can never be reduced to zero. If we ultimately need to retreat, this process has a long lead time to be most effective.

2.9 Concluding Remarks

The chapter illustrates that understanding the impacts of global and RSLR crosses many disciplines and embraces natural, social, and engineering sciences, and major gaps in that understanding remain. The success or failure of adaptation in general, and protection in particular, is an important issue which deserves more attention and has lead to what Nicholls and Tol (2006) termed the "optimistic" and "pessimistic" views of the importance of sea-level rise. The pessimists tend to focus on high rises in sea level and extreme events like Katrina, and view our ability to adapt as being rather limited, resulting in alarming impacts, including widespread human displacement from coastal areas. The optimists tend to focus on lower rises in sea level and stress a high ability to protect and high benefit-cost ratios in developed areas and wonder what all the fuss is about.

The optimists have empirical evidence to support their views that sea-level rise is not a big problem in terms of the subsiding megacities that are also thriving. Importantly, these analyses suggest that improved protection under rising sea levels is more likely and rational than is widely assumed. Hence the common assumption of a widespread retreat from the shore in the face of global sea-level rise is not inevitable, and coastal societies will have more choice in their response than is often assumed. However, the pessimists also have evidence to support their view. First, the widely used SRES socioeconomic scenarios are optimistic about future economic growth and its more equitable distribution: lower growth and greater concentration of wealth in parts of the world may mean less damage in monetary terms, but it will also lead to a lower ability to protect. Secondly, the benefit-cost approach implies a proactive attitude to protection, while historical experience shows most protection has been a reaction to actual or near disaster. Therefore, high rates of sea-level rise may lead to more frequent coastal disasters, even if the ultimate response is better protection. Thirdly, disasters (or adaptation failures) such as Hurricane Katrina could trigger coastal abandonment, a process that has not been analyzed to date. This could have a profound influence on society's future choices concerning coastal protection as the pattern of coastal occupancy might change radically. A cycle of decline in some coastal areas is not inconceivable, especially in future world scenarios where capital is highly mobile and collective action is weaker. As the issue of sea-level rise is so widely known, disinvestment from coastal areas may be triggered even without disasters actually occurring: for example, the economies of small islands may be highly vulnerable if investors become cautious (Barnett and Adger 2003). Lastly, the commitment to sea-level rise beyond 2100 raises questions about the appropriate balance between protection and retreat: while we can protect now, should we protect in the long-term? As retreat and accommodation have long lead times, the benefits are greatest if implementation occurs soon, but we do not appear to be systematically exploiting these options as yet. For all these reasons, adaptation may not be as successful as some assume, especially if rises in global sea level are at the higher end of the range of predictions.

Thus the optimists and the pessimists both have arguments in their favor. Sea-level rise is clearly a threat, which demands a response. The commitment to sea-level rise means that an ongoing adaptation response is essential through the 21st century and beyond. However, mitigation can reduce the commitment to sea-level rise, most particularly the potential Greenland and West Antarctic contributions. Scientists need to better understand these threats, including the implications of different mixtures of adaptation and mitigation, as well as the need to engage with the coastal and climate policy process so that these scientific perspectives are heard. Importantly, it has been recognized that a combination of mitigation (to reduce the risks of a large global rise in sea level) and adaptation (to the inevitable global rise) appears to be the most appropriate course of action, as these two policies are more effective when combined than when followed independently, and together they address both immediate and longer-term concerns (Nicholls et al. 2007a).

Acknowledgments

Susan Hanson, University of Southampton, helped prepare the figures and references.

References

Adger W.N., Agrawala S., Mirza M.M.Q., Conde C., O'Brien K., Pulhin J. et al. (2007) Assessment of adaptation practices, options, constraints and capacity. In: *Climate Change 2007: Impacts, Adaptation and Vulnerability. Contribution of Working Group II to the Fourth Assessment Report of the Intergovernmental Panel on Climate Change* (Parry M.L., Canziani O.F., Palutikof J.P., van der Linden P.J. and Hanson C.E., eds). Cambridge University Press, Cambridge.

Anthoff D., Nicholls R.J., Tol R.S.J. and Vafeidis A.T. (2006) *Global and Regional Exposure to Large Rises in Sea-Level: A Sensitivity Analysis*. Working Paper 96. Tyndall Centre for Climate Change Research, University of East Anglia, Norwich.

Aráujo I. and Pugh D.T. (2008) Sea levels at Newlyn 1915–2005: analysis of trends for future flooding risks. *Journal of Coastal Research* **24**(sp3), 203–12.

Barnett J. and Adger W.N. (2003) Climate dangers and atoll countries. *Climatic Change* **61**, 321–37.

Barras J., Beville S., Britsch D., Hartley S., Hawes S., Johnston J. et al. (2003) *Historical and Projected Coastal Louisiana Land Changes: 1978–2050*. Open File Report 03-334 39. US Geological Survey.

Barth M.C. and Titus J.G. (eds) (1984) *Greenhouse Effect and Sea Level Rise: A Challenge for this Generation*. Van Nostrand Reinhold, New York.

Bijlsma L., Ehler C.N., Klein R.J.T., Kulshrestha S.M., McLean R.F., Mimura N. et al. (1996) Coastal zones and small islands. In: *Impacts, Adaptations and Mitigation of Climate Change: Scientific-Technical Analyses*. The Second Assessment Report of the Intergovernmental Panel on Climate Change, Working Group II (Watson R.T., Zinyowera M.C. and Moss R.H., eds). Cambridge University Press, Cambridge.

Bird E.C.F. (1985) *Coastline Changes: A Global Review*. John Wiley and Sons, New York.

Bird E.C.F. (2000) *Coastal Geomorphology: An Introduction*. Wiley and Sons, Chichester.

Bosello F., Roson R. and Tol R.S.J. (2007) Economy-wide estimates of the implications of climate change: sea level rise. *Environmental & Resource Economics* **37**, 549–71.

Cahoon D.R., Day J.W. and Reed D.J. (1999) The influence of surface and shallow subsurface soil processes on wetland elevation: a synthesis. *Current Topics in Wetland Biogeochemistry* **3**, 72–88.

Cahoon D.R., Hensel P.F., Spencer T., Reed D.J., McKee K.L. and Saintilan N. (2006) Coastal wetland vulnerability to relative sea-level rise: wetland elevation trends and process controls. In: *Wetlands as a Natural Resource, Vol. 1: Wetlands and Natural Resource Management*. Springer Ecological Studies Series (Verhoeven J., Whigham D., Bobbink R. and Beltman B., eds). Springer Verlag, Berlin.

Capobianco M., Devriend H.J., Nicholls R.J. and Stive M.J.F. (1999) Coastal area impact and vulnerability assessment: The point of view of a morphodynamic modeller. *Journal of Coastal Research* **15**, 701–16.

Coleman J.M., Huh O.K. and Braud D. (2008) Wetland loss in world deltas. *Journal of Coastal Research* **24**, 1–14.

Cowell P.J., Stive M.J.F., Niedoroda A.W., De Vriend H.J., Swift D.J.P., Kaminsky G.M. and Capobianco M. (2003a) The coastal tract. Part 1: A conceptual approach to aggregated modelling of low-order coastal change. *Journal of Coastal Research* **19**, 812–27.

Cowell P.J., Stive M.J.F., Niedoroda A.W., Swift D.J.P., De Vriend H.J., Buijsman M.C. et al. (2003b) The coastal tract. Part 2: Applications of aggregated modelling of lower-order coastal change. *Journal of Coastal Research* **19**, 828–48.

Cowell P.J., Thom B.J., Jones R.A., Everts C.H. and Simanovic D. (2006) Management of uncertainty in predicting climate-change impacts on beaches. *Journal of Coastal Research* **22**, 232–45.

Crossland C.J., Kremer H.H., Lindeboom H.J., Marshall Crossland J.I. and Le Tissier M.D.A. (eds) (2005) *Coastal Fluxes in the Anthropocene*. Springer, Berlin.

Darwin R.F. and Tol R.S.J. (2001) Estimates of the economic effects of sea level rise. *Environmental and Resource Economics* **19**, 113–29.

Dawson R.J., Hall J.W., Bates P.D. and Nicholls R.J. (2005) Quantified analysis of the probability of flooding in the Thames Estuary under imaginable worst case sea-level rise scenarios. Special edition on Water and Disasters. *International Journal of Water Resources Development* **21**, 577–91.

Dawson R.J., Dickson M., Nicholls R.J., Hall J.W., Walkden M., Stansby P. et al. (2007) *Integrated Analysis of Risks of Coastal Flooding and Cliff Erosion under Scenarios of Long Term Change*. Tyndall Centre for Climate Change Research, University of East Anglia, Norwich.

De La Vega-Leinert A.C. and Nicholls R.J. (2001) *Proceedings of the SURVAS Overview Workshop on the Future of Vulnerability and Adaptation Studies*. 28–30 June 2001, Enfield.

Dickson M.E., Walkden M.J.A. and Hall J.W. (2007) Systemic impacts of climate change on an eroding coastal region over the twenty-first century. *Climatic Change* **84**, 141–66.

Dixon T.H., Amelung F., Ferretti A., Novali F., Rocca F., Dokka R. et al. (2006) Subsidence and flooding in New Orleans. *Nature* **441**, 587–8.

Ehler C.N., Cicin-Sain B., Knecht R., South R. and Weiher R. (1997) Guidelines to assist policy makers and managers of coastal areas in the integration of coastal management programs and national climate-change action plans. *Ocean & Coastal Management* **37**(1), 7–27.

Emery K.O. and Aubrey D.G. (1991) *Sea Levels, Land Levels and Tide Gauges.* Springer Verlag, New York.

Ericson J.P., Vörösmarty C.J., Dingman S.L., Ward L.G. and Meybeck M. (2006) Effective sea-level rise and deltas: causes of change and human dimension implications. *Global and Planetary Change* **50**, 63–82.

Essink G. (1996) *Impact of Sea Level Rise on Groundwater Flow Regimes. A Sensitivity Analysis for the Netherlands.* Delft Technical University of Technology, Delft.

Essink G. (2001a) Improving fresh groundwater supply – problems and solutions. *Ocean & Coastal Management* **44**, 429–49.

Essink G. (2001b) Salt water intrusion in a three-dimensional groundwater system in the Netherlands: a numerical study. *Transport in Porous Media* **43**, 137–58.

Evans E., Ashley R.M., Hall J., Penning-Rowsell E., Saul A., Sayers P. et al. (2004) *Foresight Future Flooding Scientific Summary; Volumes I and II.* Office of Science and Technology, London.

Fankhauser S. (1995) Protection versus retreat: estimating the costs of sea-level rise. *Environment and Planning A* **27**, 299–319.

Fankhauser S. and Tol R.S.J. (2005) On climate change and economic growth. *Resource and Energy Economics* **27**, 1–17.

Fankhauser S., Smith J.B. and Tol R.S.J. (1999) Weathering climate change: some simple rules to guide adaptation decisions. *Ecological Economics* **30**, 67–78.

Few R., Ahern M., Matthies F. and Kovats S. (2004) *Floods, Health and Climate Change: A Strategic Review.* Working Paper 63. Tyndall Centre for Climate Change Research, University of East Anglia, Norwich.

Fletcher C.A. and Spencer T. (eds) (2005) *Flooding and Environmental Challenges for Venice and its Lagoon: State of Knowledge.* Cambridge University Press, Cambridge.

Flynn T.J., Walesh S.G., Titus J.G. and Barth M.C. (1984) Implications of sea level rise for hazardous waste sites in coastal floodplains. In: *Greenhouse Effect and Sea Level Rise: A Challenge for this Generation* (Barth M.C. and Titus J.G., eds). Van Nostrand Reinhold, New York.

Ganopolski A., Petoukhov V., Rahmstorf S., Brovkin V, Claussen M., Eliseev A. and Kubatzki C. (2001) CLIMBER-2: a climate system model of intermediate complexity. Part II: model sensitivity. *Climate Dynamics* **17**, 735–51.

Gardiner S., Hanson S., Nicholls R.J., Zhang Z., Jude S.R., Jones A. et al. (2007) The Habitats Directive, coastal habitats and climate change – case studies from the south coast of the U.K. In: *Proceedings of the ICE International Conference on Coastal Management 2007* (McInnes R., ed.), pp. 193–202. Thomas Telford, London.

Gibbons S.J.A. and Nicholls R.J. (2006) Island abandonment and sea-level rise: An historical analog from the Chesapeake Bay, USA. *Global Environmental Change – Human and Policy Dimensions* **16**, 40–7.

Gornitz V., Couch S. and Hartig E.K. (2002) Impacts of sea-level rise in the New York City metropolitan area. *Global and Planetary Change* **32**, 61–88.

Grossi P. and Muir-Wood R. (2006) *Flood Risk in New Orleans: Implications for Future Management and Insurability*. Risk Management Solutions (RMS), London.

Hartig E.K., Gornitz V., Kolker A., Muschacke F. and Fallon D. (2002) Anthropogenic and climate-change impacts on salt marshes of Jamaica Bay, New York City. *Wetlands* **22**, 71–89.

Hay J.E. and Mimura N. (2005) Sea-level rise: Implications for water resources management. *Mitigation and Adaptation Strategies for Global Change* **10**, 717–37.

Hinkel J. and Klein R.J.T. (2007) Integrating knowledge for assessing coastal vulnerability to climate change. In: *Managing Coastal Vulnerability: An Integrated Approach*. (McFadden L., Nicholls R. and Penning-Rowsell E.C., eds). Elsevier Science, Amsterdam.

Hoozemans F.M.J., Marchand M. and Pennekamp H.A. (1993) *A Global Vulnerability Analysis: Vulnerability Assessment for Population, Coastal Wetlands and Rice Production on a Global Scale*, 2nd edn. Delft Hydraulics, Delft.

Hull C.H.J. and Titus J.G. (eds) (1986) *Greenhouse Effect, Sea Level Rise, and Salinity in the Delaware Estuary*. US Environmental Protection Agency and Delaware River Basin Commission, Washington DC.

IGES (2007) *Sustainable Groundwater Management in Asian Cities: A Final Report of Research on Sustainable Water Management Policy*. IGES Freshwater Resources Management Project, Hayama, Japan. Institute for Global Environmental Strategies (IGES), Hayama.

IPCC CZMS (1990) *Strategies for Adaptation to Sea Level Rise*. Report of the Coastal Zone Management Subgroup, Response Strategies Working Group of the Intergovernmental Panel on Climate Change. Ministry of Transport, Public Works and Water Management, The Hague.

IPCC (2007) *Climate Change 2007: Synthesis Report*. Contribution of Working Groups I, II and III to the Fourth Assessment Report of the Intergovernmental Panel on Climate Change (Core Writing Team, Pachauri R.K and Reisinger A., eds). Cambridge University Press, Cambridge.

Jacob K., Gornitz V. and Rosenzweig C. (2007) Vulnerability of the New York City metropolitan area to coastal hazards, including sea-level rise: Inferences for urban coastal risk management and adaptation policies. In: *Managing Coastal Vulnerability* (McFadden L., Nicholls R.J. and Penning-Rowsell E., eds). Elsevier, Oxford.

Kearney M.S. and Stevenson J.C. (1991) Island land loss and marsh vertical accretion rate: Evidence for historical sea-level changes in Chesapeake Bay. *Journal of Coastal Research* **7**, 403–15.

Keller K., Tol R.S.J., Toth F.L. and Yohe G.W., (guest editors) (2008) Special Issue: Abrupt Climate Change near the Poles. *Climatic Change* **91**, 1–209.

Klein R.J.T. and Nicholls R.J. (1999) Assessment of coastal vulnerability to climate change. *Ambio* **28**, 182–7.

Klein R.J.T., Nicholls R.J. and Mimura N. (1999) Coastal adaptation to climate change: can the IPCC Technical Guidelines be applied? *Mitigation and Adaptation Strategies for Global Change* **4**, 51–64.

Klein R.J.T., Aston J., Buckley E.N., Capobianco M., Mizutani N., Nicholls R.J. et al. (2000) Coastal adaptation. In: *IPCC Special Report on Methodological and Technological Issues in Technology Transfer* (Metz B., Davidson O.R., Martens J.W., Van Rooijen S.N.M. and Van Wie McGrory L.L., eds). Cambridge University Press, Cambridge.

Klein R.J.T., Nicholls R.J., Ragoonaden S., Capobianco M., Aston J. and Buckley E.N. (2001) Technological options for adaptation to climate change in coastal zones. *Journal of Coastal Research* **17**, 531–43.

Kremer H.H., Le Tisser M.D.A., Burbridge P.R., Talaue-McManus L., Rabalais N.N., Parslow J. et al. (2004) *Land-Ocean Interactions in the Coastal Zone: Science Plan and Implementation Strategy*. IGBP Report 51/IHDP Report 18 60. GBP Secretariat, Stockholm.

Leatherman S.P. (1992) Coastal land loss in the Chesapeake Bay Region: An historical analogy approach to global climate analysis and response. In: *Regions and Global Warming: Impacts and Response Strategies* (Schmandt J., ed.). Oxford University Press, Oxford.

Lee, M. (2001) Coastal defence and the habitats directive: predictions of habitat change in England and Wales. *The Geographical Journal* **167**, 39–56.

McCarthy J., Canziani O.F., Leary N.A., Dokken D.L. and White K.S. (2001) *Climate Change 2001: Impacts, Adaptation and Vulnerability*. Contribution of the Working Group II to the Third Assessment Report of the Intergovernmental Panel on Climate Change (IPCC). Cambridge University Press, Cambridge.

McFadden L., Spencer T. and Nicholls R.J. (2007) Broad-scale modelling of coastal wetlands: what is required? *Hydrobiologia* **577**, 5–15.

McGranahan G., Balk D. and Anderson B. (2007) The rising tide: assessing the risks of climate change and human settlements in low elevation coastal zones. *Environment and Urbanization* **19**, 17–37.

Meehl G.A., Stocker T.F., Collins W.D., Friedlingstein P., Gaye A.T., Gregory J.M. et al. (2007) Global Climate Projections. In: *Climate Change 2007: The Physical Science Basis*. Contribution of Working Group 1 to the Fourth Assessment Report of the Intergovernmental Panel on Climate Change (Solomon S., Qin D., Manning M., Marquis M., Averyt K., Tignor M.M.B. et al., eds), pp. 747–845. Cambridge University Press, Cambridge.

Milliman J.D. and Haq B.U. (eds) (1996) *Sea Level Rise and Coastal Subsidence: Causes, Consequences and Strategies*. Kluwer Academic, Dordrecht.

Milliman J.D., Broadus J.M. and Gable F. (1989) Environmental and economic implications of rising sea level and subsiding deltas: the Nile and Bengal examples. *Ambio* **18**, 340–345.

Mimura N. (2000) *Distribution of Vulnerability and Adaptation in the Asia and Pacific Region. Proceedings of the APN/SURVAS/LOICZ Joint Conference on the Coastal Impacts of Climate Change and Adaptation in the Asia-Pacific Region*, APN and Ibaraki University, pp. 21–25.

Mimura N. and Nobuoka H. (1996) *Verification of the Bruun Rule for the Estimation of Shoreline Retreat Caused by Sea-Level Rise. Proceedings of Coastal Dynamics '95*, pp. 607–16. ASCE, Gdansk.

Myers N. (2002) Environmental refugees: a growing phenomenon of the 21st century. *Philosophical Transactions of the Royal Society B* **357**, 609–63.

National Research Council (1987) *Responding to Changes in Sea Level: Engineering Implications*. National Academy Press, Washington DC.

National Research Council (1995) *Science Policy and the Coast: Improving Decision Making*. National Academy Press, Washington DC.

Neumann J.E., Yohe G., Nicholls R.J. and Manion M. (2001) Sea level rise and its effects on global resources. In: *Climate Change: Science, Strategies and Solutions* (Claussen E., ed.). Brill, Boston.

Nicholls R.J. (1995) Coastal megacities and climate change. *GeoJournal* **37**, 369–79.

Nicholls R.J. (2004) Coastal flooding and wetland loss in the 21st century: changes under the SRES climate and socio-economic scenarios. *Global Environmental Change – Human and Policy Dimensions* **14**, 69–86.

Nicholls R.J. (2007) *Adaptation Options for Coastal Areas and Infrastructure: An Analysis for 2030*. Report to the United Nations Framework Convention on Climate Change. United Nations, Bonn.

Nicholls R.J. and Mimura N. (1998) Regional issues raised by sea-level rise and their policy implications. *Climate Research* **11**, 5–18.

Nicholls R.J. and Klein R.J.T. (2005) Climate change and coastal management on Europe's coast. In: *Managing European Coasts: Past, Present and Future* (Vermaat J.E., Ledoux L., Turner K., Salomons W. and Bouwer L., eds). Springer, Environmental Science Monograph Series, London.

Nicholls R.J. and Lowe J.A. (2006) Climate stabilisation and impacts of sea-level rise. In: *Avoiding Dangerous Climate Change* (Schellnhuber H.J., Cramer W., Nakicenovic N., Wigley T.M.L. and Yohe G., eds). Cambridge University Press, Cambridge.

Nicholls R.J. and Tol R.S.J. (2006) Impacts and responses to sea-level rise: a global analysis of the SRES scenarios over the twenty-first century. *Philosophical Transactions of the Royal Society A* **364**, 1073–95.

Nicholls R.J., Hoozemans M.J. and Marchand M. (1999) Increasing flood risk and wetland losses due to global sea-level rise: regional and global analyses. *Global Environmental Change* **9**, S69–87.

Nicholls R.J., Wong P.P., Burkett V.R., Codignotto J.O., Hay J.E., McLean R.F., Ragoonaden S. and Woodroffe C.D. (2007a) Coastal systems and low-lying areas. In: *Climate Change 2007: Impacts, Adaptation and Vulnerability. Contribution of Working Group II to the Fourth Assessment Report of the Intergovernmental Panel on Climate Change* (Parry M.L., Canziani O.F.,

Palutikof J.P., van der Linden P.J. and Hanson C.E., eds). Cambridge University Press, Cambridge.

Nicholls R.J., Hanson S., Herweijer C., Patmore N., Hallegatte S., Corfee-Morlot J. et al. (2007b) *Ranking Port Cities with High Exposure and Vulnerability to Climate Extremes – Exposure Estimates*. Environmental Working Paper No. 1. Organisation for Economic Co-operation and Development (OECD), Paris.

Nicholls R.J., Klein R.J.T. and Tol R.S.J. (2007c) Managing coastal vulnerability and climate change: a national to global perspective. In: *Managing Coastal Vulnerability* (McFadden L., Nicholls R.J. and Penning-Rowsell E., eds). Elsevier, Oxford.

Nordhaus W.D. (1991) To slow or not to slow. *Economics Journal* **5**, 920–37.

Nurse L., Sem G., Hay J.E., Suarez A.G., Wong P.P., Briguglio L. and Ragoonaden S. (2001) Small island states. In: *Climate Change 2001: Impacts, Adaptation and Vulnerability* (McCarthy J.J., Canziani O.F., Leary N.A., Dokken D.J. and White K.S., eds). Cambridge University Press, Cambridge.

Pernetta J.C. (1992) Impacts of climate change and sea-level rise on small island states: national and international responses. *Global Environmental Change* **2**, 19–31.

Phien-Wej N., Giao P.H. and Nutalaya P. (2006) Land subsidence in Bangkok, Thailand. *Engineering Geology* **82**, 187–201.

Rahmstorf S., Cazenave A., Church J.A., Hansen J.E., Keeling R.F., Parker D.E. and Somerville R.C.J. (2007) Recent climate observations compared to projections. *Science* **316**, 709.

Ranjan S.P., Kazama S. and Sawamoto M. (2006) Effects of climate and land use changes on groundwater resources in coastal aquifers. *Journal of Environmental Management* **80**, 25–35.

Rodolfo K.S. and Siringan F.P. (2006) Global sea-level rise is recognised, but flooding from anthropogenic land subsidence is ignored around northern Manila Bay, Philippines. *Disaster Management* **30**, 118–39.

Rosenzweig C., Casassa G., Karoly D.J., Imeson A., Liu C., Menzel A. et al. (2007) Assessment of observed changes and responses in natural and managed systems. In: *Climate Change 2007: Impacts, Adaptation and Vulnerability*. Contribution of Working Group II to the Fourth Assessment Report of the Intergovernmental Panel on Climate Change (Parry M.L., Canziani O.F., Palutikof J.P., van der Linden P.J. and Hanson C.E., eds). Cambridge University Press, Cambridge.

Sachs J.D., Mellinger A.D. and Gallup J.L. (2001) The geography of poverty and wealth. *Scientific American* **284**, 70–75.

Scavia D., Field J.C., Boesch D.F., Buddemeier R.W., Burkett V., Cayan D.R. et al. (2002) Climate Change Impacts on U.S. Coastal and Marine Ecosystems. *Estuaries* **25**, 149–64.

Sherif M.M. and Singh V.P. (1999) Effect of climate change on sea water intrusion in coastal aquifers. *Hydrological Processes* **13**, 1277–87.

Smith J.B. and Tirpak D.A. (eds) (1989) *The Potential Effects of Global Climate Change on the United States*. US Environmental Protection Agency, Washington DC.

Sorensen R.M., Weisman R.N. and Lennon G.P. (1984) Control of erosion, inundation, and salinity intrusion. In: *Greenhouse Effect and Sea Level Rise: A Challenge for this Generation* (Barth M.C. and Titus J.G., eds.). Van Nostrand Reinhold, New York.

Stern N. (2006) *The Economics of Climate Change. The Stern Review*. Cambridge University Press, Cambridge.

Sterr H. (2008) Assessment of vulnerability and adaptation to sea-level rise for the coastal zone of Germany. *Journal of Coastal Research* **24**, 380–93.

Stive M.J.F. (2004) How important is global warming for coastal erosion? An editorial comment. *Climatic Change* **64**, 27–39.

Stive M.J.F., Roelvink J.A. and De Vriend H.J. (1990) *Large-Scale Coastal Evolution Concept. Proceedings of the 22nd International Conference on Coastal Engineering, Delft, ASCE, New York*, pp. 1962–74.

Stive M.J.F., Cowell P. and Nicholls R. J. (2009) Beaches, cliffs and deltas. In *Geomorphology and Global Environmental Change* (Slaymaker O., Spencer T. and Embleton-Hamann C., eds). Cambridge University Press, Cambridge, pp. 158–79.

Sugiyama M., Nicholls R.J. and Vafeidis A. (2008) *Estimating the Economic Cost of Sea-Level Rise*. Report No. 156. MIT Joint Program on the Science and POLICY of Global Change, Boston.

Syvitski J.P.M. (2008) Deltas at risk. *Sustainability Science* **3**, 23–32.

Syvitski J.P.M., Vörösmarty C.J., Kettner A.J. and Green P. (2005) Impact of humans on the flux of terrestrial sediment to the global coastal ocean. *Science* **308**, 376–80.

Ten Brinke W.B.M., Bonnink B.A. and Ligtvoet W. (2008) The evaluation of flood risk policy in the Netherlands. *Water Management, Proceedings of the Institution of Civil Engineers* **161**, 181–8.

Thorne C., Evans E. and Penning-Rowsell E. (eds) (2007) *Future Flooding and Coastal Erosion Risks*. Thomas Telford, London.

Titus J.G. (ed.) (1988) *Greenhouse Effect and Sea-Level Rise – the Cost of Holding Back the Sea*. US Environmental Protection Agency, Washington DC.

Titus J.G., Kuo C.Y., Gibbs M.J., Laroche T.B., Webb M.K. and Waddell J.O. (1987) Greenhouse effect, sea level rise, and coastal drainage systems. *Journal of Water Resources Planning and Management* **113**, 216–25.

Titus J.G., Park R.A., Leatherman S.P., Weggel J.R., Greene M.S., Mausel P.W. et al. (1991) Greenhouse effect and sea level rise: the cost of holding back the sea. *Coastal Management* **19**, 171–204.

Tol R.S.J. (2002a) Estimates of the damage costs of climate change. Part I: Benchmark estimates. *Environmental and Resource Economics* **21**, 47–73.

Tol R.S.J. (2002b) Estimates of the damage costs of climate change. Part II: Dynamic estimates. *Environmental and Resource Economics* **21**, 135–60.

Tol R.S.J. (2007) The double trade-off between adaptation and mitigation for sea level rise: an application of FUND. *Mitigation and Adaptation Strategies for Global Change* **12**, 741–53.

Tompkins E.L., Nicholson-Cole S.A., Hurlston L., Boyd E., Hodge G.B., Clarke J., Gray G., Trotz N. and Varlack L. (2005) *Surviving Climate Change in Small Islands: A Guidebook*. Tyndall Centre for Climate Change Research, University of East Anglia, Norwich.

Tribbia J. and Moser S.C. (2008) More than information: what coastal managers need to plan for climate change. *Environmental Science & Policy* **11**, 315–28.

Turner R.K., Doktor P. and Adger W.N. (1995) Assessing the costs of sea-level rise. *Environment and Planning A* **27**, 1777–96.

Valiela I. (2006) *Global Coastal Change*. Blackwell, Malden, MA.

Van Goor M.A., Zitman T.J., Wang Z.B. and Stive M.J. (2003) Impact of sea-level rise on the morphological equilibrium state of tidal inlets. *Marine Geology* **202**, 211–27.

Van Koningsveld M., Mulder J.P.M., Stive M.J.F., Vandervalk L. and Vanderweck A.W. (2008) Living with sea-level rise and climate change: A case study of the Netherlands. *Journal of Coastal Research* **24**, 367–79.

Vellinga P. and Leatherman S.P. (1989) Sea level rise, consequences and policies. *Climatic Change* **15**, 175–89.

von Storch H. and Woth K. (2008) Storm surges: perspectives and options. *Sustainability Science* **3**, 33–43.

Walkden M. and Dickson M. (2008) Equilibrium erosion of soft rock shores with a shallow or absent beach under increased sea level rise. *Marine Geology* **251**, 75–84.

Warrick R.A., Barrow E.M. and Wigley T.M.L. (eds) (1993) *Climate and Sea Level Change: Observations, Projections, Implications*. Cambridge University Press, Cambridge.

Willows R.I. and Connell R.K. (eds) (2003) *Climate Adaptation. Risk, Uncertainty and Decision Making*. UK Climate Impacts Programme (UKCIP), Oxford.

Woodroffe C.D., Nicholls R.J., Saito Y., Chen Z. and Goodbred S.L. (2006) Landscape variability and the response of Asian megadeltas to environmental change. In: *Global Change Implications for Coasts in the Asia-Pacific Region* (Harvey N., ed.). Springer, London.

Woodworth P.L. and Blackman D.L. (2004) Evidence for systematic changes in extreme high waters since the mid-1970s. *Journal of Climate* **17**, 1190–7.

Yohe G. (1990) The cost of not holding back the sea – toward a national sample of economic vulnerability. *Coastal Management* **18**, 403–31.

Yohe G. and Schlesinger M. (1998) Sea level change: the expected economic cost of protection or abandonment in the United States. *Climatic Change* **38**, 447–72.

Yohe G.W., Neumann J.E., Marshall P. and Ameden H. (1996) The economic costs of sea level rise on US coastal properties. *Climatic Change* **32**, 387–410.

Zhang K., Douglas B.C. and Leatherman S.P. (2000) Twentieth-century storm activity along the U.S. east coast. *Journal of Climate* **13**, 1748–61.

Zhang K.Q., Douglas B.C. and Leatherman S.P. (2004) Global warming and coastal erosion. *Climatic Change* **64**, 41–58.

3 A First-Order Assessment of the Impact of Long-Term Trends in Extreme Sea Levels on Offshore Structures and Coastal Refineries

Ralph Rayner and Bev MacKenzie

3.1 Introduction

One of the projected impacts of global warming is a progressive rise in mean sea level throughout the 21st century and beyond. Historical observations of mean sea level show a general upward trend. Recent work on historical records indicates that the rate of change is increasing (Church 2001; Church and White 2006). Church and White (2006) identify a global mean rise of 195 mm over the period 1870–2004, a 20th-century rate of sea-level rise of 1.7 ± 0.3 mm $year^{-1}$ and a significant acceleration of sea-level rise of 0.013 ± 0.006 mm $year^{-2}$. At this rate of acceleration, the rise over the period 1990–2100 would be in the range 280–340 mm.

The Intergovernmental Panel on Climate Change (IPCC) Third Assessment Report (IPCC 2001) projects a rise of 90–880 mm between 1990 and 2100 (depending on future carbon dioxide-emission scenarios) with a mid estimate of 480 mm (Church et al. 2001). It is evident from both the analysis of historical records and the IPCC report that the level of uncertainty in future projections of mean sea level is large. However, it is now generally accepted that present trends will be sustained or may further accelerate if Greenland and Antarctica become significant sources of freshwater input (Church et al. 2001) as detected in recent observations (Chapter 7).

Even greater uncertainty is associated with predicting future changes in extreme storm surges and wave heights. For both of these parameters future trends are an indirect consequence of possible long-term trends in atmospheric pressure fields and winds as well as changes in mean sea level. In particular, trends in the severity

Understanding Sea-Level Rise and Variability, 1st edition. Edited by John A. Church, Philip L. Woodworth, Thorkild Aarup & W. Stanley Wilson.

and frequency of storm events associated with generating large surges and extreme waves are very difficult to predict.

The sometimes devastating impact of extreme sea levels resulting from the combination of high tides, surges, and storm waves is well known. The severe coastal flooding around the North Sea in 1953 and the frequent coastal floods in the low-lying areas of Bangladesh are examples of the many events which have resulted in loss of life and extensive damage to infrastructure and the environment. The recent catastrophic effects of Hurricanes Rita and Katrina have heightened awareness of property damage and loss of life that can result from coastal inundation. These two events have also provoked an intense debate about the relative contributions of chance, natural cycles, and anthropogenic change (Cooper and Stear 2006).

It has become increasingly important to take account of probable long-term trends in sea level, extreme surges, and extreme waves in ensuring the protection of life and property in coastal regions. A number of authors have sought to determine the socioeconomic and environmental impact of increased risk of coastal flooding due to rising sea levels and changes in storm severity and frequency. Examples are the work of Titus et al. (1991) and more recently Nicholls (2003), Nicholls and Lowe (2004), and Woodworth et al. (2004). The work of these authors has concentrated on the general impact on coastal regions.

Long-term trends in sea level and in the frequency and magnitude of extreme waves and surges have a specific impact on individual facilities such as ports and harbors, coastal process plants, and offshore oil- and gas-production platforms (Figures 3.1 and 3.2). The economic impact on facilities of these kinds has received limited attention.

Here we consider the impact of long-term sea-level trends on the design of new offshore structures and coastal refineries and the possible additional costs of protecting existing installations.

Figure 3.1 The Henry Hub natural-gas facility surrounded by floodwaters in the aftermath of Hurricane Rita, Sunday September 25, 2005. (Photo credit: Associated Press/David J. Phillip.)

Figure 3.2 An oil platform construction facility surrounded by floodwaters in the aftermath of Hurricane Rita, Sunday September 25, 2005. (Photo credit: Associated Press/David J. Phillip.)

3.2 Design Considerations

Part of the established approach to the design of offshore and coastal installations is the estimation of environmental conditions that might occur over their design life. This includes determination of criteria for extremes of sea level. With varying degrees of sophistication, such design extremes are based on statistical estimation of the magnitude of the maximum sea level that might be expected to occur at least once in a specified return period.

Extremes are calculated based on the analysis of time histories of measured or modeled sea levels and the application of probabilistic techniques to extrapolate observed or modeled extreme values to return periods which are generally well in excess of the length of the data set. Probability distribution functions such as Weibull, Fisher Tippett, and Gumbel are used to extrapolate to these longer return periods (Kotz and Nadarajah 2000).

When estimating extremes of sea level, time histories of sea-surface elevation are broken down into a tidal and a non-tidal component. Maximum astronomical tidal elevations are derived using tidal predictions based on harmonic analysis of the tidal component of the data. The extreme surge component of sea-surface elevation is estimated based on probabilistic extrapolation of the non-tidal component using one or more probability distribution functions. Maximum tide and extreme surge elevations are then simply added or, more usually, combined based on joint probability of occurrence. A detailed account of the techniques used for the estimation of extreme tide and surge elevations and the use of joint probability techniques for their combination is provided by Pugh (1987, 2004).

This established approach to the estimation of extreme sea-surface elevations may be applied by means of specific calculations for individual facilities or through use of preset design standards and design codes applied on a regional basis.

Return periods are selected based on the expected life of the structure and the economic, safety, and environmental risks associated with inundation of sensitive plant, structural failure, or failed protection measures. It is through selection of different return periods that the trade-off between economic, safety, and environmental impacts and the costs of construction are assumed to be taken into account. For example, offshore oil- and gas-production structures have typically been designed for a 100-year return period and coastal nuclear power stations for a return period of 10 000 years or more.

The obvious limitation of this process for estimating extreme values is that it relies on nature being ergodic: it assumes that the time history used as the basis for the analysis is statistically representative of conditions over the generally far longer period for which extreme values are to be derived. This explicit assumption of stationarity is not valid when there is a long-term and accelerating rise in sea level or if a single event from a class of events which are absent in the observed past occurs (for example, impact of a hurricane at a location where no hurricanes have previously been observed).

There are some 6000 offshore oil- and gas-production facilities that have been designed on the basis of extreme sea-level criteria derived using this established approach. Almost all new offshore structures continue to be designed on the basis that long-term trends in environmental parameters, whether natural or anthropogenic, will be accommodated in the inherent conservatism of safety factors built into structural design. Similarly, the environmental design criteria for major coastal installations such as refineries and liquid-natural-gas processing and reception facilities have, until recently, taken no account of any long-term trends in environmental parameters.

Hurricanes Rita and Katrina have been something of a wake up call for the limitations of this approach. In a number of cases the design criteria for offshore structures in the Gulf of Mexico were exceeded during these two events. The level of damage to offshore and coastal installations and coastal infrastructure was significant. The full economic impact has yet to be determined but will certainly run into billions of dollars. In response to this wake-up call regulatory authorities, insurers, and offshore operators are increasingly seeking revisions to design codes and to the basis for calculating design criteria (Versowski et al. 2006).

3.3 Impact of Long-Term Trends in Extreme Sea Levels

At the design stage the impact of taking into account any long-term trends in extreme sea levels is predominantly economic as, over the design life of typical installations, long-term trends are unlikely to necessitate an entirely different design concept or make the proposed project impossible to engineer. For example, one can consider a fixed offshore production facility, one that is supported by a rigid structure attached to the seabed, such as a jacket or gravity platform. An increase in the extreme sea-surface elevations included in its design, to take account of long-term trends, would typically necessitate a higher elevation for parts of the structure (with consequent impact on steelwork or concrete costs) rather than a completely different design concept.

The move to offshore oil and gas production in ever deeper water has resulted in a large reduction in the use of fixed structures, although approximately 100 fixed structures per year continue to be installed for shallow-water production. In water depths greater than approximately 300 m, production takes place mostly from floating structures (which are tethered or anchored to the seabed). Such structures are not significantly impacted by long-term changes in mean sea-surface elevation.

For new coastal installations such as refineries and natural-gas processing and storage facilities, the predominant concern arising from sea level trends is the possibility of an increased risk of flooding over the design life of the facility. Where such facilities are protected by artificial sea defenses a significant increase in extreme sea level or extreme waves will necessitate increasing the design elevation of the defenses. Where location prevents use of artificial defenses it may be necessary to increase the elevation of the entire plant.

For existing installations, such as the approximately 6000 fixed offshore production facilities, 200 coastal refineries, and 70 natural-gas import and export facilities around the world, the principal impacts of exceeding environmental design criteria are possible damage to the facility and income losses associated with any disruption to operations. With operating facilities there are of course also potential impacts in terms of loss of life and harm to the environment.

Winds and waves associated with Hurricances Rita and Katrina destroyed 114 fixed offshore platforms with 52 other platforms sustaining damage (Minerals Management Service 2005). In addition, a relatively new deepwater floating production structure, the perhaps appropriately named tension-leg platform Typhoon, was capsized. Loss of production, either as a result of evacuation and shutdown of facilities, or because of structural damage, was highly significant. From June through November 2005, an estimated 100 million barrels of crude oil and 525 billion cubic feet of gas were left in the ground due to hurricane activity (Balint and Orange 2006). This shut-in production had a significant affect on prices for refined product. There was, fortunately, no loss of life in any of these offshore incidents, in stark contrast to events on the Louisiana coast.

Following Rita and Katrina there has been an understandable increase in the level of interest in the issue of hurricane frequency and intensity from offshore operators, regulators, offshore standards organizations, and insurers. As well as an increased interest in the specific case of hurricanes in the Gulf of Mexico these two high-profile events have sparked a more general awareness that long-term trends in environmental parameters might impact on the design of new maritime installations and the protection and operation of existing facilities.

3.4 Evaluating the Economic Impact

A rigorous approach to the evaluation of the potential economic impact of long-term trends in sea level, storm surges, and extreme waves on the protection of proposed and existing coastal and offshore installations requires the following stages:

- identification of installations at risk;
- determination of the cost of protection of these installations against possible long-term trends over their expected design life;
- determination of the likely damage costs that might arise if additional protection measures are not implemented.

As discussed above, offshore oil- and gas-production facilities are unlikely to be significantly impacted by projected trends in sea level over the remainder of the century given that the majority of fixed structures are likely to reach the end of their lives in the coming few decades and that most new installations are based on floating production concepts which are not impacted by rising sea levels. A much greater concern for new and existing floating production systems is the impact of forces associated with any adverse long-term trends in extreme winds or waves.

Our research has confirmed that estimates of long-term trends in sea level are now factored into the design process for most new coastal installations. This is achieved by adding the projected increase in mean sea level over the life of the facility to the extreme sea level derived from the analysis of historical data. Alternatively, some facilities are being designed in such a way that the costs of future modification to take account of unexpected sea-level rise are minimized. However, existing coastal facilities are vulnerable; their flood-protection measures have been designed without any consideration of long-term trends in sea level and they are located at sites likely to be occupied for many decades into the future.

Here we derive a first estimate of additional protection costs for existing coastal refinery sites under sea-level-rise scenarios of 25 and 50 cm. Table 3.1 indicates the number of coastal and estuarine refinery sites by region. This analysis is based on data provided by the Midwest Publishing Company (2006). We have worked with engineers in a major multinational refining company to determine a representative range of costs of additional protection measures for a typical medium-sized refining installation. Their analysis indicates that provision of additional flood protection for such a site would be in the range of US$10–20 million for a 25-cm sea-level rise and $20–30 million for a 50-cm rise.

If we assume that this range of costs is applicable to all of the existing refinery sites identified in Table 3.1 then we arrive at a first estimate of global additional protection cost associated with long-term sea-level rise of $2–4 billion for a 25-cm rise and $4–6 billion for a 50-cm rise. Clearly this first estimate is an

Table 3.1 Numbers of active coastal and estuarine refinery locations.

Region	Number of sites
Europe	66
North America	34
South America	20
Asia	66
Oceania	7
Africa	6
Total	199

approximation as it takes no account of the differences between specific sites in terms of scale and degree of exposure to damage as a result of sea-level rise. However, it gives an indication of the range of cost likely to be associated with probable sea-level rise scenarios.

Further work is now being undertaken to refine this first-order estimate based on a more detailed evaluation of the configuration of each refining site. This work will also extend the analysis to other coastal installations such as natural-gas processing and reception facilities. As well as considering additional protection costs this further work will also seek to evaluate the potential damage costs of site flooding.

3.5 Conclusions

High-impact events, such as hurricanes and major coastal floods, together with an increasing acceptance that both natural and anthropogenic long-term changes are occurring, are forcing maritime industries to pay more attention to how long-term trends will affect their operations.

In this study, we have shown that the impact of sea-level rise on new offshore structures is likely to be small. This is because the majority of new developments are likely to be in deep water where the floating structures typically employed are unaffected by long-term changes in sea level. It is also the case that much of the stock of existing shallow-water fixed installations will reach the end of their design lives before long-term trends in sea level have any impact on their operation.

For coastal installations, such as refineries and gas-processing facilities, long-term trends in sea level are of greater significance over the expected life of new and existing infrastructure. Given that new coastal installations are being designed on the basis of trends in sea level consistent with those currently projected as a mean case by the IPCC then these facilities are likely to be adequately protected unless rates of increase in sea level accelerate further. Existing coastal refinery sites are likely to be occupied for periods over which trends in rising sea level will become significant. Our preliminary work has identified the probable range of cost associated with protecting such facilities.

In comparison with the more general costs of protection of critical coastal infrastructure (roads, railways, power distribution, ports, etc.), these costs are small. For example, it is estimated that a more than 1-m rise in sea level (made up of a combination of sea-level rise and local subsidence) projected for the Gulf Coast region between Alabama and Houston (the location of a number of major refining installations) over the next 50–100 years would permanently flood nearly a third of the region's roads as well as putting more that 70% of the region's ports at risk (CCSP 2008). The costs of adaptation to this level of impact will be orders of magnitude greater than those associated with protecting individual refinery sites.

References

Balint S.W. and Orange D. (2006) *Future of the Gulf of Mexico after Katrina and Rita. Proceedings of the Offshore Technology Conference, Houston, 1–4 May 2006.*

CCSP (2008) *Impacts of Climate Change and Variability on Transportation Systems and Infrastructure: Gulf Coast Study, Phase I.* A Report by the U.S. Climate Change Science Program and the Subcommittee on Global Change Research (Savonis M.J., V.R. Burkett and J.R. Potter, eds). Department of Transportation, Washington DC.

Church J.A. (2001) How fast are sea levels rising? *Nature* **294**, 802–3.

Church J.A. and White N.J. (2006) A 20th century acceleration in global sea-level rise. *Geophysical Research Letters* **33**, L01602.

Church J.A., Gregory J.M., Huybrechts P., Kuhn M., Lambeck K., Nhuan M.T. et al. (2001) Changes in sea level. In: *Climate Change 2001: The Scientific Basis.* Contribution of Working Group 1 to the Third Assessment Report of the Intergovernmental Panel on Climate Change (Houghton J.T., Ding Y., Griggs D.J., Noguer M., van der Linden P.J., Dai X. et al., eds), pp. 639–94. Cambridge University Press, Cambridge.

Cooper C. and Stear J. (2006) *Hurricane Climate in the Gulf of Mexico. Proceedings of the Offshore Technology Conference, Houston, 1–4 May 2006.*

IPCC (2001) *Climate Change 2001: The Scientific Basis.* Contribution of Working Group 1 to the Third Assessment Report of the Intergovernmental Panel on Climate Change (Houghton J.T., Ding Y., Griggs D.J., Noguer M., van der Linden P.J., Dai X. et al., eds). Cambridge University Press, Cambridge.

Kotz S. and Nadarajah S. (2000) *Extreme Value Distributions, Theory and Applications.* Imperial College Press, London.

Midwest Publishing Company (2006) *Refining and Gas Processing Industry Worldwide Directory.* Midwest Publishing Company, Tulsa, OK.

Minerals Management Service (2005) *Potential Obstructions to Mariners and Offshore Operators*. US Department of the Interior Minerals Management Service Safety Alert no. 234, 9 December 2005.

Nicholls R. (2003) *Case Study on Sea Level Impacts*. OECD Workshop on the Benefits of Climate Policy: Improving Information for Policy Makers, ENV/EPOC/GSP(2003)9/FINAL. OECD, Paris.

Nicholls R.J. and Lowe J.A. (2004) Benefits of mitigation of climate change for coastal areas. *Global Environmental Change* **14**, 229–44.

Pugh D.T. (1987) *Tides, Surges and Mean Sea-Level: a Handbook for Engineers and Scientists*. Wiley, Chichester.

Pugh D.T. (2004) *Changing Sea Levels, Effects of Tides, Weather and Climate*. Cambridge University Press, Cambridge.

Titus J.G., Park R.A., Leatherman S.P., Weggel J.R., Greene M.S., Mausel P.W. et al. (1991) Greenhouse effect and sea level rise: the cost of holding back the sea. *Coastal Management* **19**, 171–204.

Versowski P., Rodenbusch G., O'Conner P. and Prins M. (2006) *Hurricane Impact Reviewed Through API. Proceedings of the Offshore Technology Conference, Houston, 1–4 May 2006.*

Woodworth P.L., Gregory J.M. and Nicholls R.J. (2004) Long term sea level changes and their impacts. In: *The Sea*, vol. **13** (Robinson A.R., McCarthy J. and Rothschild, B.J., eds), pp. 717–52. Harvard University Press, Cambridge, MA.

4 Paleoenvironmental Records, Geophysical Modeling, and Reconstruction of Sea-Level Trends and Variability on Centennial and Longer Timescales

Kurt Lambeck, Colin D. Woodroffe, Fabrizio Antonioli, Marco Anzidei, W. Roland Gehrels, Jacques Laborel, and Alex J. Wright

4.1 Introduction

The level of the sea does not remain constant. It changes at varying rates, geographically and over time. Changes of sea level have affected human civilizations in the past (as preserved in legends, such as that of Atlantis), and anticipated accelerated sea-level rise as a consequence of enhanced global warming (Church et al. 2001; Bindoff et al. 2007) seems certain to have serious ramifications in the future.

Understanding past sea-level change plays an important role in determining the underlying causes, and also permits the extrapolation of past sea levels to locations and epochs for which there are no instrumental data. In addition, the Earth has a memory of past events, and the pattern of relative sea-level change, today and in the future, will continue to respond to past events. This occurs for several reasons, including crustal response to past glacial retreats and re-advances, coastal subsidence from past sediment deposition, from present tectonic upheavals the underlying processes of which have their origins in the remote past, and from

Understanding Sea-Level Rise and Variability, 1st edition. Edited by John A. Church, Philip L. Woodworth, Thorkild Aarup & W. Stanley Wilson.

delays in the adjustment of the oceans to changing thermal or salinity regimes. In the context of future sea-level rise in response to global warming, it is important to also be aware of the inexorable pace at which many Earth-surface and oceanographic processes, such as thermal expansion of the oceans and the response of ice sheets, occur. As a consequence, sea-level changes will continue for centuries, even if greenhouse gas concentrations in the atmosphere are stabilized now.

In order to understand present and future sea-level variability we must understand the past record over a range of timescales, and it is instructive to review how our understanding of the processes involved has improved over the past two centuries. In this chapter, we will summarize what is known about past sea-level change prior to the instrumental record, concentrating particularly on changes in ocean volume that have occurred over the past few thousand years since the termination of deglaciation of the major ice sheets. These changes are driven by global factors, including climate change and thermal expansion of the oceans, or regional responses, including glacio- and hydro-isostasy following deglaciation, or local responses to tectonic movements or deformations including volcanic and sediment loading. The roles of each of these adjustments are important in the face of future changes in ocean volume, whether those are driven by thermal expansion, ice-melt, or by land movements. Geographical and temporal variability in sea-level change will be described, together with the potential sources of new high-resolution and high-accuracy information on paleo-sea-level change. In particular, the past offers the opportunity to decipher what other high-frequency contributions to relative sea level occur that may mask the ocean-volume signal. All ages given in this chapter are in calibrated years, unless otherwise stated.

4.2 Past Sea-Level Changes

4.2.1 Historical Context

Our ability to understand present and future change in global sea level is largely determined by our knowledge of how sea level has fluctuated in the past, and its relationship to changes in climate. In the early 19th century there was general acceptance of the idea that the Biblical Flood could explain the anomalous geological deposits in the landscape. Fortunately, we have progressed from that with the work by James Hutton (Playfair 1802), who proposed the uniformitarianism principle that the past is the key to the present and that massive geological evolution can occur through the continual operation of everyday processes. Charles Lyell was a major advocate of the uniformitarian approach and was aware that the relative position of sea level had changed. Indeed, the frontispiece of his *Principles of Geology* (Lyell 1832) pictured the Roman columns, bored by marine mollusks during temporary submergence, at Pozzuoli (termed by Lyell the Temple of Serapis; see below), a site that continues to stimulate research on sea-level change (Morhange et al. 2006; Figure 4.1a). In the mid 19th century came the

Figure 4.1 (a) Remains of the Roman market at Pozzuoli in the Phlaegrean Fields, showing the columns on which marine borings can be seen up to 7 m above present sea level, indicating submergence followed by uplift. Recent radiocarbon dating of these boring organisms indicates local relative sea-level highstands at three periods: during the 5th century AD, the early Middle Ages, and before the 1538 eruption of Monte Nuovo (Morhange et al. 2006) (photo credit: M. Anzidei). (b) A former raised shoreline (arrow) at Scilla in Calabria, southern Italy, that can be related to a sea stand during the Last Interglacial (marine oxygen isotope stage 5e, MIS-5e) (Antonioli et al. 2006). Its present elevation is approximately 125 m (photo credit: F. Antonioli). (c) A flight of emergent reef terraces on the Huon Peninsula, Papua New Guinea, where the Last Interglacial shoreline (arrow indicates the front of the reef) has been elevated to more than 100 m above sea level as a result of rapid co-seismic uplift (photo credit: A. Tudhope). (d) A sequence of tidal notches on a cliff at Iraion, near Corinth, resulting from seismic uplift events during the past 6000 years. The upper notch is approximately 3 m above present sea level (photo credit: K. Lambeck). (e) Partial excavation of the Roman-age harbor of Phalasarna, uplifted about 6.5 m during the AD 365 Crete earthquake. Moorings and biological remains are well preserved and provide a precise measure of the sea level change at this site (photo credit: M. Anzidei).

major realization that there had been a series of ice ages, with recognition by Agassiz (1840) that anomalous boulders (erratics) in the Swiss Alps and deposits in Scotland were evidence of past glacial activity. There was considerable speculation on the number and causes of ice ages over the ensuing century.

That the Earth's orbital motion might be the cause of ice ages was first postulated by Joseph Adhémar in 1842, and the 'orbital theory' and its link to global climate changes was written up in detail by Croll (1864, 1875). The Earth's orbit experiences three periodic variations: variation in its eccentricity (the shape of the ellipse), its obliquity (angle of tilt of the orbital plane), and the precession of the equinoxes (the rotation of the orbital plane). The mathematical consequences of these known periodicities were calculated meticulously by Milankovitch (1941). This became known as the astronomical theory of ice ages, with the important consequence that it implied a series of ice ages. Milankovitch had shown by 1914 that changes in orbital eccentricity and the precession of the equinoxes could produce sufficient effects on the heat balance at different latitudes to influence the size of the ice sheets. He later incorporated the influence of changes in tilt of the Earth's axis. The combined 100 000-year eccentricity, 41 000-year obliquity, and 22 000-year precession rhythms result in significant heating differences at latitudes 55–65° north of the equator where much of the land in the Northern Hemisphere is concentrated (Berger 1992).

By the beginning of the 20th century it was perceived that there had been a series of glaciations in the Alps (Penck and Brückner 1909). At the same time, it was recognized that shorelines identified at varying elevations around the world were related to past ice ages (Daly 1910, 1915, 1925). Fossil shorelines from around the Mediterranean Sea provided evidence for changes of sea level (Figure 4.1b), and it was thought that these were correlated with the sequence of glaciations, although in the absence of any means of dating a particular shoreline it was presumed that the highest was the oldest and that sea level had lowered progressively during each successive interglacial (Zeuner 1945). Isostatic uplift associated with melting of the glaciers was also recognized (Daly 1934; Gutenberg 1941).

It was the analysis of sediments from the ocean floor that was to provide one set of results that indicated the periodic fluctuation of ocean volume in relation to ice ages. Following the pioneering work of David Ericson, Goesta Wollin, Maurice Ewing, and Cesare Emiliani, the analysis of oxygen isotope ratios in the tests of foraminifera preserved in deep-sea cores was shown to record synchronous ocean/ice volume oscillations throughout the Quaternary (Hays et al. 1976). Initially it was considered that the record of oxygen isotope variations shown by foraminifera reflected temperature dependence, but the oxygen isotope concentration in ocean water is itself related to ocean volume as the ocean is enriched in ^{18}O during glaciations. More detailed analysis of cores from key sites (Shackleton and Opdyke 1973; Imbrie et al. 1984, 1992, 1993; Shackleton 1987) confirmed the periodicities and the orbital control on climate, particularly on ice and ocean volumes, is now well accepted (Alley et al. 2005).

4.2.2 Relative Sea Level and the Causes of Sea-Level Change

Sea level as recorded in the geological record by geomorphological and ecological shoreline indicators, or by archaeological evidence and historical documents, is a relative measurement. It records the shifting relationship between the ocean surface and land. Changes in ocean volume result in an overall change in global mean sea level, termed 'eustatic' by Suess (1888). But it is important to also take account of changes in the elevation of the land. These are most striking in seismically active areas where there is rapid uplift. For example, the series of raised coral reef terraces on the Huon Peninsula (Figure 4.1c) occur because this site is experiencing rapid uplift at rates of up to 4 mm/year (Chappell et al. 1996). Co-seismic uplift that accompanied the Sumatran earthquakes in 2004 and 2005 led to uplift of coral reefs of several meters and to levels at which the corals could no longer survive (Meltzner et al. 2006). Co-seismic uplift is also indicated by notches along the coast in the Gulf of Corinth (Figure 4.1d) and elsewhere in Greece such as Rhodes (Villy et al. 2002), or along the eastern coast of Sicily (Antonioli et al. 2003). In some instances this can be precisely verified from uplifted archaeological remains such as the Roman-age harbor of Phalasarna in western Crete (Pirazzoli et al. 1992; Shaw et al. 2008) (Figure 4.1e). In addition to local tectonic uplift, there are also locations experiencing tectonic subsidence (Dokka 2006) including land movements produced by active volcanism. Striking examples from Italy include the submerged Roman city of Baia (see Figure 4.8e, below) in the Phlaegrean Fields (Parascandola 1947; Dvorak and Mastrolorenzo 1991; Morhange et al. 1999, 2006) and the wharf at Basiluzzo in the volcanic arc of the Aeolian Islands (Tallarico et al. 2003).

Besides these tectonic movements, land and relative sea-level movements also occur as a result of the exchange of water between the continents and the oceans. The growth and decay of ice sheets is the most obvious example of this and the associated relative sea-level change is typically of the order of 100–150 m over time intervals of about 10 000 years. But change in surface and ground water on the continents can also have a perceptible effect on sea level at the millimeters-per-year scale (Church et al. 2001). In response to a redistribution of water between ice sheets, ground and surface water, and oceans, sea-level change will comprise a number of interrelated factors: (1) changes in ocean volume, (2) radial displacement of the land surface by the changing load, (3) changes in the gravitational potential as a result of the deformation of the planet and the redistribution of mass on its surface, (4) changes in the shape of the ocean basins, and (5) redistribution of water within these basins such that the ocean surface remains an equipotential surface. The combined surface deformation and geoid changes are referred to as the glacial isostatic adjustment (GIA) of the ocean surface. They are the dominant causes of sea-level change during glacial cycles, are global and continue after ice volumes have stabilized.

Thus changes in sea level are driven by a combination of climate and tectonic forcing and the geological record shows that this has occurred on all timescales (Lambeck et al. 2002a). Sea levels have been changing over millions of years and are recorded, though imperfectly, by the stratigraphy of sedimentary basins (Figure 4.2a). The volume of water in the ocean basins may increase gradually through the addition of new water (juvenile water) from within the Earth's rocks, although this occurs at a negligible rate on timescales of less than 10^6 years. The prime control at this long timescale appears to be plate tectonics; changes in the rate of spreading at mid-ocean ridges, the associated cooling of oceanic crust, and changes in ocean-basin configurations (Pitman 1978). Sequence stratigraphy, based on seismic records of the major sedimentary basins, has enabled the compilation of long-term patterns of sea-level variation (Haq et al. 1987; Miller et al. 2005) although often these records are also contaminated by tectonic signals (Cloetingh et al. 1987).

The generally accepted pattern of sea-level change over the last glacial cycle is shown in Figure 4.2b, comprising a sea level close to present (but probably several meters above) during the Last Interglacial (marine oxygen isotope substage 5e (MIS-5e) at ≈125 000 years ago), followed by two interstadials of slightly lower sea level (substages 5c and 5a), and then a series of progressively lower oscillations culminating in the glacial maximum around 25 000 years ago. It is important, however, to recognize that there are a number of global, regional, and local factors (tectonics, coastal subsidence, ocean circulation, responses to volcanic and sedimentary loadings, and the GIAs) that influence the record of relative sea-level change observed at any point on the globe. There may also be changes of several meters of amplitude persisting for hundreds of years, in response to sub-orbital variations (Esat and Yokoyama 2006). Several abrupt postglacial changes in sea level have been detected in response to injections of meltwater from terrestrial ice sheets but whereas there is agreement on the existence of some of the meltwater pulses (MWPs), such as MWP-1A at approximately 14 000 years ago of approximately 14 m amplitude (Fairbanks 1989; Blanchon and Shaw 1995), there is less consensus on the timing of earlier events or on the source of the meltwater (Yokoyama et al. 2000; Lambeck et al. 2002b; Liu and Milliman 2004; Shennan et al. 2005; Bassett et al. 2005; Peltier 2005). There is also a growing body of evidence in support of a later and smaller rapid and global sea-level rise (≈5 m) approximately 7600 years ago (MWP-2) (Blanchon and Shaw 1995; Bird et al. 2007; Yu et al. 2007). The rates of rise during these events are usually less well determined because of uncertainties in dating their onset and termination but are of the order of several tens of millimeters per year and significantly greater than anything observed during post-glacial phases.

Climate-induced changes in sea level result from ocean temperature increase and from redistribution of heat within the oceans as well as from long-term fluctuations in atmospheric pressure and wind fields. Thus climate oscillations, such as the El Niño Southern Oscillation (ENSO), that show global teleconnections may also be expressed in sea-level fluctuations on regional scales. Changes observed over recent decades result primarily from climate-related thermal expansion of ocean water (termed steric, Chapter 6).

The observation of relative sea-level change therefore contains information on (1) land movements, (2) mass redistribution or geoid changes, and (3) changes

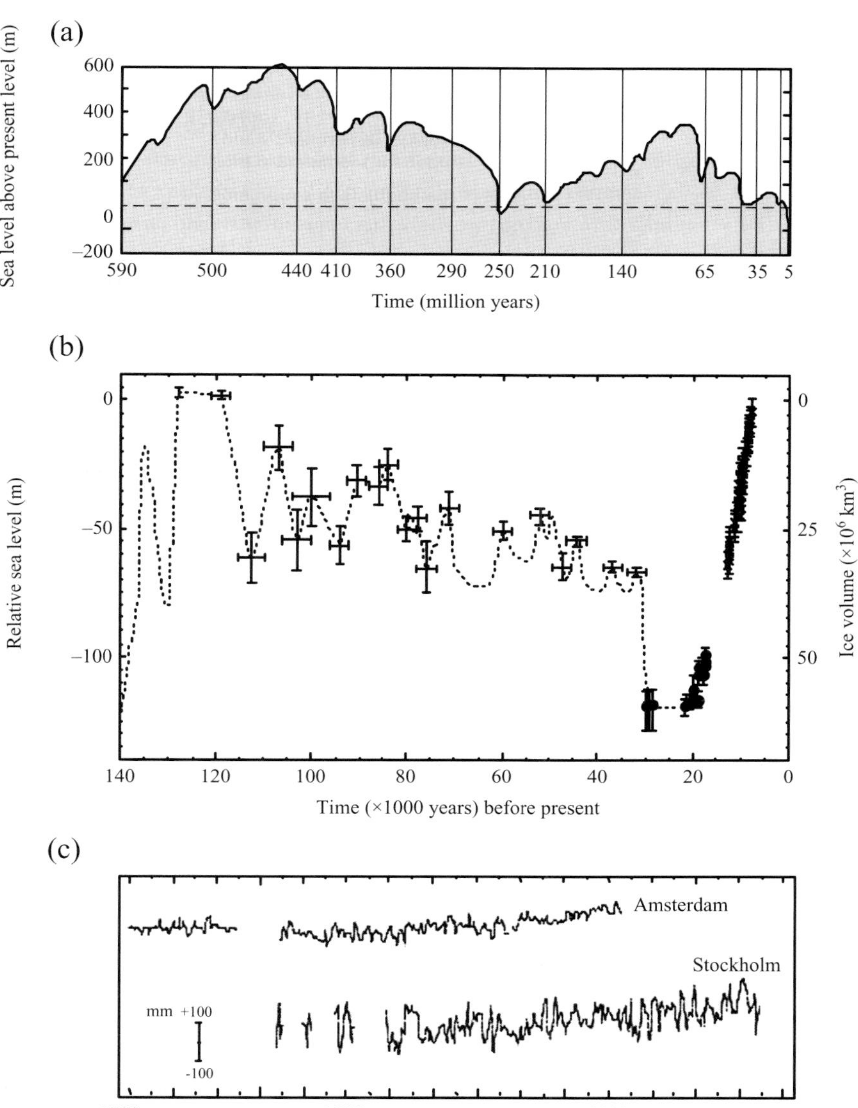

Figure 4.2 Sea-level change at different timescales. (a) The past 600 million years based on sequence stratigraphy analysis of sedimentary basins (after Hallam 1984). This change is driven primarily by plate tectonic factors. (b) The last glacial cycle (≈140 000 years) derived from fossil shoreline evidence from the Huon Peninsula reefs, Papua New Guinea and from sediment cores in the Bonaparte Gulf of Western Australia (Lambeck and Chappell 2001). This result is representative of far-field continental margin sites. It shows the Last Interglacial (MIS-5e) at 130 000–120 000 years ago followed by a series of progressively lower fluctuations to the Last Glacial Maximum at approximately 25 000 years ago, and the rapid post-glacial sea-level rise to about 7000 years ago. (c) Sea level from historical tide-gauge records at Amsterdam and Stockholm (after Church et al. 2001). The latter record has been detrended to remove the glacial isostatic rise in local sea level of about 4 mm/year.

in ocean volume and in the distribution of water within the ocean basins. Within the context of understanding the relation between sea level and climate change we are primarily interested in (3) but all practical outcomes are concerned with the changing disposition of land and sea and the paleoenvironmental and geological records indicate that we cannot avoid (1) and (2). Paleo-sea-level records provide the geological context for the interpretation of the shorter-term pattern of sea-level change observed with modern geodetic instrumentation (Chapter 10).

4.2.3 Quaternary Sea-Level Fluctuations

The paleo-record provides a relatively detailed picture of sea-level position and timing over the Quaternary, particularly of recent interglacial and interstadial highstands (Lambeck and Chappell 2001). Fossil reefs, emerged and forming terraces elevated above modern shorelines on coasts that are experiencing tectonic uplift, have been especially important for research since the advent of suitable dating techniques to discriminate units of Holocene and Late Pleistocene age (initially conventional radiocarbon, and then uranium-series disequilibrium dating methods, supplemented by other techniques such as electron spin resonance). The reef sequences on Barbados, and the more extensive series on the Huon Peninsula on the north coast of Papua New Guinea, contain corals, now elevated by uplift, and dates on these corals have been used to constrain successively more refined estimates of the age of past shorelines (Mesolella et al. 1969; Bloom et al. 1974; Chappell 1974). Three such terraces occur above sea level on Barbados; more occur on the rapidly uplifted Huon Peninsula in Papua New Guinea (Figure 4.1c). On the basis of dating of the Huon Peninsula terraces from several sites at which there have been different rates of uplift, a history of sea-level positions has been developed (Figure 4.2b), and it has been possible to link this record with that of oxygen isotopes from deep sea cores (Chappell and Shackleton 1986; Chappell et al. 1996).

The increased refinement of the U-series dating measurements has enabled more detailed discrimination, initially of separate interglacial reefs, and recently of sub-orbital fluctuations such as interstadials, and cycles within an interglacial (Chappell 2002; Potter et al. 2004; Thompson and Goldstein 2005; Esat and Yokoyama 2006). Initial application of U-series dating by Veeh (1966) showed that there was widespread evidence from shorelines on stable islands, such as Bermuda, Bahamas, Seychelles, and continental margins, such as Western Australia, for a Last Interglacial shoreline around 125 000 years old at heights averaging 6 m above present sea level (ranging 2–10 m). However, several of these island sites, such as Oahu in the Hawai'ian Islands and the makatea islands in the Cook Islands, are now considered to have undergone flexural adjustments in response to loading of the ocean floor by adjacent volcanic islands (Lambeck 1981; Nakiboglu et al. 1983). Reefs associated with MIS interglacial stage 7 are also widely recognized (Harmon et al. 1978; Woodroffe et al. 1991), and reef limestones associated with interglacials MIS-9 and MIS-11 have been dated, but there is less certainty about sea-level height at these earlier times (Hearty et al. 1999; Stirling et al. 2001).

The record of sea-level highstands derived from radiometric thermal ionization mass spectrometry dating of corals from Pleistocene reefs provides chronological support for control on the timing of ice-sheet growth and ocean volume variations through the superimposition of variations in the eccentricity, obliquity and precession of the Earth's orbit, supporting the Milankovitch hypothesis. The coral records also offer the prospect of resolving millennial-scale, sub-orbital oscillations overprinted on the longer-term oscillations. Such variations of approximately 10–15 m in amplitude have been observed during glacial periods, 55–20 ka (thousand years ago; Esat and Yokoyama 2006; Yokoyama et al. 2001a), and also appear within interglacial reef terraces such as those on Huon Peninsula, Oahu, and Barbados during MIS-5 (Potter et al. 2004; Thompson and Goldstein 2005).

4.2.4 Sea-Level Variations Since the Last Glacial Maximum

At the Last Glacial Maximum (LGM), around 25–20 ka, sea level was of the order of 130 m lower than present as a result of the volume of water that was locked up in ice sheets (Lambeck et al. 2002b; Peltier and Fairbanks 2006). Reconstruction of the shoreline at this time was initially derived largely from morphological evidence such as submerged cliffs and notches, and deposits of shallow-water shells (Jongsma 1970; Veeh and Veevers 1970). Analysis of cores from the continental shelf around Australia has provided a more detailed chronology and depth control on this lowstand (Ferland et al. 1995; Yokoyama et al. 2000, 2001b). Reefs are generally not good indicators of this lowstand, either because the material is buried, or because it is now in deep water. Once corrected for differential glacio-hydro isostasy and vertical tectonics, there is generally good agreement between the sea-level curves from diverse sites for the ensuing period of rapid sea-level rise following the LGM (Figure 4.3). The initial reconstruction was based on drilling of offshore reefs around Barbados (Fairbanks 1989). A similar pattern of rise was determined at almost the same time by drilling through the Holocene reef on the Huon Peninsula in New Guinea (Chappell and Polach 1991). Subsequently, comparable long records of post-glacial sea-level rise have been derived from

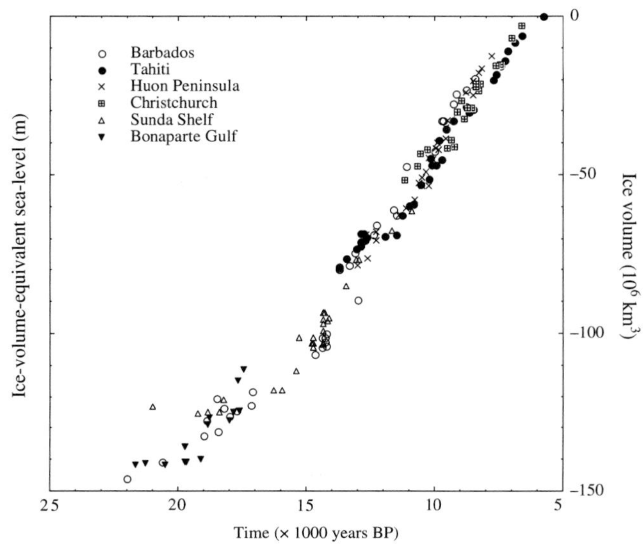

Figure 4.3 Ice-volume-equivalent sea function during and after the LGM inferred from coral, sedimentary, and paleontological data. The records from the individual sites have been corrected for the glacio-hydro isostatic effects. The equivalent grounded ice volumes (including ice grounded on the shelves) are shown by the scale on the right-hand side. The error bars are not shown here (see Lambeck et al. 2002b). The anomalous point at the LGM from the Sunda Shelf is based on a bulk sediment which is believed to be too old (K. Statteger, personal communication).

Mayotte in the Comoros Islands (Colonna et al. 1996) and Tahiti in the Society Islands (Montaggioni et al. 1997), with only a slightly shorter record from the Abrolhos Islands off the coast of Western Australia (Eisenhauer et al. 1993). Independent corroboration of the sea-level pattern comes from dating of muddy shoreline deposits off northwestern Australia and on the Sunda Shelf (Yokoyama et al. 2000; De Deckker and Yokoyama 2009: Hanebuth et al. 2000, 2009). Periods during which relative sea level was stable long enough may also be marked by the erosion of notches; notches at several distinct levels (7, 15, 25, 35, 55, 65, and 90 m) have been reported from around the world (Laborel et al. 1999), but it remains difficult to correlate these in the absence of age control because sea levels that formed at a time when ocean volume was constant will now not be at the same position everywhere.

The timing and source of individual meltwater pulses need to be further resolved, as do their consequences for coastal landforms. The peak of the last glaciation appears to have been terminated by rapid polar melting around 19 ka with little evidence that reefs in existence at that time were able to keep pace with the rising sea (Yokoyama et al. 2000; Clark et al. 2004). At least three phases of subsequent reef generation are recognizable: 17.5–14.7 ka, 13.8–11.5 ka, and the most recent Holocene phase separated by meltwater pulses (IA around 14 ka and IB around 11.5 ka). The Younger Dryas, a short return to glacial conditions triggered by meltwater disruption of warm water in the northern Atlantic, is another example of an excursion from the general pattern of climate change (Figure 4.3). Whether or not there was a further meltwater pulse (MWP-2), and the rapidity of its rise, have been debated. It was inferred that several Caribbean reefs appeared not to have been able to keep pace with rapid rates of sea-level rise around 7600 years ago (Blanchon and Shaw 1995) but reef growth did not appear to undergo disruption on those coasts that are uplifting (Montaggioni 2005). Recent observations from Singapore (Bird et al. 2007) and from the Baltic (Yu et al. 2007) add further evidence for a rapid rise prior to reaching present level around 7500 years ago.

Glacio-isostasy-induced geographic variability of relative sea level is important at all time periods and for all locations (Figure 4.4). They are most significant near former ice margins in Norway and Scotland (near-field sites) where it can be seen particularly well as emergent terraces (Gray 1983). Only in some cases has it been possible to date these platforms directly (Stone et al. 1996) but it has been possible to provide some constraint on their age and relationship to sea level by dating the separation from the sea of tidal embayments (isolation basins) at elevations now below and above the platform. Such results in Norway indicate that differential rebound can reach 2 m/km during the Younger Dryas (e.g. Anundsen 1985; Svendsen and Mangerud 1987). But even far from the former ice sheets (far-field sites) the spatial gradients of the sea-level response can be significant, for example at the time of the LGM of approximately 1 m/20 km across the broad northwestern margin of Australia (Yokoyama et al. 2000) or between the Sunda Shelf sites of Hanebuth et al. (2000) (Lambeck et al. 2002b). Along the eastern margin of Australia the spatially differential response is more subtle, of

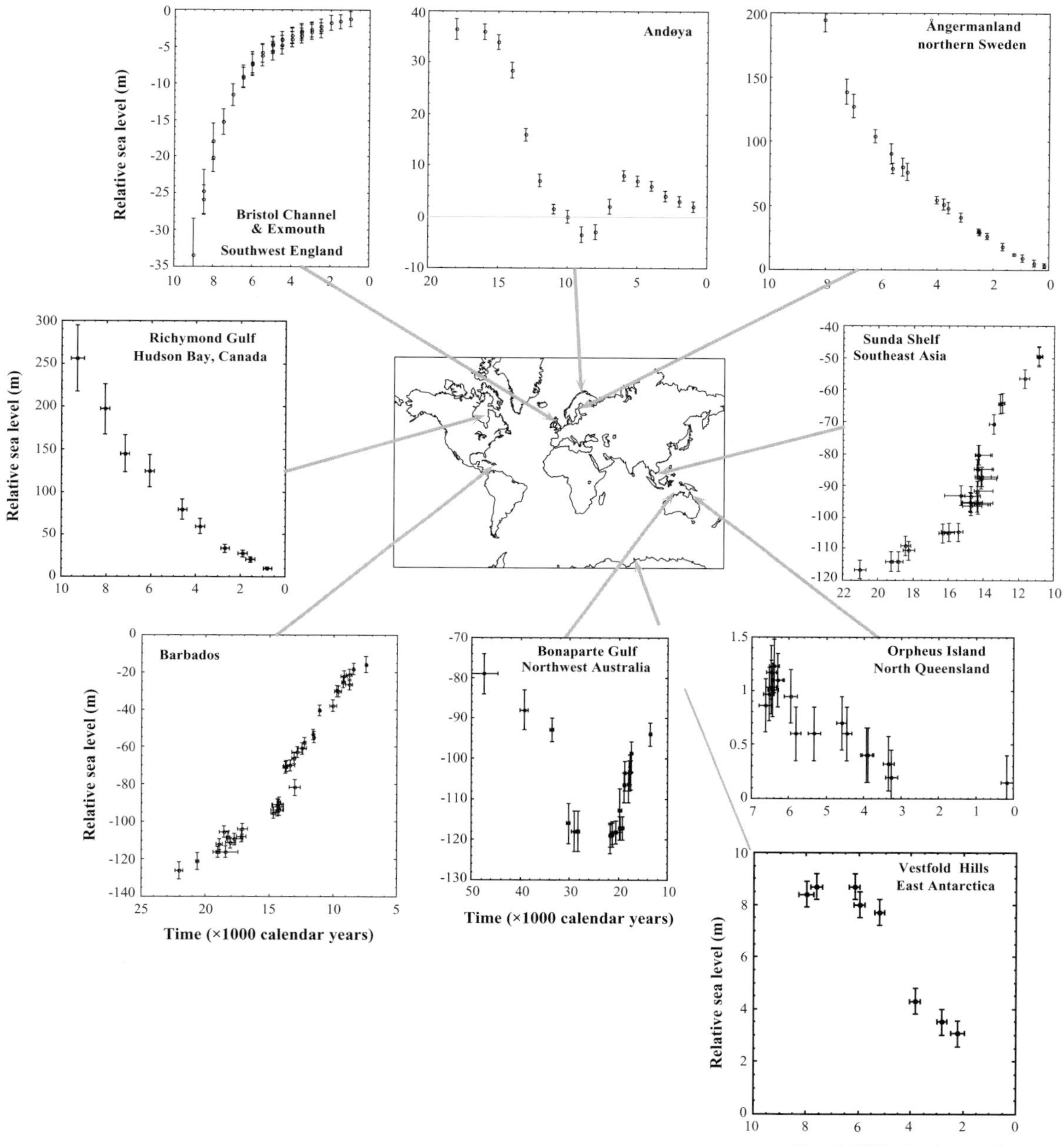

Figure 4.4 Observed sea level at different places round the world illustrating the spatial variability that occurs in response to ice-volume changes during the last glacial cycle. Note particularly the contrasts between near-field sites (e.g. Ångerman and Hudson Bay), ice-marginal sites (e.g. Åndoya, Vestfold Hills in East Antarctica), intermediate-field sites (Bristol Channel), and far-field sites (Queensland).

the order of 1–3 m at 6000 years ago (Nakada and Lambeck 1989) but greater than the observational accuracies of many sea-level indicators for this time, and important for establishing the isostatic response function.

Thus in most instances sea-level information from different localities should not be combined into a single sea-level curve unless it is first demonstrated that the expected spatial variability is less than the observational uncertainty. In the absence of a complete observational record to relate information from one location to the next, geophysical prediction or interpolation models are required that are based on a knowledge of the ice history during the glacial cycle and of the Earth response to loading on timescales of hundreds to thousands of years, and for evaluating any tectonic contributions.

Early attempts by coastal geologists and geomorphologists to determine Holocene sea-level history involved a global compilation of radiocarbon dates. For example, Fairbridge (1961) plotted radiocarbon ages for many shoreline indicators from around the tropics onto a single age-depth diagram. He interpreted the evidence to indicate a series of oscillations in sea level over the past several millennia, but the evidence for periods of sea level above its present level came primarily from Australia, whereas that for sea level remaining below present level came from Florida and the Caribbean. Since the 1970s, it has been increasingly accepted that there has been a significant difference between the pattern of sea-level change in the Pacific and that in the West Indies; the sea-level envelope which best describes the Pacific region reached a level close to present around 7000 years ago with evidence of a mid- Holocene sea-level highstand (Thom and Chappell 1975; Adey 1978). However, Caribbean sites are not free from isostatic signals, and the gradual rise of sea level relative to islands in the West Indies reflects ongoing isostatic adjustment to melting of the Laurentide Ice Sheet. This disparity between the characteristic Pacific and Caribbean sea-level envelopes became apparent in the two atlases of sea-level curves (Bloom 1977; Pirazzoli 1991) produced during research programs of the International Geological Correlation Programme (IGCP) and the International Quaternary Association (INQUA) (Pirazzoli 1996).

There is widespread evidence from the Indo-Pacific region that the sea reached a level close to present at least 7000 years ago, and that it has been slightly higher since then in much of the region and then slowly fell to its present position. The clearest evidence comes from regions where massive corals, termed microatolls (see below), are found in their position of growth (as opposed to corals dislodged and moved by storms; see below) such as on the Great Barrier Reef (Chappell 1983), and in the eastern Indian and central Pacific Oceans (Pirazzoli and Montaggioni 1988; Woodroffe et al. 1990a, 1990b; Woodroffe and McLean 1998). High-resolution records from salt marsh deposits along the northwestern Atlantic seaboard also suggest that century-scale sea-level fluctuations during the Late Holocene did not exceed 10–20 cm (e.g. van de Plassche et al. 1998, 2001; Gehrels 1999; Gehrels et al. 2005). Nevertheless, there has been a resurgence of the debate as to whether there might have been a series of fluctuations of sea level during past millennia using data from both Australia (Larcombe et al. 1995; Harris 1999;

Baker and Haworth 2000a, 2000b; Baker et al. 2004) and southeast Asia (Tjia 1996; Woodroffe and Horton 2005).

In the western Atlantic, construction of a sea-level curve for the West Indies has been based primarily on dating of staghorn coral, *Acropora palmata*, which dominates the reef crest and grows most prolifically within water depths of 0–5 m on the reef front, and there is no evidence for sea level having been higher than present during the Holocene (Lighty et al. 1982, Toscano and Macintyre 2003). However, there continues to be a practice of combining data from different Caribbean localities into a single sea-level curve (Fairbanks 1989; Toscana and Macintyre 2003) despite evidence that the glacial isostatic response is variable across the region (Lambeck et al. 2002b; Gischler 2006). In contrast, the Brazilian coastline shows Holocene emergence similar to that seen in other far-field continental margin areas (Angulo et al. 2006), consistent with the isostatic predictions (Milne et al. 2005).

4.3 Sea-Level Indicators

Paleo-sea-level indicators consist of various types, including sedimentary (such as beachrock), erosional (such as notches), ecological (such as accretionary biotherms constructed by coralline algae), microfossils (such as diatoms, testate amoebae, and foraminifera), and archaeological (such as fish tanks). There have been several substantial reviews of methodologies on the reconstruction of sea level from such proxies (van de Plassche 1986; Devoy 1987; Pirazzoli 1996; Elias 2007).

Sea-level curves, or more usually sea-level envelopes, need to be constructed with due consideration for the uncertainties that surround both the determination of vertical relationships between the position of the sea-level indicator and mean sea level, and for the errors associated with the age determination. Most sea-level indicators refer to some level within the tidal range other than mean sea level. This may be close to mean high water, as is the case for salt marsh markers (Horton et al. 1999), or mean low water springs (MLWS), as for coral microatolls (Smithers and Woodroffe 2000; Hopley et al. 2007). Or it may be that the observation is indicative only, and gives simply a limiting value. For example, a coral inferred to be in growth position but not showing evidence that it was intertidal and without any further stratigraphic information, indicates merely that MLWS occurred above this position by an amount corresponding to the growth depth range of the particular coral species. Thus the formation position with respect to mean sea level and the past tidal range need to be known when different indicators are combined into a single sea-level curve or sea-level envelope for a particular locality. In addition, there are positional uncertainties from survey error and from disturbances of the marker since formation, due to, for example, bioturbation, compaction, or storm activity. Age uncertainties include not only the intrinsic instrumental uncertainties but also the nature of the dated materials and any post-depositional disturbance or contamination of the record. For example, bulk sediment ages may be contaminated by old carbon from runoff or reworked and re-deposited sand grains (Woodroffe et al. 2007), as well as possible contamination by younger carbon rootlets that have penetrated the dated layer.

Geological, biological, and archaeological proxy indicators can sometimes be used in combination. For example, a coral reef is a geological structure, the interpretation of which is based primarily on ecological characteristics of the organisms making it up. Detailed analysis of muddy shoreline sediments often makes use of microfossil assemblages, sometimes of multiproxy data (Patterson et al. 2000; Plater et al. 2000). Even in the case of archaeological evidence, there is scope for cross-correlation, for example using marine mollusk growth or borings on archaeological structures (Morhange et al. 2001). Each of these lines of evidence requires some dating method to provide a constraint on age, and positional information on the elevation relative to some reference point. For all indicators this preferably involves correlation with a modern equivalent.

4.3.1 Geological and Biological Sea-Level Indicators

The most visibly distinct evidence for sea-level change is morphological, for example the erosional terraces or depositional shorelines shown in Figures 4.1b and c. Erosional notches, cut into hard rock, are also good indicators of former sea-level positions (Figure 4.1d), and can include structural notches that are controlled by rock strength, and abrasion notches that are the result of wave erosion. These can occur at one of several different elevations, in some cases being excavated at the highest point that waves reach, or at other elevations in the intertidal range, or even subtidal (Pirazzoli 1994; Laborel et al. 1999). Tidal notches are the most useful in determining former sea level; these occur as mid-littoral erosional features and are particularly distinctive on limestone coasts, in many cases with the deepest point, or vertex, coinciding with mean sea level (Spencer 1988). In more exposed settings, surf benches may form, termed trottoirs in the Mediterranean; these are erosional in origin, but generally have an accretionary lip. The reliability of erosional coastal features as indicators of former sea level, including potholes, tafoni, and honeycombing, depends on identification and understanding of the processes that were active in development of the features (Pirazzoli 1996).

Organisms that occupy only a very narrow vertical range within the intertidal zone are optimal for reconstructing relative sea-level change. On hard substrates, such as cliffs, breakwaters, and seawalls, there tends to be a distinct hiatus in composition of benthic organisms although it is often possible to discriminate mid-littoral from supralittoral (above) and subtidal (below). There are also organisms that provide evidence of submersion (Laborel et al. 1994). These include boring mollusks such as *Lithophaga lithophaga*, and species of encrusting vermetids (*Vermetus triqueter*) and fixed oysters (*Chama* and *Spondylus*). The mid-littoral, dominated by barnacles (*Balanus*, *Elminius*, *Tetraclita*, and *Chthamalus*), oysters (*Chama* and *Ostrea*), and mussels (*Mytilus*), is submerged frequently as opposed to the supralittoral, which is not (Figures 4.5a and b). A distinct zonation occurs, but its elevation depends on tide and wave characteristics. The frondose coralline rhodophyte *Lithphyllum lichenoides* has been used to reconstruct former sea level along the southern coast of France, and at other sites in the Mediterranean (Laborel and Laborel-Deguen 1994). The vermetid *Dendropoma petraeum* can form reefs up to 10 m wide with a depth range of over 0.4 m. Along the Sicilian coast Antonioli et al. (1999) identified platform-type *Dendropoma* reefs that are similar to coral-fringing reefs (Figure 4.5c), and which support the reconstruction of a relative sea-level curve over the past 700 years. Other indicators may be more subtle, for example a line below the lower cupulae of mid-littoral limpet erosion, and the reconstructed sea level is consequently less precise.

Figure 4.5 Biological indicators of shoreline position. (a) A typical zonation on rocky coasts in the Mediterranean showing the distinct hiatus between organisms characterizing the supralittoral, mid-littoral, and sublittoral (photo credit: J. Laborel). (b) Similar biological differentiation of supralittoral, mid-littoral and sublittoral on an irregular rocky shore (photo credit: J. Laborel). (c) Vermetid reef in Sicily (photo credit: F. Antonioli). (d) Serpulid overgrowth on submerged stalagmites in Argentarola Cave, Italy (photo credit: F. Antonioli). (e) Large flat-topped coral microatolls of a massive species of *Porites*, the living polyps around the margin contrast with the dead upper surface. This example is from Orpheus Island and is actually moated behind a recently formed boulder ridge on an older reef platform. (photo credit: K. Lambeck). (f) A fossil microatoll from Orpheus Island on the inner Great Barrier Reef, Australia. This coral grew around 6000 years until it reached the upper level to which corals can grow and then continued growth laterally. That level is nearly a meter above the modern equivalent limit to coral growth (photo credit: A. Lambeck).

In optimum conditions, fossil specimens of these fixed biological indicators offer a vertical precision, when related to their modern equivalents, of as high as ±5 cm in low-tidal-range environments (Pirazzoli et al. 1996; Morhange et al. 2001). Precision decreases on coasts with larger tidal range, or with more complex topography or wave conditions, or where preservation state of the fossil populations is poor. In Australia, the polychaete tubeworm, *Galeolaria caespitosa*, which secretes durable calcareous tubes within the intertidal to an upper limit between mid-tide and high neap tide, has been recognized as a useful fixed biological indicator that could be used to monitor sea-level changes (Bird 1988). Radiocarbon dating of fossil tubeworms provided evidence that Holocene sea level had been higher than present in northern New South Wales (Flood and Frankel 1989; Baker and Haworth 1997). Baker and Haworth (2000a) and Baker et al. (2001a) indicate that *Galeolaria* can be used to identify former sea level to a precision of ±0.5 m on sheltered rocky coasts, and that this may be increased when supplemented using the interface between *Galeolaria* and barnacles, limpets, or other mollusks occupying the upper intertidal (Baker et al. 2001b). In northeastern Queensland the oysters *Saccostrea cucullata* and *Crassostrea amasa* form conspicuous encrustations from mean sea level to mean high water neaps, the exact elevation varying with exposure (Hopley et al. 2007).

Cave deposits offer a further means of determining past sea-level behavior from areas now below water. Some data on sea-level lows have been derived from submerged speleothems, which can only have grown when sea level was below the depths at which they are found (Li et al. 1989; Lundberg et al. 1990). Some submerged stalagmites studied in Italy are of particular interest because they exhibit several contiguous layers corresponding to periods of low sea level. During highstands, the cave was flooded and the speleothems were covered by overgrowths formed by colonies of serpulids (Figure 4.5d). This marine and continental archive (sampled to water depths of 52 m) has provided some discrete points on the sea-level curve from approximately 250 ka before present (BP) up to the present for the stable coast in central Italy (Bard et al. 2002, Antonioli et al. 2004; Dutton et al. 2009).

Most corals grow in shallow water, although some species can grow in depths of more than 100 m in clear waters. Thus fossil corals, whose age has been determined by radiometric dating, in most cases are indicative, rather than definitive, of sea-level position, even if they are in their growth position (Figure 4.6). Some species of corals have a relatively narrow depth constraint; for example, *Acropora palmata* grows on the crest of Caribbean coral reefs and is most abundant in water depths of less than 5 m, making it a useful sea-level indicator although there are places where it grows in deeper water. Other corals, for example *Porites*, can grow to an upper level that experiences exposure at lowest tides (generally somewhere between mean low water springs, MLWS and mean low water neaps) (Figure 4.6a) and form microatolls that can provide precise constraints on the paleo mean low water neap level.

Coral microatolls are disc-shaped corals that have grown into the intertidal zone and whose upward growth is limited by exposure at low tide (Scoffin and

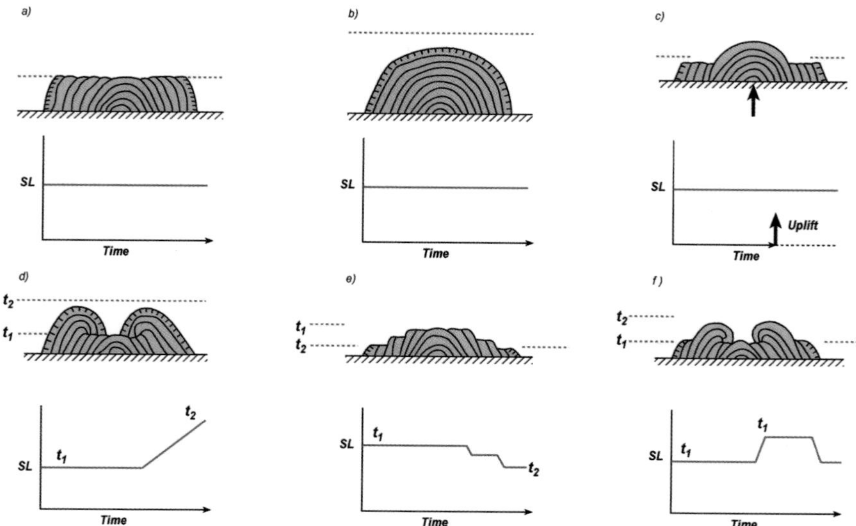

Figure 4.6 Schematic illustration of the response of coral, particularly the upper surface of microatolls, to changes of sea level or uplift of the land, as shown by annual banding within the coral skeleton. (a) If sea level (SL) remains constant from year to year, a massive coral that has grown up to sea level continues to grow outwards but with its upper surface constrained at water level. The inner part of the upper surface is partly protected by the outer rim which will usually be the only living part of the colony. (b) If the coral does not reach water level it adopts a domed growth form and is not constrained by water level. Its upper surface could be several meters below sea level. (c) If the coast undergoes uplift (or sea level falls), a coral previously not limited by sea level may be raised above its growth limit and the exposed upper surface will die, but with continued lateral growth at a lower elevation. (d) If the water level increases, a coral previously constrained by exposure at low tides can resume vertical growth and begin to overgrow the formerly dead upper surface. (e) If water level falls episodically then the microatoll adopts the form of a series of terracettes. (f) If there are fluctuations of water level with a periodicity of several years then the upper surface of the microatoll consists of a series of concentric undulations. Such a pattern can be seen on microatolls from reef flats in the central Pacific where El Niño results in interannual variations in sea level (after Woodroffe and McLean 1990).

Stoddart 1978). Microatolls are dead on the central part of the upper surface but continue to grow laterally around their outer margin (Figures 4.5e and 4.6a) and they are therefore sensitive to small changes in sea level. Typically, microatolls comprise a single colony of massive *Porites*, though several other genera, particularly *Goniastrea*, can also adopt the microatoll form and grow to several meters in diameter (Figure 4.5f). Microatolls in their growth position are fixed biological sea-level indicators and can be used to indicate previous limits to coral growth (Woodroffe and McLean 1990; Smithers and Woodroffe 2000). The elevation of radiocarbon-dated fossil microatolls compared with their living counterparts indicates that the near-equatorial sea level has either been stable or has fallen throughout Late Holocene time at rates that vary from location to location (Chappell 1983; Woodroffe et al. 1990b; Pirazzoli and Montaggioni 1988; Nunn and Peltier 2001; Hopley et al. 2007).

Because corals form annual bands that can be detected using X-radiography or fluorescence, banding within microatolls can confirm that growth has been primarily lateral and that the growth limit has been constrained by the tide level. The banding can also indicate periods during which the limit to coral growth has been temporarily raised or lowered, and the upper surface of individual coral colonies can track sea level and provide records of interannual sea-level variations at decadal (vertical precision of the order of ±5 cm) to millennial (vertical precision of the order of ±25 cm) scale that have been used for detailed reconstructions of Holocene sea level in those parts of the Indo-Pacific region which experienced a mid-Holocene sea-level highstand (Woodroffe 2005).

However, several cautions need to be exercised in using microatolls to reconstruct sea level. First, storms and other disturbances can re-align corals or impound them in moats giving a highly localized pattern of water level change (Hopley and Isdale 1977). Second, microatolls are related not to mean sea level but to some low water level that is related to the physiological response of the polyps to exposure. On open ocean atolls where the tidal range is minimal, microatolls may form without moating and small water-level changes may be recorded. However, moating is common on reef flats on the Great Barrier Reef within temporary pools forming amongst reef-flat rubble. In these circumstances gross sea-level changes may be recorded but not any subtle smaller changes (Hopley et al. 2007).

The dead upper surface of some microatolls can undergo erosion so the record will deteriorate (for example, fossil microatolls from the central Pacific do not appear to preserve ENSO-related undulations like those shown in Figure 4.6f). It is also important to remember that there is a time lag in the coral's response to water-level rise because of the time required for the coral to grow vertically, and thus trends are smoothed, and coral does not provide the instantaneous record that modern instrumental records do. Nevertheless, microatolls are an important source of additional data for sea level at geological and historical timescales as well as for other land-sea adjustments such as seismic deformation.

Sedimentary indicators of sea level include beach facies. On tropical and subtropical coasts, beach sands may be cemented into beachrock with clear preservation of the bedding parallel with the beachface. Numerous beachrock deposits have been preserved at different depths (0–50 m) along the Sardinia-Corsica continental shelf with remarkable continuity (Demuro and Orrù 1998), and these have been radiocarbon dated to establish a local relative sea-level curve for the past 10 000 years. Despite the distinctive preservation of the former beach, cementation appears possible at a range of vertical elevations within and above the tidal range, and beachrock does not seem to give a particularly precise indicator of the height of sea level (Stoddart and Scoffin 1983; Kelletat 2006; Hopley et al. 2007). The level of cementation may itself be constrained by the water table, and distinctive carbonate cements, within a range of carbonate reef facies, can be used as a sea-level indicator (Montaggioni and Pirazzoli 1984).

Muddy shoreline deposits have been widely used outside the tropics. Salt marshes occur in the upper intertidal zone on sheltered temperate shores, and the range of paleoecological techniques developed for salt marshes have to a lesser extent been applied to mangrove environments in the tropics. At the broadest level the intercalation of mud and peat has been interpreted to determine past sea level. Detailed reconstruction of sea-level tendency (i.e. the increase or decrease of marine influence indicated by stratigraphic changes) has been possible from peat-mud contacts following an approach advocated by Shennan (1986) and consistent sea-level tendencies over larger regions (positive or negative) can be indicative of regional sea-level change.

The most effective methods for detailed analysis of salt marsh stratigraphy involve microfossils. Such analysis, based on foraminifera, was pioneered by Scott and Medioli (1978) in the macrotidal region around the Bay of Fundy, and has been widely applied since. This has led to high-resolution sea-level reconstructions using salt marsh foraminifera (e.g. Gehrels et al. 1996, 2005). These types of salt marsh proxy records are based on the principle that marsh foraminiferal assemblages record the degree to which the salt marsh surface is able to maintain its position relative to a rising sea level. Higher marsh (less frequently inundated) foraminifera assemblages occur when marsh accretion is faster than the rate of sea-level rise; lower marsh assemblages appear when sea level rises faster than the rate at which marsh sediments accumulate. This method works best in regions where relative sea level is rising, so that accommodation space is created in which the salt marshes can accrete vertically (Haslett et al. 1997). The procedure involves measuring the height at which modern foraminifera are distributed across the salt-marsh surface (Figure 4.7) relative to a tidal datum (most commonly used is the mean high water level). A simple 'transfer function' is then produced, which quantifies the relationship between modern foraminiferal assemblage distributions and marsh-surface elevation (Edwards et al. 2004a, 2004b). The height at which fossil marsh sediments were deposited is determined from foraminifera preserved and retrieved from a sample core of sediment. The elevation at which the fossil foraminifera lived (their 'indicative meaning') is calculated by applying the transfer function which is based on the vertical distribution of the modern counterparts (e.g. Horton et al. 1999; Gehrels 2000). This results in a reconstruction of changes in the height of the paleo-marsh surface (PMS) relative to paleo-mean high water. These height changes are a combination of variations in the rate of sea-level rise and in the rate at which sediments accumulated. Therefore, if the accumulation history can be determined by dating PMS indicators, it is possible to produce a sea-level reconstruction (generally paleo-mean high water) from salt-marsh biostratigraphy.

Detailed stratigraphic investigations based on plant remains and benthic foraminifera have been conducted along the east coast of North America (van de Plassche et al. 1998, 2001; Gehrels 1999; Gehrels et al. 2005; Donnelly et al. 2004; Donnelly 2006), and with extension into Iceland (Gehrels et al. 2006). Age constraints on cores are generally obtained by radiometric techniques such as radiocarbon dating. These analyses can be extended into historical times with shorter-lived isotopes, such as ^{210}Pb, or marker layers, such as ^{137}Cs, pollution horizons and pollen. At best, salt-marsh sediments can produce proxy records with a precision of ±10 years in the 19th century and ±5 years in the 20th century,

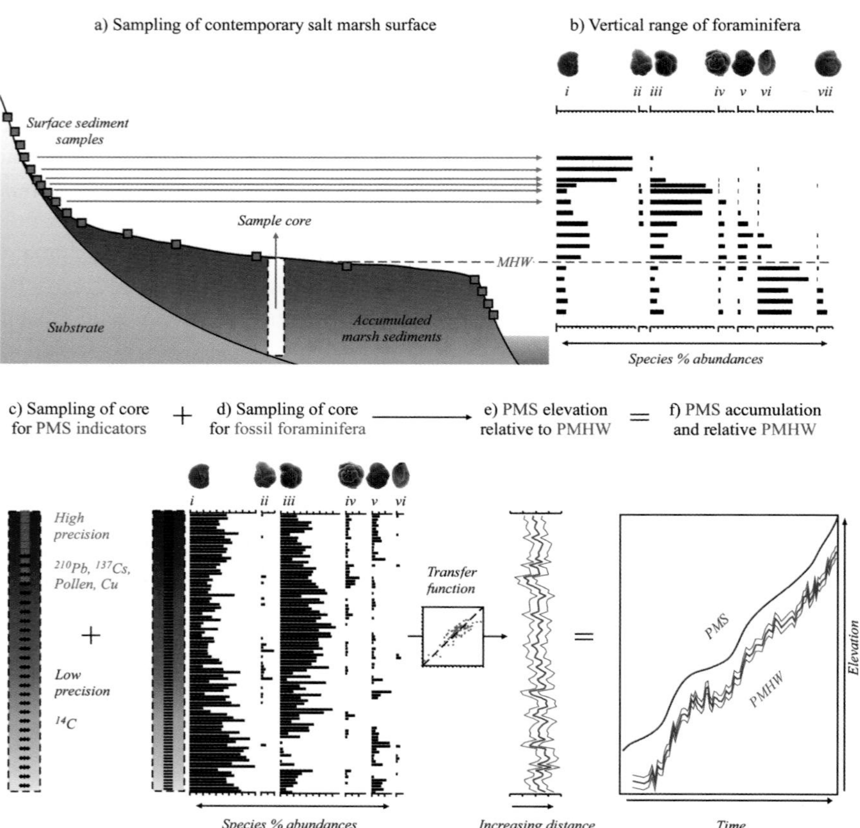

Figure 4.7 Schematic illustration of the steps involved in reconstructing sea-level changes using salt-marsh microfossils. The contemporary surface distribution (a) of foraminifera species (b) is related to elevation of the marsh surface above a tidal datum (MHW, mean high water), and subsequently represented by a transfer function. Paleo-marsh surface (PMS) indicators are sampled in a sediment core (c), with dating control of the upper core provided by techniques such as ^{210}Pb or ^{137}Cs and of the lower core by ^{14}C. The core is also analyzed for fossil foraminifera species abundances (d), which are interpreted in terms of PMS height (e) relative to paleo-mean high water (PMHW) using the transfer function arrived on the basis of steps (a) and (b). These combine (f) to reconstruct the PMS accumulation history and rate of relative PMHW rise. (Source: A.J. Wright, unpublished PhD thesis in preparation.)

depending on sedimentation rate and sampling resolution. However, for much of the past millennium the resolving power is probably on the order of 50 years, given the limitations of the radiocarbon dating technique. A vertical precision of individual paleo-sea-level estimates of ±5–20 cm is generally possible depending on site-specific factors such as tidal range.

Salt-marsh proxy records for recent times can be validated by comparison with tide-gauge records, providing independent evidence of a significant increase in the rate of sea-level rise in the 20th century compared to preceding centuries (Donnelly et al. 2004; Gehrels et al. 2005, 2006). Examples of such proxy sea-level reconstructions spanning the five past centuries are given in Chapter 5. Along coastlines with small tidal ranges, salt-marsh proxy methods are capable of resolving sub-centennial sea-level fluctuations as small as 10 cm (Gehrels et al. 2005).

Mangrove sediments provide an indication of shoreline position in sheltered tropical locations, and mangrove pollen has been used to reconstruct the ecology of former mangrove-dominated shorelines (Woodroffe 1990, 1995). Preliminary investigations have been undertaken concerning the use of other microfossils in these extensive tropical wetlands (Haslett 2001; Horton et al. 2003; Woodroffe et al. 2005). In all studies from muddy shorelines it is important to take into consideration the probable rate of compaction of these fine sediments. It has been shown that there has been considerable compaction (a meter or more) in muddy estuarine environments around northwestern Europe (Allen 2000a, 2000b), and sea-level reconstructions need either to be produced using basal samples (resting on a non-compactable substrate), or with allowance for compaction through a comparison between basal and non-basal sediments of similar age (Gehrels 1999).

4.3.2 Archaeological Sea-Level Indicators

Coastal settlements and ports constructed in antiquity provide important insights into sea-level changes during past millennia, and reconstructions of historical sea-level change using archaeological coastal installations have been particularly effective in the Mediterranean Sea, whose coastlines preserve remnants of human activity since the LGM. The first pioneering results using geophysical interpretations from archaeological indicators for the sea-level change estimation were published by Flemming (1969), Caputo and Pieri (1976), and Pirazzoli (1976).

Unfortunately, despite the large number of archaeological remains in the Mediterranean, as well as significant examples from other regions, such as the Bronze Age carvings from a former coastal dwelling site now inland in southwestern Sweden (Figure 4.8a), only a small fraction can be used to obtain precise information on their former relationship to sea level. Limitations arise as a result of their uncertain use, poor preservation, or because they were built in geologically unstable areas (for examples soft sediments at the mouth of rivers). Two types of relationship provide evidence of former sea level; first, those structures or artefacts which must have been terrestrial but have now been flooded, and second those coastal structures that were built with a precise relation to water level and which no longer function because of a change in sea level. The flooded paintings of Cosquer cave (Southern France) (Clottes and Courtin 1994; Clottes et al. 1997) belong to the first category and provide the oldest known archaeological constraints on sea-level change in the Mediterranean (Lambeck and Bard 2000). The second category is particularly valuable and includes slipways, fish tanks, piers and harbor constructions generally built after 2500 years BP. The Roman fish tanks, generally dated from between the first century BC and the first century AD, are particularly valuable (Figures 4.8c and d). Quarries carved along the coastlines and located nearby fish tanks, and harbors or villas of similar age, can provide additional data on past water level, as well as information on their functional elevation above sea level, although these are less precise indicators (Flemming 1978). Wells, such as those submerged off the coast of Israel (Figure 4.8b), indicate gradual sea-level rise since 8000 years ago (Sivan et al. 2001, 2004). Likewise, now-submerged sites of human activity can provide limiting information. One example is an Early Neolithic burial site at −8.5 m in northern Sardinia

(a)
(b)
(c)
(d)
(e)
(f)

where the archaeological style of a nearby settlement gives an age of approximately 7600 years BP at which time local sea level was more than 8.5 m lower than present (Antonioli et al. 1996). Other examples include submerged mosaics associated with the sunken city of Baia (Figure 4.8e) and the submerged breakwater at Capo Malfatano in Sardinia (Figure 4.8f).

Functional elevation is defined from the elevation of specific architectural parts of an archaeological structure with respect to the local mean sea level at that location and at the time of its construction, and provides the basis for determining sea-level change. It depends on the type of structure, its use and the local tide amplitudes. The minimum elevation of particular structures above the local highest tides can also be defined. So far, pier surfaces can be estimated at an elevation of at least 1 m above sea level, bollards at 0.7–1.0 m, and up to 2.0 m for large harbors such as at Leptis Magna (Lybia), depending on the ship size. Slipways can be found at several sites in the Mediterranean, such as Sicily, Cyprus, and Turkey, and it is apparent that they will function only if they have enough dry length to house the ship out of the water and sufficient depth of water at the bottom of the slipway for the ship to enter (Blackman 1973).

Roman fish tanks have constructional elements that bear directly on sea level at the time of construction (Figures 4.8c and d). Their distribution, limited to the Mediterranean Sea, is mainly along the western Italian peninsula, with a few examples in the Adriatic Sea and in the other areas of the Mediterranean, such as northern Africa, Spain, Greece, and Israel. Their construction and use is well

Figure 4.8 (*Opposite*) Archaeological evidence of former sea level. (a) Bronze Age carvings from a former coastal dwelling site now inland in southwestern Sweden. This is an area of uplift and the present elevation provides a limiting estimate of the total uplift since the time of carving (photo credit: K. Lambeck). (b) A pre-pottery Neolithic submerged well at Atlit Yam, radiocarbon-dated from settlement refuse at between 9500 and 8200 calibrated years BP, in water depth of 10 m, off the coast of Israel. The base of the well provides a limiting estimate of sea-level change since it was abandoned by rising saline groundwater (Sivan et al. 2001, with permission from Elsevier) (photo credit: the Israel Antiquities Authority and E. Galili). (c) Fish tanks constructed during Roman times at Briatico, Calabria (southern Italy). The tanks have been cut in the sandstone and are interconnected and connected to the sea by a series of canals in which the flow is tidally controlled via a series of sluices (photo credit: M. Anzidei). (d) A fixed sluice gate in a Roman fish tank from the Ventotene Islands. For it to function successfully it must have a well-defined position with respect to the tidal range which is about 20 cm here. These gates, generally made of lead or stone, consist of (1) a horizontal stone surface that defines the threshold and is cut by a groove to receive the gate, (2) two vertical posts with grooves to guide the vertical movement of the gate, (3) an upper stone slab with horizontal slot to extract the gate, and (4) the gate itself with small holes to permit water exchange and prevent the escape of fish. This example records a relative sea-level change of 1.5 ± 0.2 m since 2000 years ago (Lambeck et al. 2004a) (photo credit: M. Anzidei, with permission from Elsevier). (e) The mosaics of the sunken city of Baia (Italy) where submergence is due to volcanic deformation (photo credit: A. Benini). (f) The Capo Malfatano breakwater and dock of Punic age (2300 years ago BP) at a depth of 2.1 m, Sardinia, Italy (photo credit: F. Antonioli).

documented in the writings of authors such as Plinius, Varro, and Columella, who described the use of piscinae as holding tanks for fish culture. The fish tanks were generally carved into rock to a depth of 2.7 m or less, and the basins themselves were tidally controlled and protected from wave and storm action by exterior walls. Significant sea-level markers can be identified at these installations, including channels, sluice gates with posts, sliding grooves and, in some cases, the actual sliding gates, thresholds, and footwalks (Lambeck et al. 2004a).

4.4 Geophysical Modeling of Variability in Relative Sea-Level History

In the absence of tectonic processes the principal cause of global sea-level change has been that associated with the growth and decay of ice sheets and the associated deformational and gravitational response of the earth to the changing surface loads of ice and water. A useful zero-order approximation for the resulting sea-level change is the average change over the ocean resulting from the melting of ice sheets, defined as

$$\Delta\zeta_{esl}(t) = -\frac{\rho_i}{\rho_o}\int_t \frac{1}{A_o(t)}\frac{dV_i}{dt}dt \qquad (4.1)$$

where V_i is the ice volume at time t, $A_o(t)$ is the ocean surface area at time t, and ρ_i and ρ_o are the average densities of ice and ocean water. (In the absence of other factors that lead to changes in ocean volume this *ice-volume-equivalent sea level* is equal to eustatic sea level as defined by Suess.)

Sea level at any locality φ and time t is then defined as

$$\Delta\zeta_{rsl}(\varphi,t) = \Delta\zeta_{esl}(t) + \Delta\zeta_I(\varphi,t) + \Delta\zeta_T(\varphi,t) \qquad (4.2)$$

where $\Delta\zeta_{rsl}(\varphi,t)$ represents the change of the sea surface relative to land at location φ and time t compared to its present position at time t_P. The second term, $\Delta\zeta_I(\varphi,t)$, is the (glacio-hydro) isostatic contribution, and the last term is a tectonic contribution for tectonically active areas. Both of these terms are functions of position φ and of time t. Figure 4.9 shows the definition of terms used when comparing former and present shorelines when both ocean volumes and land surfaces change.

In a simple approximation the isostatic term can be conveniently viewed as two parts: glacio-isostasy corresponding to the deformational and gravitational effects on sea level from the changing ice load, and hydro-isostasy corresponding to the concomitant changes in the ocean load. (The two terms are not independent; the water loading, for example, being dependent on geoid changes induced by the changing ice sheets.) The former is dominant in the areas of former glaciation (near-field sites) and for large ice masses the rebound dominates the ice-volume

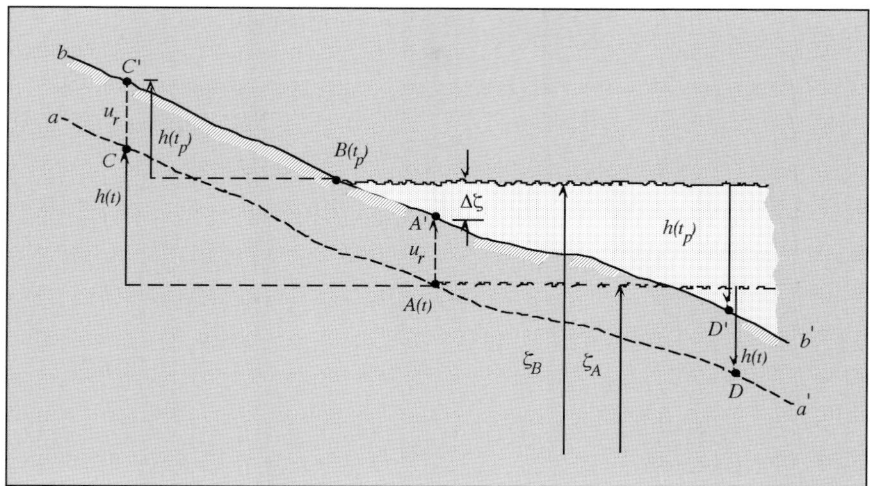

Figure 4.9 Definition of relative sea-level change. At time t the land surface is at a–a', the shoreline is at A(t), and the ocean surface is at a distance ζ_A from the center of the Earth. In the interval from t to the present t_P the land has been uplifted by u_r and the land surface is now at b–b' and the original shoreline is displaced to A′. In the same interval, the ocean volume has increased so that the ocean surface is now at a distance ζ_B from the Earth's center of mass and the shoreline is now at B(t_P). The observed relative change in sea level is the position of A′ relative to B(t_P) or $\Delta\zeta = u_r - (\zeta_B - \zeta_A)$. Offshore (D) or onshore (C) the paleo water depth or terrain elevation at a past time t is $h(t) = h(t_P) - \Delta\zeta$.

equivalent change such that the relative change since the onset of deglaciation is one of a falling sea level. The amplitudes and timing of the sea-level change will be functions of both the ice history and the mantle rheology. At sites far from the former ice sheets (far-field sites), the hydro-isostatic signal is the dominant departure from the globally averaged change, particularly at continental margins, where it results in a slowly falling sea level of the order 0.2–0.4 mm/year during the Late Holocene. The consequences of the glacio-isostatic contribution is also seen at these distant sites, largely through the increased holding capacity of the ocean basins as the broad deformational bulges around the ice sheets subside. This effect, sometimes termed ocean siphoning (Mitrovica and Peltier 1991), together with local hydro-isostatic adjustments, means that much of the far-field actually experienced a relatively higher sea level around 6000 years ago and the pattern of relative sea-level change has seen a gradual fall since that time (Farrell and Clark 1976; Nakada and Lambeck 1987; Mitrovica and Milne 2002; Mitrovica 2003).

At the intermediate sites, the post-LGM glacio-isostatic signal is mainly one of a rising sea level as the broad peripheral bulge around the ice sheet, shaped by mantle flow and changes in gravity, slowly subsides. This signal dominates the water-loading effect along the North Atlantic coastlines and for most of the Mediterranean, with the result that here sea levels have continued to rise slowly for the past 6000–7000 years. The relative importance of the two isostatic contributions produces the regional variability emphasized by Farrell and Clark (1976) and seen observationally in Figure 4.4 and illustrated schematically in Figure 4.10.

In a first approximation the spatially variable $\Delta\zeta_I(\varphi, t)$ is a function of the distance from the ice sheets (the glacio-isostatic part) but short wavelengths, of the order of the lithosphere's effective elastic thickness, also occur far from the ice sheets due to the changes in water loading (the hydro-isostatic part). Both

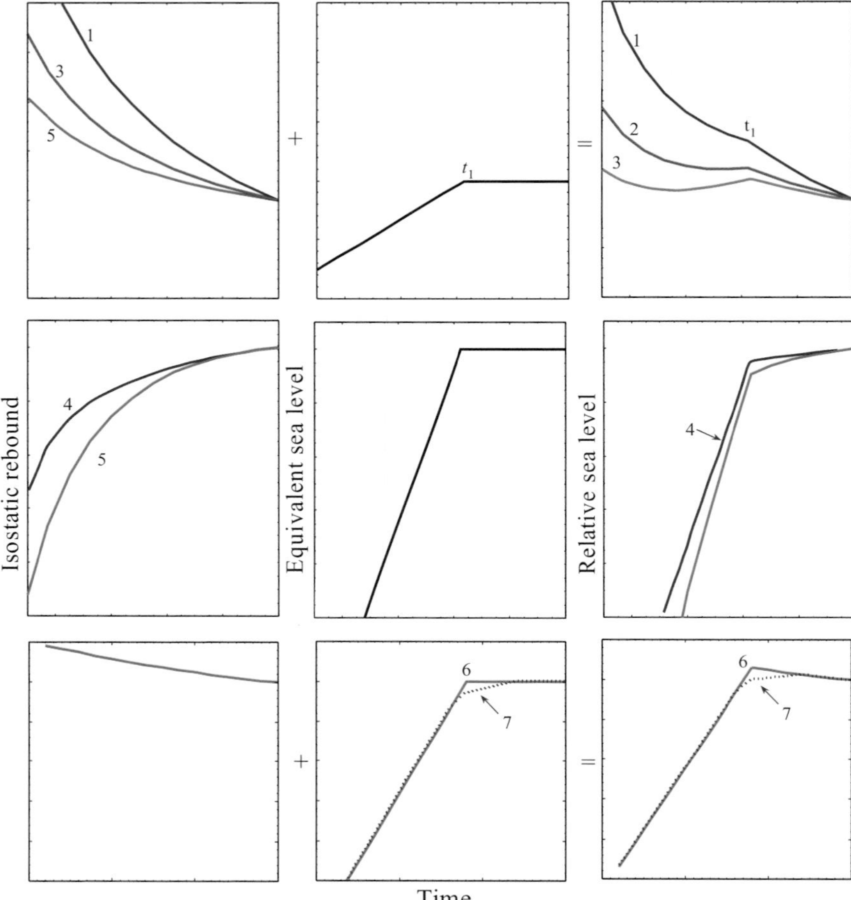

Figure 4.10 Schematic construction of the spatial variability in sea level introduced by the Earth's response to deglaciation. The left-hand panels indicate the crustal rebound, the central panels give the ice-volume-equivalent sea-level (esl) function, and the right-hand panels are the sum of the two. Note that the scales are not quantified and that the vertical scales differ in the three rows of panels. Gravitational effects are not considered, nor is the meltwater realistically distributed. (a) For sites within the former ice margin where the crustal rebound (left) exceeds the esl change (center) and the relative sea level is falling throughout (compare with the Ångerman result, Figure 4.4). If the cessation of global melting is abrupt at time t_1 a small cusp will appear in the relative sea-level curve at the time of cessation. The rebound from sites further from the center of rebound will become progressively smaller (see Åndoya, Figure 4.4) and near the ice margin (e.g. location 3) the rebound and eustatic signals may be of comparable amplitude but opposite sign. (b) At sites outside the former ice margin on a broad peripheral zone the rebound signal is one of subsidence, and the isostatic contribution is of the same sign as the esl change with an amplitude that depends on distance from the ice margin (see Bristol Channel, Figure 4.4) (site 4 lies closer to this margin than 5 which lies near the zone of maximum subsidence). (c) For the far-field continental margin sites the dominant signal is a subsidence of the sea floor induced by the water loading of the ocean floor (not just the shelf) with a concomitant fall in the water surface. The coastal zone is dragged down less than the offshore sea floor and sea level at the coast will appear to be falling. Once major melting has ceased the relaxation of the sea floor dominates and results in the characteristic highstand in relative sea level at approximately 6000 years ago (compare with Orpheus Island, Figure 4.4). For a site in a deeply indented bay, the down dragging effect of the coastal zone is less and the fall in sea level will appear greater by an amount that is rheology dependent. Two esl functions are shown in the central panel of (c), one in which all melting ceased at t_1 (curve 6) and the other in which a small amount of melting occurred until more recent time (curve 7). The resulting reduction in the highstand is effectively rheology-independent.

parts are predictable if the growth and decay history of the continental ice is known and the patterns of the changes are well established and calibrated against observational data. In contrast, the changes wrought by tectonic processes, of shorter wavelength and more episodic than the glacially driven change, tend to be less predictable and if the aim is to understand the climate-driven processes, then tectonically active areas should be avoided or, alternatively, the analysis should be designed such that the tectonic component can be removed.

As the emphasis here is on establishing constraints on changes in ocean volume during the current interglacial, it is the post-glacial observational data from far- and intermediate-field locations that will be most important. This is because the model predictions will be least sensitive to uncertainties in the details of the ice models, with the isostatic corrections being primarily a function of the amounts and rates at which water has been added into the oceans. Nevertheless the isostatic corrections will be dependent on both the assumed rheological model and on assumptions about the ice-volume-equivalent sea-level function. Separation of the two can, however, be achieved (Nakada and Lambeck 1989; Lambeck 2002).

Writing (4.2) as

$$\Delta\zeta_{\mathrm{I}}(\varphi,t)=\Delta\zeta_{\mathrm{I-g}}(\varphi,t)+\Delta\zeta_{\mathrm{I-h}}(\varphi,t) \tag{4.3}$$

where $\Delta\zeta_{\mathrm{I\text{-}g}}(\varphi, t)$ is the glacio-isostatic part and $\Delta\zeta_{\mathrm{I\text{-}h}}(\varphi, t)$ the hydro-isostatic part, the sea levels at two locations φ_1, φ_2 are (ignoring tectonics)

$$\Delta\zeta(\varphi_1,t)=\Delta\zeta^{0}_{\mathrm{esl}}(t)+\delta\Delta\zeta^{0}_{\mathrm{esl}}(t)+\Delta\zeta_{\mathrm{I-g}}(\varphi_1,t)+\Delta\zeta_{\mathrm{I-h}}(\varphi_1,t)$$

$$\Delta\zeta(\varphi_2,t)=\Delta\zeta^{0}_{\mathrm{esl}}(t)+\delta\Delta\zeta^{0}_{\mathrm{esl}}(t)+\Delta\zeta_{\mathrm{I-g}}(\varphi_2,t)+\Delta\zeta_{\mathrm{I-h}}(\varphi_2,t)$$

The $\Delta\zeta^{0}_{\mathrm{esl}}$ represents the contribution to the ice-volume function based on the adopted ice model and the $\delta\zeta^{0}_{\mathrm{esl}}(t)$ is a corrective term to allow for any limitations in this model. The difference is

$$\begin{aligned}\Delta\zeta(\varphi_1,t)-\Delta\zeta(\varphi_2,t)&=\{\Delta\zeta_{\mathrm{I-g}}(\varphi_1,t)-\Delta\zeta_{\mathrm{I-g}}(\varphi_2,t)\}+\{\Delta\zeta_{\mathrm{I-h}}(\varphi_1,t)-\Delta\zeta_{\mathrm{I-h}}(\varphi_2,t)\}\\&\approx\{\Delta\zeta_{\mathrm{I-h}}(\varphi_1,t)-\Delta\zeta_{\mathrm{I-h}}(\varphi_2,t)\}\end{aligned} \tag{4.4}$$

if the two sites are far from the ice margins and sufficiently close to each other for the glacio-isostatic parts to be small and comparable. This differential observation is primarily a function of the mantle response to the water loading and only to second order of $\Delta\zeta_{\mathrm{esl}}(t)$. Hence it provides an estimate of the rheological parameters which are then used to improve the $\Delta\zeta_{\mathrm{I}}(\varphi_1,t)$ as well as an improved estimate for the ice-volume-equivalent sea level through

$$\delta\zeta^{0}_{\mathrm{esl}}(t)=\Delta\zeta(\varphi_1,t)-\{\Delta\zeta^{0}_{\mathrm{esl}}(t)+\Delta\zeta_{\mathrm{I}}(\varphi_1,t)\} \tag{4.5}$$

Usually an iterative process is used to arrive at the solution in which the far-field and near-field solutions are progressively improved to yield better ice models and rheological parameters and, with a number of provisos, we can estimate the change in ocean volume through time. For much modeling, the nominal globally averaged pattern of sea-level change adopted has been that shown in Figure 4.10 based on an assumption that ice melt was complete at t_i approximately 7000 years ago and that the volume of the ocean has remained constant since then. But such an assumption needs further assessment for any high-resolution modeling of fluctuations in ice sheet and mountain glacier ice volumes during the current interglacial.

Geophysical modeling of the isostatic response of the Earth to surface load redistribution can be undertaken at global or at regional scales. In the present context we are mostly concerned with improving the estimates for the ocean-volume change during the present interglacial and since the time that sea levels globally approached their present level about 7000–6000 years ago. For this we prefer to carry out regional and or local analyses rather than global solutions. The far-field sites will be most important because of their insensitivity to ice-sheet details but where it has been possible to establish reliable ice models the data from near- and intermediate-field are also important. The reasons for this include an ability to take into consideration, to a first order, the possibility that there is lateral variation in the mantle response function; the possibility of focusing on single data types rather than on merging different data which may bear different relations to sea level; and a ready ability to compare the regional results for consistency or for establishing whether there are indeed regional variations in the sea-level change that are of a non-glacial nature. This provides the framework within which the regional differences in sea-level behavior can be understood. However, it is important to recognize that there are several other factors that contribute to the relative sea-level trend that may be observed at any point, and one way to assess these other factors is to compare the observational paleoenvironmental proxy data for former shorelines with the modeled output. Several regional examples are used below to illustrate these points.

4.5 Regional Case Studies

4.5.1 Far-Field: Eastern Australian

The detail of sea-level change in eastern Australia during the past 6000 years has been controversial, with vigorous debate about whether a Holocene highstand can be identified, and its amplitude, timing, and geographical extent (Gill and Hopley 1972; Thom et al. 1972; Thom and Roy 1985; Young et al. 1993). The evidence is best examined through a combination of field observations and geophysical modeling. This region is relatively stable tectonically, but has recorded the spatial variability of isostatic components during the last deglaciation (Nakada and Lambeck 1989; Lambeck 2002).

There is now widespread acceptance that coral evidence from the Great Barrier Reef region indicates that local sea level was higher than present in the mid-Holocene (Hopley et al. 2007). The significance of flat-topped intertidal coral microatolls as precise indicators of former sea levels was recognized during the 1973 Great Barrier Reef expedition, and numerous microatolls have subsequently been surveyed and dated (McLean et al. 1978; Scoffin and Stoddart 1978; Stoddart and Scoffin 1979; Hopley 1982; Chappell 1983; Chappell et al. 1983; Zwartz 1993). The observations from Orpheus Island (Figure 4.11a) are representative and indicate that here sea levels have fallen gradually over the past 6000 years without

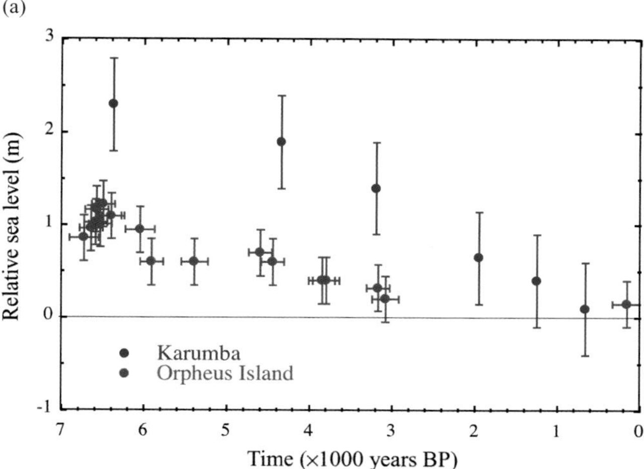

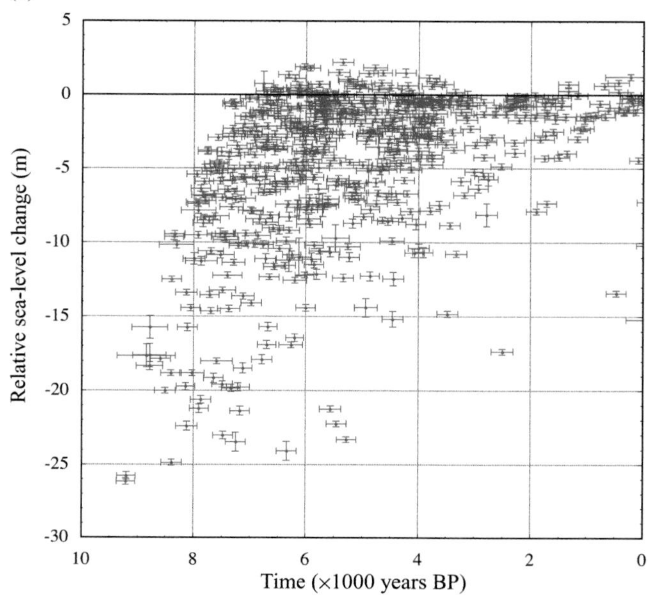

Figure 4.11 (a) Relative sea level for the past 7000 years at Orpheus Island on the inner Great Barrier Reef in northern Queensland, Australia, based on microatoll data by Chappell et al. (1983) and additional data collected by K. and A. Lambeck and D. Zwartz in 1989–91, and for Karumba in the Gulf of Carpentaria based on sedimentological evidence (Chappell et al. 1983). The two locations lie about 500 km apart and the differences in amplitude are a consequence of different hydro-isostatic effects; the Karumba site lies further from the water load added into the Pacific Ocean (Nakada and Lambeck 1989). (b) A compilation by D. Zwartz and D. Hopley of all fossil coral age/depth relationships across the Great Barrier Reef corrected for differential isostatic signals between the site and a reference site near the center of all sites. All depths have been reduced to mean sea level. Microatoll data are included and the upper envelope gives an estimate of mean sea level at the reference site (from Lambeck 2002).

evidence for oscillations exceeding the observational noise level of approximately 0.25 m. It has been common practice to combine microatoll data from different sites across the Great Barrier Reef into a single sea-level record but this ignores the spatial variability that is created mainly by the hydro-isostatic response. For example the amplitude of the mid-Holocene highstand is predicted to vary by approximately 1.5 m across the reef with coastal highstands higher than offshore island highstands (Nakada and Lambeck 1989; Lambeck and Nakada 1990).

While the absolute values for the isostatic corrections are dependent on model parameters, the differential values for nearby sites are much less so: corrections from ocean volume change will be common to all sites and the glacio-isostatic correction varies only slowly across the region. Thus a better practice is to reduce the observations from a region to a common site by applying a differential isostatic correction – the difference between the predicted correction at the observation site and that at the reference site – before combining the data into a single curve or envelope. Figure 4.11b shows the results for corals sampled from reef cores across the Great Barrier Reef, including microatoll data. Most of these cores give limiting values only: sea level must be above the sampling depth at the time of growth. All have been reduced to a common location and the upper limit provides the best estimate for sea-level change at this reference locality and supports the earlier interpretation of the microatoll data that if sea-level oscillations have occurred during the past 6000 years they are small compared with observational uncertainties.

As indicated by equation 4.4 the spatial difference in the relative sea level is a function of rheological parameters while the amplitudes are functions of these parameters as well as of the ice volumes. Thus a separation of the Earth and ice parameters can be achieved. This approach has led to the inference that ice models in which all melting ceased at approximately 7000 years ago (≈6000 radiocarbon years BP) lead to highstands that are larger than observed but that replicate well the observed spatial variability. The implication is that ocean volumes have continued to increase during the past 6000 years, as indicated in Figure 4.11 for two data sets from Northern Australia, as well as from a larger data set for the Australian margin (Lambeck 2002). This evidence suggests that ocean volumes have increased, such as to give an ice-volume-equivalent sea-level rise of approximately 3 m, with much of this rise having occurred between about 6000 and 2000 years ago. Again, within the time and height resolution of the data, particularly for the two high-resolution records from Orpheus Island and from Karumba (Figure 4.11), there is no evidence to support that large fluctuations occurred superimposed on this trend.

There have, however, been some recent studies that have returned to the Fairbridge (1961) model of Holocene sea-level oscillations, and ignored the importance of the isostatic contributions, using evidence in eastern Australia (Larcombe et al. 1995; Baker and Haworth 2000b; Baker et al. 2001b; Lewis et al. 2008). For example, Larcombe et al. (1995) postulated a major oscillation around 8000 years ago prior to sea level reaching present level. The evidence on which this was based was reviewed by Harris (1999), who demonstrated that there were insufficient grounds for such an interpretation: the adopted vertical errors of 0.2 m record uncertainty in the depth at which material was recovered from cores but does not take into consideration the relationship of the coral to mean sea level which is much greater than this. Furthermore, the coral dates came from very different locations, particularly from different distances from the coast and the resulting differential isostatic contributions were not taken into consideration which, once done, removes much of the purported signal (Figure 4.11).

Based on several types of fixed biological indicators, primarily worm tubes from along the New South Wales coast of Australia, Baker and Haworth have inferred sea-level oscillations of approximately 1 m during the Holocene (Baker et al. 2001a). Although ages on fossil tubeworm remains provide support for a mid-Holocene highstand on the New South Wales coast (Baker and Haworth 2000a), the evidence is fragmentary from a few sites clustered in central New South Wales and cannot be accurately related to mean sea level. They also included oyster data from Queensland sites which were compared with the tubeworm results without consideration of the differential isostatic responses to water loading which can amount to as much as 1 m depending on exact location (Nakada and Lambeck 1989). Microatolls can be more precisely related to tidal datum and the careful analysis of Chappell et al. (1983) does not support oscillations in sea level during the Late Holocene. We consider it premature to revise the Late Holocene sea-level history.

As indicated above, the far-field hydro-isostatic corrections are not strongly dependent on the source of the water added into the oceans. But while the Australian sites lie far from the northern ice sheets, the more southern locations occur within the expected intermediate field of Antarctic rebound. Thus if there has been a substantial reduction in Antarctic volume some north–south dependence in the sea-level signal can be expected (Nakada and Lambeck 1988). This is because there will be a long-wavelength signal emanating from Antarctica in which southern Australian Holocene highstands are reduced relative to those in the north, as is indeed suggested by the fragmentary evidence from both New Zealand and eastern Australia (Lambeck and Nakada 1990). Generally the available evidence from the southern latitudes is less satisfactory: evidence for Holocene highstands does occur along part of the coast (Belperio et al. 2002) but there is no coral record and the indicators are from sources that are less securely related to mean sea level. A more concerted effort is required to quantify the north–south gradient of past sea level along the Australian east coast.

4.5.2 *The Mediterranean Sea*

The Mediterranean Sea is beyond the immediate influence of the ice sheets and the glacio-isostatic signal for the last 7000 years is one of rising sea level that mostly dominates the hydro-isostatic contribution and only in the eastern extremities are the two predicted to be equal in magnitude but opposite in sign.

Thus, in the absence of tectonics, Late Holocene sea levels for much of the Mediterranean will be characterized by slowly rising values.

A positive feature of the Mediterranean is that it is an area of low tidal amplitudes and restricted wave activity such that the record of past sea levels has a higher probability of being preserved than elsewhere, and an accurate reduction of sea-level indicators to a consistent datum should be possible. There is also a good observational database, or potential database, including geological, biological, and archaeological sources, to address some of the specific issues about Late Holocene sea level. However, a negative feature is that tectonic movements are important in many localities and will complicate the pattern of sea-level change expected from glacial cycles alone. Also, being a nearly closed basin, variations in sea level from salinity or thermal adjustments may be important on timescales of decades and centuries.

There have been many studies of Mediterranean sea level, both with the view of establishing the average rate of change and with estimating tectonic uplift rates (e.g. Flemming 1969; 1978; Caputo and Pieri 1976; Labeyrie et al. 1976; van Andel 1989; Laborel et al. 1994). The isostatic contributions were mostly neglected in these studies and their importance, including their variability across the basin were subsequently modeled by Lambeck (1995, 1996), Lambeck et al. (2004b) and Lambeck and Purcell (2005). Tectonically stable areas, within the Mediterranean basin can be identified from either the geological or geophysical records or from the position of the Last Interglacial shoreline, the Tyrrhenian shoreline, which is clearly identified by its morphological and litho-stratigraphic markers and the associated Senegalese fauna (Issel 1914; Antonioli et al. 2006). For example, parts of the Italian peninsula, Sardinia and northwest Sicily can be considered to be stable on the Holocene timescale but sites in southern Italy, such as Calabria and northeast Sicily are uplifting at rates of up to 2.4 mm/year (Antonioli et al. 2006). The evidence for uplift is often direct, such as the sequences of erosion notches seen near Corinth (Figure 4.1d) (Flemming and Webb 1986; Pirazzoli 1994; Villy et al. 2002), and the evidence here is more useful for estimating magnitudes and repeat times of seismic events than for the study of sea level. But where Last Interglacial shorelines are absent, uplift/subsidence rates for Holocene time can still be calculated if the glacio-hydro-isostatic model has been calibrated against the data from the known tectonically stable areas (e.g. Monaco et al. 2004; Orrù et al. 2004) and then applied to the areas of suspected tectonic activity.

The assumption of uniform rates of tectonic vertical movement is not always valid as exemplified by the examples of Basiluzzo and Pozzuoli, respectively located in the volcanic arc of the Aeolian Islands and the centre of the Phlaegrean Fields volcanic complex in Italy. At the latter site the Roman columns have long been known for their marine borings at elevations up to 7 m above sea level (Figure 4.1a) (Morhange et al. 1999). Recent radiocarbon dating of *in situ Lithophaga* and other mollusk shells in the perforations of these columns, as well as of *in situ* corals and mollusks from nearby cave and cliff sites, indicates that local sea level reached 7 m above present during three periods since Roman times; in the fifth century AD, early Middle Ages between approximately 700 and 900 AD, and between approximately 1300 and 1500 AD before the 1538 eruption of Monte Nuovo (Morhange et al. 2006). At other sites there is evidence for long term changes in uplift rates with the average rates for the past approximately 120 000 years being up to twice those for the past 10 000 years (Antonioli et al. 2006).

Archaeological evidence provides many records of patterns of change across the region, for example, along the coast of Tunisia (Slim et al. 2004) at the buried port of Troy (Kraft et al. 2003), at the ancient harbor of Marseilles (Morhange et al. 2001), or the submerged urban quarters at Tyre where a seawall 3 m high indicates >3 m of submergence, compared with a submerged quarry indicating relatively little sea-level change at Sidon (Marriner et al. 2006). Sivan et al. (2001) show that on the coast of Israel, sea level rose to reach a level close to present around 2000 years ago, with evidence of changes of <30–40 cm since. Evidence includes Neolithic pottery, wells and rock-cut pools, in addition to abrasion platforms and accretionary vermetid reefs (Sivan et al. 2004).

The Roman fish tank data from the more stable areas of the central Mediterranean have provided the most accurate estimates for paleo-sea levels because the relation of structural features to the tidal range is well established and it indicates an average relative sea-level change along the Tyrrhenian coast near Rome of -1.35 ± 0.07 m at 2000 years ago (Lambeck et al. 2004a), although elsewhere, such as in Calabria, the change has been greater because of tectonic uplift (Anzidei et al. 2006). Most of this change from the stable region can be attributed to ongoing isostatic signals. When corrected for this and when compared with present rates of change estimated from nearby tide-gauge sites, the change in sea level attributed to change in ocean volume since the Roman Period is a rise of 0.13 ± 0.09 m, which is indicative of an onset of the modern sea-level rise having occurred only about 100 years ago (Lambeck et al. 2004a).

One feature of the isostatic response within the Mediterranean basin is that the spatial variability is significant because of the variability in the coastline geometry and because of variable distance from the former ice sheets. This is illustrated for two sites in Figure 4.12: the vermetid and coralline algae data from the French

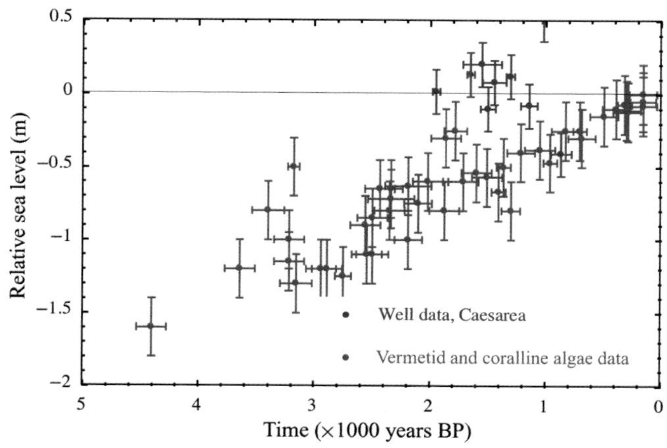

Figure 4.12 Relative sea-level observations from two localities in the Mediterranean Sea. The vermetid and coralline algae data are from the Mediterranean coast between Marseilles and Nice (Laborel et al. 1994) where the differential isostatic effects are small but not wholly negligible (Lambeck and Bard 2000). The well data are from Caesarea (Sivan et al. 2004). The different responses at the two locations reflect the increasing distance from the former ice loads.

Mediterranean coast between Marseilles and Nice (Laborel et al. 1994) and the well data from Israel (Sivan et al. 2004). The comparison shows some significant difference that can be attributed to the differences in isostatic contribution at the two sites. Thus the differential techniques used for the Australian region can also be used with the same advantage here to separate out the isostatic and ocean volume contributions. In all cases, whether from Greece, Italy, Israel, or Mediterranean France, the results point to there having been an ongoing increase in ocean volume until about 3000 years ago, such as to raise global sea level by about 3 m between 7000 and approximately 3000 years BP, with a near-constant ocean volume over the past approximately 2000 years (Lambeck and Purcell 2005).

4.5.3 Ice Margin Sites: The Baltic Sea

Well within the former ice margins the relative sea-level change since the time of deglaciation has been one of falling levels because crustal rebound exceeds the ocean volume contribution. Beyond the ice margin the change has been one of a rising sea level because it is the latter contribution that dominates. Thus there will be an intermediate zone, one that is time-dependent, in which the glacio- and hydro-isostatic contributions cancel each other such that the actual sea-level change will provide a direct measure of the change in ocean volume (e.g. Lambeck et al. 1998b), or where the sum of all contributions is zero. A locality where the latter condition is approximated occurs in southwestern Norway, in Jæren, where the relative sea level has oscillated within a few meters about its present position for about the last 15 000 years (Bird and Klemsdal 1986). But while such sites are theoretically ideal for separating out ocean volume and isostatic corrections the data quality is often unsatisfactory because of the long time interval in which overprinting of the record occurs.

The southern Baltic Sea does provide one environment where accurate and high-resolution records of sea level have been preserved. Along the coast of Blekinge and Småland in southeastern Sweden, in particular, isolation basin studies have provided quite detailed results for local and regional sea-level change from the time of deglaciation to the present (Svensson 1991; Berglund et al. 2005). The comprehensive data set for sea-level change has permitted quite detailed rebound models to be developed that reflect the broad features of the rebound seen in geological and geodetic data across Scandinavia as a whole (Lambeck et al. 1998a,b; Lambeck 1999; Milne et al. 2001). We consider here only the Littorina data (≈7000 years or younger) from the Baltic basin extending from the Gulf of Finland to Denmark within and to the south of the Younger Dryas moraine zone. In this region any ice-history and Earth-model limitations result only in long-wavelength perturbations in the predicted values such that systematic shorter-period fluctuations should be indicative of unmodeled oscillations in the assumed ice volume function that, in this case, assumes that there has been a rise in sea level after 7000 years BP. Figure 4.13a illustrates the

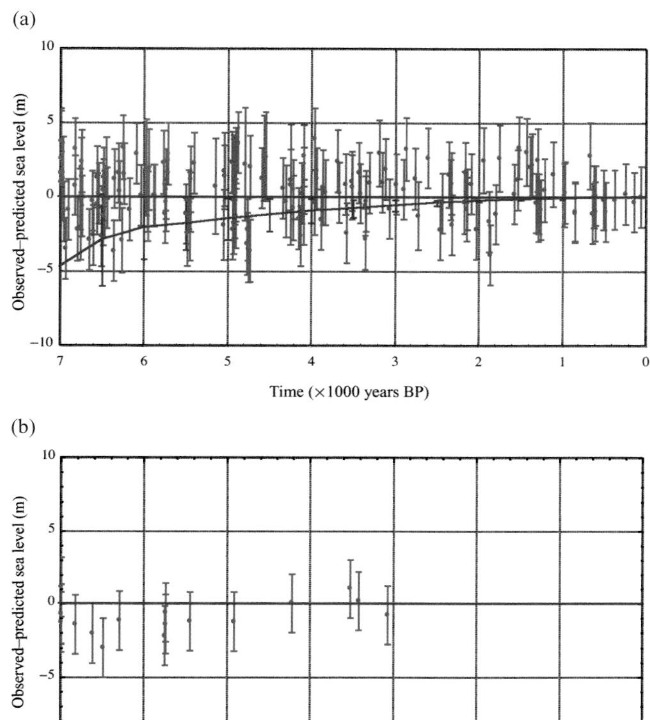

Figure 4.13 (a) Observed less predicted sea-level estimates from within the Baltic Basin south of the Younger Dryas marginal moraine zone in Finland, Åland, Sweden, and the western Baltic. The predicted values are based on the 2006 Australian National University (ANU) iterations for the Scandinavian and more distant ice sheets and for mantle rheology. The observations all correspond to isolation basin analyses for which the radiocarbon ages are consistent with pollen results. The equivalent sea-level function corresponding to the adopted ice sheets is shown in green (data from N.-O. Svensson and K. Lambeck, unpublished). (b) Same as (a) but for the restricted data from Blekinge, corrected for differential isostatic corrections between the individual basin sites and the reference locality corresponding to the mean position of the basins (data from Yu 2003, and Berglund et al. 2005).

results, with the predictions based on the most recent solution for Scandinavia (Lambeck 2006).

In at least one location it has been possible to extract more detailed constraints on ocean volume change. This is the case for the evidence for Blekinge, southern Sweden, where the basins have been selected to lie approximately on contours of constant isostatic rebound such that the differential isostatic corrections are very small (Berglund et al. 2005; Yu 2003) (Figure 4.13b). For both the local and regional data sets and within observational uncertainties (1) the results are in agreement with the rebound model and assumed ice-volume function and (2) there is no need to invoke globally occurring fluctuations in excess of approximately 1 m during the Late Holocene.

4.6 Discussion and Conclusions

Taken together, the above analyses indicate that while most sea-level data will contain isostatic signals, they can be used to extract information about changes

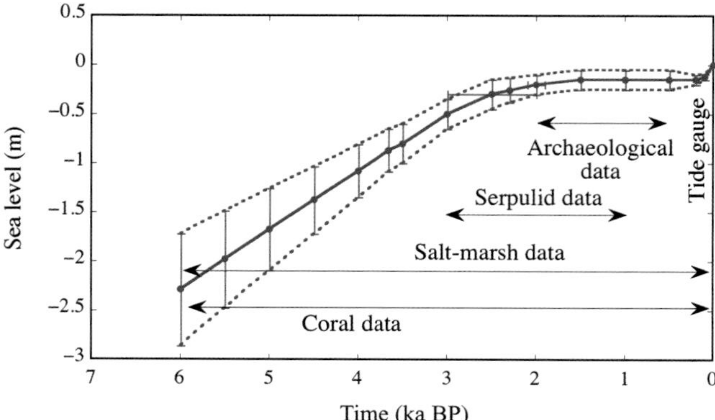

Figure 4.14 Current best estimates, including uncertainty estimates, of the eustatic sea-level change over the past 6000 years as inferred from geological and archaeological data and from the tide-gauge data for the past century.

in ocean volume that is largely independent from the details of the isostatic model assumed. The principal conclusions reached for the last 6000–7000 years, the time when sea level reached or approached its present level for the first time since the last interglacial, are:

- there has been a slow increase in ocean volume from about 7000 to 2000 or 3000 years ago such as to raise levels globally by approximately 2–3 m;
- ocean volumes have remained nearly constant from 2000–3000 years ago until the onset of the modern rise about 100 years ago; and
- there is no convincing evidence for oscillations in global ocean volume, or for periods of rapid increase in volume, during the past 6000–7000 years.

Figure 4.14 illustrates what we consider is our best current estimate of the globally averaged sea level change for this period, including the instrumentally recorded rise for the past approximately 100 years.

These results do not identify the origin of the volume increase during the latter part of the Holocene but the requisite amounts are not inconsistent with evidence of local or regional ice retreat within this time interval. Antarctic contributions, particularly from West Antarctica, cannot be ruled out in view of model results that show that this latter ice sheet is especially vulnerable to rising sea level and that the change in Holocene grounding line lags behind the sea-level rise (Huybrechts 2003). Nor are these results inconsistent with observational evidence. Stone et al. (2003), for example, have concluded from cosmogenic ages of glacial deposits that the West Antarctic Ice Sheet is still adjusting to Holocene temperature and sea level and that it continued to supply water to the oceans after the Northern Hemisphere ice sheets stabilized or disappeared. They also infer that melting from the Marie Byrd Land coast alone could have contributed 0.2–0.3 m to global sea level since 7000 years BP, and if this is representative of the West Antarctic Ice Sheet as a whole then total changes of 2–3 m are plausible. Observational evidence for reduction in mountain glacier volume indicates contributions to sea-level rise of the order 0.3–0.4 mm/year for the past century which, if representative of the longer period, would contribute some 2 m to global sea level.

4.6.1 Improvements in Data Sets and Modeling

In order to improve the resolution for the past ice volume changes, the following requirements are essential:

- sea-level data from localities that are largely free from major isostatic and tectonic contributions;
- if the first point cannot be achieved, to use analysis methods and data sets that effectively separate the isostatic and ocean-volume signals;
- high-resolution, high-precision data from markers that bear a consistent relation between formation position and mean sea level and that have a modern counterpart;
- data sets that bridge the interval between the Holocene geological evidence and the historical record;
- data sets that permit distinction between local, regional, and global change;
- sea-level data that address specific questions about the past ice sheets;
- improvements in ice and Earth models for the glacial adjustment modeling.

Pure Ocean Volume Sea-Level Markers

While the isostatic response is position- and time-dependent, there will be circumstances when (compare with equation 4.1)

$$\Delta\zeta_{\mathrm{rsl}}(\varphi,t) \cong \Delta\zeta_{\mathrm{esl}}(t)$$

or

$$\Delta\zeta_{\mathrm{I-g}}(\varphi,t) + \Delta\zeta_{\mathrm{I-h}}(\varphi,t) = 0$$

and the observations should provide a direct measure of changes in ocean volume. Ocean islands far from the former ice margins were believed to be such sites (Bloom 1967) but, as described above, these are now known to also contain isostatic information. Nevertheless there are circumstances where the glacio- and hydro-isostatic contributions cancel each other and one example is a zone through Denmark and northern Germany, around the Scandinavian Ice Sheet, where a number of long tide-gauge records exist (Lambeck et al. 1998b). The location of this zone is Earth- and ice-model-dependent but for a range of models that are consistent with geological sea-level data from within the ice margins a broad zone exists for recent times within which

$$\Delta\zeta_{\mathrm{rsl}}(\varphi,t) = \Delta\zeta_{\mathrm{esl}}(t) \pm 3\sigma_{\mathrm{rsl}} \tag{4.6}$$

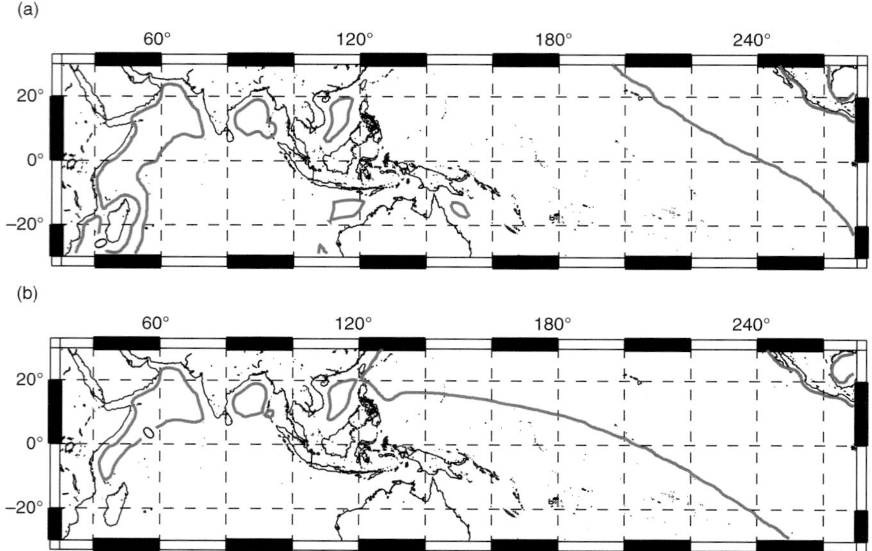

Figure 4.15 Contours identifying locations where $\Delta\zeta_{rsl}(\varphi, t) = \Delta\zeta_{esl}(t)$ in the equatorial Pacific and Indian Oceans at two epochs: (a) during the LGM 20 000 years ago and (b) 10 000 years ago. Sea-level data from locations near these contours would provide a direct measure of the ice-volume-equivalent sea level. The actual locations of the zero contours will show some model-parameter dependence.

where σ_{rsl} is the observational accuracy. The zone may shift with time but in general it will be possible to find in-principle locations where this condition is satisfied at any epoch. Whether there is potential for geological observational data from such locations is another matter. Figure 4.15 illustrates far-field examples of where globally the condition is met for two epochs, based on the (late 2006) Australian National University (ANU) ice- and Earth models. At the LGM, for example, isostatic-free conditions are predicted for several near-equatorial island groups, including the northern Marianas, Christmas (Kiritimati) and Easter Island in the Pacific and Comores and Aldabra in the western Indian Ocean. Note that the location of the zero contour evolves with time and at 10 000 years ago, for example, the Hawai'ian Islands approach the dip-stick condition (Bloom 1967), as does Mauritius in the Indian Ocean.

Differential Methods for Separating Contributions to Sea-Level Change

The analysis of the Australian data, in particular, has demonstrated the value of using differential methods to separate the isostatic and ocean volume contributions. Predictions for other locations illustrate that sites can be found with similar characteristics and one example is illustrated in Figure 4.16 for the western Indian Ocean where significant spatial variability is predicted mainly because of the hydro-isostatic dependence on shoreline geometry of, and distance from, the African and Madagascar coasts. Other locations where this approach has considerable potential is across the Sunda Shelf where considerable spatial variability is predicted (Lambeck et al. 2002b) consistent with observational evidence (Hanebuth et al. 2000). Further information from both regions would

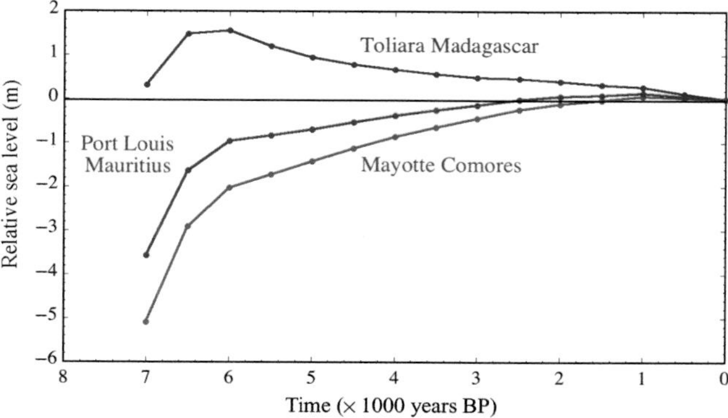

Figure 4.16 Predicted sea-level response for three locations within an approximately 1000-km-sided triangle in the western Indian Ocean based on the 2006 ANU ice and Earth models and for the past 7000 years. Late Holocene highstands are predicted only for the Madagascar coast. Data for this period from these locations would be important for estimating both the regional mantle structure and any variations in ocean volume.

be constructive, both for ice-volume studies and for evaluation of whether lateral variation in mantle viscosity is likely to be important.

High-Precision, High-Resolution Data

Requirements for improved observational data sets include:

- sea-level markers with a well-defined relationship to mean sea level within at least the region of study and preferably globally;
- the ability to measure the positions of these indicators with respect to their modern positions and the validity of the assumption that the tidal range has remained unchanged since formation age or that the changes are known;
- ability for precise dating of the age of formation of the sea-level indicators;
- indicators that are sensitive to both sea-level rise and fall.

Corals generally will not be useful because of their ability to grow in water depths of several meters to 100 m or more. But, as discussed above, where they form microatolls at levels that are related to a specific point in the tidal range, they can be used to track changes in sea level with very considerable precision subject to the cautions discussed above.

Isolation basin analyses from slowly uplifting areas can also provide high precision information on changes in ocean volume in areas of low tidal range, provided that the basins are closely clustered such that differential isostatic variations can be reliably modeled, as illustrated in Figure 4.13b. Estimation of changes in ocean volume from single locations requires confidence in the isostatic models but isostatic signals over the past few thousand years will not exhibit high-frequency changes such that this evidence can place constraints on magnitudes of oscillations from other causes.

Other data that can be exploited further are the accretionary biological constructions of coralline algae that are particularly important in the Mediterranean. The age of these records rarely extends beyond a few thousand years but it has been demonstrated that they have the potential to track sea level with considerable precision (Laborel et al. 1994) (Figure 4.12). In so far as the Mediterranean is a region in which the spatial variability of the isostatic response is very considerable, this data type is potentially important for testing models of this variability. Also, the algal constructions span the time interval of the more important archaeological data, and in some regions they coexist, such that it becomes possible to evaluate the relative merits of the two data types.

Finally, an important and recent source of information comes from the salt-marsh analyses, as both high-precision, high-resolution records of sea-level change for bridging the geological and instrumental records (see below) and for improving our understanding of the spatial variability of Late Holocene sea-level change.

One of the limitations of past work has been the resolution of the dating methods, often based on a combination of radiocarbon and pollen analyses, but other dating systems, notably high-precision $^{238}U/^{230}Th$ (McCulloch and Mortimer 2008) and ^{14}C (Marshall et al. 2007), and possibly ^{210}Pb and ^{137}Cs, will become increasingly important for the more recent periods.

Bridging the Geological and Instrumental Data Sets

Although a few of the longer tide-gauge records suggest a mid-19th-century acceleration in sea-level rise, for example those from Amsterdam, Liverpool, and Stockholm (Church et al. 2001) (Figure 4.2c), most are of too short a duration to be adequately compared with paleoecological proxies of sea-level change. Generally the accuracy of the latter proxies is such that signals characteristic of the past century would barely rise out of the noise levels of the data for the earlier period. However, there is potential for bridging the gap between the two data types (Goodwin et al. 2000), most important of which so far are the reconstructions of sea-level changes for the past 1000 years from former salt-marsh surfaces.

Salt-marsh proxy records, constructed as shown in Figure 4.7, can be validated by comparison with tide-gauge records (Donnelly et al. 2004; Gehrels et al. 2005, 2006) as discussed in the following chapter. Sea-level variations during the Little Ice Age and the Medieval Warm Period were probably too small to be resolved by salt-marsh records (Gehrels et al. 2005). The absence of pre-20th-century sea-level fluctuations in proxy records from North America suggests that the 20th-century sea-level acceleration was unprecedented in (at least) the past millennium.

As discussed above, coral microatolls may also be useful for bridging the two timescales. Thus Smithers and Woodroffe (2001) have demonstrated synchronous broad fluctuations of sea level in the eastern Indian Ocean, of about 20 years' length, with an amplitude of 5–10 cm, using two microatolls from different parts of the reef flat around Cocos. These microatolls showed an average rate of sea-level rise of only 0.35 mm/year over the past century, considerably lower than the global mean rate for this period. Elsewhere in the Pacific Ocean the undulations on the top of microatolls have been shown to be related to the ENSO phenomenon (Woodroffe and McLean 1990; Spencer et al. 1997).

Local, Regional, and Global Variations

In the ocean volume model for the past 7000 years (Figure 4.14) it has been assumed that the increase in ocean volume has been continuous because of the absence of convincing and global evidence that would indicate otherwise. Addition of ice into the oceans will result in a globally synchronized sea-level rise although the amplitude may vary spatially and an increase in sea level as a series of discrete and episodic events could be expected, as has been suggested for a somewhat earlier epoch, at approximately 8000–7500 years BP, by Blanchon and Shaw (1995), Liu and Milliman (2004), or Bird et al. (2007). A global fall in ocean volume during interglacial conditions is more difficult to envisage because it requires rapid withdrawal of water from the ocean: a global fall of approximately 0.5 m requires an equally rapid growth of a continent-based ice sheet of about $2 \times 10^5\,\text{km}^3$ in volume which corresponds approximately to the British ice sheet at the time of the LGM.

There is evidence in some homogeneous and consistent records for oscillations in sea level during the past 7000 years but it has not been demonstrated whether this is a global phenomenon or restricted to certain areas. In the Baltic, for example, small-amplitude (<1 m) oscillations have been noted in many locations (Hyvärinen 1988; Berglund et al. 2005), particularly for the period between about 7000 and 4000 years BP, but there is as yet no clear correlation of these features within either the basin, primarily because of uncertainties in the age control, or with similar changes outside of the Baltic Basin. Baltic levels are strongly correlated with meteorological conditions (Ekman 1998; Andersson 2002) and it is possible that what is recorded in the isolation basins is atmospherically forced (Berglund et al. 2005), although the evidence for the past two centuries indicates that the atmospheric forcing of sea level does not exceed a few decimeters (Ekman 1998; Andersson 2002). As discussed above, the coral evidence from the Great Barrier Reef (Chappell et al. 1983), or from Abrolhos Islands of Western Australia (Collins et al. 2006), does not require oscillations during the Late Holocene and suggests that what is seen in the Baltic is a regional rather than global signal.

Sea-Level Data that Address Specific Questions about the Past Ice Sheets

Considerable uncertainties remain about the knowledge of the melting history of the ice sheets that have contributed to the sea-level history since the time of the last glacial melting. Models for the pre-LGM ice sheets do not generally impact greatly on the rebound models for the post-LGM interval, particularly for far-field sites, provided that the total changes in ice volume are consistent with those inferred from the far-field sea-level data for the period since the last interglacial.

For all ice sheets, locations can be identified from which new data would be desirable to improve particular aspects of the ice model. For example, for the Scandinavian Ice Sheet improved data from the Jærand and Vest Agder regions of southwestern Norway, particularly for the earlier part of the last deglaciation, is desirable for establishing the ice thickness over this region and its extent onto the North Sea. Likewise additional data from the eastern sector is desirable to improve understanding of the ice history between the Gulf of Finland and the White Sea. But two ice sheets deserve particular attention: Antarctica and Greenland.

The LGM limits and post-LGM history of the Antarctic Ice Sheet remain poorly constrained by inversions of sea-level and rebound data although the analysis by Nakada et al. (2000), and Zwartz et al. (1997) have shown the usefulness of such analyses to place regional constraints on the ice cap. The potential for new geological observations of relative sea-level change from the Antarctic margins is, however, limited and examination of changes in ice elevation using cosmogenic dating of exposed rock surfaces, erratics and marginal moraines is likely to be more profitable (Stone et al. 2003) as would be dating of the last ice retreat across the shelf (Anderson et al. 2002). Geodetic observations of crustal rebound and gravity changes will also play an important role in future analyses but with the proviso that such observations will also contain information on more recent changes in ice volume and separation of this from the past changes requires independent knowledge of the ice history (see Chapter 10 of this volume).

The Greenland Ice Sheet is more amenable to analyses of sea-level data (Fleming and Lambeck 2004; Tarasov and Peltier 2006) although much of the older field data for sea-level change is of relatively low quality, being mostly limiting indicators rather than measures of actual sea-level change (Weidick et al. 2004; Fleming and Lambeck 2004). However, more recent results by Long et al. (1999, 2003), Bennike et al. (2002, and Sparrenbom et al. (2006a, 2006b) have demonstrated that from isolation basin analyses it is possible to obtain high-quality data from the coast and inner fjords and that these results are pointing to a Greenland Ice Sheet that in the south extended out to the shelf edge (Sparrenbom 2006).

Modeling Improvements

The available evidence indicates that the amplitude of ocean volume changes during the current interglacial are likely to be of a similar magnitude or less than other perturbations in sea level. To obtain improved estimates of both the amplitudes and rates of this change, therefore, requires either an observational strategy that eliminates these other factors or an improved understanding of the other contributions. The tectonic contributions at these scales are the most problematical because the range of driving mechanisms is considerable and their geophysical modeling is still only little more than order of magnitude analyses. The assumption of tectonic stability if the Last Interglacial shorelines are close to present-day sea level is usually a satisfactory one except in cases where the vertical movements are cyclic, as for the Mediterranean examples discussed above (Figures 4.1a, 4.8c and e). Where the Last Interglacial shorelines lie far from present sea level, the usual practice is to correct for vertical tectonics by assuming uniform rates of uplift or subsidence but the evidence from some of the Italian localities suggests that this is not always valid (Antonioli et al. 2006). In the case of the uplifted Huon terraces this assumption can be tested by comparing results from different parts of the coast with different uplift rates and by comparing rates estimated from the uplifted mid-Holocene reef with results from the Last Interglacial as well as with older interglacial reefs, and the evidence points to a general validity of the assumption of uniform uplift rates when averaged over about a thousand years. But on shorter timescales the vertical movements are likely to be episodic, as illustrated by the Iraion example in Figure 4.1d, and as may also have occurred on the Huon Peninsula (Chappell et al. 1996).

The other major departures from eustasy during the interglacial are the isostatic contributions. The physics here are well understood but there remain a number of issues that need addressing. These fall into three categories: numerical methods, assumptions about the mantle rheology, and ice-sheet input parameters.

While it has been convenient to discuss the isostatic response as the sum of glacio- and hydro-isostatic contributions the two are not decoupled, as discussed above, and iterative solutions are usually introduced which allow for this coupling and, at the same time, for the migration of shorelines and ice-grounding lines. There has been some recent debate about this (Peltier 2002; Lambeck et al. 2002b, 2003; Mitrovica 2003) with Mitrovica concluding that the theory developed at ANU and used here "is technically sound" (see also Mitrovica and Milne 2003 and Chapter 10 of this volume). Partial comparisons have been carried out between the ANU code and that developed by Mitrovica and Milne and no differences have been identified that impact on the inferences drawn here.

A publicly available sea-level solution code is provided by Spada and Stocchi (2007) but this does not yet include some of the higher-order complications of the complete solution.

Most models assume linear response of the planet to loading, primarily for the reason that this appears to be satisfactory at the level of current observational accuracy and because of mathematical convenience. Likewise, for much the same reasons, most models assume lateral uniformity in mantle response, although there is some evidence that lateral variation in at least the effective lithospheric thickness and effective upper mantle viscosity may be important (Lambeck and Chappell 2001). There remains unresolved debate about the correct formulation of the internal boundary conditions; whether or not the position of the seismically defined phase-transformation boundaries respond to stress changes by radial deformation (Fjeldskaar and Cathles 1984; Dehant and Wahr 1991; Johnston et al. 1997). Again for reasons of expediency, most models assume that the kinetics of the phase transformation are slow so as not to be significant on the timescale considered here. Solutions with different assumptions about this behavior indicate that this trade-off occurs mainly between viscosity and ice thickness (Johnston et al. 1997; Lambeck et al. 1996): that when the ice-Earth model parameter combination is calibrated against the same observational data, solutions with slow and fast kinetics of mineralogical phase transformations give statistically equivalent solutions and hence are equally valid for interpolation purposes although the physical consequences may be different.

The major limitation in the present models is believed to be from the limitations of the ice sheets. Ice margin locations as a function of time are reasonably well known on the continents for the time from the LGM to the end of deglaciation (Anderson et al. 2002; Boulton et al. 2001; Dyke et al. 2002; Mangerud et al. 2002) but there is less certainty about the ice retreat history across the continental shelves. Also the ice thickness is usually not directly observed and the assumed values are model dependent. Thus in solutions of the sea-level equation aspects of both the ice history and of the Earth rheology must be considered as unknowns and strategies have to be developed to separate the parameters (Lambeck et al. 1998a). Some correlation between the two classes of parameters will often remain, for example, between ice thickness and upper-mantle viscosity or between upper-mantle viscosity and lithospheric thickness, and different combinations of parameters may be equally effective for interpolation between the observational data. For the ocean-volume analysis this should not be of significant consequence if the analysis is restricted to far-field locations.

4.6.2 Summary of Recommendations

Paleo-sea level provides a reference surface for quantifying rates of vertical tectonic movements, their episodicity as well as their longer-term average rates. But it will require the highest-quality records in order to provide a means of evaluating climate-forced signals and determining whether there are global signals that can be attributed to regional climate indicators such as the Little Ice Age or the Medieval Climatic Optimum, or regional changes in sea level associated with long-term changes in atmospheric circulation such as in the North Atlantic Oscillation (NAO). Future research should include the following tasks:

- identify data types and potential records for the extraction of precise and accurate local relative sea-level signals for the past 6000 years, to address:
  - longer-term change, including ocean-volume increase, and
  - variability of sea level on millennial to decadal timescales;
- correlate records from different environments and different geographical areas to establish whether identified changes are global, regional, or local;
- compare the principal field methodologies and modeling approaches, and assess their effectiveness at establishing isostatic and other unresolved vertical motion contributions to sea-level changes.

To address issues concerning the choice of isostatic response functions, new or additional field data on past sea level are required from a wider range of geographical areas and different epochs. Some characteristics for new sites and data have been discussed above but the list is not exclusive. High-resolution, high-precision data from far-field sites during the current interglacial are clearly important for establishing the benchmarks for separating geological background signals from possible 'greenhouse' signals. But this background record is a function of the earlier glacial cycle and new data to improve our ability to quantify this effect are also required. This includes, in particular, data from near the margins of Antarctica (Zwartz et al. 1998) and Greenland (Long et al. 2003; Sparrenbom et al. 2006a, 2006b) to establish constraints on their past histories but also data from near the interiors of the former continental-based ice sheets which in most instances will be restricted to geodetic data of crustal rebound or gravity change. Additional data sets from different geological provinces are also desirable in order to improve solutions for lateral variability in the mantle response to the variable surface loads of ice and water.

The quest for the average global rate of sea-level change has been a goal for much of the early paleo-sea-level research, but the geological record and its paleoenvironmental complement, together with archaeological evidence, demonstrate unequivocally that most places in the world do not experience this global rate of change and that regional and local departures from it can be at least as important, often more important, and scientifically more interesting. The paleo-expression of this spatial and temporal record is what is required to assess past and present trends and to predict future trends for planners and managers at the local level, to ensure sustainable use of the world's coasts in the future.

Acknowledgments

We thank the reviewers, Dr D. Hopley, Dr N. Harvey, and Dr P. Woodworth as well as two anonymous referees for constructive comments on the manuscript. K.L thanks the Australian National University, the Australian Research Council, and the ACE-CRC for research support.

References

Adey W.H. (1978) Coral reef morphogenesis: a multidimensional model. *Science* **202**, 831–7.

Agassiz L. (1840) On glaciers and the evidence of their having once existed in Scotland, Ireland and England. *Proceedings of the Geological Society of London* **3**, 327–32.

Allen J.R.L. (2000a) Morphodynamics of Holocene salt marshes: a review sketch from the Atlantic and Southern North Sea coasts of Europe. *Quaternary Science Reviews* **19**, 1155–1231.

Allen J.R.L. (2000b) Holocene coastal lowlands in NW Europe: autocompaction and the uncertain ground. In: *Coastal and Estuarine Environments: Sedimentology, Geomorphology and Geoarchaeology* (Pye K. and Allen J.R.L., eds), Special Publication, pp. 239–52. Geological Society, London.

Alley R.B., Clark P.U., Huybrechts P. and Joughin I. (2005) Ice-sheets and sea-level changes. *Science* **310**, 456–60.

Anderson J.B., Shipp S.S., Lowe A.L., Wellner J.S. and Mosola A.B. (2002) The Antarctic Ice Sheet during the Last Glacial Maximum and its subsequent retreat history: a review. *Quaternary Science Reviews* **21**, 49–70.

Andersson H.C. (2002) Influence of long-term regional and large-scale atmospheric circulation on the Baltic sea level. *Tellus* **54A**, 76–88.

Angulo R.J.A., Lessa G.C.B. and De Souza M.C.A. (2006) A critical review of mid- to late-Holocene sea-level fluctuations on the eastern Brazilian coastline. *Quaternary Science Reviews* **25**, 486–506.

Antonioli F., Ferranti L. and Lo Schiavo F. (1996) The submerged Neolithic burials of the Grotta Verde at Capo Caccia (Sardinia, Italy) implication for the Holocene sea-level rise. *Memorie Descrittive del Servizio Geologico Nazionale* **52**, 329–36.

Antonioli F., Chemello R., Improta S. and Riggio S. (1999) The Dendropoma (Mollusca Gastropoda, Vermetidae) intertidal reef formations and their paleoclimatological use. *Marine Geology* **161**, 155–70.

Antonioli F., Kershaw S., Rust D. and Verrubbi V. (2003) Holocene sea-level change in Sicily, and its implications for tectonic models: new data from the Taormina area, NE Sicily. *Marine Geology* **196**, 53–71.

Antonioli F., Bard E., Silenzi S., Potter E.K. and Improta S. (2004) 215 kyr history of sea level based on submerged speleothems. *Global and Planetary Change* **43**, 57–68.

Antonioli F., Ferranti L., Lambeck K., Kershaw S., Verrubbi V. and Dai Pra G. (2006) Late Pleistocene to Holocene record of changing uplift rates in southern Calabria and eastern Sicily (southern Italy, Central Mediterranean Sea). *Tectonophysics* **422**, 23–40.

Anundsen K. (1985) Changes in shore-level and ice-front position in Late Weichsel and Holocene, southern Norway. *Norsk geografisk tidsskrift* **39**, 205–25.

Anzidei M., Esposito A., Antonioli F., Benini A., Tertulliani A. and Del Grande C. (2006) I movimenti verticali nell'area di Briatico: evidenze da indicatori archeologici marittimi nell'area del terremoto del 1905. In: *8 settembre 1905, terremoto in Calabria* (Guerra I. and Bavaglio A., eds), pp. 301–21. Università della Calabria, Regione Calabria.

Baker R.G.V. and Haworth R.J. (1997) Further evidence from relic shell crust sequences for a late Holocene higher sea level for eastern Australia. *Marine Geology* **141**, 1–9.

Baker R.G.V. and Haworth R.J. (2000a) Smooth or oscillating late Holocene sea-level curve? Evidence from cross-regional statistical regressions of fixed biological indicators. *Marine Geology* **163**, 353–65.

Baker R.G.V. and Haworth R.J. (2000b) Smooth or oscillating late Holocene sea-level curve? Evidence from palaeo-zoology of fixed biological indicators in east Australia and beyond. *Marine Geology* **163**, 367–86.

Baker R.G.V., Haworth R.J. and Flood P.G. (2001a) Inter-tidal fixed indicators of former Holocene sea levels in Australia: a summary of sites and a review of methods and models. *Quaternary International* **83–85**, 257–73.

Baker R.G.V., Haworth R.J. and Flood P.G. (2001b) Warmer or cooler late Holocene marine palaeoenvironments?: interpreting southeast Australian and Brazilian sea-level changes using fixed biological indicators and their $\delta^{18}O$ composition. *Palaeogeography, Palaeoclimatology, Palaeoecology* **168**, 249–72.

Baker R.G.V., Haworth R.J. and Flood P.G. (2004) An oscillating Holocene sea-level? Revisiting Rottnest Island, Western Australia and the Fairbridge eustatic hypothesis. *Journal of Coastal Research* **42**, 3–14.

Bard E., Antonioli F. and Silenzi S. (2002) Sea-level during the penultimate interglacial period based on submerged stalagmite from Argentarola Cave (Italy). *Earth and Planetary Science Letters* **196**, 135–46.

Bassett S.E., Milne G.A., Mitrovica J.X. and Clark P.U. (2005) Ice sheet and solid earth influences on far-field sea-level histories. *Science* **309**, 925–8.

Belperio A.P., Harvey N. and Bourman R.P. (2002) Spatial and temporal variability in the Holocene highstand around southern Australia. *Sedimentary Geology* **150**, 153–69.

Bennike O., Björck S. and Lambeck K. (2002) Estimates of South Greenland late-glacial ice limits from a new relative sea level curve. *Earth and Planetary Science Letters* **197**, 171–86.

Berger A.L. (1992) Astronomical theory of paleoclimates and the last glacial-interglacial cycle. *Quaternary Science Reviews* **11**, 571–80.

Berglund B.E., Sandgren P., Barnekow L., Hannon G., Hui J., Skog G. and Yu S-Y. (2005) Early Holocene history of the Baltic Sea, as reflected in coastal sediments in Blekinge, southeastern Sweden. *Quaternary International* **130**, 111–39.

Bindoff N., Willebrand J., Artale V., Cazenave A., Gregory J.M., Gulev S. et al. (2007) Observations: ocean climate change and sea level. In: *Climate Change 2007: The Physical Science Basis*. Contribution of Working Group 1 to the Fourth Assessment Report of the Intergovernmental Panel on Climate Change (Solomon S., Qin D., Manning M., Marquis M., Averyt K., Tignor M.M.B. et al., eds), pp. 385–432. Cambridge University Press, Cambridge.

Bird E.C.F. (1988) The tubeworm *Galeolaria caespitosa* as an indicator of sea level rise. *Victorian Naturalist* **105**, 98–105.

Bird E.C.F. and Klemsdal T. (1986) Shore displacement and the origin of the lagoon at Brusand, southwestern Norway. *Norsk Geografisk Tidsskrift* **40**, 27–35.

Bird M.I., Fifield L.K., Teh T.S., Chang C.H., Shirlaw N. and Lambeck K. (2007) An inflection in the rate of early mid-Holocene sea-level rise: a new sea-level curve from Singapore. *Estuarine Coastal and Shelf Science* **71**, 523–36.

Blackman D.J. (1973) Evidence of sea level change in ancient harbours and coastal installations. *Marine Archaeology, Colston Papers* **23**, 114–17.

Blanchon P. and Shaw J. (1995) Reef drowning during the last deglaciation: Evidence for catastrophic sea-level rise and ice-sheet collapse. *Geology* **23**, 4–8.

Bloom A.L. (1967) Pleistocene shorelines: a new test of isostasy. *Geological Society of America Bulletin* **78**, 1477–94.

Bloom A.L. (1977) *Atlas of Sea-Level Curves*. IGCP 61. Cornell University, Ithaca.

Bloom A.L., Broecker W.S., Chappell J.M.A., Matthews R.K. and Mesolella K.J. (1974) Quaternary sea level fluctuations on a tectonic coast, new ^{230}Th/^{234}U dates for the Huon Peninsula, New Guinea. *Quaternary Research* **4**, 185–205.

Boulton G.S., Dongelmans P., Punkari M. and Broadgate M. (2001) Palaeoglaciology of an ice sheet through a glacial cycle: the European ice sheet through the Weichselian. *Quaternary Science Reviews* **20**, 591–625.

Caputo M. and Pieri L. (1976) Eustatic variation in the last 2000 years in the Mediterranean. *Journal of Geophysical Research* **81**, 5787–90.

Chappell J. (1974) Geology of coral terraces, Huon Peninsula, New Guinea: a study of quaternary tectonic movements and sea-level changes. *Geological Society of America Bulletin* **85**, 553–70.

Chappell J. (1983) Evidence for smoothly falling sea levels relative to north Queensland, Australia, during the past 6000 years. *Nature* **302**, 406–8.

Chappell J. (2002) Sea level changes forced ice breakouts in the Last Glacial cycle: new results from coral terraces. *Quaternary Science Reviews* **21**, 1229–40.

Chappell J. and Shackleton N.J. (1986) Oxygen isotopes and sea level. *Nature* **324**, 137–40.

Chappell J. and Polach H. (1991) Post glacial sea level rise from a coral record at Huon Peninsula, Papua New Guinea. *Nature* **349**, 147–9.

Chappell J., Chivas A., Wallensky E., Polach H.A. and Aharon P. (1983) Holocene palaeo- environment changes, central to north Great Barrier Reef inner zone. *BMR Journal Australian Geology and Geophysics* **8**, 223–35.

Chappell J., Omura A., Esat T., McCulloch M., Pandolfi J., Ota Y. and Pillans B. (1996) Reconciliation of late Quaternary sea levels derived from coral terraces at Huon Peninsula with deep sea oxygen isotope records. *Earth and Planetary Science Letters* **141**, 227–36.

Church J.A., Gregory J.M, Huybrechts P., Kuhn M., Lambeck K., Nhuan M.T. et al. (2001) Changes in Sea Level. In: *Climate Change 2001: The Scientific Basis*. Contribution of Working Group 1 to the Third Assessment Report of the Intergovernmental Panel on Climate Change (Houghton J.T., Ding Y., Griggs D.J., Noguer M., van der Linden P.J., Dai X. et al., eds), pp. 639–94. Cambridge University Press, Cambridge.

Clark P.U., McCabe A.M., Mix A.C. and Weaver A.J. (2004) Rapid rise of sea level 19,000 years ago and its global implications. *Science* **304**, 1141–4.

Cloetingh S., Lambeck K. and McQueen H. (1987) Apparent sea-level fluctuations and a palaeostress field for the North Sea region. In: *Petroleum Geology of North West Europe* (Brooks J. and Glennie K., et al.), pp. 49–57. Graham & Trotman, London.

Clottes J. and Courtin J. (1994) *La Grotte Cosquer*. Editions du Seuil, Paris.

Clottes J., Courtin J., Collina-Girard J., Arnold M. and Valladas H. (1997) News from Cosquer Cave: climatic studies recording, sampling, dates. *Antiquity* **71**, 321–6.

Collins L.B., Zhao J.-X. and Freeman H. (2006) A high-precision record of mid-late Holocene sea-level events from emergent coral pavements in the Houtman

Abrolhos Islands, southwest Australia. *Quaternary International* **145-6**, 78–85.

Colonna M., Casanova J., Dullo W.C. and Camoin G. (1996) Sea-level changes and ^{18}O record for the past 30,000 yr from Mayotte Reef, Indian Ocean. *Quaternary Research* **46**, 335–9.

Croll J. (1864) On the physical cause of the change of climate during geological epochs. *Philosophical Magazine* **28**, 121–37.

Croll J. (1875) *Climate and time in their geological relations*. Daldy, Isbister & Co., London.

Daly R.A. (1910) Pleistocene glaciation and the coral reef problem. *American Journal of Science* **30**, 297–308.

Daly R.A. (1915) The glacial-control theory of coral reefs. *Proceedings of the American Academy of Arts and Science* **51**, 155–251.

Daly R.A. (1925) Pleistocene changes of level. *American Journal of Science* **10**, 281–313.

Daly R.A. (1934) *The Changing World of the Ice Age*. Yale University Press, New Haven.

De Deckker P. and Yokoyama Y. (2009) Micropalaeontological evidence for Late Quaternary sea-level changes in Bonaparte Gulf, Australia, *Global and Planetary Change* **66**, 85–92.

Dehant V. and Wahr J.M. (1991) The response of a compressible, non-homogenous Earth to internal loading: Theory. *Journal of Geomagnetism and Geoelectricity* **43**, 157–78.

Demuro S. and Orrù P. (1998) Il contributo delle beach-rock nello studio della risalita del mare olocenico. Le beach-rock post-glaciali della Sardegna nord-orientale. *Il Quaternario* **11**, 19–39.

Devoy R.J.N. (ed.) (1987) *Sea Surface Studies: a Global View*. Croom Helm, London.

Dokka R.K. (2006) Modern-day tectonic subsidence in coastal Louisiana. *Geology* **34**, 281–4.

Donnelly J.P. (2006) A revised Late Holocene sea-level record for northern Massachusetts, USA. *Journal of Coastal Research* **22**, 1051–61.

Donnelly J.P., Cleary P., Newby P. and Ettinger R. (2004) Coupling instrumental and geological records of sea level change: Evidence from southern New England of an increase in the rate of sea level rise in the late 19th century. *Geophysical Research Letters* **31**, L05203.

Dutton A., Bard E., Antonioli F., Esat T.M., Lambeck K. and McCulloch M.T. (2009) Phasing and amplitude of sea-level and climate change during the penultimate interglacial. *Nature Geoscience* **2**, 355–9.

Dvorak J. and Mastrolorenzo G. (1991) The mechanisms of recent vertical crustal movements in Campi Flegrei caldera, southern Italy. *Geological Society of America Special Paper* 263.

Dyke A.S., Andrews J.T., Clark P.U., England J.H., Miller G.H., Shaw J. and Veillette J.J. (2002) The Laurentide and Innuitian ice sheets during the Last Glacial Maximum. *Quaternary Science Reviews* **21**, 9–31.

Edwards R.J., Wright A.J. and van de Plassche O. (2004a) Surface distributions of salt-marsh foraminifera from Connecticut, USA: Modern analogues for high resolution sea level studies. *Marine Micropalaeontology* **51**, 1–21.

Edwards R.J., van de Plassche O., Gehrels W.R. and Wright A.J. (2004b) Assessing sea-level data from Connecticut, USA, using a foraminiferal transfer function for tide level. *Marine Micropalaeontology* **51**, 239–55.

Eisenhauer A., Wasserburg G.J., Chen J.H., Bonani G., Collins L.B., Zhu Z.R. and Wyrwoll K.H. (1993) Holocene sea-level determination relative to the Australian continent: U/Th (TIMS) and 14C (AMS) dating of coral cores from the Abrolhos Islands. *Earth and Planetary Science Letters* **114**, 529–47.

Ekman M. (1998) Long-term changes of interannual sea-level variability in the Baltic Sea and related changes of winter climate. *Geophysica* **34**, 131–40.

Elias S.A. (ed.) (2007) Sea level studies. In: *Encyclopedia of Quaternary Science* **4**, pp. 2967–3095. Elsevier, Amsterdam.

Esat T.M. and Yokoyama Y. (2006) Growth patterns of the last ice age coral terraces at Huon Peninsula. *Global and Planetary Change* **54**, 216–24.

Fairbanks R.G. (1989) A 17,000-year glacio-eustatic sea level record: influence of glacial melting rates on the Younger Dryas event and deep-ocean circulation. *Nature* **342**, 637–42.

Fairbridge R.W. (1961) *Eustatic Changes in Sea Level, Physics and Chemistry of the Earth*, pp. 99–185. Pergamon Press, New York.

Farrell W.E. and Clark J.A. (1976) On postglacial sea level. *Geophysical Journal of the Royal Astronomical Society* **46**, 647–67.

Ferland M.A., Roy P.S. and Murray-Wallace C.V. (1995) Glacial lowstand deposits on the outer continental shelf of southeastern Australia. *Quaternary Research* **44**, 294–9.

Fjeldskaar W. and Cathles L.M. (1984) Measurement requirements for glacial uplift detection of nonadiabatic density gradients in the mantle. *Journal of Geophysical Research* **89**, 10115–24.

Fleming K. and Lambeck K. (2004) Constraints on the Greenland Ice Sheet since the Last Glacial Maximum from sea-level observations and glacial-rebound models. *Quaternary Science Reviews* **23**, 1053–77.

Flemming N.C. (1969) Archaeological evidence for eustatic change of sea level and earth movements in the Western Mediterranean during the last 2000 years. *Geological Society of America* Special paper **109**, 1–125.

Flemming N.C. (1978) Holocene eustatic changes and coastal tectonics in the northeast Mediterranean; implications for models of crustal consumption. *Philosophical Transactions of the Royal Society of London* **289**, 405–58.

Flemming N.C. and Webb C.O. (1986) Tectonic and eustatic coastal changes during the last 10,000 years derived from archeological data. *Zeitschrift für Geomorphologie* (suppl. 62), 1–29.

Flood P.G. and Frankel E. (1989) Late Holocene higher sea level indicators from eastern Australia. *Marine Geology* **90**, 193–5.

Gehrels W.R. (1999) Middle and late Holocene sea-level change in eastern Maine reconstructed from foraminiferal saltmarsh stratigraphy and AMS ^{14}C dates on basal peats. *Quaternary Research* **52**, 350–9.

Gehrels W.R. (2000) Using foraminiferal transfer functions to produce high-resolution sea-level records from saltmarsh deposits, Maine, USA. *The Holocene* **10**, 367–76.

Gehrels W.R., Belknap D.F. and Kelley J.T. (1996) Integrated high-precision analyses of Holocene relative sea-level changes: lessons from the coast of Maine. *Geological Society of America Bulletin* **108**, 1073–88.

Gehrels W.R., Kirby J., Prokoph A., Newnham R., Achterberg E., Evans H., Black S. and Scott D. (2005) Onset of recent rapid sea level rise in the western Atlantic Ocean. *Quaternary Science Reviews* **24**, 2083–2100.

Gehrels W.R., Marshall W.A., Gehrels M.J., Larsen G., Kirby J.R., Eriksson J., Heinemeier J. and Shimmield T. (2006) Rapid sea level rise in the North Atlantic Ocean since the first half of the 19th century. *The Holocene* **16**, 948–64.

Gill E.D. and Hopley D. (1972) Holocene sea levels in eastern Australia – a discussion. *Marine Geology* **12**, 223–33.

Gischler E. (2006) Comment on "Corrected western Atlantic sea-level curve for the last 11,000 years based on calibrated ^{14}C dates from Acropora palmata framework and intertidal mangrove peat" by Toscano and Macintyre, *Coral Reefs* **22**, 257–70 (2003), and their response in *Coral Reefs* **24**, 187–90 (2005). *Coral Reefs* **25**, 273–9.

Goodwin I., Harvey N., van de Plassche O., Oglesby R. and Oldfield F. (2000) Prospects for resolving climate-induced sea level fluctuations over the last 2000 years. *EOS Transactions American Geophysical Union* **81**, 311–12.

Gray J.M. (1983) The measurement of relict shoreline altitudes in areas affected by glacio-isostasy, with particular reference to Scotland. In: *Shorelines and Isostasy* (Smith D.E. and Dawson A.G., eds). Academic Press, London.

Gutenberg B. (1941) Changes in sea level, postglacial uplift, and mobility of the earth's interior. *Geological Society of America Bulletin* **52**, 721–72.

Hallam A. (1984) Pre-Quaternary sea-level changes. *Annual Review of Earth and Planetary Sciences* **12**, 205–43.

Hanebuth T., Stattegger K. and Grootes P.M. (2000) Rapid flooding of the Sunda Shelf: a late-glacial sea-level record. *Science* **288**, 1033–5.

Hanebuth T.J.J., Stattegger K. and Bojanowski A. (2009) Termination of the Last Glacial Maximum sea-level lowstand: The Sunda-Shelf data revisited. *Global and Planetary Change* **66**, 76–84.

Haq B.U., Hardenbol J. and Vail P. (1987) Chronology of fluctuating sea levels since the Triassic. *Science* **235**, 1156–67.

Harmon R.S., Schwarcz H.P. and Ford D.C. (1978) Late Pleistocene sea level history of Bermuda. *Quaternary Research* **9**, 205–18.

Harris P.T. (1999) Sequence architecture during the Holocene transgression: an example from the Great Barrier Reef shelf, Australia – comment. *Sedimentary Geology* **125**, 235–9.

Haslett S.K. (2001) The palaeoenvironmental implications of the distribution of intertidal foraminifera in a tropical Australian estuary: a reconnaissance study. *Australian Geographical Studies* **39**, 67–74.

Haslett S.K., Davies P. and Strawbridge F. (1997) Reconstructing Holocene sea-level change in the Severn Estuary and Somerset Levels: the foraminifera connection. *Archaeology in the Severn Estuary* **8**, 29–40.

Hays J.D., Imbrie J. and Shackleton N.J. (1976) Variations in the earth's orbit: pacemaker of the Ice Ages. *Science* **194**, 1121–32.

Hearty P.J., Kindler P., Cheng H. and Edwards R.L. (1999) A +20 m middle Pleistocene sea-level highstand (Bermuda and the Bahamas) due to partial collapse of Antarctic ice. *Geology* **27**, 375–8.

Hopley D. (1982) *The Geomorphology of the Great Barrier Reef: Quaternary Development of Coral Reefs*. Wiley, New York.

Hopley D. and Isdale P. (1977) Coral micro-atolls, tropical cyclones and reef flat morphology: a North Queensland example. *Search* **8**, 79–81.

Hopley D., Smithers S.G. and Parnell K.E. (2007) *The Geomorphology of the Great Barrier Reef: Development, Diversity and Change*. Cambridge University Press, Cambridge.

Horton B.P., Edwards R.J. and Lloyd J.M. (1999) UK intertidal foraminiferal distributions: implications for sea-level studies. *Marine Micropalaeontology* **36**, 205–23.

Horton B.P., Larcombe P., Woodroffe S.A., Whittaker J.E., Wright M.R. and Wynn C. (2003) Contemporary foraminiferal distributions of a mangrove environment, Great Barrier Reef coastline, Australia: implications for sea-level reconstruction. *Marine Geology* **198**, 225–43.

Huybrechts P. (2003) Antarctica: modelling. In: *Mass Balance of the Cryosphere: Observations and Modelling of Contemporary and Future Changes* (Bamber J.L. and Payne A.J., eds), pp. 491–523. Cambridge University Press, Cambridge.

Hyvärinen H. (1988) Definition of the Baltic stages. *Annales Academiae Scientiarum Fennicae Ser. A III* **148**, 7–11.

Imbrie J., Hays J.D., Martinson D.G., McIntyre A., Mix A.C., Morley J.J. et al. (1984) The orbital theory of Pleistocene climate: support from a revised chronology of the marine ^{18}O record. In: *Milankovitch and Climate: Understanding the Response to Astronomical Forcing* (Berger A.L., Imbrie J., Hays J., Kukla G. and Saltzman B., eds), pp. 269–305. Reidel, Dordrecht.

Imbrie J., Boyle E.A, Clemens S.C., Duffy A., Howard W.R., Kukla G. et al. (1992) On the structure and origin of major glaciation cycles. I. Linear responses to Milankovitch forcing. *Paleoceanography* **7**, 701–38.

Imbrie J., Berger A., Boyle E.A, Clemens S.C., Duffy A., Howard W.R. et al. (1993) On the structure and origin of major glaciation cycles 2. The 100,000-year cycle. *Paleoceanography* **8**, 699–735.

Issel A. (1914) Lembi fossiliferi quaternari e recenti nella Sardegna meridionale. *Acc. Naz. dei Lincei* **23** (ser. 5), 759–70.

Johnston P., Lambeck K. and Wolf D. (1997) Material versus isobaric internal boundaries in the Earth and their influence on postglacial rebound. *Geophysical Journal International* **129**, 252–68.

Jongsma D. (1970) Eustatic sea level changes in the Arafura Sea. *Nature* **228**, 150–1.

Kelletat D. (2006) Beachrock as sea-level indicator? Remarks from a geomorphological point of view. *Journal of Coastal Research* **22**, 1558–64.

Kraft J.C., Rapp G.R., Kayan I. and Luce J.V. (2003) Harbor areas at ancient Troy: sedimentology and geomorphology complement Homer's *Iliad*. *Geology* **31**, 163–6.

Labeyrie M.J., Lalou C., Monaco A. and Thommeret J. (1976) Chronologie des niveaux eustatiques sur la côte du Roussillon de – 33 000 ans BP à nos jours. *Comptes Rendus de l'Académie des Sciences* **282** (série D), 349–52.

Laborel J. and Laborel-Deguen F. (1994) Biological indicators of relative sea-level variations and of co-seismic displacements in the Mediterranean region. *Journal of Coastal Research* **10**, 395–415.

Laborel J., Morhange C., Lafont R., Campion J., Laborel-Deguen F. and Sartoretto S. (1994) Biological evidence of sea-level rise during the last 4500 years on the rocky coasts of continental southwestern France and Corsica. *Marine Geology* **120**, 203–23.

Laborel J., Morhange C., Collina-Girard J. and Laborel-Deguen F. (1999) Shoreline bioerosion, a tool for the study of sea level variations during the Holocene. *Danish Geological Journal* **45**, 144–8.

Lambeck K. (1981) Lithospheric response to volcanic loading in the southern Cook Islands. *Earth and Planetary Science Letters* **55**, 482–96.

Lambeck K. (1995) Late Pleistocene and Holocene sea-level change in Greece and south-western Turkey: a separation of eustatic, isostatic and tectonic contributions. *Geophysical Journal International* **122**, 1022–44.

Lambeck K. (1996) Sea-level change and shore-line evolution in Aegean Greece since Upper Palaeolithic time. *Antiquity* **70**, 588–611.

Lambeck K. (1999) Shoreline displacement in southern-central Sweden and the evolution of the Baltic Sea since the last maximum glaciation. *Journal of the Geological Society of London* **156**, 465–86.

Lambeck K. (2002) Sea-level change from mid-Holocene to recent time: an Australian example with global implications. In: *Glacial Isostatic Adjustment and the Earth System* (Mitrovica J.X. and Vermeersen L.L.A., eds), pp. 33–50. American Geophysical Union, Washington DC.

Lambeck K. (2006) Hyperbolic tangents are no substitute for simple classical physics. *GFF*, **128**, 349–50.

Lambeck K. and Nakada M. (1990) Late Pleistocene and Holocene sea-level change along the Australian coast. *Palaeogeography, Palaeoclimatology, Palaeoecology (Global and Planetary Change Section)* **89**, 143–76.

Lambeck K. and Bard E. (2000) Sea-level change along the French Mediterranean coast since the time of the Last Glacial Maximum. *Earth and Planetary Science Letters* **175**, 202–22.

Lambeck K. and Chappell J. (2001) Sea level change through the last glacial cycle. *Science* **292**, 679–86.

Lambeck K. and Purcell A. (2005) Sea-level change in the Mediterranean Sea since the LGM: model predictions for tectonically stable areas. *Quaternary Science Reviews* **24**, 1969–88.

Lambeck K., Johnston P., Smither C. and Nakada M. (1996) Glacial rebound of the British Isles – III. Constraints on mantle viscosity. *Geophysical Journal International* **125**, 340–54.

Lambeck K., Smither C. and Johnston P. (1998a) Sea-level change, glacial rebound and mantle viscosity for northern Europe. *Geophysical Journal International* **134**, 102–44.

Lambeck K., Smither C. and Ekman M. (1998b) Tests of glacial rebound models for Fennoscandinavia based on instrumented sea- and lake-level records. *Geophysical Journal International* **135**, 375–87.

Lambeck K., Esat T.M. and Potter E.K. (2002a) Links between climate and sea levels for the past three million years. *Nature* **419**, 199–206.

Lambeck K., Yokoyama Y. and Purcell A. (2002b) Into and out of the Last glacial Maximum Sea Level change during Oxygen Isotope Stages 3-2. *Quaternary Science Reviews* **21**, 343–60.

Lambeck K., Purcell A., Johnston P., Nakada M. and Yokoyama Y. (2003) Water-load definition in the glacio-hydro-isostatic sea-level equation. *Quaternary Science Reviews* **22**, 309–18.

Lambeck K., Anzidei M., Antonioli F., Benini A. and Esposito E. (2004a) Sea level in Roman time in the central Mediterranean and implications for modern sea level rise. *Earth and Planetary Science Letters* **224**, 563–75.

Lambeck K., Antonioli F., Purcell A. and Silenzi S. (2004b) Sea level change along the Italian coast for the past 10,000 yrs. *Quaternary Science Reviews* **23**, 1567–98.

Larcombe P., Carter R.M., Dye J., Gagan M.K. and Johnson D.P. (1995) New evidence for episodic post-glacial sea-level rise, central Great Barrier Reef, Australia. *Marine Geology* **127**, 1–44.

Lewis S.E., Wüst R.A.J., Webster J.M. and Shields G.A. (2008) Mid-late Holocene sea-level variability in eastern Australia. *Terra Nova* **20**, 74–81.

Li W.-X., Lundberg J., Dickin A.P., Ford D.C., Schwarcz H.P., McNutt R. and William D. (1989) High precision mass spectrometric uranium-series dating of cave deposits and implications for paleoclimate studies. *Nature* **339**, 534–6.

Lighty R.G., Macintyre I.G. and Stuckenrath R. (1982) *Acropora palmata* reef framework: a reliable indicator of sea level in the western Atlantic for the past 10,000 years. *Coral Reefs* **1**, 125–30.

Liu J.P. and Milliman J.D. (2004) Reconsidering meltwater pulses 1A and 1B: global impact of rapid sea-level rise. *Journal of Ocean University of China* **3**, 183–90.

Long A.J., Roberts D.H. and Wright M.R. (1999) Isolation basin stratigraphy and Holocene relative sea-level change on Arveprinsen Ejland, Disko Bugt, West Greenland. *Journal of Quaternary Science* **14**, 323–45.

Long A.J., Roberts D.H. and Rasch M. (2003) New observations on the relative sea level and deglacial history of Greenland from Innaarsuit, Disko Bugt. *Quaternary Research* **60**, 162–71.

Lundberg J., Ford D.C., Schwarcz H.P., Dickin A.P. and Li W.-X. (1990) Dating sea level in caves. *Nature* **343**, 217–18.

Lyell C. (1832) *Principles of Geology*. Murray, London.

Mangerud J., Astakhov V. and Svendsen J.I. (2002) The extent of the Barents-Kara Ice Sheet during the Last Glacial Maximum. *Quaternary Science Reviews* **21**, 111–19.

Marriner N., Morhange C., Doumet-Serhal C. and Carbonel P. (2006) Geoscience rediscovers Phoenicia's buried harbors. *Geology* **34**, 1–4.

Marshall W.A., Gehrels W.R., Garnett M.H., Freeman S.P.H.T., Maden C. and Xu S. (2007) The use of "bomb spike" calibration and high-precision AMS ^{14}C analyses to date salt-marsh sediments deposited during the past three centuries. *Quaternary Research* **68**, 325–37.

McCulloch M.T. and Mortimer G.E. (2008) Applications of the ^{238}U-^{230}Th decay series to dating of fossil and modern corals using MC-ICPMS. *Australian Journal of Earth Sciences* **55**, 955–65.

McLean R.F., Stoddart D.R., Hopley D. and Polach H.A. (1978) Sea level change in the Holocene on the northern Great Barrier Reef. *Philosophical Transactions of the Royal Society of London A* **291**, 167–86.

Meltzner A.J., Sieh K., Abrams M., Agnew D.C., Hudnut K.W., Avouac J.P. and Natawidjaja D.H. (2006) Uplift and subsidence associated with the great Aceh-Andaman earthquake of 2004. *Journal of Geophysical Research* **111**, B02407.

Mesolella K.J., Matthews R.K., Broecker W.S. and Thurber D.L. (1969) The astronomical theory of climatic change: Barbados data. *Journal of Geology* **77**, 250–74.

Milankovitch M. (1941) *Kanon der erdbestrahlung und seine anwendung auf das eiszeitenproblem 132*. Special Publication. Royal Serbian Academy, Belgrade.

Miller K.G., Koniz M.A., Browning J.V., Wright J.D, Mountain G.S., Katz M.E. et al. (2005) The Phanerozoic record of global sea-level change. *Science* **310**, 1293–8.

Milne G.A., Davis J.L., Mitrovica J.X., Scherneck H.-G., Johansson J.M., Vermeer M. and Koivula H. (2001) Space-geodetic constraints on glacial isostatic adjustment in Fennoscandia. *Science* **291**, 2381–5.

Milne G.A., Long A.J. and Bassett S.E. (2005) Modelling Holocene relative sea-level observations from the Caribbean and South America. *Quaternary Science Reviews* **24**, 1183–1202.

Mitrovica J.X. (2003) Recent controversies in predicting post-glacial sea-level change. *Quaternary Science Reviews* **22**, 127–33.

Mitrovica J.X. and Peltier W.R. (1991) On post-glacial geoid subsidence over the equatorial oceans. *Journal of Geophysical Research* **96**, 20053–71.

Mitrovica J.X. and Milne G.A. (2002) On the origin of late Holocene sea-level highstands within equatorial ocean basins. *Quaternary Science Reviews* **21**, 2179–90.

Mitrovica J.X. and Milne G.A. (2003) On post-glacial sea level: I. General theory, *Geophysical Journal International* **154**, 253–67.

Monaco C., Antonioli F., De Guidi G., Lambeck K., Tortorici L. and Verrubbi V. (2004) Holocene tectonic uplift of the Catania Plain (eastern Sicily). *Quaternaria Nova* **8**, 171–85.

Montaggioni L.F. (2005) History of Indo-Pacific coral reef systems since the last glaciation: development patterns and controlling factors. *Earth Science Reviews* **71**, 1–75.

Montaggioni L.F. and Pirazzoli P.A. (1984) The significance of exposed coral conglomerates from French Polynesia (Pacific Ocean) as indications of recent sea-level changes. *Coral Reefs* **3**, 29–42.

Montaggioni L.F., Cabioch G., Camoin G.F., Bard E., Laurenti A.R., Faure G., Déjardin P. and Récy J. (1997) Continuous record of reef growth over the past 14 k.y. on the mid-Pacific island of Tahiti. *Geology* **25**, 555–8.

Morhange C., Bourcier M., Laborel J., Giallanella C., Goiran J., Crimaco L. and Vecchi L. (1999) New data on historical relative sea level movements in Pozzuoli, Phlaegrean Fields, southern Italy. *Physics and Chemistry of the Earth* **24**, 349–54.

Morhange C., Laborel J. and Hesnard A. (2001) Changes of relative sea level during the past 5000 years in the ancient harbour of Marseilles, Southern France. *Palaeogeography, Palaeoclimatology, Palaeoecology* **166**, 319–29.

Morhange C., Marriner N., Laborel J., Micol T. and Oberlin C. (2006) Rapid sea-level movements and noneruptive crustal deformation in the Phlegrean Fields caldera, Italy. *Geology* **34**, 93–6.

Nakada M. and Lambeck K. (1987) Glacial rebound and relative sea-level variations: a new appraisal. *Geophysical Journal of the Royal Astronomical Society* **90**, 171–224.

Nakada M. and Lambeck K. (1988) The melting history of the late Pleistocene Antarctic ice sheet. *Nature* **333**, 36–40.

Nakada M. and Lambeck K. (1989) Late Pleistocene and Holocene sea-level change in the Australian region and mantle rheology. *Geophysical Journal of the Royal Astronomical Society* **96**, 497–517.

Nakada M., Kimuru R., Okuno J., Moriwaki K., Miura H. and Maemoku H. (2000) Late Pleistocene and Holocene melting history of the Antarctic ice sheet derived from sea-level variations. *Marine Geology* **167**, 85–103.

Nakiboglu S.M., Lambeck K. and Aharon P. (1983) Postglacial sea levels in the Pacific: implications with respect to deglaciation regime and local tectonics. *Tectonophysics* **91**, 335–58.

Nunn P.D. and Peltier W.R. (2001) Far-field test of the ICE-4G model of global isostatic response to deglaciation using empirical and theoretical Holocene sea-level reconstructions for the Fiji Islands, southwestern Pacific. *Quaternary Research* **55**, 203–14.

Orrù P.E., Antonioli F., Lambeck K. and Verrubbi V. (2004) Holocene sea-level change in the Cagliari coastal plain (southern Sardinia, Italy). *Quaternaria Nova* **8**, 193–212.

Parascandola A. (1947) *I fenomeni bradisismici del Serapeo di Pozzuoli.* Privately published, Naples.

Patterson R.T., Hutchinson I., Guilbault J.P. and Clague J.J. (2000) A comparison of the vertical zonation of diatom, foraminifera, and macrophyte assemblages in a coastal marsh: implications for greater paleo-sea level resolution. *Micropaleontology* **46**, 229–44.

Peltier W.R. (2002) Comments on the paper of Yokoyama et al. (2000), entitled "Timing of the Last Glacial Maximum from observed sea level minima". *Quaternary Science Reviews* **21**, 409–14.

Peltier W.R. (2005) On the hemispheric origins of meltwater pulse 1a. *Quaternary Science Reviews* **24**, 1655–71.

Peltier W.R. and Fairbanks R.G. (2006) Global glacial ice volume and Last Glacial Maximum duration from an extended Barbados sea level record. *Quaternary Science Reviews* **25**, 3322–37.

Penck A. and Brückner E. (1909) *Die Alpen im Eiszeitalter*. C.H. Tauchnitz, Leipzig.

Pirazzoli P.A. (1976) Sea level variation in the northwest Mediterranean during Roman times. *Science* **194**, 519–21.

Pirazzoli P.A. (1991) *World Atlas of Holocene Sea-Level Changes*. Elsevier Oceanography Series 58. Elsevier, Amsterdam.

Pirazzoli P.A. (1994) Tectonic coasts. In: *Coastal Evolution: Late Quaternary Shoreline Morphodynamics* (Carter R.W.G. and Woodroffe C.D., eds), pp. 451–76. Cambridge University Press, Cambridge.

Pirazzoli P.A. (1996) *Sea-Level Changes: the Last 20,000 Years*. Wiley, Chichester.

Pirazzoli P.A. and Montaggioni L.F. (1988) Holocene sea-level changes in French Polynesia. *Palaeogeography, Palaeoclimatology, Palaeoecology* **68**, 153–75.

Pirazzoli P.A., Ausseil-Badie J., Giresse P., Hadjidaki E. and Arnold M. (1992) Historical environmental changes at Phalasarna harbour, West Crete. *Geoarchaeology* **7**, 371–92.

Pirazzoli P.A., Laborel J. and Stiros S.C. (1996) Coastal indicators of rapid uplift and subsidence: examples from Crete and other eastern Mediterranean sites. *Zeitschrift für Geomorphologie N.F. Supplement – Band* **102**, 21–35.

Pitman W.C. (1978) Relationship between eustacy and stratigraphic sequences of passive margins. *Geological Society of America Bulletin* **89**, 1369–1403.

Plater A.J., Horton B.P., Haworth E.Y., Rutherford M.M., Zong Y., Wright M.R. and Appleby P.G. (2000) Holocene tidal levels and sedimentation rates using a diatom-based palaeo-environmental reconstruction: the Tees estuary, northeastern England. *The Holocene* **10**, 441–52.

Playfair J. (1802) *Illustrations of the Huttonian Theory of the Earth*. Dover, New York.

Potter E.-K., Esat T.M., Schellmann G., Radtke U., Lambeck K. and McCulloch M.T. (2004) Suborbital-period sea-level oscillations during Marine Isotope Substages 5a and 5c. *Earth and Planetary Science Letters* **225**, 191–204.

Scoffin T.P. and Stoddart D.R. (1978) The nature and significance of micro atolls. *Philosophical Transactions of the Royal Society of London B* **284**, 99–122.

Scott D.B. and Medioli F.S. (1978) Vertical zonations of marsh foraminifera as accurate indicators of former sea-levels. *Nature* **272**, 528–31.

Shackleton N.J. (1987) Oxygen isotopes, ice volume and sea level. *Quaternary Science Reviews* **6**, 183–90.

Shackleton N.J. and Opdyke N.D. (1973) Oxygen isotope and paleomagnetic stratigraphy of equatorial Pacific core V28-238: oxygen isotope temperatures and ice volumes on a 10^5 and 10^6 year scale. *Quaternary Research* **3**, 39–55.

Shaw B., Ambraseys N.N., England P.C., Floyd M.A., Gorman G.J., Higham T.F.G. et al. (2008) Eastern Mediterranean tectonics and tsunami hazard inferred from the AD 365 earthquake. *Nature Geoscience* **1**, 268–76.

Shennan I. (1986) Flandrian sea-level changes in the Fenland. II: Tendencies of sea-level movement, altitudinal changes, and local and regional factors. *Journal of Quaternary Science* **1**, 155–79.

Shennan I., Hamilton S., Hillier C. and Woodroffe S. (2005) A 16,000-year record of near-field relative sea-level changes, northwest Scotland, United Kingdom. *Quaternary International* **133–4**, 95–106.

Sivan D., Lambeck K., Galili E. and Raban A. (2001) Holocene sea-level changes along the Mediterranean coast of Israel, based on archaeological observations and numerical model. *Palaeogeography, Palaeoclimatology, Palaeoecology* **167**, 101–17.

Sivan D., Lambeck K., Toueg R., Raban A., Porath Y. and Shirman B. (2004) Ancient coastal wells of Caesarea Maritima, Israel, an indicator for sea level changes during the last 2000 years. *Earth and Planetary Science Letters* **222**, 315–30.

Slim H., Trusset P., Paskoff R. and Oueslati A. (2004) *Le littoral de la Tunisie. Etude Géoarcheologique et Historique*. CNRS Editions, Paris.

Smithers S.G. and Woodroffe C.D. (2000) Microatolls as sea-level indicators on a mid-ocean atoll. *Marine Geology* **168**, 61–78.

Smithers S.G. and Woodroffe C.D. (2001) Coral microatolls and 20th century sea level in the eastern Indian Ocean. *Earth and Planetary Science Letters* **191**, 173–84.

Spada G. and Stocchi P. (2007) SELEN: a Fortran 90 program for solving the "Sea Level Equation". *Computers and Geosciences* **33**, 538–62.

Sparrenbom C.J. (2006) *Constraining the Southern Part of the Greenland Ice Sheet Since the Last Glacial Maximum from Relative Sea-Level Changes, Cosmogenic Dates and Glacial-Isostatic Adjustment Models*. LUNDQUA Thesis 56, Department of Geology, Lund University.

Sparrenbom C.J., Bennike O., Björck S. and Lambeck K. (2006a) Relative sea-level changes since 15000 cal. yr BP in the Nanortalik area, southern Greenland. *Journal of Quaternary Science* **21**, 29–48.

Sparrenbom C.J., Bennike O., Björck S. and Lambeck K. (2006b) Holocene relative sea-level changes in the Qaqortoq area, southern Greenland. *Boreas* **35**, 171–87.

Spencer T. (1988) Limestone coastal morphology: the biological contribution. *Progress in Physical Geography* **12**, 66–101.

Spencer T., Tudhope A.W., French J.R., Scoffin T.P. and Utanga A. (1997) Reconstructing sealevel change from coral microatolls, Tongareva (Penrhyn)

Atoll, northern Cook Islands. *Proceedings of the 8th International Coral Reef Symposium* **1**, 489–94.

Stirling C.H., Esat T.M., Lambeck K., McCulloch M.T., Blake S.G., Lee D.-C. and Halliday A.N. (2001) Orbital forcing of the marine isotope stage 9 interglacial. *Science* **291**, 290–3.

Stoddart D.R. and Scoffin T.P. (1979) Microatolls: review of form, origin and terminology. *Atoll Research Bulletin* **224**, 1–17.

Stoddart D.R. and Scoffin T.P. (1983) Phosphate rock on coral reef islands. In: *Chemical Sediments and Geomorphology* (Goudie A.S. and Pye K., eds), pp. 369–400. Academic Press, London.

Stone J., Lambeck K., Fifield L.K., Evans J.M. and Cresswell R.G. (1996) A Lateglacial age for the Main Rock Platform, western Scotland. *Geology* **24**, 707–10.

Stone J.O., Balco G.A., Sugden D.E., Caffee M.W., Sass L.C., Cowdery S.G and Siddoway C. (2003) Holocene deglaciation of Marie Byrd Land, West Antarctica. *Science* **299**, 99–102.

Suess E. (1888) *Das Antlitz der Erde.* F. Tempsky, Vienna.

Svendsen J.I. and Mangerud J. (1987) Late Weichselian and Holocene sea-level history for a cross-section of western Norway. *Journal of Quaternary Science* **2**, 113–32.

Svensson N.-O. (1991) Late Weichselian and Early Holocene shore displacement in the central Baltic Sea. *Quaternary International* **9**, 7–26.

Tallarico A., Dragoni M., Anzidei M. and Esposito A. (2003) Modeling long-term ground deformation cooling of a magma chamber, Aeolian Islands, Italy. *Journal of Geophysical Research* **108**(B12), 2568.

Tarasov L. and Peltier W.R. (2006) A calibrated deglacial drainage chronology for the North American continent: evidence of an Arctic trigger for the Younger Dryas. *Quaternary Science Reviews* **25**(7–8), 659–88.

Thom B.G. and Chappell J. (1975) Holocene sea levels relative to Australia. *Search* **6**, 90–3.

Thom B.G. and Roy P.S. (1985) Relative sea levels and coastal sedimentation in southeast Australia in the Holocene. *Journal of Sedimentary Petrology* **55**, 257–64.

Thom B.G., Hails J.R., Martin A.R.H. and Phipps C.V.G. (1972) Post-glacial sea levels in eastern Australia – a reply. *Marine Geology* **12**, 233–42.

Thompson W.G. and Goldstein S.L. (2005) Open-system coral ages reveal persistent suborbital sea-level cycles. *Science* **308**, 401–4.

Tjia H.D. (1996) Sea-level changes in the tectonically stable Malay-Thai peninsula. *Quaternary International* **31**, 95–101.

Toscano M.A. and Macintyre I.G. (2003) Corrected western Atlantic sea-level curve for the last 11,000 years based on calibrated ^{14}C dates from *Acropora palmata* framework and intertidal mangrove peat. *Coral Reefs* **22**, 257–70.

van Andel T.H. (1989) Late Quaternary sea-level changes and archaeology. *Antiquity* **63**, 733–46.

van de Plassche O. (ed.) (1986) *Sea-Level Research.* Geo Books, Norwich.

van de Plassche O., van der Borg K. and de Jong A.F.M. (1998) Sea level-climate correlation during the past 1400 yr. *Geology* **26**, 319–22.

van de Plassche O., Edwards R.J., van der Borg K. and de Jong A.F.M. (2001) ^{14}C wiggle-match dating in high-resolution sea-level research. *Radiocarbon* **43**, 391–402.

Veeh H.H. (1966) Th 230/U238 and U234/U238 ages of Pleistocene high sea level stand. *Journal of Geophysical Research* **71**, 3379–86.

Veeh H.H. and Veevers J.J. (1970) Sea level at −175 m off the Great Barrier Reef 13,600 to 17,000 years ago. *Nature* **226**, 536–7.

Villy A.K., Tsoulos N. and Stiros S.C. (2002) Coastal uplift, earthquakes and active faulting of Rhodes island (Aegean Arc): modelling based on geodetic inversion. *Marine Geology* **186**, 299–317.

Weidick A., Kelly M. and Bennike O. (2004) Late Quaternary development of the southern sector of the Greenland Ice Sheet, with particular reference to the Qassimiut lobe. *Boreas* **33**, 284–99.

Woodroffe C.D. (1990) The impact of sea-level rise on mangrove shorelines. *Progress in Physical Geography* **14**, 483–520.

Woodroffe C.D. (1995) Response of tide-dominated mangrove shorelines in northern Australia to anticipated sea-level rise. *Earth Surface Processes and Landforms* **20**, 65–85.

Woodroffe C.D. (2005) Late Quaternary sea-level highstands in the central and eastern Indian Ocean: a review. *Global and Planetary Change* **49**, 121–38.

Woodroffe C.D. and McLean R.F. (1990) Microatolls and recent sea level change on coral atolls. *Nature* **344**, 531–4.

Woodroffe C.D. and McLean R.F. (1998) Pleistocene morphology and Holocene emergence of Christmas (Kiritimati) Island, Pacific Ocean. *Coral Reefs* **17**, 235–48.

Woodroffe C.D., Stoddart D.R., Spencer T., Scoffin T.P. and Tudhope A. (1990a) Holocene emergence in the Cook Islands, South Pacific. *Coral Reefs* **9**, 31–9.

Woodroffe C.D., McLean R.F., Polach H. and Wallensky E. (1990b) Sea level and coral atolls: Late Holocene emergence in the Indian Ocean. *Geology* **18**, 62–6.

Woodroffe C.D., Short S., Stoddart D.R., Spencer T. and Harmon R.S. (1991) Stratigraphy and chronology of Pleistocene reefs and the southern Cook Islands, South Pacific. *Quaternary Research* **35**, 246–63.

Woodroffe C.D., Samosorn B., Hua Q. and Hart D.E. (2007) Incremental accretion of a sandy reef island over the past 3000 years indicated by component-specific radiocarbon dating. *Geophysical Research Letters* **34**, L03602.

Woodroffe S.A. and Horton B.P. (2005) Holocene sea-level changes in the Indo-Pacific. *Journal of Asian Earth Sciences* **25**, 29–43.

Woodroffe S.A., Horton B.P., Larcombe P. and Whittaker J.E. (2005) Intertidal mangrove foraminifera from the central Great Barrier Reef shelf, Australia: implications for sea-level reconstruction. *Journal of Foraminiferal Research* **35**, 259–70.

Yokoyama Y., Lambeck K., De Deckker P., Johnston P. and Fifield K. (2000) Timing of the Last Glacial Maximum from observed sea-level minima. *Nature* **406**, 713–16.

Yokoyama Y., De Deckker P., Lambeck K., Johnston P. and Fifield L.K. (2001a) Sea-level at the Last Glacial Maximum: evidence from northwestern Australia to constrain ice volumes for oxygen isotope stage 2. *Palaeogeography, Palaeoclimatology, Palaeoecology* **165**, 281–97.

Yokoyama Y., Purcell A., Lambeck K. and Johnston P. (2001b) Shore-line reconstruction around Australia during the Last Glacial Maximum and Late Glacial Stage. *Quaternary International* **83**, 9–18.

Young R.W., Bryant E.A., Price D.M., Wirth L. and Pease M. (1993) Theoretical constraints and chronological evidence of Holocene coastal development in central and southern New South Wales, Australia. *Geomorphology* **7**, 317–29.

Yu S.-Y. (2003) *The Littorina Transgressions in Southeastern Sweden and its Relation to Mid-Holocene Climate Variability*. Lundqua Thesis 51, Department of Geology, Lund.

Yu S.-Y., Berglund B.E., Sandgren P. and Lambeck K. (2007) Evidence for a rapid sea-level rise 7600 yr ago. *Geology* **35**(10), 891–4.

Zeuner F.E. (1945) *The Pleistocene Period: its Climate Chronology and Faunal Successions*. Ray Society, London.

Zwartz D.P. (1993) *The Recent History of the Antarctic Ice Sheet: Constraints from Sea-Level Change*. PhD Thesis, Australian National University.

Zwartz D., Lambeck K., Bird M. and Stone J. (1997) Constraints on the former Antarctic Ice Sheet from sea-level observations and geodynamics modeling. In: *The Antarctic Region: Geological Evolution and Processes* (Ricci C., ed.), pp. 821–8. Terra Antarctica, Siena.

Zwartz D., Bird M., Stone J. and Lambeck K. (1998) Holocene sea-level change and ice-sheet history in the Vestfold Hills, East Antarctica. *Earth and Planetary Science Letters* **155**, 131–45.

Modern Sea-Level-Change Estimates

Gary T. Mitchum, R. Steven Nerem, Mark A. Merrifield, and W. Roland Gehrels

5.1 Introduction

Why are we interested in global sea-level rise? As discussed in Chapters 2 and 3, one answer is that low-lying island nations and some coastal communities on continental coastlines might be facing increased risks of inundation and coastal erosion. Such enhanced flood risks have major implications for protection of coastal infrastructure and preservation of the coastal natural environment. Coastal flooding probabilities will also change in response to the frequency of extreme sea-level events, such as storm surges, and this topic is treated in Chapter 11.

In addition, we maintain that the study of sea-level rise is important because observations of ocean volume change can place constraints on climate model simulations, which is important because such models provide the only predictions we presently have concerning how the Earth system might evolve in coming decades in response to increasing greenhouse gases. We also argue that these ocean volume change estimates, when coupled to ocean density change estimates, are potentially an important integral constraint on estimates of the global ice-mass budget.

Global mean sea level can change for two reasons. The shape of the ocean basins can change (Chapter 10), or the volume of the water in the oceans can change. Globally averaged sea level is to first order a proxy for ocean volume changes, although corrections to satellite altimeter and tide-gauge data are needed to allow for the changing shape of the oceans for the most accurate estimates of ocean-volume change. The volume of the ocean changes via a change in the mass of the ocean, or by a change in the average density of the ocean. The amount of mass of the ocean is driven by the exchange of water with the continents, primarily through the melting of grounded ice due to a warming climate (Chapter 7). Changes in the average density of the ocean are largely a result of variations in the temperature of the ocean, which is affected by atmospheric warming (Chapter 6). Globally averaged sea level is thus an integral of changes in the Earth's heat budget. Precise estimates of the rate of global mean-sea-level change, and any

Understanding Sea-Level Rise and Variability, 1st edition. Edited by John A. Church, Philip L. Woodworth, Thorkild Aarup & W. Stanley Wilson.

change in this rate, thus provide constraints on global climate models, and those models should be able to adequately represent observed sea-level change.

There is a more subtle point, however. One of our main concerns is with understanding the observed sea-level changes. This task requires the partitioning of the observed sea-level change into contributions from ocean-mass addition and oceanic warming (ignoring for a moment the change in the shape of the ocean basins). Closing this budget has proved to be difficult, and perhaps a new perspective is needed. We suggest that if we can make precise estimates of the total sea-level change, then similarly precise estimates of the ocean-density changes or of the ocean mass addition (i.e. primarily the ice melt rate, but including land-ocean-mass exchanges) might enable us to obtain the least well-known term by difference.

This objective requires first that we be able to make precise estimates of the rate of globally averaged sea-level change over any timescale of interest; for example, the 20th century or recent decades. In this chapter we review efforts to do this, and we make recommendations of what must be done to further this goal.

To this end, we will focus in this review on measurements from tide gauges over the last century or so, and we will then turn to the measurements made by satellite altimeters over the past decade. We will also attempt to determine what improvements in the observation system might result in the precision we need to constrain both the climate models and the ice-mass budget. However, we note that tide gauges and altimetry are in fact not the only sources of recent sea-level information; geological and archeological measurements (Chapter 4) can provide estimates of millennial- and centennial-timescale sea-level changes which overlap with the ones we will discuss. In addition, there are evolving paleo-oceanographic estimates of sea-level change that provide information on centennial timescale changes. Therefore, we will start this chapter by presenting proxy sea-level information from the paleo-oceanographic evidence which will be seen to complement that from tide-gauge data in many ways. We will then return to our main focus on the results from tide gauges and satellite altimetry.

5.2 Estimates from Proxy Sea-Level Records

There are few centennial-scale records of mean-sea-level change based on tide-gauge data. The records from Brest in western France (starting in 1807; see Wöppelmann et al. 2006, 2008) and Swinoujscie in Poland (1811) are among the longest observations from tide gauges extending back to the early 19th century. The Permanent Service for Mean Sea Level (PSMSL) also gives information on records from Stockholm (since 1774), which can claim to be the longest, continuous record in Europe (Ekman 1988), and lists other records for Amsterdam (1700–1925), Liverpool (since 1768), and Kronstadt (1773–1993), while Pouvreau (2008) has recently reviewed tide-gauge measurements from Brest as far back as 1679. However, many of these historical records contain large gaps and are based on only high water readings and the inference of mean-sea-level change from them requires careful analysis.

The only reliable 19th-century tide-gauge measurements for the Southern Hemisphere are from Fremantle (starting 1897) and Sydney (1886) in Australia. Data archaeology, the process of re-discovering historical information in its many forms, has recently uncovered a record from Port Arthur in Tasmania dating from 1841–2 (Pugh et al. 2002; Hunter et al. 2003) and other examples of recovery of historical information are in progress. However, the poor coverage of tide-gauge sea-level records, both geographically and temporally, means that there are difficulties in estimating 19th- and 20th-century rates of sea-level rise on a global scale, although recent attempts have been made, as will be discussed below.

Therefore, it is clear that other techniques and data sources capable of providing insight into long-term sea-level change must be analyzed alongside information from the tide-gauge instrumental records.

Proxy records of relative sea-level change from salt marshes have been shown to be capable of providing valuable 'substitute tide-gauge data' at locations where tide-gauge records are not available. In addition, they can have a temporal coverage comparable to the longest tide-gauge records (depending upon the location), extending back in time to cover the pre-industrial era. The character of the data is such that it provides a form of low-pass filter of much the interannual variability seen in tide-gauge records.

For example, recent investigations in the North Atlantic (Donnelly et al. 2004; Gehrels et al. 2005, 2006) are presented in Figure 5.1. Donnelly et al. (2004) measured the accretion rate of a salt marsh in eastern Connecticut, USA, and used this rate as a substitute for sea-level rise by dating the base of the salt marsh along a contact with an underlying glacial erratic boulder, using radiocarbon, pollen, and geochemical (pollution) markers. Each dated point provides a former sea-level position (Figure 5.1). The salt-marsh sediments contained the fossil remains of plants which could be accurately related to mean high water based on distributions of the plants in modern salt marshes. The eastern Connecticut proxy record (Figure 5.1d) extends to the middle of the 19th century and shows that the rate of relative sea-level rise was 1.0 ± 0.2 mm/year between AD 1300 and 1850. By comparison, the tide-gauge record of New York (since 1856) shows a linear rate of rise of 2.8 mm/year, leading to the conclusion that during the late 1800s the rate of sea-level rise in eastern Connecticut almost tripled. The increase in rate of 1.8 mm/year (i.e. from 1.0 to 2.8 mm/year) is approximately equal to estimates of global averaged sea-level rise for the 20th century (see below).

Salt-marsh proxy records from Nova Scotia (Gehrels et al. 2005) and Iceland (Gehrels et al. 2006) are based on vertical accretion rates in single cores or exposures, rather than on encroachment rates along a sloping surface. These records use the fossil remains of foraminifera to establish the relationship between the fossil sediments and sea level by using modern distributions of salt-marsh foraminifera as a guide (see Chapter 4, Figure 4.7). With this method, each foraminiferal sample provides a former sea-level position. Its age is based on an age model determined from the dated levels of the core. The chronology of the records is provided by a variety of dating methods, including radiocarbon, stable

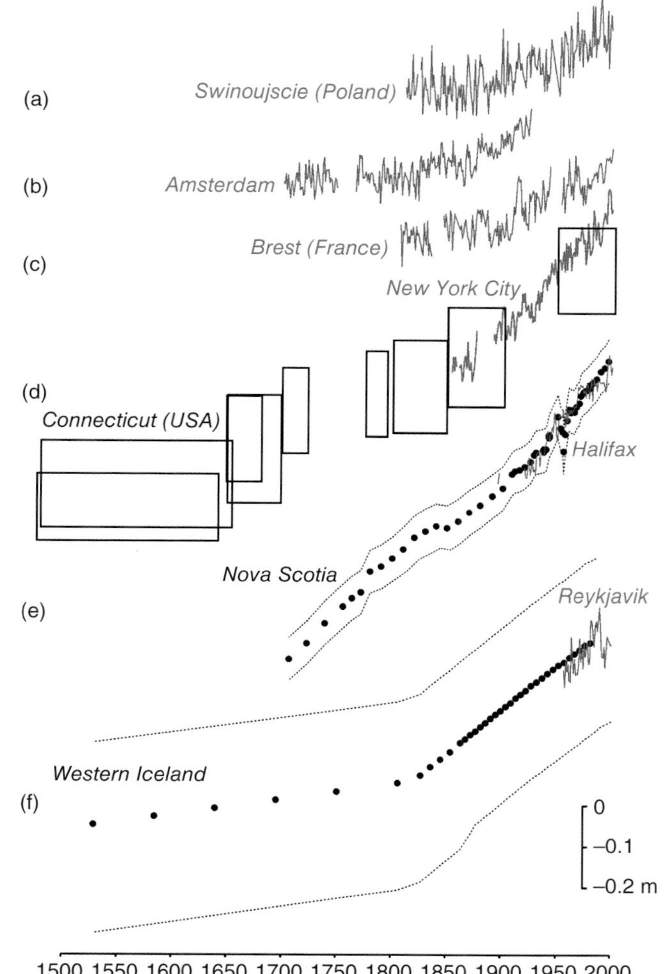

Figure 5.1 (a–c) Three long sea-level records. (d) Proxy sea-level reconstruction from eastern Connecticut, USA. Boxes represent dated levels of the base of a salt-marsh sequence, by radiocarbon, pollen, and metal analyses. Their dimensions reflect age and height uncertainties. The New York City tide gauge provides observations since 1856. Adapted from Donnelly et al. (2004). (e) Proxy sea-level reconstruction from Chezzetcook, Nova Scotia, eastern Canada, and the tide-gauge record from Halifax. (f) Proxy sea-level reconstruction from Vidarholmi, western Iceland, and the tide-gauge record from Reykjavik. Dots represent former sea-level positions derived from fossil foraminifera sampled in an exposed salt-marsh section. Adapted from Gehrels et al. (2006). Adapted from Gehrels et al. (2005). In (e) and (f) age and height uncertainties are reflected by the dotted lines.

isotopes of lead, lead concentrations, lead-210, cesium-137, pollen, and paleomagnetism. Possible sediment compaction can be assessed by bulk-density measurements. These records offer reconstructions up to the present day and therefore allow a direct comparison of the proxy sea-level records with the trend of nearby tide-gauge records for the period when instrumental and proxy records overlap.

The proxy record from Nova Scotia (Figure 5.1e) is comparable with the Donnelly et al. (2004) reconstruction from Connecticut. It shows a 19th-century rate of relative sea-level rise of 1.6 ± 0.1 mm/year, compared to a 20th-century rate (at this site) of 3.2 ± 0.1 mm/year. Again, the increase in rate of 1.6 mm/year (from 1.6 to 3.2 mm/year) is approximately equal to estimates of global averaged sea-level rise for the 20th century (see below). This implied acceleration is also seen in tide-gauge data and is discussed further below. The increase in the rate of sea-level rise is estimated to have occurred between 1900 and 1920 and is in agreement with the Halifax tide-gauge record (Gehrels et al. 2005 and see also Gehrels et al. 2002 for earlier work in the Gulf of Maine). However, in Iceland (Figure 5.1f), the onset of rapid sea-level rise is estimated to have occurred between 1800 and 1840, earlier than in the western Atlantic (Gehrels et al. 2006) but in reasonable agreement with the observations at Swinoujscie, Amsterdam, and Brest (Figure 5.1a–c).

More recently, Gehrels et al. (2008) have found evidence for an acceleration in New Zealand, from 0.3 ± 0.3 mm/year prior to 1900 to 2.8 ± 0.5 mm/year during the 20th century. This increase in rate exceeds the global averaged rate of 20th-century sea-level rise. As for the European Atlantic coast, Leorri et al. (2008) have provided proxy information from salt marshes in northern Spain, together with validation by the longer tide-gauge records in the region, including evidence for late-19th-century acceleration. These proxy records together provide evidence very suggestive of a middle-to-late 19th-century acceleration in sea-level rise in the North Atlantic Ocean.

As more proxy measurements such as these are assembled for many parts of the world, they are likely to provide an increasingly valuable complement to the instrumental records from tide gauges and extend those records to the centennial and longer scale. Although it is unlikely that such proxy records will ever have a truly global distribution, the additional information that they provide could provide useful inputs to knowledge of global change.

5.3 Estimates of Global Sea-Level Change from Tide Gauges

5.3.1 Tide-Gauge Data and Main Findings

Tide-gauge time series typically include a quasi-linear trend, which is a measure of the slow variation of the water level relative to the adjacent land. A number of studies have considered how to use these records to estimate global sea-level trends over the past century, and to determine whether these trends are accelerating in accordance with climate model predictions. The main uncertainties in these estimations involve how to extract global averages from an unevenly distributed set of tide gauges, how to estimate secular trends and accelerations in the presence of sea-level variability with long temporal and spatial decorrelation scales, and how to account for vertical land motion in the tide-gauge records.

The primary resource for tide-gauge studies of global sea-level change is the Revised Local Reference (RLR) database of monthly averaged time series maintained by PSMSL (Woodworth and Player 2003). The RLR data are screened for reference datum stability over time relative to benchmarks on the nearby land. However, corrections for the movement of the benchmarks themselves are not applied. The geographic coverage of long-record stations is poor, however, with the majority located in the Northern Hemisphere in Europe, North America, and

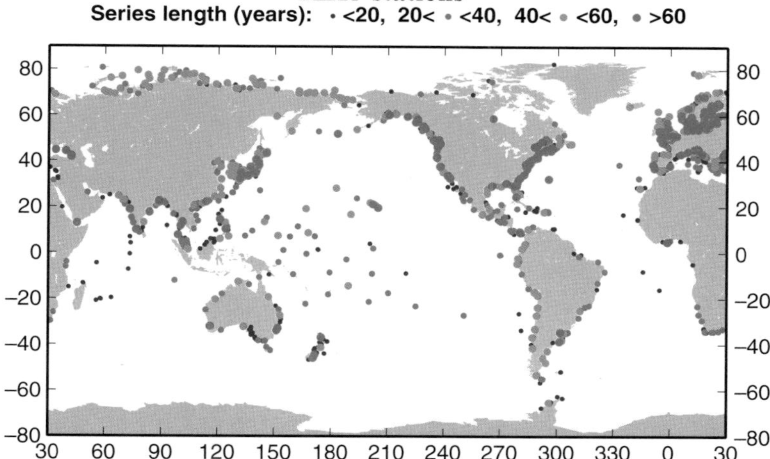

Figure 5.2 Time-series record lengths for RLR stations in the PSMSL database.

Japan (Figure 5.2). Also, the RLR stations are located predominantly along continental coasts, with relatively few long records available from island stations in the ocean interior. The uneven spatial coverage poses a major challenge for computing global averaged sea level.

Until a few years ago, most tide-gauge-based estimates of global sea-level change were computed as the average of linear trends obtained from a small number of RLR time series (e.g. Peltier and Tushingham 1991; Douglas 1991). The underlying assumptions were that the tide gauges measure a common ocean trend, which can be estimated with statistical significance given a suitably long time series (>60–70 years according to Douglas 2001) to minimize the impact of low-frequency variability in each record, and that the main remaining error is uncorrected vertical land motion at each station. The process of maximizing the number of stations with long record length while minimizing the number in tectonically active regions results in approximately 10 geographic groupings of stations (Douglas 2001). Land-motion corrections have relied primarily on models of glacial isostatic adjustment (GIA) (e.g. Tushingham and Peltier 1991; Peltier 2001).

However, there have recently been attempts to derive estimates of global sea-level change using so-called "reconstruction" techniques. For example, Church et al. (2004) and Church and White (2006) "reconstructed" a global sea-level time series from 1870 onwards, using empirical orthogonal functions (EOFs) of ocean variability determined from altimeter data since 1993 as basis functions, together with coastal tide-gauge information. First-differences of annual mean tide-gauge data were used to account for datum uncertainty. This approach was first developed by Chambers et al. (2002) in a study of interannual and decadal variations in global sea level from altimeter data. The authors note that the method allows for a larger number of tide-gauge stations to be used in the computation, in contrast to the Douglas et al. "average of trends" approach. Church et al. (2004)

estimated a linear global sea-level rise of 1.8 ± 0.3 mm/year for the period 1950–2000, similar to the previous results of Douglas (1997) and others, found similar decadal variations as Chambers et al. (2002), and noted that these fluctuations limit the ability to detect a net global sea-level acceleration over the 50-year record. Berge-Nguyen et al. (2008) later applied similar techniques using basis functions derived from altimetry, ocean temperature information, or ocean models. All of these techniques essentially attempt to provide forms of spatial averaging of the tide-gauge data, in some ways similar to the spherical harmonic parameterizations of sea-level change by Nakiboglu and Lambeck (1991), with the lowest mode taken as the global-average rate, but with a more plausible oceanographic parameterization. Jevrejeva et al. (2006) applied an alternative averaging technique with the use of a "virtual station" method of combining information from the global set of tide-gauge records.

What has been learned from this body of work? The main reviews of this subject have been included in the Third Assessment Report of the Intergovernmental Panel on Climate Change (IPCC; Church et al. 2001) and in the Fourth Assessment Report (Bindoff et al. 2007). A consensus seems to have been achieved that the 20th-century rise in global sea level was closer to 2 than 1 mm/year, with values around 1.7 mm/year having been obtained recently for the past century (e.g. Church and White 2006) or past half-century (e.g. Church et al. 2004; Holgate and Woodworth 2004).

Concerning acceleration in the rate of global sea-level change, the acceleration between the 19th and 20th centuries seen in the individual long European tide-gauge records (e.g. Woodworth 1999), and also in the salt-marsh data discussed above, has also been found in the reconstructions. For example, the Church et al. (2004) analysis was extended to include relative sea-level data from 1870 to 2001 by Church and White (2006), who reported a significant acceleration in global sea-level rise (0.013 ± 0.006 mm/year^2), with the main increase in the trend occurring during the first half of the 20th century. An alternative approach was taken by Jevrejeva et al. (2006), who use Monte Carlo Singular Spectrum Analysis on RLR data from 12 ocean regions in an attempt to separate 2–30-year fluctuations from a longer-term nonlinear global sea-level trend. The global sea-level trend is computed as a mean over all regions. The method again highlights the variations in global sea-level trends over periods of around 30 years, and yields a similar 20th-century linear trend estimate (1.8 mm/year) as in older and newer studies (e.g. Douglas 1997; Church et al. 2004). Their reconstruction also suggested a 19th–20th-century acceleration, which as suggested in a later paper could have occurred even earlier (Jevrejeva et al. 2008).

However, this simplified view of a 19th–20th-century acceleration leading to a 20th-century sea-level rise is far from the whole story. For example, in earlier studies which were largely confined to 20th-century data (Woodworth 1990; Douglas 1992), the authors concluded that a *deceleration* had taken place, largely because of a reduction in trend since about 1960 (see also Holgate 2007). To put these different acceleration analyses into perspective, we show (Figure 5.3) the Church and White (2006) reconstruction along with four other, generally

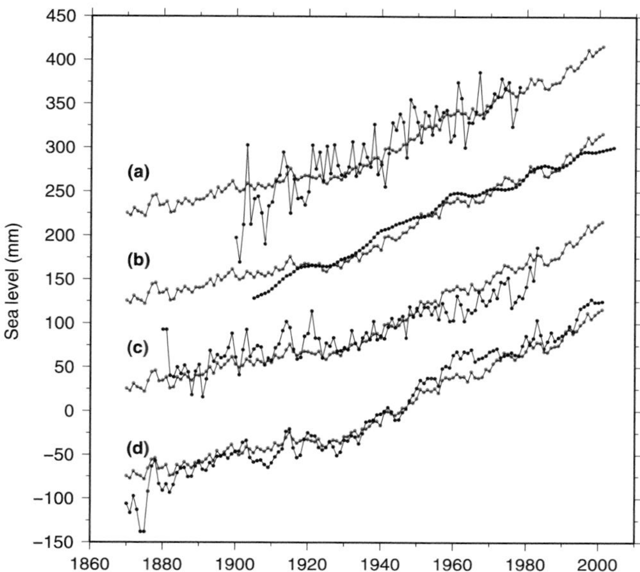

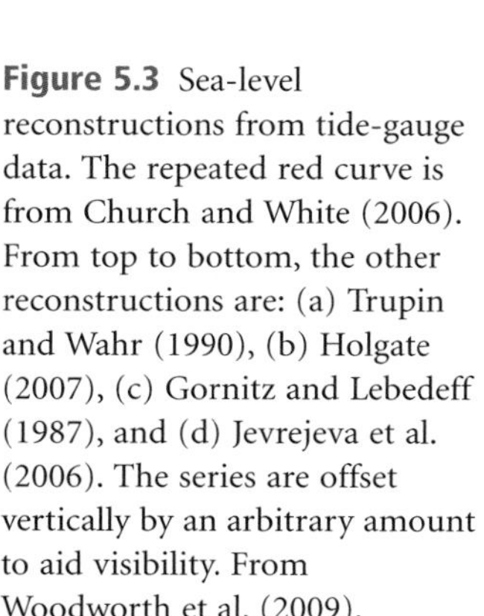
Figure 5.3 Sea-level reconstructions from tide-gauge data. The repeated red curve is from Church and White (2006). From top to bottom, the other reconstructions are: (a) Trupin and Wahr (1990), (b) Holgate (2007), (c) Gornitz and Lebedeff (1987), and (d) Jevrejeva et al. (2006). The series are offset vertically by an arbitrary amount to aid visibility. From Woodworth et al. (2009).

earlier, estimates of global sea-level change (Woodworth et al. 2009). First, we consider that the scatter between these different reconstructions is disconcerting given that all of them use essentially the same basic PSMSL data set and vary primarily in the analysis methodology. More importantly, though, regardless of the analysis method chosen, it is clear that different results on trends and accelerations are obtained depending on the epoch (i.e. time period) used to estimate them.

From the preceding and from a more detailed discussion by Woodworth et al. (2009), there is evidence from the tide-gauge data that there was an acceleration between the 19th and 20th centuries and that there were larger rates in some periods (such as in the decades after 1930 and since the late 1980s) and lower rates in others (such as the 1960s). Regarding the apparent acceleration, the paucity of tide gauges in the 19th century must be kept in mind and we suggest that further research is needed to put realistic error bars on the 19th–20th-century rate difference. Concerning the decadal variations in the rate, Merrifield et al. (2009) have suggested that some of these variations might be due to spatial sampling limitations of the tide-gauge network. Their work was by no means conclusive, but does indicate the need for further research.

5.3.2 *Spatial Variability in Trends*

The emergence of global altimeter data sets and reconstructions of upper ocean heat content based on historic hydrographic data provided insight into the spatial patterns associated with interannual and lower-frequency sea-level variations

(Cabanes et al. 2001). The dominant sea-level signal at these timescales is associated with ocean volume redistribution, and not net volume change. Because of their global spatial coverage, altimeters effectively average out these redistributions when estimating global sea level. In the averaging approach of Douglas and others described above, trends at each tide-gauge station are computed before any attempt is made to remove the redistribution signal. The trend errors owing to the redistribution signals would ideally be considerably reduced in any average of individual trends, but because of the priority placed on long record lengths there are relatively few stations or station groupings that can be used in the average.

Cabanes et al. (2001) suggested that this undersampling problem could be a major problem in the determination of any global average trend, casting doubt on the rates reported above. In particular, they suggested that apparent warming of the North Atlantic sub-tropical gyre could have contributed disproportionately to rates of sea-level change observed on the US east coast where there are many long records that contribute to the global average. Miller and Douglas (2004, 2006) and Lombard et al. (2005) have since re-examined this suggestion and convincingly rejected this possibility. Nevertheless, the Cabanes et al. (2001) study was important in alerting the community to the importance of spatial variability in sea-level change and its possible role in the global average.

Another type of spatial bias was discussed by Holgate and Woodworth (2004) who speculated that there may have been differences in coastal versus open ocean rates of change over the past decade. To our knowledge this remains an open issue, although White et al. (2005) argued that while the Holgate and Woodworth result applied for the 1993–2003 period it did not apply to other periods. Prandi et al. (2009) have also discounted the possibility over a longer period. Holgate and Woodworth also pointed to the high rates of sea-level rise observed during the 1990s, a topic to which Merrifield et al. (2009) have recently returned, pointing to the importance of rising sea levels in the tropics and southern latitudes in this period. They caution that trend estimates that are weighted heavily toward Northern Hemisphere gauges may misrepresent the recent acceleration. Of course, whether the apparent rise in recent years is the start of a long-term acceleration associated with climate change remains to be seen (see also Rahmstorf et al. 2007).

A most intriguing type of spatial bias was discussed by Miller and Douglas (2007). They argued that the apparent 20th-century acceleration seen at the eastern boundary of the North Atlantic Ocean could have been an artifact of the spin-down of the sub-tropical gyre, the strength of which was represented by air pressure measured at the centre of the gyre. A similar inference was made for data from the eastern boundary of the North Pacific, using data from the San Francisco tide gauge. The study was limited to some extent as it attempted to make conclusions on century-timescale change based on data only from 1880 onwards. Nevertheless, its conclusions give an important reminder that long-term redistribution of water remains a factor to be considered alongside the processes usually considered responsible for climate-driven sea-level change.

The spatial variations in sea level may cause a difficulty for researchers interested only in global averages, but the variations can provide important insight on the reasons for change and what changes may occur in the future. Such research is sometimes called "fingerprinting" and is discussed further below from an altimeter data perspective and in Chapter 10 from the point of view of geophysical change. Sea-level "fingerprints" can occur due to ocean processes as well as ones due to, for example, ice-sheet melting, and the deciphering of the many signals represented in spatial variations of sea-level change represents a major research challenge (e.g. Marcos and Tsimplis 2007; Stammer 2008).

5.3.3 The Importance of Precise Vertical Land-Movement Information

Both the "average of trends" approach of Douglas and others and the "global sea-level reconstruction" approach of Church and White and others are sensitive to land-motion errors. Lacking direct land-motion estimates, the most common approach is to employ some form of GIA correction at each tide-gauge station (Peltier 2001). Efforts are underway to apply direct land-motion estimates using a variety of techniques, most notably continuous Global Positioning System (GPS; see Haines et al. 2003; Caccamise et al. 2005; Snay et al. 2007; Wöppelmann et al. 2007, 2009), Doppler Orbitography and Radiopositioning Integrated by Satellite (DORIS; see Cazenave et al. 1999), and absolute gravity (Teferle et al. 2006) measurements. However, here we will present some evidence to suggest that direct land measurements may have a significant impact in reducing tide-gauge trend errors relative to what has been achieved using GIA models.

We select a subset of RLR stations with a relatively even geographic distribution. In this case we consider the Global Climate Observing System (GCOS) set of stations (Figure 5.4). For each GCOS station, we follow the approach of Nerem and Mitchum (2002) and compute an estimate of land motion at the tide gauge as the trend of the difference between tide-gauge sea level (TG) and the nearest altimeter sea-level grid point to the tide gauge (ALT; using the gridded altimeter product provided by the University of South Florida for 1993–2005). TG records with less than 80% common overlap with their corresponding ALT records are not included in the comparison, and neither are stations that are outside the range of the altimeter data set (poleward of ≈66° latitude). In addition, a few TG records with obvious datum issues are excluded. The TG-ALT trends exhibit a range of positive and negative values scattered approximately about zero (Figure 5.4). A comparison of the TG-ALT trends with GIA trends obtained from the ICE4G-VM2 model (Peltier 2001) shows that GIA rates equatorward of approximately 40° latitude, where the majority of GCOS stations are located, are generally much weaker than the TG-ALT rates (Figure 5.4). In other words, a GIA correction to the tide-gauge data would have little impact on the TG-ALT trend estimates in this latitude range. Seen another way, this might be evidence for substantial land-motion signals that are not captured by GIA physics.

This discrepancy in rates has a number of ramifications for tide-gauge error budgets. First, many of the longest tide-gauge records are at high latitudes and so

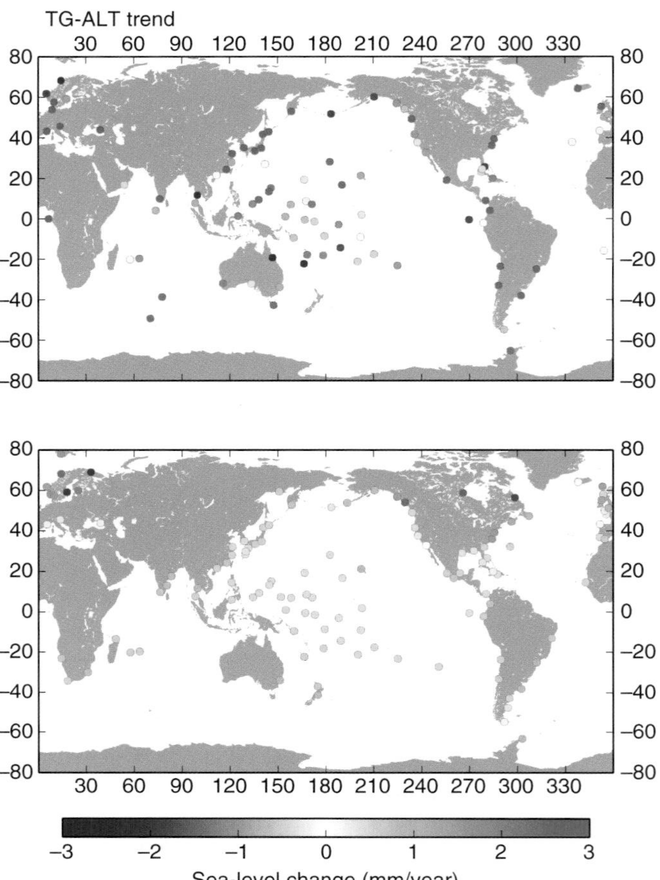

Figure 5.4 (Top) The linear trend of the sea-level difference computed using GCOS tide-gauge records (TG) and the nearest gridded altimeter data point (ALT). The trend is used as an estimate of vertical land motion following Nerem and Mitchum (2002). A positive trend corresponds to rising sea level relative to land. (Bottom) The component of sea-level rise at each tide gauge owing to GIA land motion as simulated by the ICE4G-VM2 model (Peltier 2001). A positive trend corresponds to a downward trend in land motion.

the conclusion reached in previous tide-gauge trend studies (e.g. Douglas 1997) that GIA corrections are important for obtaining reliable global sea-level rates is not inconsistent with our result. Second, large TG-ALT trends are found at stations that would not have been considered tectonically active in the average of trend studies. Third, the assumption that the scatter in previous tide-gauge rate estimates is consistent with the scatter in GIA model rates appears to be overstated, unless the analysis is heavily weighted toward high-latitude stations. It is likely that land motion contributes errors at the ±2 mm/year level (the scatter of the TG-ALT rates is of this order), but not because of poor modeling of GIA. Fourth, the trends are scattered about zero and so it is possible that land-motion errors will average out to some degree as long as a large number of tide-gauge stations is used in the analysis (see Mitchum 1998). This would be a concern, however, for studies extending further back in time, such as Church and White (2006) when there were fewer and fewer stations available. Fifth, these findings appear to be consistent with Nakiboglu and Lambeck (1991) who noted that GIA corrections had little impact on global sea-level rates determined using spherical harmonic reconstructions of tide-gauge trends. Conversely, it would seem likely that direct land-motion estimates might have a significant effect on such an analysis.

These results are suggestive rather than conclusive, yet appear to illustrate the potential impact that direct land-motion estimates might have in correcting the global tide-gauge data base. We conclude that the correction of land motion in tide-gauge records remains the primary concern for reducing recent global sea-level trend uncertainties. Also, Mitchum (2000) has shown that the uncertainty in land-motion rates at the individual tide-gauge stations is also the limiting error in altimeter calibrations, and hence in the altimeter-derived estimates of sea-level change, which is the subject of the remainder of this chapter.

5.4 Estimates of Global Sea-Level Change from Satellite Altimetry

Reliable estimates of global mean-sea-level rise from satellite altimetry first became possible with the launch of the TOPEX/Poseidon radar altimeter satellite (T/P) in 1992. While there were altimeter missions prior to the launch of T/P (Seasat, Geosat, ERS-1), T/P was the first to provide all of the key elements necessary to precisely measure changes in global mean sea level: a precision dual-frequency altimeter, a microwave radiometer, and a suite of instruments for precise orbit determination. Although the T/P mission recently came to an end, the Jason-1 satellite was launched in 2001 to continue the precision sea-level measurements along the T/P 10-day repeat groundtrack. This was since followed by Jason-2 in June 2008.

Figure 5.5 shows the 15-year time series of 10-day estimates of global mean sea level from these satellites. Each 10-day point has a precision of 4–5 mm. The time series shows a well-defined seasonal variation caused by the annual exchange of water with the continents plus an annual variation in thermal expansion. Figure 5.6 shows the same curve after the removal of seasonal variations, which reveals substantial interannual variability in global mean sea level, most notably an approximately 15-mm rise and fall during the 1997–1998 El Niño Southern Oscillation (ENSO) event (Nerem et al. 1999) and a more recent drop during the current La Niña conditions. The cause of this ENSO signal is still open to debate, although there is a suggestion that it is related to water-mass exchange with the continents because the signal is not observed in estimates of thermosteric sea-level change.

The secular trend of global mean sea level over the entire time series is 3.2 mm/year. GIA causes a small secular subsidence of the ocean basins (changing their volume) as mantle material moves back into previously glaciated regions. Modeling suggests that estimates of global mean sea level should be increased by 0.3 mm/year to account for GIA (Peltier 2001). The current best estimate of

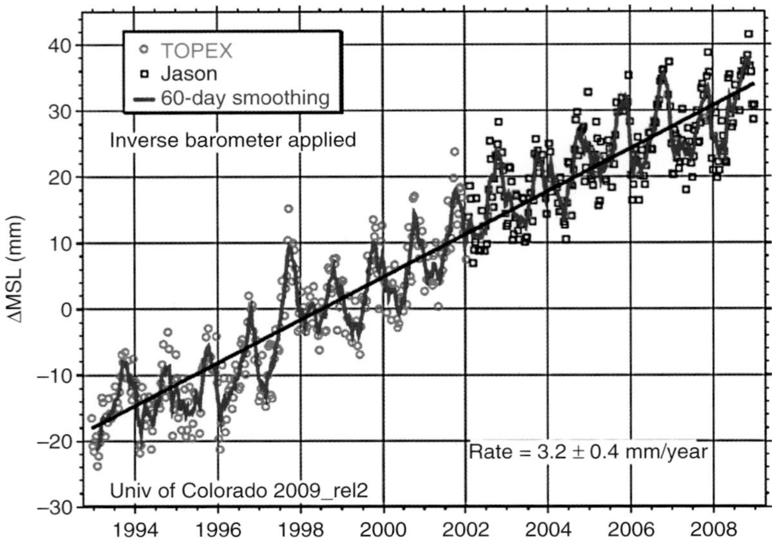

Figure 5.5 Ten-day estimates of global mean-sea-level change (ΔMSL) from the TOPEX and Jason satellites.

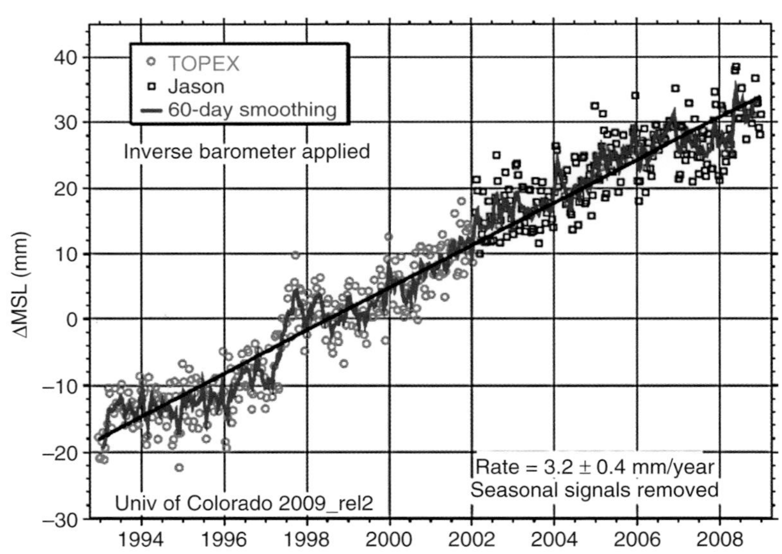

Figure 5.6 Same as Figure 5.5, except that seasonal variations are removed.

the average rate of sea-level rise over the last 15 years is thus 3.5 ± 0.4 mm/year. The error bar is composed of three components: (1) the formal error of the least squares fit, (2) an inflation factor to account for serial correlation of the residuals, and (3) error in the instrument calibration. It is this last term that dominates the error budget.

The T/P and Jason missions benefit from a number of dedicated *in situ* calibration sites designed to evaluate the measurements and monitor the instrument

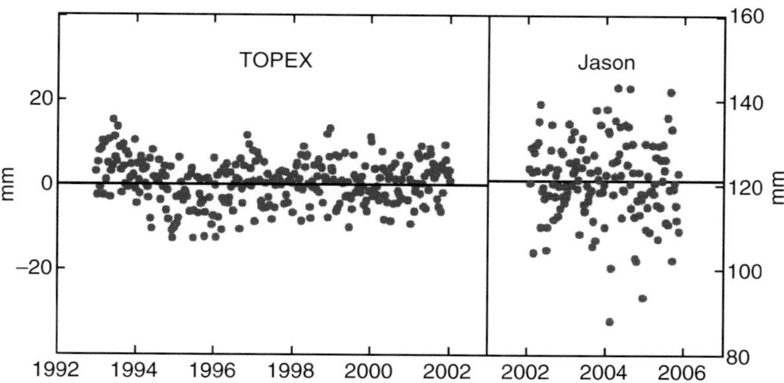

Figure 5.7 Time series of TOPEX and Jason differences relative to the global tide-gauge network. Note the difference in the values on the vertical axes. The Jason mean is 121 mm relative to TOPEX. The cause of this bias is presently unknown. The TOPEX scatter is 5.3 mm; the Jason scatter is 9.3 mm.

performance over time. However, the global tide-gauge network has provided better sensitivity to changes in the instrument behavior. Essentially, the instrument performance is monitored by differencing tide-gauge sea-level measurements with altimeter sea-level measurements (averaged in the vicinity of each tide gauge), and then averaging these differences over the global network of tide gauges available in each 10-day repeat cycle (Mitchum 2000). A sample of a time series of these calibration measurements is shown in Figure 5.7. This procedure has successfully detected a number of small errors in the altimeter data over the years, but when corrections are developed for these errors, the remaining instrument drift is not statistically different from zero. The error in the tide-gauge-derived drift calibration is dominated by errors in the knowledge of the vertical motion of the tide gauges (Mitchum 2000), which are attached to the Earth's crust and thus can move in response to GIA and other phenomena (local subsidence, tectonic activity, etc.). Reducing the error bar on the drift estimate can only be accomplished by measuring the vertical motion of the tide gauges via geodetic techniques such as GPS, DORIS, etc.

There are two changes in the altimetric time series that are particularly important to address. The first is when the TOPEX altimeter was switched to its redundant electronics (Side A to Side B) at the beginning of 1999. The second is the switch in the time series from TOPEX to Jason-1 (mid-2002 for the data presented here). Both of these events (and the transition from Jason-1 to Jason-2 and future transitions) have the potential to introduce biases in the sea-level measurements, and it is thus particularly important to accurately calibrate potential offsets, which are the altimetric analog of shifts in a tide-gauge datum.

What has been learned from this measurement record? While the record is relatively short, it suggests that there has been an increase in the rate of sea-level rise as compared to the historical tide-gauge record. Part of the recently observed rise (≈0.5 mm/year) may be due to the recovery of sea level after the cooling effects of the eruption of Mt Pinatubo in 1991 (Church et al. 2005). There may be residual effects of decadal variability, but the most recent IPCC consensus report (Bindoff et al. 2007) suggests that there has been a recent change in the long-term rate of sea-level rise. As discussed above, acceleration over the 19th–20th-century period are more difficult to assess, but the changes in the past decade are more reliably determined. Recent assessments suggest roughly half of the observed rise is due to ocean temperature change (Willis et al. 2004; Bindoff et al. 2007; but see also Chapter 6 of this book), with the remainder due to the influx of water from the continents, most likely melting of mountain glaciers and polar ice.

The T/P and Jason-1 missions provide sea-level measurements within ±66° latitude and portions of the polar regions are not observed. Surface temperatures have risen much more in the polar regions than elsewhere on Earth, and it is possible that part of the warming signal in sea level is missed. In addition, the melting of the ice sheets is expected to cause a distinct "fingerprint" in the sea-level rise maps that is most easily observed in the polar regions (Mitrovica et al. 2001; see also Chapters 9 and 10). Note also that these fingerprints are due to the effect of mass loading and gravitational changes and are not due to oceanic mass simply remaining near the ice-melt regions. One reason these regions are not covered by current precision altimeter missions is that the higher-inclination orbits significantly degrade the east–west accuracy of sea-level gradient measurements required for some ocean circulation studies. Nevertheless, a number of missions (ERS-1, ERS-2, Envisat, etc.) have flown in sun-synchronous orbits (98.6° inclination) that provide coverage to ±81.4° latitude. While much good science has been done with these missions, sun-synchronous orbits do not have good tidal-aliasing characteristics for purposes of monitoring long-term sea-level change (solar tides alias to zero frequency). In addition, the precision of the orbit determination and other environmental corrections is generally poorer than for T/P and Jason-1. In fact, estimates of long-term sea-level change from these missions have relied on ties to T/P and Jason-1 within ±66° latitude in order to reduce the larger errors present in the measurements. Finally, it should be noted that there is no evidence that omitting regions outside ±66° latitude significantly affects the global mean-sea-level curve in Figure 5.5. In fact, truncating the data to ±55° latitude has no significant effect on the curve (Nerem 1995).

The current 15-year time series is still too short to estimate the long-term acceleration of sea-level rise. Knowledge of the acceleration is important for evaluating climate models and predicting future sea levels. As discussed in Chapter 12 of this book, the Jason-2/Ocean Surface Topography Mission (OSTM) was launched in June 2008 and flies the same 10-day repeating groundtrack as T/P and Jason-1, and, at the time of writing, data from this mission are becoming available and being added to the earlier altimetric data. In addition, a Jason-3 mission has now been approved. While other satellite altimeter missions are planned, they are generally less accurate and are targeted for sun-synchronous orbits, which are undesirable for monitoring long-term sea-level change because of their tidal-aliasing characteristics. Therefore the Jason-3 and subsequent missions are critical to continue this fundamental climate change data set.

Preliminary discussions on subsequent missions have suggested lowering the altitude from 1336 to 800 km in order to reduce spacecraft costs and to be in a more benign radiation environment. This would mean abandoning the T/P/Jason-1/Jason-2 groundtrack. Studies have suggested a 78° inclination orbit with a near 20-day repeat would meet the requirements of many different communities. For long-term sea-level change, calibration of the measurements from the new orbit relative to the existing time series is a major concern. We believe that the global tide-gauge network could accomplish this task with sufficient accuracy, provided that the orbit is *not* sun-synchronous, and that it is *not* a non-repeating orbit. Although the tide-gauge calibration is in principle capable of dealing with a non-repeating orbit, this has never been implemented in an operational fashion. Also, questions arise concerning the mean sea surface and tidal models for such a mission. We conclude, then, that neither a sun-synchronous nor non-repeating orbits are desirable characteristics for a mission that has climate science as part of its objectives. Coverage to ±78° latitude would provide a better chance of detecting the sea-level fingerprint of polar ice melt (Mitrovica et al. 2001) without significantly compromising the climate objectives of the mission. This assessment is biased to the determination of global mean-sea-level change. For studies of oceanic variability other arguments for maintaining the T/P/Jason groundtrack have been advanced. For example, predictions of ENSO, which are of extreme economic value, might be compromised without a T/P/Jason-class instrument (e.g. Périgaud and Cassou 2000; Périgaud et al. 2006). These considerations are beyond our scope, but should be carefully considered, and we also note that the global sea-level change estimations would also most likely be best accomplished with a continuation of the T/P/Jason series of altimetry missions.

5.5 Recommendations

Based on these findings, we offer the following recommendations. It is our opinion that adopting these will allow continued success in monitoring sea-level change, and will also allow continued reduction in the error bars of future analyses. The global tide-gauge network (the Global Sea Level Observing System (of the Intergovernmental Oceanographic Commission); GLOSS) and the satellite altimetry missions should both be considered operational entities, meaning that continuation is a given, and expansion of both components of the sea-level-observing system should be made at every opportunity.

- Sustain the Jason class of high-accuracy satellite altimeters for the foreseeable future through the implementation of a Jason-3 and follow-on high-accuracy missions with equivalent performance. The best solution is to continue in the T/P groundtrack, which has proven to provide high-quality sea-level change estimates. If this must be changed, *a priori* studies of accuracy and precision must be made. A sun-synchronous or non-repeating orbit will not meet the requirements for climate studies.
- Continued maintenance and enhancement of the global tide-gauge network is essential, along with continued exploration of opportunities for sea-level data archaeology and the development of useful sea-level proxies. These data provide the historical context for analyses from altimetry and the tide-gauge data are also essential to the altimeter calibration problem. Improving spatial distribution is

critical, particularly in the Southern Hemisphere. The GLOSS Core Network and the GCOS Climate subset of that network provide the blueprint for future network development.

- Address the land-motion problem at the tide gauges. Complement GIA model estimates with direct geodetic measurements at each site. In addition, although this has not been discussed in detail here, the larger issue of placing all of these measurements into a consistent, high-quality global reference frame (Chapter 9) is important.
- Continued study of the best offsets between different altimetric missions (e.g. TOPEX Side A to Side B and TOPEX to Jason-1 and to subsequent missions) is needed.
- Study of the variations in the rate of sea-level change over the past century should be continued. At present there is good consensus on the 20th-century rate and its error bar, but results on acceleration are conflicting. The mass and density redistribution signals that do not contribute to global sea-level change need to be carefully investigated.

Acknowledgments

We thankfully acknowledge our colleagues (Trevor Baker, John Church, Eric Leuliette, Laury Miller, Claire Périgaud, and Wilton Sturges) who provided valuable comments on an early version of this chapter, and the reviewers (Anny Cazenave, Simon Holgate, Svetlana Jevrejeva, and Philip Woodworth) who spent valuable time helping us to substantially improve the final version.

References

Berge-Nguyen M., Cazenave A., Lombard A., Llovel W., Viarre J. and Cretaux J.F. (2008) Reconstruction of past decades sea level using thermosteric sea level, tide gauge, satellite altimetry and ocean reanalysis data. *Global and Planetary Change* **62**, 1–13.

Bindoff N., Willebrand J., Artale V., Cazenave A., Gregory J.M., Gulev S. et al. (2007) Observations: ocean climate change and sea level. In: *Climate Change 2007: The Physical Science Basis*. Contribution of Working Group 1 to the Fourth Assessment Report of the Intergovernmental Panel on Climate Change (Solomon S., Qin D., Manning M., Marquis M., Averyt K., Tignor M.M.B. et al., eds), pp. 385–432. Cambridge University Press, UK.

Cabanes C., Cazenave A. and LeProvost C. (2001) Sea level rise during past 40 years determined from satellite and in situ observations. *Science* **294**, 840–2.

Caccamise D.J., Merrifield M.A., Bevis M., Foster J., Firing Y.L., Schenewerk M.S. et al. (2005) Sea level rise at Hawaii: GPS Estimates of Differential Land Motion. *Geophysical Research Letters* **32**, L03607.

Cazenave A., Dominh K., Ponchaut F., Soudarin L., Cretaux J.F. and Le Provost C. (1999) Sea level changes from TOPEX-Poseidon altimetry and tide gauges, and vertical crustal motions from DORIS. *Geophysical Research Letters* **26**, 2077–80.

Chambers D.P., Melhaff C.A., Urban T.J., Fuji D. and Nerem R.S. (2002) Low-frequency variations in global mean sea level: 1950–2000. *Journal of Geophysical Research* **107**, 3026.

Church J.A. and White N.J. (2006) A 20th century acceleration in global sea-level rise. *Geophysical Research Letters* **33**, L01602.

Church J.A., Gregory J.M, Huybrechts P., Kuhn M., Lambeck K., Nhuan M.T. et al. (2001) Changes in sea level. In: *Climate Change 2001: The Scientific Basis.* Contribution of Working Group 1 to the Third Assessment Report of the Intergovernmental Panel on Climate Change (Houghton J.T., Ding Y., Griggs D.J., Noguer M., van der Linden P.J., Dai X. et al., eds), pp. 639–94. Cambridge University Press, Cambridge.

Church J.A., White N.J., Coleman R., Lambeck K. and Mitrovica, J.X. (2004) Estimates of the regional distribution of sea-level rise over the 1950 to 2000 period. *Journal of Climate* **17**, 2609–25.

Church J.A., White N.J. and Arblaster J.M. (2005) Significant decadal-scale impact of volcanic eruptions on sea level and ocean heat content. *Nature* **438**, 74–7.

Donnelly J.P., Cleary P., Newby P. and Ettinger R. (2004) Coupling instrumental and geological records of sea level change: evidence from southern New England of an increase in the rate of sea level rise in the late 19th century. *Geophysical Research Letters* **31**, L05203.

Douglas B.C. (1991) Global sea level rise. *Journal of Geophysical Research* **96**, 6981–92.

Douglas B. (1992) Global sea level acceleration. *Journal of Geophysical Research* **97**, 12699–796.

Douglas B.C. (1997) Global sea-rise: a redetermination. *Surveys in Geophysics* **18**, 279–92.

Douglas B. (2001) Sea level changes in the era of the recording tide gauge, In: *Sea Level Rise: History and Consequences* (Douglas B., Kearney M. and Leatherman S., eds), pp. 37–64. Academic Press, London.

Ekman M. (1988) The world's longest continued series of sea level observations. *Pure and Applied Geophysics* **127**, 73–7.

Gehrels W.R., Belknap D., Black S. and Newnham R. (2002) Rapid sea level rise in the Gulf of Maine, USA, since AD 1800. *The Holocene* **12**, 383–9.

Gehrels W.R., Kirby J., Prokoph A., Newnham R., Achterberg E., Evans H., Black S. and Scott D. (2005) Onset of recent rapid sea level rise in the western Atlantic Ocean. *Quaternary Science Reviews* **24**, 2083–2100.

Gehrels W.R., Marshall W.A., Gehrels M.J., Larsen G., Kirby J.R., Eriksson J. et al. (2006) Rapid sea level rise in the North Atlantic Ocean since the first half of the 19th century. *The Holocene* **16**, 948–64.

Gehrels W.R., Hayward B.W., Newnham R.M. and Southall, K.E. (2008) A 20th century sea-level acceleration in New Zealand. *Geophysical Research Letters* **35**, L02717.

Gornitz V. and Lebedeff S. (1987) Global sea level changes during the past century. In: *Sea Level Fluctuation and Coastal Evolution* (Nummedal D., Pilkey O.H. and Howard J.D., eds), pp. 3–16. SEPM Special Publication No. 41. Society for Economic Paleontologists and Mineralogists, Tulsa, OK.

Haines B.J., Dong D., Born G.H. and Gill S.K. (2003) The Harvest Experiment: Monitoring Jason-1 and TOPEX/POSEIDON from a California Offshore Platform. *Marine Geodesy* **26**, 239–59.

Holgate S.J. (2007) On the decadal rates of sea level change during the twentieth century. *Geophysical Research Letters* **34**, L01602.

Holgate S.J. and Woodworth P.L. (2004) Evidence for enhanced coastal sea level rise during the 1990s. *Geophysical Research Letters* **31**, L07305.

Hunter J., Coleman R. and Pugh D. (2003) The sea level at Port Arthur, Tasmania, from 1841 to the present. *Geophysical Research Letters* **30**, 1401.

Jevrejeva S., Grinsted A., Moore J.C. and Holgate S. (2006) Nonlinear trends and multiyear cycles in sea level records. *Journal of Geophysical Research* **111**, C09012.

Jevrejeva S., Moore J.C., Grinsted A. and Woodworth, P.L. (2008) Recent global sea level acceleration started over 200 years ago? *Geophysical Research Letters* **35**, L08715.

Leorri E., Horton B.P. and Cearetta A. (2008) Development of a foraminifera-based transfer function in the Basque marshes, N. Spain: implications for sea-level studies in the Bay of Biscay. *Marine Geology* **251**, 60–74.

Lombard A., Cazenave A., Le Traon P.-Y. and Ishii M. (2005) Contribution of thermal expansion to present-day sea-level change revisited. *Global and Planetary Change* **47**, 1–16.

Marcos M. and Tsimplis M.N. (2007), Forcing of coastal sea level rise patterns in the North Atlantic and the Mediterranean Sea. *Geophysical Research Letters* **34**, L18604.

Merrifield M.A., Merrifield S.T. and Mitchum G.T. (2009) An anomalous recent acceleration of global sea level rise. *Journal of Climate* **22**, 5772–81.

Miller L. and Douglas B.C. (2004) Mass and volume contributions to 20th century global sea level rise. *Nature* **438**, 406–9.

Miller L. and Douglas B.C. (2006) On the rate and causes of twentieth century sea-level rise. *Philosophical Transactions of the Royal Society A* **364**, 805–20.

Miller L. and Douglas B.C. (2007) Gyre-scale atmospheric pressure variations and their relation to 19th and 20th century sea level rise. *Geophysical Research Letters* **34**, L16602.

Mitchum G.T. (1998) Monitoring the stability of satellite altimeters with tide gauges. *Journal of Atmospheric and Oceanic Technology* **15**, 721–30.

Mitchum G.T. (2000) An improved calibration of satellite altimetric heights using tide gauge sea levels with adjustment for land motion. *Marine Geodesy* **23**, 145–66.

Mitrovica J.X., Tamisiea M.E., Milne G.A. and Davis J.L. (2001) Recent mass balance of polar ice sheets inferred from patterns of global sea-level change. *Nature* **409**, 1026–9.

Nakiboglu S.M. and Lambeck K. (1991) Secular sea level change. In: *Glacial Isostasy, Sea Level and Mantle Rheology* (Sabadini R., Lambeck K. and Boschi E., eds), pp. 237–58. Kluwer Academic Publishers, Dordrecht.

Nerem R.S. (1995) Measuring global mean sea level variations using TOPEX/Poseidon altimeter data. *Journal of Geophysical Research* **100**, 25135–51.

Nerem R.S. and Mitchum G.T. (2002) Estimates of vertical crustal motion derived from differences of TOPEX/POSEIDON and tide gauge sea level measurements. *Geophysical Research Letters* **29**(19), 1934.

Nerem R.S., Chambers D.P., Leuliette E., Mitchum G.T. and Giese B.S. (1999) Variations in global mean sea level during the 1997–98 ENSO event. *Geophysical Research Letters* **26**, 3005–8.

Peltier W.R. (2001) Global glacial isostatic adjustment and modern instrumental records of relative sea level history. In: *Sea Level Rise: History and Consequences* (Douglas B.C., Kearney M.S. and Leatherman S.P., eds), pp. 61–95. Academic Press, London.

Peltier W.R. and Tushingham A.M. (1991) Influence of glacial isostatic adjustment on tide gauge measurements of secular sea level change. *Journal of Geophysical Research* **96**, 6779–96.

Périgaud C. and Cassou C. (2000) Importance of oceanic decadal trends and westerly wind bursts for forecasting El Nino. *Geophysical Research Letters* **27**, 389–92.

Périgaud C., Boulanger J.-P. and Illig, S. (2006) Importance of TOPEX/Poseidon/Jason data to improve the coupled ocean-atmosphere modeling of El Nino. In: *Proceedings of Fifteen Years of Progress in Radar Altimetry Symposium*. Venice, Italy 13–18 March 2006 (European Space Agency, ed.). http://earth.esa.int/venice06/abstractbook_venice06.pdf.

Pouvreau N. (2008) *Trois cents ans de mesures marégraphiques en France: outils, méthodes et tendances des composantes du niveau de la mer au port de Brest*. PhD thesis, University of La Rochelle.

Prandi P., Cazenave A. and Becker M. (2009) Is coastal mean sea level rising faster than the global mean? A comparison between tide gauges and satellite altimetry over 1993–2007. *Geophysical Research Letters* **36**, L05602.

Pugh D., Hunter J., Coleman R. and Watson C. (2002) A comparison of historical and recent sea level measurements at Port Arthur, Tasmania. *International Hydrographic Review* **3**, 1–20.

Rahmstorf S., Cazenave A., Church J.A., Hansen J.E., Keeling R.F., Parker D.E. and Somerville R.C.J. (2007) Recent climate observations compared to projections. *Science* **316**, 709.

Snay R., Cline M., Dillinger W., Foote R., Hilla S., Kass W. et al. (2007) Using global positioning system-derived crustal velocities to estimate rates of absolute sea level change from North American tide gauge records. *Journal of Geophysical Research* **112**, B04409.

Stammer D. (2008) Response of the global ocean to Greenland and Antarctic ice melting. *Journal of Geophysical Research* **113**, C06022.

Teferle F.N., Bingley R.M., Williams S.D.P., Baker T.F. and Dobson A.H. (2006) Using continuous GPS and absolute gravity to separate vertical land movements and changes in sea level at tide gauges in the UK. *Philosophical Transactions of the Royal Society of London A* **364**, 917–30.

Trupin A.S. and Wahr J.M. (1990) Spectroscopic analysis of global tide gauge sea level data. *Geophysical Journal International* **100**, 441–53.

Tushingham A.M. and Peltier W.R. (1991) Ice-3G: A new global model of late Pleistocene deglaciation based upon geophysical predictions of postglacial relative sea level. *Journal of Geophysical Research* **96**, 4497–4523.

White N.J., Church J.A. and Gregory J.M. (2005) Coastal and global averaged sea level rise for 1950 to 2000. *Geophysical Research Letters* **32**, L01061.

Willis J.K., Roemmich D. and Cornuelle B. (2004) Interannual variability in upper ocean heat content, temperature, and thermosteric expansion on global scales. *Journal of Geophysical Research* **109**, C12036.

Woodworth P.L. (1990) A search for accelerations in records of European mean sea level. *International Journal of Climatology* **10**, 129–43.

Woodworth P.L. (1999) High waters at Liverpool since 1768: the UK's longest sea level record. *Geophysical Research Letters* **26**, 1589–92.

Woodworth P.L. and Player R. (2003) The Permanent Service for Mean Sea Level: an update to the 21st century. *Journal of Coastal Research* **19**, 287–95.

Woodworth P.L., White N.J., Jevrejeva S., Holgate S.J., Church J.A. and Gehrels W.R. (2009) Evidence for the accelerations of sea level on multi-decade and century timescales. *International Journal of Climatology* **29**, 777–89.

Wöppelmann G., Pouvreau N. and Simon B. (2006) Brest sea level record: a time series construction back to the early eighteenth century. *Ocean Dynamics* **56**, 487–97.

Wöppelmann G., Martin Miguez B., Bouin M.-N. and Altamimi Z. (2007) Geocentric sea-level trend estimates from GPS analyses at relevant tide gauges world-wide. *Global and Planetary Change* **57**, 396–406.

Wöppelmann G., Pouvreau N., Coulomb A., Simon B. and Woodworth P. (2008) Tide gauge datum continuity at Brest since 1711: France's longest sea-level record. *Geophysical Research Letters* **35**, L22605.

Wöppelmann G., Letetrel C., Santamaria A., Bouin M.-N., Collilieux X., Altamimi Z., Williams S.D.P. and Miguez, B.M. (2009) Rates of sea-level change over the past century in a geocentric reference frame. *Geophysical Research Letters* **36**, L12607.

6 Ocean Temperature and Salinity Contributions to Global and Regional Sea-Level Change

John A. Church, Dean Roemmich, Catia M. Domingues, Josh K. Willis, Neil J. White, John E. Gilson, Detlef Stammer, Armin Köhl, Don P. Chambers, Felix W. Landerer, Jochem Marotzke, Jonathan M. Gregory, Tatsuo Suzuki, Anny Cazenave, and Pierre-Yves Le Traon

6.1 Introduction

The oceans are a central component of the climate system, storing and transporting vast quantities of heat. Indeed, more than 90% of the heat absorbed by the Earth over the last 50 years as a result of global warming is stored in the ocean (Bindoff et al. 2007). Understanding how the ocean heat content varies in space and time is central to understanding and successfully predicting climate variability and change.

As the oceans warm, they expand and sea level rises. The amount of expansion depends on the quantity of heat absorbed and on the water temperature (greater expansion in warm water), pressure (greater expansion at depth), and, to a smaller extent, salinity (greater expansion in saltier water). A 1000-m column of sea water expands by about 1 or 2 cm for every 0.1°C of warming. Both the temperature (thermosteric) and salinity (halosteric) contributions (or their combined impact on density (and volume), the steric contribution) are important for regional changes in sea level, but the thermosteric contribution is the dominant factor in globally averaged changes. Ocean thermal expansion, or thermosteric sea-level rise, was a major contributor to 20th-century sea-level rise and is projected to continue during the 21st century and for centuries into the future (Bindoff et al. 2007; Meehl et al. 2007). The close connection between ocean thermosteric sea-level rise and ocean heat-content changes means that understanding sea-level rise will contribute significantly to our understanding of the Earth's total climate system.

Understanding Sea-Level Rise and Variability, 1st edition. Edited by John A. Church, Philip L. Woodworth, Thorkild Aarup & W. Stanley Wilson.

Despite its importance, even the sign of global averaged thermosteric sea-level change was unknown as recently as the 1980s because, in most of the world's oceans, the sampling noise due to mesoscale eddies and interannual variability was too great to determine statistically significant 30-year trends (Barnett 1983). Roemmich (1990) and Joyce and Robbins (1996) described regional, multidecadal increases in steric sea level off southern California and near Bermuda, where long time-series observations were of sufficient duration and temporal resolution for significant results. On basin scales there were a few repeated hydrographic transects, such as at 24° and 36°N in the Atlantic (Roemmich and Wunsch 1984) and at 28° and 43°S in the southwest Pacific (Bindoff and Church 1992), that showed multidecadal warming. Levitus (1990) described decadal variability in the North Atlantic that was at least in part a result of changes in surface winds (Hong et al. 2000; Sturges and Hong 2001; Ezer et al. 1995). In a summary of observational comparisons, Church et al. (2001) found widespread indications of thermal expansion of the order of 1 mm/year, particularly in the subtropical gyres. However, a truly global thermosteric sea-level change estimate was not possible from these widely spaced observations.

A recognition of the importance of the role of the oceans in climate resulted in new global assessments of ocean heat content and thermal expansion. Widespread use of the expendable bathythermograph (XBT) to obtain upper-ocean temperature profiles, beginning in the late 1960s, led to better understanding of the space and timescales of oceanic variability (e.g. Bernstein and White 1979) and of the sampling requirements for global studies (White 1995). The Global Oceanographic Data Archeology and Rescue Project sponsored by the Intergovernmental Oceanographic Commission made an important contribution in assembling and making available the historical data (Levitus et al. 2005d). The World Ocean Circulation Experiment (WOCE) in the 1990s (Siedler et al. 2001) provided the first and highest possible quality, global, top-to-bottom survey of ocean temperature and salinity, as well as repeated transects at a number of key locations. WOCE argued for global altimetric measurements and was also responsible for developing the technology of autonomous profiling floats (Davis et al. 2001), and thus for enabling a global array of profiling floats (the Argo Project; Gould et al. 2004); Argo implementation began in 2000. WOCE obtained about 10 000 high-quality shipboard temperature/salinity profiles over a 7-year period. Argo now collects about the same number of temperature/salinity profiles autonomously, from the sea surface to about 2000 m every month at a fraction of the cost of the WOCE hydrographic survey.

In this chapter, we will assess direct observational estimates of steric sea-level rise for the second half of the 20th century; this is the longest period for which near-global data sets are available (section 6.2). The "era of satellite altimetry", beginning in late 1992 with the launch of the TOPEX/Poseidon radar altimeter satellite, is qualitatively different from the earlier period because of the WOCE global hydrographic survey, repeated XBT sampling along commercial shipping routes, and high-precision satellite altimeters. The latter allows near-global sea-level changes to be measured globally in parallel with the *in situ* measurements of thermosteric sea level. We will also consider the recent record of the Argo Project, with near-global coverage since about 2004.

In addition to the much improved observational database, data-assimilation techniques for combining observations and models that are standard tools for atmospheric reanalyses are now being applied to the ocean. This approach helps overcome the inadequate data distribution and allows synthesis of all available data in one consistent estimate of the evolving ocean. The first generation of these results has recently become available (section 6.3). It is also possible to infer ocean steric changes as the difference between changes in ocean volume (estimated from sea level) and changes in ocean mass (estimated by satellite gravity observations; section 6.4).

Since the 1960s and 1970s, global ocean and coupled atmosphere–ocean general circulation models (AOGCMs) have been developed. These models have improved rapidly as computing power has increased, improved parameterizations, numerical techniques and ocean data sets have been developed and assembled and interest in climate issues has exploded. These models are the basis for the projections of global averaged steric sea-level rise and the regional distribution of sea-level rise during the 21st century and beyond (section 6.5). Note that geophysical processes also affect the regional distribution of sea-level rise (see Chapter 10).

Significant progress in historical assessments and the modeling of steric sea-level rise has been made over the last decade. Improved *in situ* and satellite observational programs that are now operational will lead to further improvements in our understanding, assessments and projections. However, a number of significant gaps remain, including observations of the deep, abyssal, ice-covered and coastal oceans. Uncertainties remain about instrumental biases in historical data sets and there is a need for continued careful quality control of all data sets. Comparisons of observational estimates with climate models are required at global and regional scales to understand the implications for the detection, attribution, and projection of steric sea-level rise. We bring these ideas together in section 6.6 with recommendations for observational and modeling studies.

6.2 Direct Estimates of Steric Sea-Level Rise

6.2.1 *The Second Half of the 20th Century*

The data distribution is an important issue in estimating steric sea-level rise. Figure 6.1 shows the global inventory of high-quality research ship profiles (station) data to depths greater than 500 m for the 1950s, the 1990s, and for XBT profiles in the 1990s and profiling float data (the Argo Project) from 2000 to September 2007. The strong Northern-Hemisphere bias in the pre-Argo data sets is apparent, with enormous data-void regions in the two-thirds of the ocean in the Southern Hemisphere. The poor data distribution means that any global

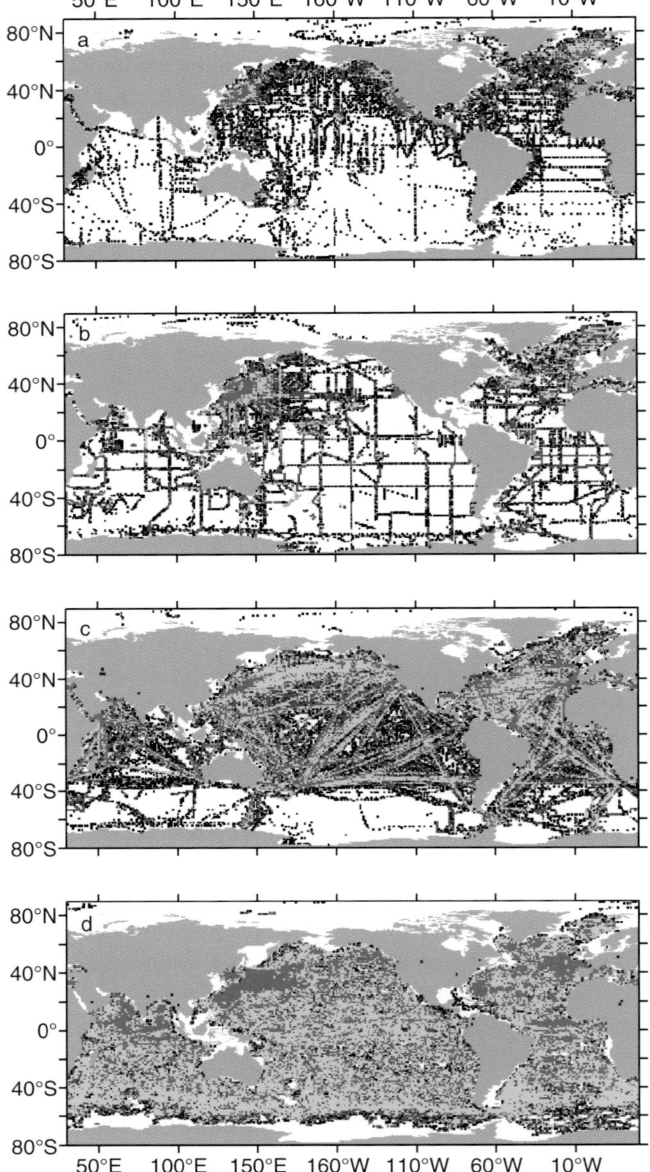

Figure 6.1 Distribution of ocean station data with temperature measurements to depths greater than 500 m. For stations (a) in the 1950s (about 46 600 total stations) and (b) in the 1990s (about 65 590 total stations); (c) XBT profiles in the 1990s (about 278 000 stations); and (d) for Argo data through September 2007 (about 368 000 stations). The number of stations in each 1-degree square is indicated by color: white = 0 stations, black = 1 station, blue = 2–5 stations, yellow = 6–20 stations, red = more than 20 stations. (Information sources: NODC and Argo Global Data Assembly Center.)

estimates will be reliant on the chosen scheme for filling data voids in space and time.

Global investigations of ocean heat storage and thermosteric sea-level rise starting from 1950 were performed by Levitus and colleagues (Levitus et al. 2000, 2005a, 2005b; Antonov et al. 2002, 2005) and by Ishii and colleagues (Ishii et al.

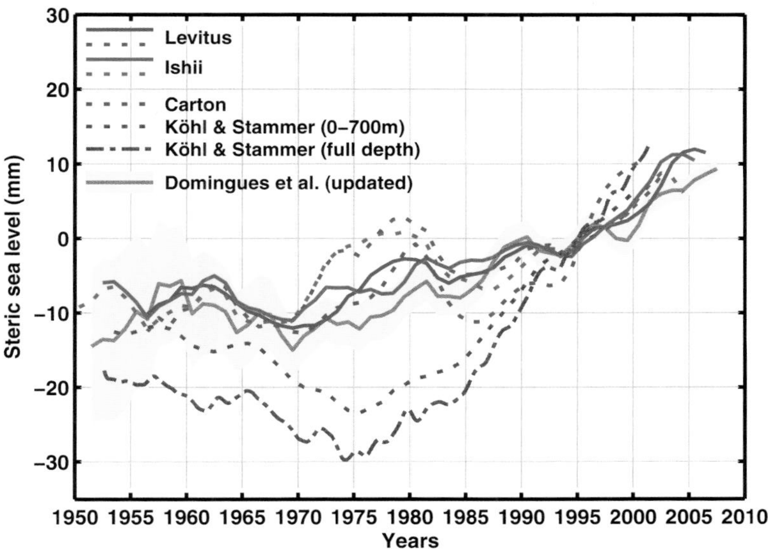

Figure 6.2 Multi-decadal direct estimates of globally averaged thermosteric sea-level changes for the upper ocean (0–700 m). The dotted lines are the thermosteric estimates of Levitus et al. (2005a) and Antonov et al. (2005; blue dotted line), Ishii et al. (2006; red dotted line), and from the Simple Ocean Data Analysis (SODA) model (Carton et al. 2005; green dotted line, to 1000 m) using XBT data not corrected for fall-rate errors. The solid lines are the equivalent curves after the XBT biases have been corrected (Levitus et al. 2009; Ishii and Kimoto 2009), with the addition of an updated estimate of Domingues et al. (2008; brown line with one standard deviation error estimates shown by the shading). The gray lines are the steric estimates from Köhl and Stammer (2008) to 700 m (gray dotted line) and full depth (gray dashed/dotted line).

2003, 2006). The World Ocean Database (WOD; Conkright et al. 2002 and later updates) forms the basis for these and other estimates, but the analyses differ in their quality-control procedures, inclusion of recent profiles, depth coverage, and analysis techniques.

A number of these results are summarized in Figure 6.2. For the 0–700 m layer and from 1955 to 2003, Levitus et al. (2005a) estimated a heat-content trend of $(0.23 \pm 0.06) \times 10^{22}$ J/year and Antonov et al. (2005) estimated the corresponding thermosteric sea-level rise of 0.33 ± 0.04 mm/year. For the same layer and time period, Ishii et al. (2006) estimated a linear trend of global-ocean heat-content increase of $(0.19 \pm 0.05) \times 10^{22}$ J/year for 1955–2003, with a corresponding rise of 0.36 ± 0.07 mm/year in thermosteric sea level. Consideration of the 0–3000 m depth range increased the estimate of Antonov et al. by approximately 20%, to 0.40 ± 0.05 mm/year for the period 1957–97.

Lombard et al. (2005; see also Levitus et al. 2005c) compared the analyses by the above two groups and found them to be similar, with much of the apparent interannual-to-decadal variability correlated with climate phenomena: the El Niño Southern Oscillation (ENSO), the Pacific Decadal Oscillation, and the North Atlantic Oscillation (NAO). While these analyses are consistent with one another, the sparse sampling, particularly in the Southern Hemisphere, and instrumental biases (see below) raise questions about their accuracy.

Optimal interpolation methods, applied in several of the above analyses, assume zero anomaly in data-void regions. Gille (2008) showed that these sampling problems are very significant in the Southern Ocean and concluded that thermosteric sea-level rise is likely to have been significantly larger than the above estimates. This conclusion is supported by tests using numerical models. Assuming zero temperature anomaly in unsampled regions resulted in small trends whereas assuming that the same average anomaly at a given time occurred in sampled and unsampled regions gave much larger trends (Gregory et al. 2004).

Instrumental biases are also a major source of uncertainty. Gouretski and Koltermann (2007) demonstrated that Mechanical Bathy Thermographs and XBTs (which dominate the historical data archive and were not designed for climate purposes) have time- and depth-dependent biases of as much as 0.4°C. Wijffels et al. (2008) and Ishii and Kimoto (2009) demonstrated that the XBT biases are dominated by inaccuracies in the depth of observations caused by errors in the estimated XBT fall rate and that these fall-rate errors change from year to year (probably as a result of small manufacturing differences from one batch to the next) but coherently around the globe, allowing an approximate correction to be applied to the historical data. One difficulty in this process is the lack of metadata specifying what XBT data were previously adjusted using the earlier Hanawa et al. (1995) fall-rate correction.

Recent Estimates of Thermosteric Sea Level

Domingues et al. (2008) addressed both instrumental and sampling biases by applying the Wijffels et al. (2008) fall-rate corrections to the global ocean data set of Ingleby and Huddleston (2007) and using a reduced-space, optimal-interpolation (RSOI) technique (Kaplan et al. 2000) to interpolate across data voids. Tests of the RSOI technique using non-eddy-resolving climate model simulations demonstrated that the errors in global averaged thermosteric sea level of the upper 700 m resulting from the spatial sampling in this database are mostly less than about 5 mm after 1960 (N.J. White et al., personal communication). The unresolved eddy variability is likely to increase these error estimates somewhat. For the period 1961–2003, the estimate of Domingues et al. (2008) of ocean thermal expansion for the upper 700 m was 0.52 ± 0.8 mm/year (i.e. a linear trend about 50% larger than earlier results; Figure 6.2). The results do not have the large decadal variability of earlier estimates (particularly the "hump" of the 1970s and 1980s). Instead they show little thermosteric rise prior to the mid-1970s, then a steady rise and variability that appears to be at least partly associated with volcanic eruptions (section 6.5).

The revised estimates of Levitus et al. (2009) and Ishii and Kimoto (2009) used different XBT bias corrections and are generally within the error bars of the results of Domingues et al. All of these analyses show little change from the 1950s to the mid-1970s and then a rise, which in the Levitus et al. and Ishii and Kimoto analyses accelerates in the mid-1990s. Over the longest time span available of 1951–2005, Ishii and Kimoto (2009) found a linear thermosteric trend of 0.29 ± 0.05 mm/year. These revised estimates are quite different from past estimates thus strengthening the argument that the 1970s peak is most likely a result of instrumental biases in earlier analyses (e.g. figure 15 in Wijffels et al. 2008). However, some smaller differences between these estimates are present;

for example the Domingues et al. series has what appears to be an anomalous minimum about 2000. With the inclusion of the most recent version of Argo data (Barker et al., unpublished work), the updated (from 2000) estimate of Domingues et al. indicates warming and expansion continues (but at a slower rate) to the end of the record. In contrast, the Levitus et al. and Ishii and Kimoto time series, which have included some corrections to early problems identified in the Argo data but not the more subtle biases identified by Barker et al., indicate a leveling off in the thermosteric sea level. Clearly further analysis and careful quality control of data is required to refine the multidecadal estimates.

Inferred Changes in Ocean Mass from Changes in Ocean Salinity

Antonov et al. (2002) and Ishii et al. (2006) estimated a global average halosteric component (expansion/contraction of ocean waters caused by changes in salinity) of 0.05 ± 0.02 and 0.04 ± 0.01 mm/year, over the 1957–94 and 1955–2003 periods, respectively. However, while the halosteric contribution is quantitatively important in the regional patterns of sea-level change, it does not contribute significantly to a global average steric change. It is instead an indication of freshening and thus of an increase in the mass of the ocean (Antonov et al. 2002; Munk 2003). This estimate of change in ocean mass is complementary to estimates inferred from changes in the storage of ice and water on land (Chapters 7 and 8) and the difference between total sea-level change and the steric component (section 6.4). The Antonov et al. (2002) estimate of the decrease of global salinity from 1957 to 1994 implied an increased mass equivalent to a sea-level rise of 1.3 ± 0.5 mm/year, if the source of the fresh water was assumed to be continental ice. However, Wadhams and Munk (2004) estimate melting of sea ice was responsible for over half of the observed freshening and thus the mass contribution to sea-level rise was estimated as about 0.6 mm/year, approximately consistent with the contributions of glaciers and ice caps, with a small contribution from the ice sheets (Chapter 7). However, measurement inaccuracies, inadequate estimates of changes in sea-ice volume and sparse ocean sampling (much poorer than for temperature) mean the quoted error estimates for freshening of the ocean are likely to be (unrealistic) lower bounds. As a result, the estimated multidecadal rates of changes in ocean mass inferred from salinity changes for the second half of the 20th century should be considered with caution. Projected increases in the cryospheric contribution to sea-level rise would lead to a larger change in ocean mass and a larger reduction in ocean salinity than during the 21st century.

6.2.2 *The "Era of Satellite Altimetry"*

The 1990s was marked by improvements in the sampling of the oceans by hydrographic and XBT networks and also the advent of high-precision satellite altimetry which provides an accurate determination of global sea-level patterns and trends. Altimetric height is highly correlated with heat content and steric height (White and Tai 1995; Gilson et al. 1998; Willis et al. 2003, 2004) and this correlation can be exploited both to assess the sampling error of the sparse *in situ* networks and to attempt to correct it (Willis et al. 2004).

Global estimates of heat content (0–750 m) and thermosteric height were completed by Willis et al. (2004) using *in situ* data alone and in combination with altimetric height. For 1993–2003 they found a thermosteric (0–750 m) sea-level rise of 1.6 ± 0.3 mm/year. From the combined *in situ* and altimetric database constructed by Guinehut et al. (2004), Lombard et al. (2006) estimated a thermosteric sea-level rise of 1.8 ± 0.2 mm/year over 1993–2003. These values are significantly larger than estimates based on *in situ* data alone (1.2 ± 0.2 mm/year; Antonov et al. 2005), possibly a result of too large a weight given to altimetry data in the Willis et al. and Lombard et al. analyses. However, it is now clear that this and indeed most of the observational estimates for the 1990s were affected by XBT fall-rate errors (Wijffels et al. 2008). Significant differences remain, even with the XBT fall-rate corrections applied. The thermosteric rise estimates for 1993–2003 are 0.8 ± 0.4 mm/year (Domingues et al. 2008), 1.3 ± 0.4 mm/year (J.K. Willis, personal communication), 1.5 ± 0.4 mm/year (Ishii and Kimoto 2009), and 1.1 ± 0.4 mm/year (Levitus et al. 2009). Further efforts are underway to agree on the best approach to correct the XBT biases.

The regional linear trends of thermosteric sea-level change for 1993–2003 for both Willis et al. (2004) and Domingues et al. (2008) are similar, with the latter being smoother because it only contains large spatial-scale variability (Figure 6.3; and Church et al. 2008). Both are similar to the regional trends in altimeter-measured sea level over the same period, demonstrating the importance of thermosteric changes for the regional distribution of sea-level rise (Figure 6.3; also see, for example, Lombard et al. 2005). The change in the tropics is largely associated with interannual variability associated with the ENSO phenomenon. The maxima in zonally integrated steric height increase at 38°N and 40°S may be a result of the increase of the atmospheric annular modes resulting in enhanced Ekman convergence and downward displacement of isopycnals at these latitudes (Roemmich et al. 2007).

One anomalous aspect of the altimetry era is that it was preceded by the explosive volcanic eruption of Mt Pinatubo in June 1991. Climate models (section 6.5; Church et al. 2005; Gregory et al. 2006; Domingues et al. 2008) indicate that sulphate aerosols injected into the stratosphere by violent volcanic eruptions such as Mt Pinatubo (and the earlier Mt Agung (1963) and El Chichon (1982) eruptions) reflect solar radiation, leading to a (rapid) cooling and thermal contraction of the oceans (about 3×10^{22} J and 5 mm) over roughly an 18-month period. The recovery of the climate system is much slower, taking decades (Church et al. 2005; Gregory et al. 2006) or even centuries (Gleckler et al. 2006a, 2006b). It is possible that the recovery from the Mt Pinatubo eruption may be partly responsible (about 0.5 mm/year) for the enhanced rate of warming and sea-level rise seen in the subsequent years (Church et al. 2005; Gregory et al. 2006).

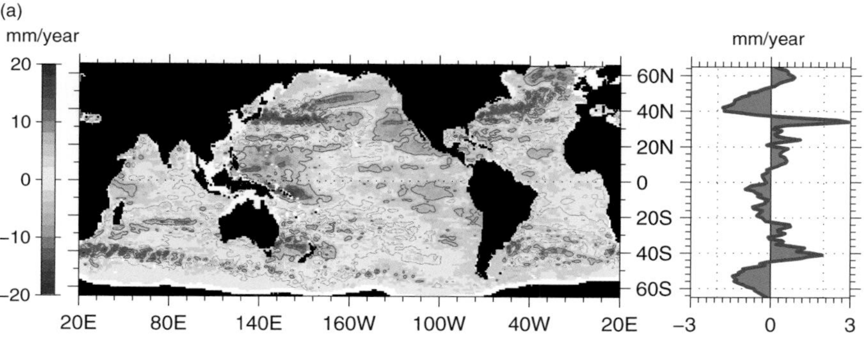

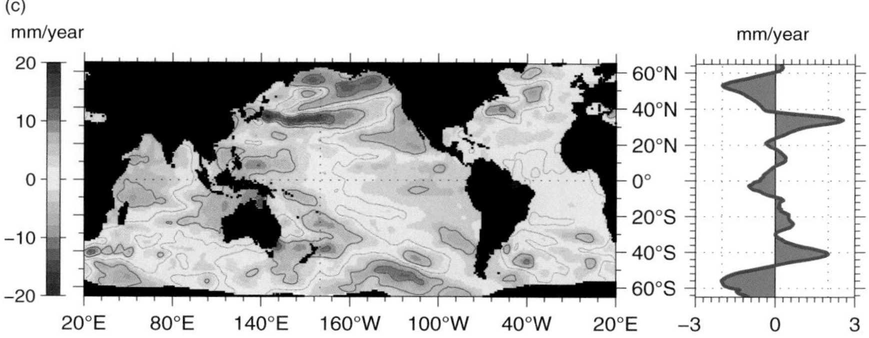

Figure 6.3 Regional distribution of thermosteric sea-level rise (mm/year) for the period 1993–2003, from the results of Willis et al. (2004; upper panel), Domingues et al. (2008) and Church et al. (2008; middle panel) and, for sea level from the satellite altimeter data (bottom panel). Panels on the right show the zonally averaged values. Note that all data are departures from the global averaged rise.

6.2.3 Progress and Gaps in the Ocean-Observing System

While the 1990s observations overcame some of the sampling problems of the previous era, significant issues remained. The WOCE survey was global, but its one-time character meant that the XBT networks (Figure 6.1) bore the main responsibility for temporal sampling. The 750-m depth range of the T-7 XBT set the limit for the heat-content estimates of Willis et al. (2004) and Domingues et al. (2008). Deeper estimates such as Antonov et al. (2005) rely on the much sparser hydrographic data sets.

Many of these problems are being significantly reduced as a result of the Argo Project (Gould et al. 2004), in which an array of autonomous floats (each with a lifetime of 3–5 years) provide temperature and salinity profiles for the upper 2000 m of the global ocean using high-quality sensors (Figure 6.1d). As of early 2008, the Argo array reached the design level of about 3000 instruments. There has been near-global coverage since 2004. Argo is now producing over 100 000 globally distributed temperature/salinity profiles per year, mostly to depths of 2000 m, and with data accuracy an order of magnitude better than XBTs but not quite the same standard as research-ship observations. It does so with better spatial coverage than the WOCE survey and it greatly reduces the Northern Hemisphere and summertime biases inherent in all previous hydrographic data sets. However, users should be aware that near-real-time Argo data are subject to only crude quality checks, and some problems identified are now being corrected in delayed-mode processing. For example, the reported upper ocean cooling since 2003 (Lyman et al. 2006) was an artifact of the transition from XBT sampling (biased warm due to fall-rate error) to Argo sampling (biased cool by the erroneous reporting of pressure in some of the Argo floats; Willis et al. 2009). Barker et al. (unpublished work) identified further inconsistencies in the Argo data (metadata, quality control, drifts in pressure sensors) that can affect thermosteric sea-level estimates. Measures are in place to eradicate the inconsistencies from the Argo data set.

In the deep and abyssal waters, below the depth of the maximum Argo float depth (2000 m) and the maximum depth of the Antonov et al. (2005) analysis (3000 m), recent results have also shown rapid warming (Fukasawa et al. 2004; Johnson and Doney 2006; Johnson et al. 2007, 2008). The warming is largest in the South Atlantic, South Pacific, and southern Indian Oceans, closer to the locations where the properties of these waters were set by their interaction with the atmosphere and the cryosphere in the high-latitude Southern Ocean. The sparse sampling precludes global estimates at this stage, but we must consider the full ocean depth (which is, on average, more than 3500 m) when estimating sea-level rise. Information about the deep and abyssal ocean is also important for understanding the climate system's sensitivity to increasing greenhouse gas concentrations. Designing and implementing an adequate deep-ocean-observing system is a high priority. Similarly, improved observations in ice-covered regions, marginal seas, and coastal areas are required.

6.3 Estimating Steric Sea-Level Change Using Ocean Syntheses

Changes in ocean conditions can also be assessed using data-assimilation techniques. The Simple Ocean Data Analysis (SODA) model (1.2 and 1.4.2; Carton et al. 2005; Carton and Giese 2008) uses an inexpensive multivariate sequential data-assimilation approach in which the model is strongly forced toward observed ocean temperature and salinity from the World Ocean Data Base (Conkright et al. 2002 and later updates) and the Global Temperature-Salinity Profile Program (Wilson 1998). Ocean dynamics and other properties are not preserved in this approach, and the resultant steric estimates from 1968 to 2001 are similar to the direct observational estimates of the original Antonov et al. and Ishii et al. (Figure 6.2) and other similar analyses (Carton and Santorelli 2008). All these results appear to be affected by the time-dependent XBT fall-rate errors.

A more sophisticated approach to estimate the evolving state of the ocean is to synthesize the available data into a dynamically consistent model using the adjoint assimilation technique (for example Köhl et al. 2007). The model then carries the information, obtained locally in space and time, forward and backward in time over many years and decades. This allows, in principle, the possibility of inferring the ocean state and its changes even in locations remote from direct observation and of rejecting potentially spurious observations that are not dynamically consistent with other data or the surface forcing.

Relevant studies include the 50-year-long syntheses (Köhl and Stammer 2008), as well as many efforts covering the last 14 years (e.g. Köhl et al. 2007; Wunsch et al. 2007; Wenzel and Schröter 2007). The time series of thermosteric sea level of Köhl and Stammer (Figure 6.2) shows an initial decrease until about 1975 and then a larger rise reaching 1.8 mm/year (1.2 mm/year over the upper 750 m) from 1992 to 2001. From 1961 to 2001, the average rise is 0.92 mm/year, with 0.66 mm/year occurring in the upper 750 m. The largest difference between these results and the direct observations is the greater fall in the first 25 years and the larger rise in subsequent decades, including in the 1990s.

Figure 6.4 shows the associated regional linear trends of the steric sea level estimated from the 50-year results for 1962 through 2001 (Köhl and Stammer 2008). The patterns suggest major changes in the Southern Ocean and smaller changes in the tropical and subtropical oceans. However, note that these are early attempts at ocean reanalysis and that significant challenges remain. For example, the results of Köhl and Stammer (2008) imply an enormous and clearly unrealistic freshwater flux out of the ocean of 10 ± 13 mm/year for 1962–2002. Although there is a small impact on steric sea level compared with regional-scale changes, until an adequate global water cycle is included in these reanalyses estimated changes of the mass of the ocean will be unrealistic. Also, estimates of large-scale integral quantities are sensitive to initial model adjustments and results obtained from simulations commencing from 1993 are quite different during the first few years from those obtained from the 50-year estimate. The long memory of the system underlines the need for ocean synthesis efforts covering decades.

There are significant differences between the various reanalysis products and between these products and the direct observational results. Ocean reanalysis has only been an active research topic since the advent of WOCE in the 1990s. Over time, increasing fidelity of such ocean synthesis efforts should lead to the best possible basis for studies of decadal sea-level and heat-content changes.

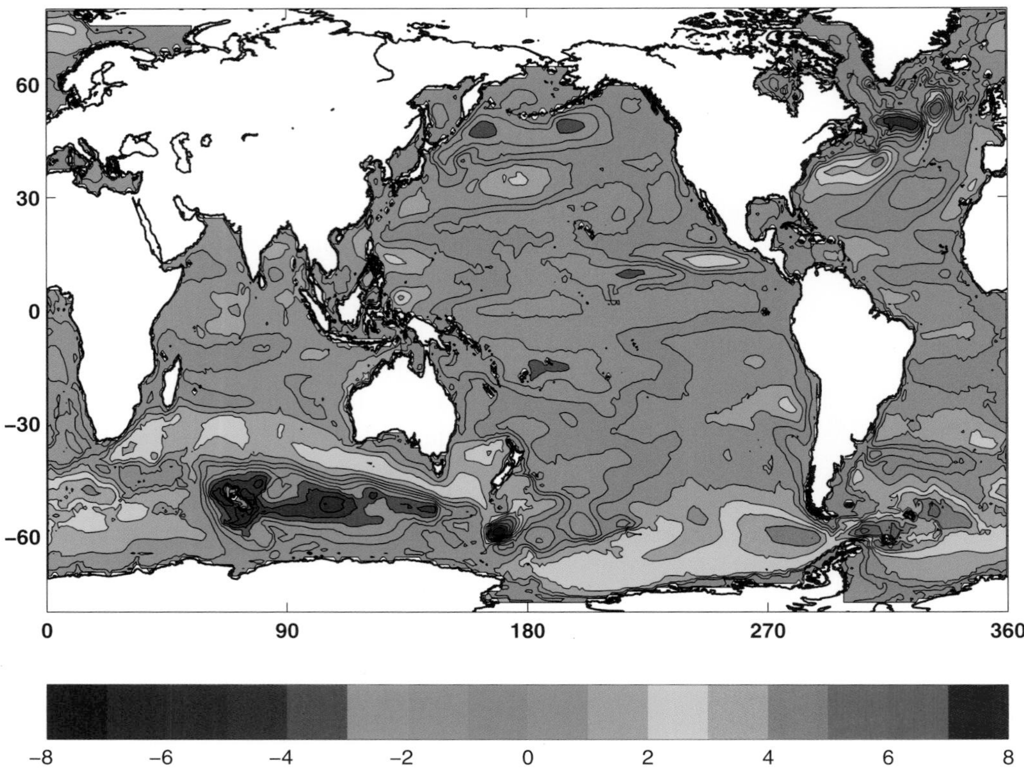

Figure 6.4 Sea-surface height trend in millimeters per year estimated from the data assimilation of Köhl and Stammer (2008) for the period 1962–2001.

6.4 Inferring Steric Sea Level from Time-Variable Gravity and Sea Level

The redistribution of water on the Earth causes temporal variations in the Earth's gravitational field (e.g. Wahr et al. 1998; Chapters 7 and 8). The Gravity Recovery and Climate Experiment (GRACE) satellite mission, launched in March 2002, has sufficient accuracy to resolve monthly movements of water on a spatial resolution of several hundred kilometers (Tapley et al. 2004).

GRACE ocean-mass estimates can be combined with sea-level measurements (of ocean volume; Chapter 5) to infer the steric component. This has been demonstrated for the seasonal variation (e.g. Chambers et al. 2004; Chambers 2006; Willis et al. 2008) and to estimate the seasonal exchange of water between the ocean and land (Figure 6.5), as previously calculated from combinations of sea-level data and hydrological models (Chen et al. 1998; Minster et al. 1999; Cazenave et al. 2000; Chapter 8). Unlike the previous studies, GRACE provides a direct

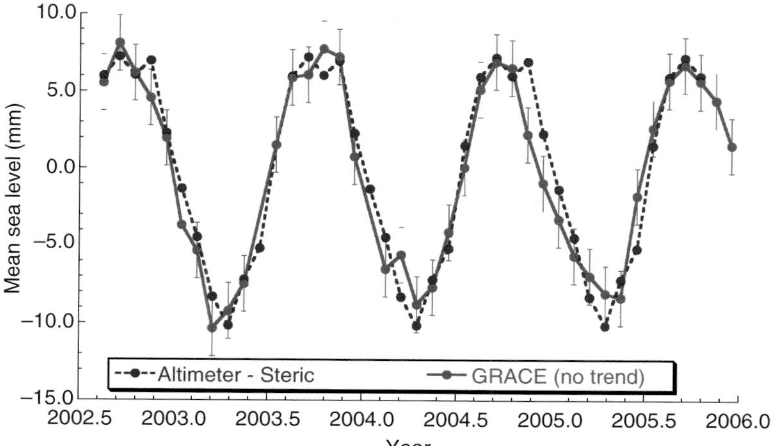

Figure 6.5 Mass component of sea level measured by GRACE (with 3-year trend removed; red), and the inferred seasonal climatological signal computed from 11 years of altimetry and steric sea level from the World Ocean Atlas 2001 (Chambers et al. 2004).

measure of changes in ocean mass. By comparing the GRACE gravity estimates with steric-corrected altimetry, Chambers et al. (2004) argued that the GRACE monthly ocean-mass estimates are accurate to about 1.8 mm equivalent sea level.

Initial attempts to estimate interannual steric sea-level variations (Lombard et al. 2007) found a positive altimeter/GRACE-derived steric sea-level trend between 2003 and 2005, in disagreement with the Lyman et al. (2006) estimates based on hydrographic data, again highlighting that the reported ocean cooling was an artifact of instrumental biases. Recently, Willis et al. (2008), Cazenave et al. (2009), and Leuliette and Miller (2009) tested the consistency of the three independent measurements of ocean mass and steric change. They used altimeter data for the total sea level, Argo data for the steric contribution, and gravity data for mass change. The seasonal signals of the detrended time series agreed within the error bars (Willis et al. 2008). However, Willis et al. found a significant inconsistency between the three measurements in the trend from mid-2003 to mid-2007, indicating a remaining systematic bias in one or more of the complementary observing systems. In contrast, Cazenave et al. used a different and larger glacial isostatic adjustment (GIA; the movement of the "solid" Earth in response to changes in the distribution of ice and water; see Chapter 10) from the Peltier and Luthcke (2009) GIA model, and managed to close the budget over the 2003–2008 period, highlighting the importance of accurate GIA estimates. Using a slightly later period (better Argo coverage), the same GIA correction as Willis et al. but different processing for the altimeter data, Leuliette and Miller (2009) also managed to close the sea-level budget. Note that all of these comparisons are for very short periods and that more rigorous and useful comparisons will be possible in several years when the time series are longer.

The long-term trends from the short GRACE data face several challenges. Although the instrumental error is only ±0.3 mm/year, significant uncertainty in the long-term rate estimate is also a result of uncertainty in GIA, geocenter changes, and interannual variability.

GIA causes an apparent decrease in the GRACE estimate of ocean mass trends that is related to movement of the "solid" Earth and unrelated to sea-level rise. The Paulson et al. (2007) GIA model, developed using the ICE-5G ice history (Peltier 2004) and a range of upper and lower mantle viscosities that produce good agreement with GRACE measurements over Hudson Bay and Fennoscandia, results in an estimated correction most likely between +1.0 and 1.5 mm/year (Willis et al. 2008), somewhat lower than the correction used by Cazenave et al. (2009) from the GIA model of Peltier and Luthcke (2009).

Ocean-mass variations measured by GRACE are sensitive to the position of the center of the Earth (geocenter) which GRACE does not measure. Seasonal geocenter variations can be estimated from other satellite tracking data (e.g. Cretaux et al. 2002), and are accurate enough to use in seasonal analysis of GRACE data (Chambers et al. 2004; Chambers 2006). Unfortunately, the geocenter interannual *trend* is poorly known as it is difficult to separate from vertical rates at the latitudinally asymmetric distribution of tracking stations used to compute the time series. Swenson et al. (2008) have recently proposed a method to estimate interannual and seasonal variations in geocenter from a combination of GRACE data and ocean models that holds promise to improve our knowledge, at least during the time of the GRACE mission.

Ocean-mass changes occur on all timescales with significant seasonal and interannual fluctuations in runoff, precipitation and evaporation. Nerem et al. (1999) estimated that seasonal exchange of water between the ocean and land-based reservoirs causes global averaged sea-level fluctuations of about ±8 mm. ENSO causes worldwide changes in precipitation and evaporation. Chambers et al. (2000), Willis et al. (2004), and Ngo-Duc et al. (2005) suggest a significant ocean-mass variation associated with the 1997 ENSO event, equivalent to about 5 mm over a 2–3 year period. These would imply the 95% confidence interval for trends of sea level for 3 years of data is about ±2.8 mm/year. Recent modeling studies (Landerer et al. 2008) suggest a somewhat smaller variation in ocean mass in response to ENSO events and would imply a smaller confidence interval.

If sufficiently long (decadal) time series are collected, these three uncertainties are small enough that GRACE and follow-on missions have the potential to measure the rate of ocean-mass change to the same order of accuracy as altimetry observes the total sea-level rise.

6.5 Modeling Steric Sea-Level Rise

Significant efforts have been made to simulate 20th-century climate and project 21st-century climate using coupled AOGCMs. These studies have been coordinated by the World Climate Research Programme and used in the Assessments of the Intergovernmental Panel on Climate Change (IPCC). The models are also used to understand the factors controlling sea-level change and to estimate projections of global mean and regional sea-level change.

Simple climate models and Earth system models of intermediate complexity are also used to estimate ocean thermal expansion. The most widely used simple climate models (MAGICC; see http://www.cgd.ucar.edu/cas/wigley/magicc/) represent the ocean as a one-dimensional (vertical) model in which heat diffuses vertically into the ocean. These models are generally tuned to represent AOGCMs or observed ocean time series. The prime value of simple climate models and Earth system models of intermediate complexity is to explore the sensitivity of sea-level-rise estimates to various climate parameters and to estimates sea-level rise (and other climate variables) for a broader range of greenhouse gas scenarios than is possible with AOGCMs (see for example Church et al. 2001). We will not discuss these models further here.

A number (although not all) of the most recent AOGCM simulations do not rely on artificial surface flux adjustments to yield a stable climate, as required in earlier simulations (e.g. Bryan 1996). As the steric sea-level anomalies depend directly on the surface fluxes, doing without artificial flux adjustments renders estimates more robust. However, many (if not all) of the models still drift at rates up to about 1 mm/year (Katsman et al. 2008) in the global average, as a result of slowly changing temperature and salinity fields of the deep ocean. The departure of the local trends from the global average can be of similar magnitude to the global averaged trends, especially at high latitudes (D. Monselesan, personal communication). To remove this residual drift, model estimates of sea-level change from natural and anthropogenic forcing of climate for the 20th and 21st centuries are usually compared with a simulation in which these time-variable forcing factors are not included (the control simulation). As the thermal expansion coefficient is a function of temperature, the thermal expansion anomalies computed from coupled AOGCMs depend (weakly) not only on the offset relative to the real ocean but also on the drift rate of the control climate. Furthermore, the inherent assumption of linear separability of climate signals and model drift when calculating anomalies of scenarios relative to the control climates may introduce (probably small) errors. Gille (2004) has also pointed to the sensitivity of global averaged sea-level change to inaccurate isopycnal and vertical diffusion parameters in models, but suggested the effect is small.

The spatial variability of steric sea-level change is linked to ocean dynamic processes, in particular to the redistribution of heat and salt horizontally and vertically through air–sea exchange and the ocean circulation. To the extent that the various AOGCMs differ in their ability to simulate the respective processes, the simulated global mean as well as regional sea level will differ, as discussed below.

Geophysical processes associated with the changing distribution of mass of the Earth system (for example, loss of mass from ice sheets and the resultant change to the Earth's gravity field and the associated crustal motion; Mitrovica et al. 2009) will also affect the regional distribution of sea-level rise. These processes must be accounted for separately to the climate factors discussed here. See Chapter 10 for further discussion of these effects.

6.5.1 Comparison of Observed and Modeled Global Averaged Thermosteric Sea-Level Rise

A number of studies (for example Levitus et al. 2001; Barnett et al. 2001, 2005; Gent and Danabasoglu 2004; Hansen et al. 2005) have shown that the observed rate of ocean warming since 1950 (and by implication thermosteric sea-level rise) is not a result of natural variability alone and that increases in greenhouse gas concentration have contributed significantly. Using 500-year simulations, Gregory et al. (2006) show that the rate of thermosteric rise is larger during the 20th century than previous centuries as a result of anthropogenic forcing and show that the combination of natural and anthropogenic forcing is critical for simulating the evolution of thermosteric sea-level rise during the 20th and 21st centuries.

For 1961–2000, the models, none of which assimilates observed ocean data, approximately reproduce the most recent thermosteric sea-level variability in the upper 700 m when the models are forced by all climate forcing factors, including time-variable volcanic and solar forcing (Figure 6.6; Domingues et al. 2008). Although there are considerable differences between the simulations, on average the multidecadal trends are smaller than observed. The models without volcanic

(a)

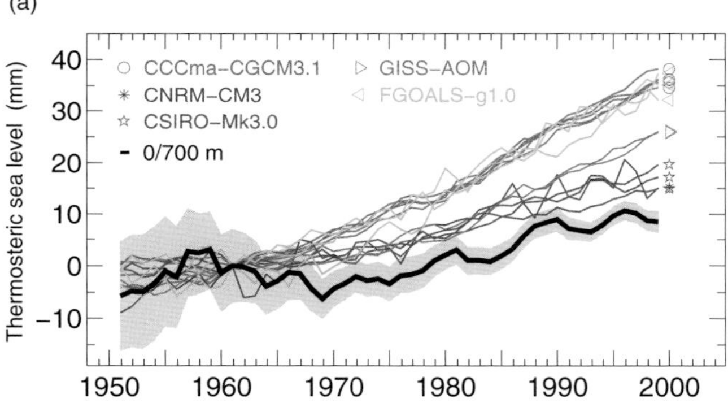

(b)

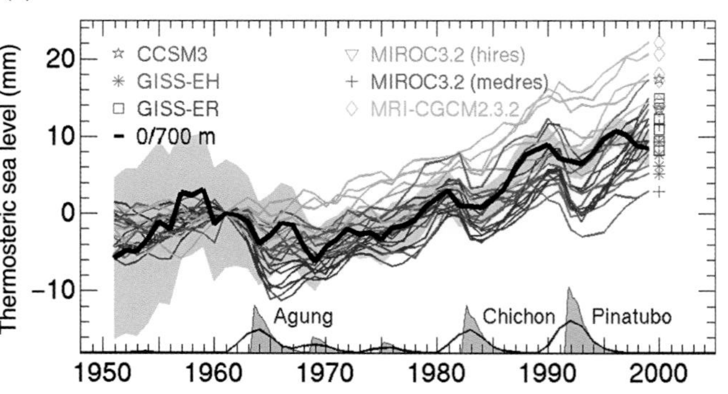

Figure 6.6 Observed and modeled global averaged ocean thermal expansion for the upper 700 m from 1950 to 2000. The models in (a) are those that include time-variable greenhouse gases, aerosols, and solar but no volcanic forcing. The models in (b) also include volcanic forcing. The observations are from Domingues et al. (2008; black with 1σ error estimates). The stratospheric loadings from major volcanic eruptions are indicated along the bottom with the brown curve a 3-year running average of these values. See Domingues et al. (2008) for more details.

forcing have less variability than observations and substantially larger trends. The comparisons clearly demonstrate the importance of volcanic eruptions (indicated on the bottom panel in Figure 6.6) for ocean heat content and thermosteric sea-level variability. The smaller trends in the models with volcanic forcing are consistent with cold anomalies persisting in the ocean for many decades after the eruptions (Delworth et al. 2005; Glecker et al. 2006a, 2006b; Gregory et al. 2006) and thus slowing the rate of 20th-century sea-level rise and partially masking any acceleration in sea-level rise (Church and White 2006). The models do not support the pronounced warming in the 1970s (Figure 6.2) and subsequent cooling in the 1980s present in the early analyses and that recent observational analyses now suggest is spurious (Wijffels et al. 2008; Domingues et al. 2008; Ishii and Kimoto 2009; Levitus et al. 2009). The approximate agreement between the modeled and observed estimates of ocean thermal expansion opens up the possibility of using the observations to constrain projected sea-level rise by either choosing between different models or the weighting of model results. This is likely to be an avenue pursued in the fifth IPCC Assessment.

6.5.2 Projections of Steric Sea-Level Change

Projected steric sea-level rise is a function of changes in atmospheric greenhouse gas concentrations and other climatic forcings and varies considerably between models. Comparing the last decade of the 21st century with 1980–2000, the global mean thermal expansion is estimated to be from 10 to 24 cm for the B1 SRES greenhouse gas scenario, 13 to 32 cm for the A1B scenario, and 17 to 41 cm for the A1FI scenario (Meehl et al. 2007). The climate sensitivity of the coupled models and their ocean heat-uptake efficiency (Raper et al. 2002) are important factors contributing to the differences between models. Gregory and Forster (2008) found the climate sensitivity and the heat-uptake efficiency were independent of each other in the models used in the 2007 IPCC Assessment. Since the ocean's thermal inertia is large, the ocean (especially the deeper layers) will continue to take up heat (resulting in steric sea-level rise) for centuries and even millennia after greenhouse gas concentrations have stabilized in the atmosphere. Contributions to steric sea level could eventually reach several times the value at the time when greenhouse gas concentrations are stabilized (Meehl et al. 2007). For example, for simulations with concentrations stabilized at the A1B level in 2100, thermal expansion during the 22nd century will be of similar magnitude to that for the 21st century and by 2300 could be in the range 30–80 cm (Meehl et al. 2005, 2007).

The steric sea-level changes in climate models and observations are not spatially uniform and some regions are projected to experience more than twice the global average rate of steric rise (e.g. Gregory et al. 2001; Gregory and Lowe 2000; Lowe and Gregory 2006; Meehl et al. 2007; Suzuki et al. 2005; Landerer et al. 2007a; Figure 6.7). Though different models predict different distributions, there are some common features. Many models show a strong sea-level rise in the Arctic

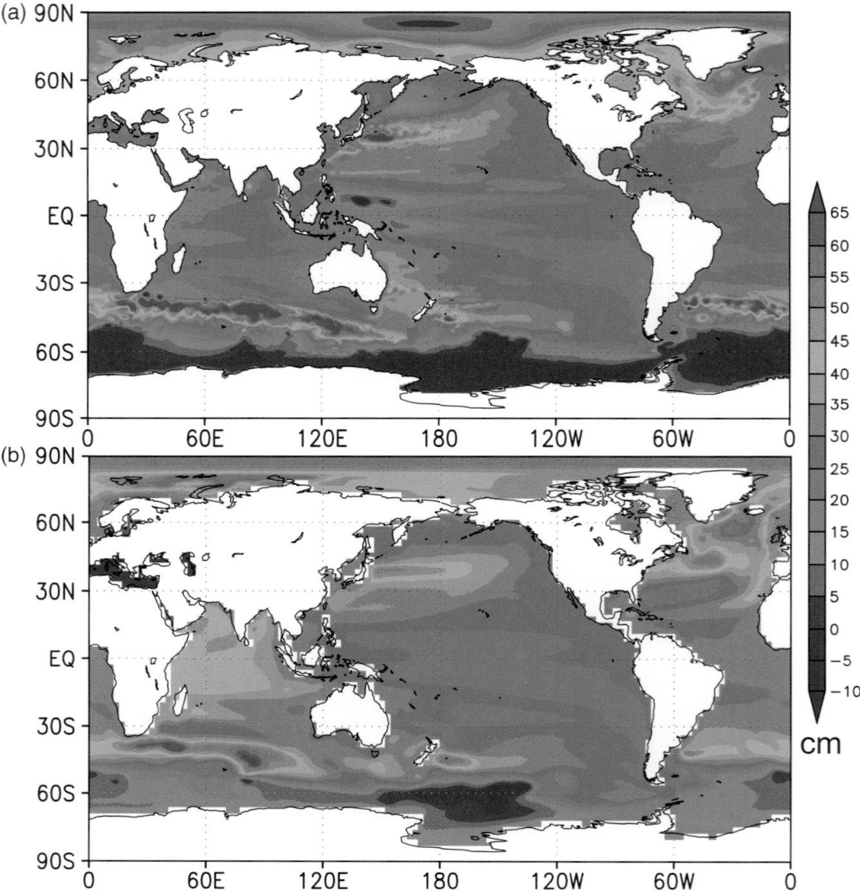

Figure 6.7 The changes in mean steric sea-surface height (2080–2100 mean minus the 1980–2000 mean) for the A1B scenario in (a) MIROC3.2_hi and (b) MIROC3.2_med (Suzuki et al. 2005). See text for abbreviations.

Ocean. Landerer et al. (2007a) found this is largely due to enhanced freshwater input from precipitation and river runoff in the northern high latitudes and a subsequent change of the density structure in this region. In the Southern Ocean, local steric sea-level rise is larger than the global average in a band running along the poleward edge of the subtropical gyre (there is a similar feature in the Northern Hemisphere) and smaller than the average south of the Antarctic Circumpolar Current. There is little improvement in agreement between models assessed in the IPCC Fourth Assessment Report (AR4) compared with the IPCC Third Assessment Report (TAR) and significant and inadequately understood differences remain between the various model projections (Pardaens et al. 2010).

6.5.3 *Higher-Resolution Model Estimates of Steric Sea-Level Rise*

Recent increases in computing power have enabled long-time integrations of higher-resolution climate models, thus permitting better (but not fully resolving) representation of bottom topography and ocean meso-scale structures such as western boundary currents, eddies, fronts, and deep-water formation. As a result, there is potential for improved representation of processes contributing to ocean thermosteric sea-level rise such as subduction of water masses, the large spatial-scale variability in ocean mixing, and deep-water formation.

The high-resolution version of the Model for Interdisciplinary Research on Climate version 3.2 (MIROC3.2_hi) has the highest resolution of all the models used in the AR4. This model (K-1 Model Developers 2004) consists of a T106 atmospheric spectral model (about 1.1° resolution at the equator) and an eddy-permitting ocean model with resolution of 0.28° zonally and 0.19° meridionally, and 48 vertical levels. Because of the greatly increased computing resources required, only one realization of the high-resolution model is available. The medium resolution version (MIROC3.2_med) has similar resolution to most other AR4 models.

The steric contribution during the 21st century is projected to be about 30 cm for the A1B scenario and 23 cm for the B1 scenario in the MIROC3.2_hi, similar to those for the MIROC3.2_med and within but near the top of the range of estimates from other models (Suzuki et al. 2005). However, the total heat flux into the ocean in the MIROC3.2_hi is larger than that in the MIROC3.2_med during the early 21st century and smaller after the middle of the 21st century. The upper ocean in the MIROC3.2_hi warms earlier than that in the MIROC3.2_med and the warming of the deeper ocean is smaller in the MIROC3.2_hi than in the MIROC3.2_med. These differences in ocean heat uptake may partly reflect the inability of coarse-resolution models to simulate the upper ocean stratification and the vertical distribution of ocean-heat uptake and/or the difference in strength of the meridional overturning circulation between the two models: the Atlantic meridional overturning circulation is weakened from 14 to 9 Sv in the MIROC3.2_hi and from 19.5 to 12.5 Sv in the MIROC3.2_med during the 21st century. It also may be related to the difference in climate sensitivity between the two models, with MIROC3.2_hi having a higher sensitivity than MIROC3.2_med.

The broad-scale features of regional sea-level rise of the coarse resolution models are similar in both models (Figure 6.7). However, in MIROC3.2_hi the representation of finer-scale features means the magnitudes of changes are more pronounced and more confined to specific areas than in the MIROC3.2_med and other coarse-resolution models. For example, in the Arctic Ocean enhanced sea-level rise is confined to the coastal region. In the Kuroshio Extension there is a reduced sea-level rise north of the Kuroshio at approximately 150°E and an enhanced sea-level rise to the south. This change is associated with the acceleration of the Kuroshio caused by changes in wind stress and the consequential spin-up of the Kuroshio recirculation (Sakamoto et al. 2005). In contrast, the Kuroshio in the MIROC3.2_med overshoots to the north, so the region of large sea-level rise in the MIROC3.2_med extends northward relative to that in the MIROC3.2_hi. Improving the resolution of these regional features and their representation on continental shelves and in semi-enclosed seas will be important in understanding the impacts of sea-level rise. As yet these issues have received little attention.

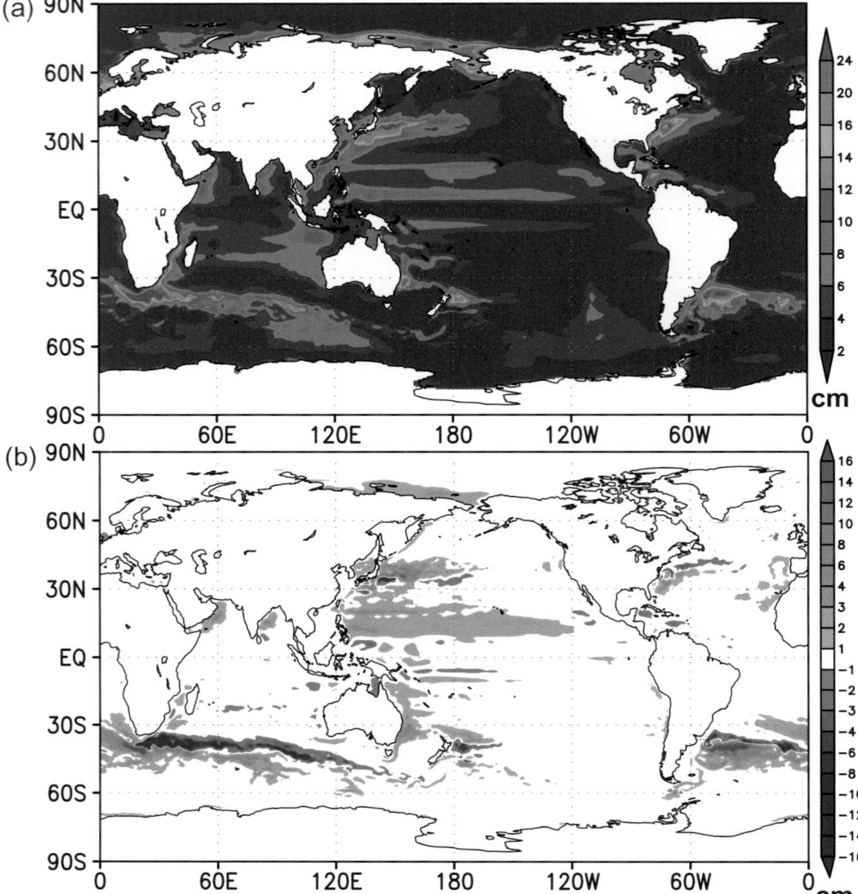

Figure 6.8 Modeled sea-surface height eddy variability. (a) The root mean square (rms) of the sea level anomaly from the 3-month running mean for the control run in MIROC3.2_hi. (b) Changes in the rms between 1980–2000 and 2080–2100 (A1B scenario) in MIROC3.2_hi (Suzuki et al. 2005).

Eddy activity is enhanced during the 21st century in MIROC3.2_hi (Figure 6.8). While the global averaged change in the rms height variability is small compared to the global averaged rise, enhanced eddy activity is confined to specific areas, and these areas overlap with the areas of large sea-level rise around some coastal regions and islands, suggesting that both changes in mean sea level and changes in eddy variability may increase the frequency and intensity of extreme sea levels in those regions during the 21st century. An example of the impact of eddies on islands is the flooding of Okinawa Island on 22 July 2001 when a warm eddy increased sea level by more than 15 cm (Tokeshi and Yanagi 2003).

6.5.4 Ocean Processes of Sea-Level Change

Coupled AOGCMs provide the opportunity to focus on the mechanisms behind global averaged and regional sea-level changes to understand better how heat enters the ocean, to understand how it is distributed, and to understand model differences (e.g. Lowe and Gregory 2006; Landerer et al. 2007a; Levermann et al. 2005; Suzuki et al. 2005).

In a series of carbon dioxide-doubling experiments, Lowe and Gregory (2006) found global mean-sea-level rise is dominated by the thermal expansion of the ocean, with changes in the salinity distribution (e.g. from changes of the hydrological cycle) having a negligible contribution to the global mean. Interior temperature changes cannot be explained solely by passive tracer transports along isopycnals with no changes in ocean circulation (Banks and Gregory 2006). Instead, the heat is redistributed by changes in the ocean circulation, and in the deeper layers there is a significant diapycnal component (Banks and Gregory 2006). The changes in regional sea level are primarily a result of density (baroclinic) changes in the ocean rather than a change in the barotropic circulation (Lowe and Gregory 2006). Temperature change is the largest contribution to these density changes and is usually positive. However, the salinity changes also make a significant local contribution, often opposing the temperature changes (see also Landerer et al. 2007a; Figure 6.9). Changes in surface wind stress were most important for determining the regional distribution of sea-level rise but changes in surface fluxes of heat and fresh water were important regionally (Lowe and Gregory 2006) and according to Köhl and Stammer (2008) buoyancy fluxes have become increasingly important for regional changes during 1992–2001. In the high-latitude Southern Ocean, the relatively small sea-level rise was related to the small thermal expansion there. In contrast, Landerer et al. (2007a) emphasized the increased wind stress, leading to a stronger Antarctic Circumpolar Current transport and a subsequent dynamic sea-surface height adjustment in the model (Figure 6.9). However, note that recent observations suggest that the strengthened winds have not resulted in a strengthening of the Antarctic Circumpolar Current (Böning et al. 2008) and in a quasi-geostrophic model the increased winds lead to an intensification of the eddy field on interannual timescales rather than a strengthening of the current (Hogg et al. 2008).

Levermann et al. (2005) find a linear scaling of 4.5–5 cm/Sv between sea level at the North American coast and the maximum North Atlantic meridional overturning circulation. However, while Landerer et al. (2007a) show that the sea-surface height difference between Bermuda and the Labrador Sea correlates highly at zero lag with the combined interannual to decadal North Atlantic gyre transport changes, changes in the overturning circulation cannot be reliably inferred from sea-surface height. Landerer et al. (2007a) report that the basin-integrated sea-surface height difference of 0.78 m between the North Atlantic and North Pacific Ocean is reduced by only 0.06 m when the North Atlantic meridional overturning circulation is reduced by 25% in their simulation, and is re-established within 100 years through a Pacific Ocean sea-surface height rise and a North Atlantic sea-surface height drop, without an analogous recovery of the North Atlantic overturning.

The vertical distribution of thermosteric and halosteric anomalies that contribute to sea-level change is very different between ocean basins. In the North Atlantic, the steric anomalies reach to depths of the North Atlantic Deep Water

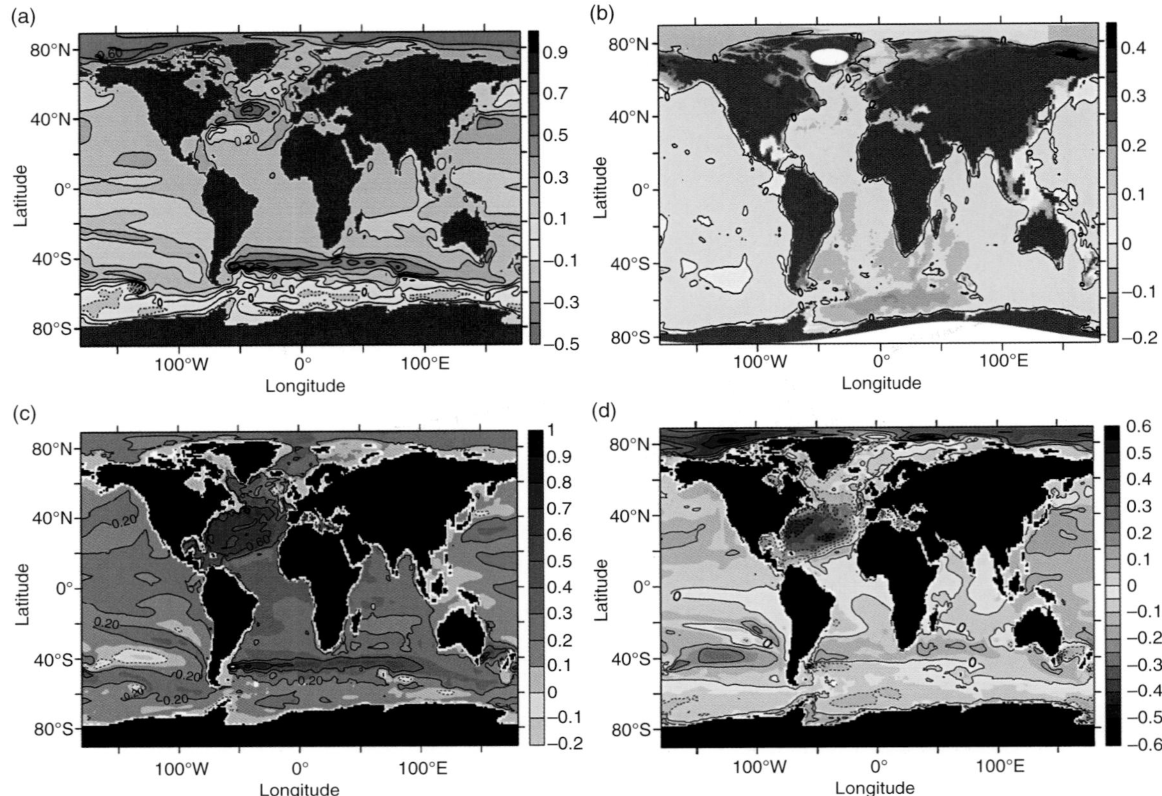

Figure 6.9 (a) Simulated sea-level changes for the period 2090–2099 relative to the pre-industrial state; (b) mass-redistribution component; (c) thermosteric contribution; and (d) halosteric contribution. No external mass source (e.g. melting glaciers) is included in the simulation (Landerer et al. 2007a, 2007b). All data are in meters. (© American Meteorological Society.)

(2000 m), whereas steric anomalies in the entire Pacific Ocean occur mainly in the upper 500 m. In the Southern Ocean, steric anomalies occur throughout the entire water column, reflecting the deep structure of the Antarctic Circumpolar Current and possibly the strong vertical exchange of buoyancy in this region (Figure 6.10; Landerer et al. 2007a). Whether or not the thermosteric and the halosteric anomalies are additive or density-compensating also varies between ocean basins and latitude bands (Levitus et al. 2005b).

Although steric sea-level changes occur at constant global ocean mass, mass redistribution within the global ocean causes bottom pressure to increase across shallow shelf areas following ocean thermal expansion (Figure 6.9). Landerer et al. (2007a, 2007b) estimate that bottom pressure increases by up to 0.4 m sea-level height equivalent in their simulation. Due to the laterally varying distribution of shelf areas, these surface mass-loading anomalies directly affect the geoid and thus

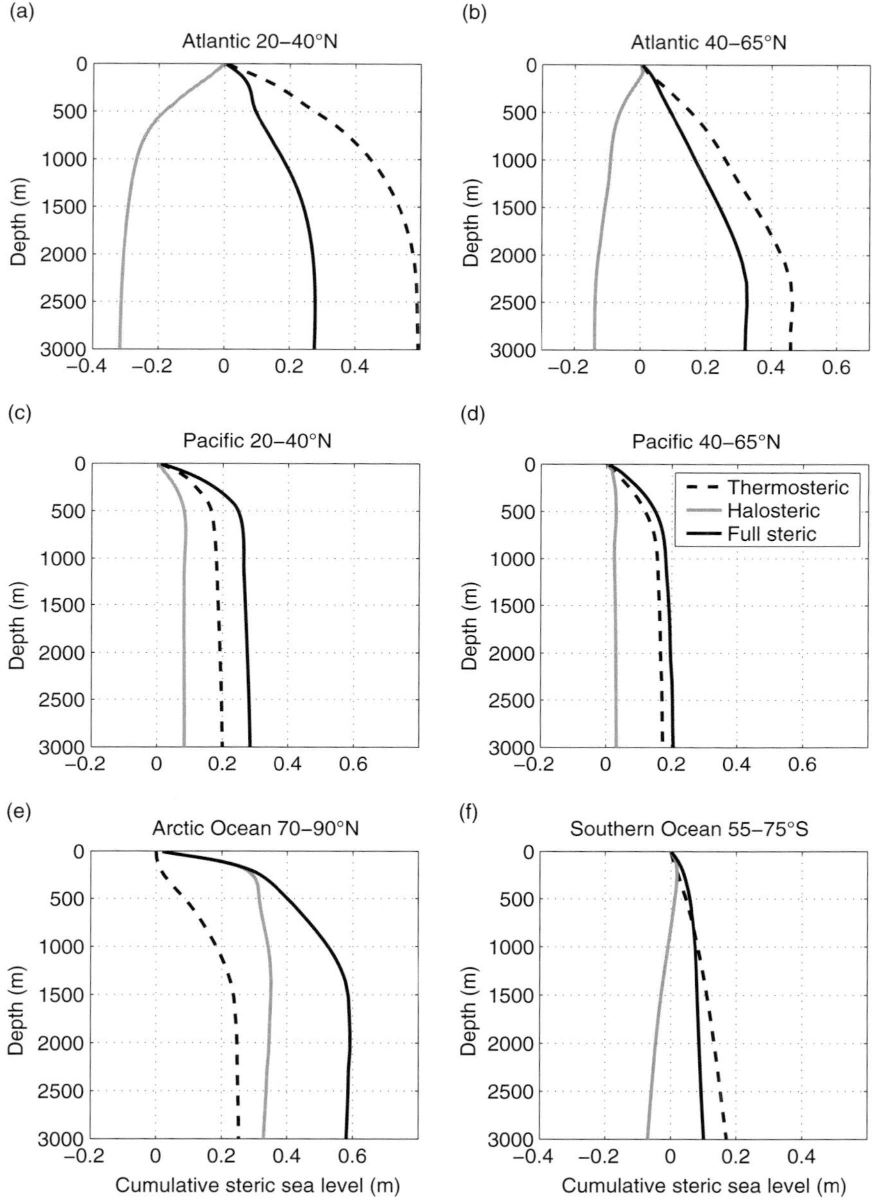

Figure 6.10 Cumulative sum of thermosteric (dashed line), halosteric (gray line), and total steric (black line) anomalies for different ocean areas. Starting at the surface, the steric anomaly from each depth layer is accumulated. Note that the abscissa has the same width in all plots (Landerer et al. 2007a). (© American Meteorological Society.)

relative sea-level rise, and also Earth orientation parameters, for example polar motion and length of day (Landerer et al. 2007b, 2009).

6.6 Conclusions and Recommendations

Significant progress has been made during the past 20 years in observing and understanding the seasonal to decadal variability and the multidecadal trends in global-ocean heat content and steric sea-level change. Adding to the historical database through data archaeology efforts has contributed enormously (Levitus et al. 2005d). However, during the 1950s and 1960s vast regions of the oceans went unsampled. Since then, greater interest in climate issues and the open ocean, and advances in measurement technology, have resulted in a tremendous evolution of the ocean-observing system. The combination of high-precision satellite altimetry and global upper-ocean observations of the Argo Project (Gould et al. 2004) means the sampling and instrumental problems of the 1950s to 1990s are now being significantly reduced for the upper-ocean heat and salt distributions.

Comparisons of different data sets and the use of different statistical approaches have helped address and partially overcome the sampling (spatial/temporal) and instrumental biases for the upper 700 m of the ocean. A recent estimate of the trend in thermosteric sea-level rise from 1961 to 2003 is larger than earlier estimates, as much as 0.52 ± 0.08 mm/year for the upper 700 m (Domingues et al. 2008). There are indications of a significant but poorly quantified deep and abyssal ocean contribution.

There has also been tremendous progress in the development of AOGCMs. The agreement between the models and observations of both the variability and trends in the upper ocean has improved (Domingues et al. 2008). However, substantial differences remain between the observations and the model estimates for the 20th century and between the various model projections for the 21st century.

There is not, as yet, convergence in assessments and projections of the regional distribution of sea-level rise (Church et al. 2004; Llovel et al. 2009; Gregory et al. 2001; Pardaens et al. 2010). An emerging issue is the importance of increased spatial resolution, including resolving continental shelves and semi-enclosed seas. The only eddy resolving model used in the last IPCC Assessment has important differences to its lower-resolution equivalent and has pointed to the potential importance of changes in the amplitude of eddy activity in some regions and its impact on ocean circulation and coastal flooding.

Our present thermosteric sea-level-observing elements of Argo for upper-ocean thermal expansion, the Jason series of high-quality satellite altimeter missions for ocean volume, and GRACE to separately observe changes in ocean mass are complementary. Argo has virtually complete coverage of the upper 1000–2000 m of the ice-free oceans. The high-quality temperature, conductivity, and pressure instruments used on Argo floats are being carefully calibrated and carry an accurate and stable thermistor. Small drifts in pressure transducers are being monitored, corrected, and reduced with improved sensors. High-quality salinity measurements are made by all floats, and in most cases are free of significant drift for 2 years or longer. Argo should ultimately (with the application of careful quality-control measures) have an error in 12-month upper-ocean global means of the order of 1 mm (Willis et al. 2004). For altimetric height, errors in multiyear global mean-sea-level trends can be as low as 0.4 mm/year (Leuliette et al. 2004). Given typical interannual variability of 1–3 mm/year for each of these quantities, the decomposition of sea level into upper-ocean steric and ocean-mass components should be adequately observed on annual, interannual, and longer timescales. Moreover, as noted above the Argo salinity measurements will provide an additional constraint on the upper ocean's freshwater budget, and consequently on ocean-mass change. Comparison of the trends of these complementary observations, as attempted by Willis et al. (2008), Cazenave et al. (2009), and Leuliette and Miller (2009), will be critically important for identifying remaining deficiencies and biases in our observation.

While significant progress has been made in observing the global oceans, clear deficiencies remain. The historical record is plagued by insufficient data, particularly in the deep and abyssal and ice-covered ocean. It is important that further efforts be made to add to the historical database through data archaeology and develop improved analysis of this data, including data assimilation. For practical reasons the Argo array is presently limited to the upper 2000 m of the oceans and to the ice-free regions, and there are relatively few floats in marginal seas. Instrument development is already mitigating these problems. Design work is needed to define the sampling requirements and techniques for observing the deep ocean. Implementation of the deep-ocean-observing system will require new resources. These deep-ocean issues are important for the very long timescales associated with steric sea-level rise and have important implications for policy decisions regarding appropriate greenhouse gas stabilization levels.

We also need to pay much greater attention to understanding the temporal and regional distribution of steric sea-level rise in both observations and models, including the presence of ocean eddies.

In addition to understanding deep-ocean steric sea-level rise and its temporal and spatial distribution, we need to understand how deep-ocean conditions impact the continental shelf and coastal sea levels. This includes both the direct effects of regional sea-level rise and also remotely forced sea-level perturbations that may be transmitted over large distances in the coastal (and equatorial) wave guides. The integrated framework for consideration of the impact of offshore phenomena on coastal sea level, including systematic observations in the coastal zone, is incomplete.

Recommendations for further observational activities and numerical model studies are:

- expand the historical database through data archaeology and improve the quality control of these data;

- sustain the Argo observational project, and extend the Argo coverage to marginal seas and ice-covered regions;
- design and implement appropriate observational strategies for the coastal region and the deep ocean (below the depth of the Argo floats);
- maintain the highest-quality satellite altimeter and time-variable gravity observations,
- improve and apply new techniques for reanalysis of the historical ocean database of 1950 to present, including statistical techniques and modern, robust data assimilation techniques;
- improve understanding, detection, and attribution of past steric sea-level rise on a range of spatial scales;
- reduce uncertainty in projections of 21st century steric sea-level rise through both model improvement and better use of observational constraints; and
- investigate high-resolution (including shelf sea/coastal) steric changes and produce better regional projections of steric sea-level change.

Acknowledgments

This chapter is a contribution to the CSIRO Climate Change Research Program and Wealth from Oceans Flagship and was supported by the Australian Government's Cooperative Research Centres Program through the Antarctic Climate and Ecosystems Cooperative Research Centre. JAC, CMD, and NJW were partly funded by the Australian Climate Change Science Program. The GECCO results of the KlimaCampus, University of Hamburg, were supported by the BMBF-funded projects "North Atlantic" and "GOCE-GRANDII". DR and JG were supported by the National Aeronautics and Space Administration (NASA) Ocean Surface Topography Science Team through JPL contract 961424, and by US Argo through NOAA grant NA17RJ1231. FWL was supported by the International Max Planck Research School on Earth System Modelling (IMPRS-ESM). We acknowledge the international modeling groups for providing their data for analysis, the Program for Climate Model Diagnosis and Intercomparison (PCMDI) for collecting and archiving the model data, the JSC/CLIVAR Working Group on Coupled Modelling (WGCM), and their Coupled Model Intercomparison Project (CMIP) and Climate Simulation Panel for organizing the model data-analysis activity, and the IPCC WG1 TSU for technical support. The IPCC Data Archive at Lawrence Livermore National Laboratory is supported by the Office of Science, US Department of Energy. DPC was supported under grants from NASA's GRACE Science Team and Interdisciplinary Science Team.

References

Antonov J.I., Levitus S. and Boyer T.P. (2002) Steric sea level variations during 1957–1994: importance of salinity. *Journal of Geophysical Research* **107**, 8013.

Antonov J., Levitus S. and Boyer T.P. (2005) Thermosteric sea level rise, 1955–2003. *Geophysical Research Letters* **32**, L12602.

Banks H.T. and Gregory J.M. (2006) Mechanisms of ocean heat uptake in a coupled climate model and the implications for tracer based predictions of ocean heat uptake. *Geophysical Research Letters* **33**, L07608.

Barnett T.P. (1983) Long-term changes in dynamic height. *Journal of Geophysical Research* **88**, 9547–52.

Barnett T.P., Pierce D.W. and Schnur R. (2001) Detection of anthropogenic climate change in the world's oceans. *Science* **292**, 270–4.

Barnett T.P., Pierce D.W., AchutaRao K.M., Gleckler P.J., Santer B.D., Gregory J.M. and Washington W.M. (2005) Penetration of human-induced warming into the World's oceans. *Science* **309**, 284–7.

Bernstein R. and White W. (1979) Design of an oceanographic network in the mid-latitude North Pacific. *Journal of Physical Oceanography* **9**, 592–606.

Bindoff N. and Church J. (1992) Warming of the water column in the southwest Pacific Ocean. *Nature* **357**, 59–62.

Bindoff N., Willebrand J., Artale V., Cazenave A., Gregory J.M., Gulev S. et al. (2007) Observations: ocean climate change and sea level. In: *Climate Change 2007: The Physical Science Basis.* Contribution of Working Group 1 to the Fourth Assessment Report of the Intergovernmental Panel on Climate Change (Solomon S., Qin D., Manning M., Marquis M., Averyt K., Tignor M.M.B. et al., eds), pp. 385–432. Cambridge University Press, Cambridge.

Böning C.W., Dispert A., Visbeck M., Rintoul S. and Schwarzkopf F.U. (2008) The Response of the Antarctic Circumpolar Current to recent climate change. *Nature Geoscience* **1**, 864–9.

Bryan K. (1996) The steric component of sea level rise associated with enhanced greenhouse warming: a model study. *Climate Dynamics* **12**, 545–55.

Carton J.A. and Giese B.S. (2008) A reanalysis of ocean climate using Simple Ocean Data Assimilation (SODA). *Monthly Weather Review* **136**, 2999–3017.

Carton J.A. and Santorelli A. (2008) Global decadal upper ocean heat content as viewed in nine analyses. *Journal of Climate* **21**, 6015–35.

Carton J.A., Giese B.S. and Grodsky S.A. (2005) Sea level rise and the warming of the oceans in the Simple Ocean Data Assimilation (SODA) ocean reanalysis. *Journal of Geophysical Research* **110**, C09006.

Cazenave A., Remy F., Dominh K. and Douville H. (2000) Global ocean mass variations, continental hydrology and the mass balance of Antarctica ice sheet at seasonal timescale. *Geophysical Research Letters* **27**, 3755–8.

Cazenave A., Dominh K., Guinehut S., Berthier E., Llovel W., Ramillien G., Ablain M. and Larnicol G. (2009) Sea level budget over 2003–2008: a reevaluation from GRACE space gravimetry, satellite altimetry and Argo. *Global Planetary Change* **65**, 83–8.

Chambers D.P. (2006) Observing seasonal steric sea level variations with GRACE and satellite altimetry. *Journal of Geophysical Research* **111**(C3), C03010.

Chambers D., Chen J., Nerem R. and Tapley B. (2000) Interannual Mean Sea Level Change and the Earth's Water Mass Budget. *Geophysical Research Letters* **27**, 3073–6.

Chambers D.P., Wahr J. and Nerem R.S. (2004) Preliminary observations of global ocean mass variations with GRACE. *Geophysical Research Letters* **31**, L13310.

Chen J.L., Wilson C.R., Chambers D.P., Nerem R.S. and Tapley B.D. (1998) Seasonal global water mass balance and mean sea level variations. *Geophysical Research Letters* **25**, 3555–8.

Church J.A. and White N.J. (2006) A 20th century acceleration in global sea-level rise. *Geophysical Research Letters* **33**, L01602.

Church J.A., Gregory J.M, Huybrechts P., Kuhn M., Lambeck K., Nhuan M.T., Qin D. and Woodworth P.L. (2001) Changes in Sea Level. In: *Climate Change 2001: The Scientific Basis*. Contribution of Working Group 1 to the Third Assessment Report of the Intergovernmental Panel on Climate Change (Houghton J.T., Ding Y., Griggs D.J., Noguer M., van der Linden P.J., Dai X. et al., eds), pp. 639–94. Cambridge University Press, Cambridge.

Church J.A., White N.J., Coleman R., Lambeck K. and Mitrovica, J.X. (2004) Estimates of the regional distribution of sea-level rise over the 1950 to 2000 period. *Journal of Climate* **17**, 2609–25.

Church J.A., White N.J. and Arblaster J.M. (2005) Significant decadal-scale impact of volcanic eruptions on sea level and ocean heat content. *Nature* **438**, 74–7.

Church J.A., White N.J., Aarup T., Wilson W.S., Woodworth P.L., Domingues C.M., Hunter J.R. and Lambeck K. (2008) Understanding global sea levels: past, present and future. *Sustainability Science* **3**, 9–22.

Conkright M.E., Antonov J.I., Baranova O.K., Boyer T.P., Garcia H.E., Gelfeld R. et al. (2002) *World Ocean Database 2001, vol. 1. Introduction [CD-ROM], NOAA Atlas NESDIS*, vol. **42** (Levitus S., ed.). Government Printing Office, Washington DC.

Crétaux J.-F., Soudarin L., Davidson F.J.M., Gennero M.-C., Berge-Nguyen M. and Cazenave A. (2002) Seasonal and interannual geocenter motion from SLR and DORIS measurements: Comparison with surface loading data. *Journal of Geophysical Research* **107**, 2374.

Davis R., Sherman J.T. and Dufour J. (2001) Profiling ALACEs and other advances in autonomous subsurface floats. *Journal of Atmospheric and Oceanic Technology* **18**, 982–93.

Delworth T.L., Ramaswamy V. and Stenchikov G.L. (2005) The impact of aerosols on simulated ocean temperature and heat content in the 20th century. *Geophysical Research Letters* **32**, L24709.

Domingues C.M., Church J.A., White N.J., Gleckler P.J., Wijffels S.E., Barker P.M. and Dunn J.R. (2008) Improved estimates of upper-ocean warming and multi-decadal sea-level rise. *Nature* **453**, 1090–3.

Ezer T., Mellor G.L. and Greatbatch R.J. (1995) On the interpentadal variability of the North Atlantic Ocean: Model simulated changes in transport, meridional heat flux and coastal sea level between 1955–59 and 1970–74. *Journal of Geophysical Research* **100**, 10559–66.

Fukasawa M., Freeland H., Perkin R., Watanabe T., Uchida H. and Nishina A. (2004) Bottom water warming in the North Pacific Ocean. *Nature* **427**, 825–7.

Gent P.R. and Danabasoglu G. (2004) Heat uptake and the thermohaline circulation in the Community Climate System Model, version 2. *Journal of Climate* **17**, 4058–69.

Gille S.T. (2004) How non-linearities in the equation of state of sea water can confound estimates of steric sea level change. *Journal of Geophysical Research* **109**, C03005.

Gille S.T. (2008) Decadal-Scale Temperature Trends in the Southern Hemisphere Ocean. *Journal of Climate* **21**, 4749–65.

Gilson J., Roemmich D., Cornuelle B. and Fu L.-L. (1998) Relationship of TOPEX/Poseidon altimetric height to the steric height and circulation in the North Pacific. *Journal of Geophysical Research* **103**, 27947–65.

Gleckler P.J., Wigley T.M.L., Santer B.D., Gregory J.M., AchutaRao K. and Taylor K.E. (2006a) Krakatoa's signature persists in the ocean. *Nature* **439**, 675.

Gleckler P.J., AchutaRao K., Gregory J.M., Santer B.D., Taylor K.E. and Wigley T.M.L. (2006b) Krakatoa lives: the effect of volcanic eruptions on ocean heat content and thermal expansion. *Geophysical Research Letters* **33**, L17702.

Gould J. and the Argo Science Team (2004) Argo profiling floats bring new era of in situ ocean observations. *EOS Transactions of the American Geophysical Union* **85**, 11 May 2004.

Gouretski V. and Koltermann K.P. (2007) How much is the ocean really warming? *Geophysical Research Letters* **34**, L01610.

Gregory J.M. and Lowe J.A. (2000) Predictions of global and regional sea level rise using AOGCMs with and without flux adjustment. *Geophysical Research Letters* **27**, 3069–72.

Gregory J.M. and Forster P.M. (2008) Transient climate response estimated from radiative forcing and observed temperature change. *Journal of Geophysical Research* **113**, D23105.

Gregory J.M., Church J.A., Boer G.J., Dixon K.W., Flato G.M., Jackett D.R. et al. (2001) Comparison of results from several AOGCMs for global and regional sea-level change 1900–2100. *Climate Dynamics* **18**, 225–40.

Gregory J., Banks H., Stott P., Lowe J. and Palmer M. (2004) Simulated and observed decadal variability in ocean heat content. *Geophysical Research Letters* **31**, L15312.

Gregory J.M., Lowe J.A. and Tett S.F.B. (2006) Simulated global-mean sea level changes over the last half-millennium. *Journal of Climate* **19**, 4576–91.

Guinehut S., Le Traon P.Y., Larnicol G. and Philipps S. (2004) Combining Argo and remote-sensing data to estimate the ocean three-dimensional temperature fields – a first approach based on simulated observations. *Journal of Marine Systems* **46**, 85–98.

Hanawa K., Rual P., Bailey R., Sy A. and Szabados M. (1995) A new depth-time equation for Sippican or TSK T-7, T-6 and T-4 expendable bathythermographs (XBT). *Deep-Sea Research, Part I* **42**, 1423–51.

Hansen J., Nazarenko L., Ruedy R., Sato M., Willis J., Del Genio A. et al. (2005) Earth's energy imbalance: confirmation and implications. *Science* **308**, 1431–5.

Hogg A.M., Meredith M.P., Blundell J.R. and Wilson C. (2008) Eddy heat flux in the southern ocean: response to variable wind forcing. *Journal of Climate* **21**, 608–20.

Hong B.G., Sturges W. and Clarke A.J. (2000) Sea level on the U.S. east coast: decadal variability caused by open-ocean wind curl forcing. *Journal of Physical Oceanography* **30**, 2088–98.

Ingleby B. and Huddleston M. (2007) Quality control of ocean temperature and salinity profiles – historical and real-time data. *Journal of Marine Systems* **65**, 158–75.

Ishii M. and Kimoto M. (2009) Reevaluation of historical ocean heat content variations with time-varying XBT and MBT depth bias. *Journal of Oceanography* **65**, 287–99.

Ishii M., Kimoto M. and Kachi M. (2003) Historical ocean subsurface temperature analysis with error estimates. *Monthly Weather Review* **131**, 51–73.

Ishii M., Kimoto M, Sakamoto K. and Iwasaki S.I. (2006) Steric sea level changes estimated from historical ocean subsurface temperature and salinity analyses. *Journal of Oceanography* **62**, 155–70.

Johnson G.C. and Doney S.C. (2006) Recent western South Atlantic bottom water warming. *Geophysical Research Letters* **33**, L14614.

Johnson G.C., Mecking S., Sloyan B.M. and Wijffels S.E. (2007) Recent bottom water warming in the Pacific Ocean. *Journal of Climate* **20**, 5365–75.

Johnson, G.C., Purkey S.G. and Bullister J.L. (2008) Warming and Freshening in the Abyssal Southeastern Indian Ocean. *Journal of Climate* **21**, 5351–63.

Joyce T.M. and Robbins P. (1996) The long-term hydrographic record at Bermuda. *Journal of Climate* **9**, 3121–31.

K-1 Model Developers (2004) *K-1 Coupled Model (MIROC) Description*. K-1 technical report 1 (Hasumi H. and Emori S., eds). Center for Climate System Research, University of Tokyo, Tokyo.

Kaplan A., Y. Kushnir Y. and Cane M.A. (2000) Reduced space optimal interpolation of historical marine sea level pressure. *Journal of Climate* **13**, 2987–3002.

Katsman C.A., Hazeleger W., Drijfhout S.S., van Oldenborgh G.J. and Burgers G. (2008) Climate scenarios of sea level rise for the northeast Atlantic Ocean: a study including the effects of ocean dynamics and gravity changes induced by ice melt. *Climatic Change* **91**, 351–74.

Köhl A. and Stammer D. (2008) Decadal sea level changes in the 50-year GECCO ocean synthesis. *Journal of Climate* **21**, 1876–90.

Köhl A., Stammer D. and Cornuelle B. (2007) Interannual to decadal changes in the ECCO global synthesis. *Journal of Physical Oceanography* **37**, 313–37.

Landerer F.W., Jungclaus J.H. and Marotzke J. (2007a) Regional dynamic and steric sea level change in response to the IPCC-A1B scenario. *Journal of Physical Oceanography* **37**, 296–312.

Landerer F.W., Jungclaus J.H. and Marotzke J. (2007b) Ocean bottom pressure changes lead to a decreasing length-of-day in a warming climate. *Geophysical Research Letters* **34**, L06307.

Landerer F.W., Jungclaus J.H. and Marotzke J. (2008) El Niño–Southern Oscillation signals in sea level, surface mass redistribution, and degree-two geoid coefficients. *Journal of Geophysical Research* **113**, C08014.

Landerer F.W. Jungclaus J.H. and Marotzke J. (2009) Long-term polar motion excited by ocean thermal expansion. *Geophysical Research Letters* **36**, L17603.

Leuliette E. and Miller L. (2009) Closing the sea level budget with altimetry, Argo and GRACE, *Geophysical Research Letters* **36**, L04608.

Leuliette E.W, Nerem R.S. and Mitchum G.T. (2004) Calibration of TOPEX/Poseidon and Jason altimeter data to construct a continuous record of mean sea level change. *Marine Geodesy* **27**, 79–94.

Levermann A., Griesel A., Hofmann M., Montoya M. and Rahmstorf S. (2005) Dynamic sea level changes following changes in the thermohaline circulation. *Climate Dynamics* **24**, 347–54.

Levitus S. (1990) Interpentadal variability of steric sea level and geopotential thickness of the North Atlantic Ocean, 1970–74 versus 1955–59. *Journal of Geophysical Research* **95**, 5233–8.

Levitus S., Antonov J.I., Boyer T.P. and Stephens C. (2000) Warming of the world ocean. *Science* **287**, 2225–9.

Levitus S., Antonov J., Wang J., Delworth T., Dixon K. and Broccoli A. (2001) Anthropogenic warming of Earth's climate system. *Science* **292**, 267–70.

Levitus S., Antonov J.I. and Boyer T.P. (2005a) Warming of the world ocean, 1955–2003. *Geophysical Research Letters* **32**, L02604.

Levitus S., Antonov J.I., Boyer T.P., Garcia H.E. and Locarnini R.A. (2005b) Linear trends of zonally averaged thermosteric, halosteric, and total steric sea level for individual ocean basins and the world ocean, (1955–1959) – (1994–1998). *Geophysical Research Letters* **32**, L16601.

Levitus S, Antonov J.I., Boyer T.P., Garcia H.E. and Locarnini R.A. (2005c) EOF analysis of upper ocean heat content, 1956–2003. *Geophysical Research Letters* **32**, L18607.

Levitus S., Sato S., Maillard C., Mikhailov N., Caldwell P. and Dooley H. (2005d) *Building Ocean Profile-Plankton Databases for Climate and Ecosystem Research.* NOAA Technical Report NESDIS 117. US Government Printing Office, Washington DC.

Levitus S., Antonov J.I., Boyer T.P., Locarnini R.A., Garcia H.E. and Mishonov A.V. (2009) Global ocean heat content 1955–2007 in light of recently revealed instrumentation problems. *Geophysical Research Letters* **36**, L07608.

Llovel W., Cazenave A., Rogel P., Lombard A. and Bergé-Nguyen M. (2009) 2-D reconstruction of past sea level (1950–2003) using tide gauge records and spatial patterns from a general ocean circulation model. *Climate of the Past Discussions* **5**, 1109–32.

Lombard A., Cazenave A., Le Traon P.-Y. and Ishii M. (2005) Contribution of thermal expansion to present-day sea-level change revisited. *Global and Planetary Change* **47**, 1–16.

Lombard A., Cazenave A., Le Traon P.-Y., Guinehut S. and Cabanes C. (2006) Perspectives on present-day sea level change: a tribute to Christian le Provost. *Ocean Modelling* **56**, 445–51.

Lombard A., Garcia D., Ramillien G., Cazenave A., Biancale R., Lemoine J.M., Flechtner F., Schmidt R. and Ishii M. (2007) Estimation of steric sea level variations from combined GRACE and satellite altimetry data. *Earth and Planetary Science Letters* **254**, 194–202.

Lowe J.A. and Gregory J.M. (2006) Understanding projections of sea level rise in a Hadley Centre coupled climate model. *Journal of Geophysical Research* **111**, C11014.

Lyman J.M., Willis J.K. and Johnson G.C. (2006) Recent cooling of the upper ocean. *Geophysical Research Letters* **33**, L18604.

Meehl G.A., Washington W.M., Collins W.D., Arblaster J.M., Hu A., Buja L.E. et al. (2005) How much more global warming and sea level rise? *Science* **307**, 1769–72.

Meehl G.A., Stocker T.F., Collins W.D., Friedlingstein P., Gaye A.T., Gregory J.M. et al. (2007) Global climate projections. In: *Climate Change 2007: The Physical Science Basis*. Contribution of Working Group 1 to the Fourth Assessment Report of the Intergovernmental Panel on Climate Change (Solomon S., Qin D., Manning M., Marquis M., Averyt K., Tignor M.M.B. et al., eds), pp. 747–845. Cambridge University Press, Cambridge.

Minster J.F., Cazenave A., Serafini Y.V., Mercier F., Gennero M.C. and Rogel P. (1999) Annual cycle in mean sea level from Topex-Poseidon and ERS-1: inferences on the global hydrological cycle. *Global and Planetary Change* **20**, 57–66.

Mitrovica J.X., Gomez N. and Clark P.U. (2009) The sea-level fingerprint of West Antarctic collapse. *Science* **323**, 753.

Munk W. (2003) Ocean freshening, sea level rising. *Science* **300**, 2041–3.

Nerem R.S., Chambers D.P., Leuliette E., Mitchum G.T. and Giese B.S. (1999) Variations in global mean sea level during the 1997–98 ENSO event. *Geophysical Research Letters* **26**, 3005–8.

Ngo-Duc T., Laval K., Polcher J. and Cazenave A. (2005) Contribution of continental water to sea level variations during the 1997–1998 El Niño–Southern Oscillation event: Comparison between Atmospheric Model Intercomparison Project simulations and TOPEX/Poseidon satellite data. *Journal of Geophysical Research* **110**, D09103.

Pardaens A.K., Gregory J.M. and Lowe J.A. (2010) A model study of factors influencing projected changes in regional sea level over the twenty-first century. *Climate Dynamics*, in press.

Paulson A., Zhong, S. and Wahr J. (2007) Inference of mantle viscosity from GRACE and relative sea level data. *Geophyical Journal International* **171**, 497–508.

Peltier W.R. (2004) Global glacial isostasy and the surface of the ice-age Earth: The ICE-5G (VM2) model and GRACE. *Annual Review of Earth and Planetary Sciences* **32**, 111–49.

Peltier W. R. and Luthcke S.B. (2009) On the origins of Earth rotation anomalies: new insights on the basis of both "paleogeodetic" data and Gravity Recovery and Climate Experiment (GRACE) data. *Journal of Geophysical Research* **114**, B11405.

Raper S.C.B., Gregory J.M. and Stouffer R.J. (2002) The role of climate sensitivity and ocean heat uptake on AOGCM transient temperature response. *Journal of Climate* **15**, 124–8.

Roemmich D. (1990) Sea level and the thermal variability of the ocean. In: *Sea Level Change*, pp. 208–17. Geophysics Study Committee, National Academy Press, Washington DC.

Roemmich D. and Wunsch C. (1984) Apparent changes in the climatic state of the deep North Atlantic Ocean. *Nature* **307**, 447–50.

Roemmich D., Gilson J., Davis R., Sutton P., Wijffels S. and Riser S. (2007) Decadal spinup of the South Pacific subtropical gyre. *Journal of Physical Oceanography* **37**, 162–73.

Sakamoto T.T., Suzuki T., Ishii M., Nishimura T., Emori S. and Hasumi H. (2005) Response of the Kuroshio and the Kuroshio Extension to global warming in a long-term climate change projection with a high-resolution climate model. *Geophysical Research Letters* **32**, L14617.

Siedler G., Church J. and Gould J. (eds) (2001) *Ocean Circulation and Climate: Observing and Modelling the Global Ocean*. Academic Press, London.

Sturges W. and Hong B.G. (2001) Decadal variability of sea level. In: *Sea Level Rise: History and Consequences*, pp. 165–80, Academic Press, New York.

Suzuki T., Hasumi H., Sakamoto T.T., Nishimura T., Abe-Ouchi A., Segawa T., Okada N., Oka A. and Emori S. (2005) Projection of future sea level and its variability in a high-resolution climate model: Ocean processes and Greenland and Antarctic ice-melt contributions. *Geophysical Research Letters* **32**, L19706.

Swenson S., Chambers D. and Wahr J. (2008) Estimating geocenter variations from a combination of GRACE and ocean model output. *Journal of Geophysical Research* **113**, B08410.

Tapley B.D., Bettadpur S., Watkins M. and Reigber C. (2004) The Gravity Recovery and Climate Experiment: mission overview and early results. *Geophysical Research Letters* **31**, L09607.

Tokeshi T. and Yanagi T. (2003) High sea level change at Naha in Okinawa Island. *Oceanography in Japan* **12**, 395–405.

Wadhams P. and Munk W. (2004) Ocean freshening, sea level rising, sea ice melting. *Geophysical Research Letters* **31**, L11311.

Wahr J., Molenaar M. and Bryan F. (1998) Time variability of the Earth's gravity field: Hydrological and oceanic effects and their possible detection using GRACE. *Journal of Geophysical Research* **103**, 30205–29.

Wenzel M. and Schröter J. (2007) The global ocean mass budget in 1993–2003 estimated from sea level change. *Journal of Physical Oceanography* **37**, 203–13.

White W. (1995) Design of a global observing system for gyre-scale upper ocean temperature variability. *Progress in Oceanography* **36**, 169–217.

White W.B. and Tai C.-K. (1995) Inferring interannual changes in global upper ocean heat storage from TOPEX altimetry. *Journal of Geophysical Research* **100**, 24943–54.

Wijffels S.E., Willis J., Domingues C.M., Barker P., White N.J., Gronell A., Ridgway K. and Church J.A. (2008) Changing expendable bathythermograph fall rates and their impact on estimates of thermosteric sea level rise. *Journal of Climate* **21**, 5657–72.

Willis J., Roemmich D. and Cornuelle B. (2003) Combining altimetric height with broadscale profile data to estimate steric height, heat storage, subsurface temperature and SST variability. *Journal of Geophysical Research* **108**, 3292.

Willis J.K., Roemmich D. and Cornuelle B. (2004) Interannual variability in upper ocean heat content, temperature, and thermosteric expansion on global scales. *Journal of Geophysical Research* **109**, C12036.

Willis J.K., Chambers D.T. and Nerem R.S. (2008) Assessing the globally averaged sea level budget on seasonal to interannual time scales. *Journal of Geophysical Research* **113**, C06015.

Willis J.K., Lyman J.M., Johnson G.C. and Gilson J. (2009) In situ data biases and recent ocean heat content variability. *Journal of Atmospheric and Oceanic Technology* **26**(4), 846–52.

Wilson J.R. (1998) *Global Temperature-Salinity Profile Programme (GTSPP) – Overview and Future.* Intergovernmental Oceanographic Commission Technical Series 49. UNESCO, Paris.

Wunsch C., Ponte R.M. and Heimbach P. (2007) Decadal trends in sea level patterns: 1993–2004. *Journal of Climate* **20**, 5889–5911.

7 Cryospheric Contributions to Sea-Level Rise and Variability

Konrad Steffen, Robert H. Thomas, Eric Rignot, J. Graham Cogley, Mark B. Dyurgerov[1], Sarah C.B. Raper, Philippe Huybrechts, and Edward Hanna

7.1 Introduction

Global mean sea level rose by approximately 1.8 mm/year over the last 50 years, increasing to approximately 3.1 mm/year during the 1990s (Church et al. 2004; Holgate and Woodworth 2004; Cazenave and Nerem 2004). Thermal expansion of ocean water accounts for approximately 0.4 mm/year of sea-level rise for the past four to five decades, rising to approximately 1.5 mm/year during the last decade (Levitus et al. 2005; Ishii et al. 2006; Willis et al. 2004; Lombard et al. 2006). Contributions from water on land are probably small, with sequestration by dams more or less balanced by release of groundwater, but uncertainties are large (Chapter 8). The most important source of the remainder is likely land ice which, if it were all to melt, would cause sea-level rise of more than 60 m: with contributions of 88 and 11% from the big ice sheets in Antarctica and Greenland respectively, and 1% from glaciers and ice caps (Table 7.1)[2]. Many glaciers and ice caps are currently retreating and have contributed a sea-level rise of 0.3–0.45 mm/year over the last 50 years, rising to 0.8 mm/year over the last decade (Dyurgerov and Meier 2005)[3]. This leaves an unexplained contribution of approximately 1 mm/year, which probably comes from the big ice sheets.

[1] Deceased.

[2] Clarification of terminology: two general categories of ice are considered as contributors to global sea-level rise. (1) The ice sheets of Antarctica and Greenland, both of which drain into the surrounding ocean via a number of ice streams or outlet glaciers. Ice sheets are sufficiently thick to cover most of the bedrock topography. (2) Glaciers and ice caps, mostly outside Antarctica and Greenland. A glacier is a mass of ice on the land flowing downhill under gravity and an ice cap is a mass of ice that typically covers a highland area.

[3] Note that estimates of sea-level rise associated with transfer of ice to the ocean are calculated by dividing the equivalent freshwater volume by ocean area. Elastic depression of the ocean bed would reduce sea-level rise estimates given in this chapter by about 6% (G. Milne, personal communication).

Understanding Sea-Level Rise and Variability, 1st edition. Edited by John A. Church, Philip L. Woodworth, Thorkild Aarup & W. Stanley Wilson.

Table 7.1 Summary of the recent mass balance of Greenland and Antarctica and our interpretation of current knowledge of the contribution to sea-level change by the main components of the cryosphere: the polar ice sheets, and small glaciers and ice caps found at most latitudes.

	Small glaciers/ice caps	**Greenland**	**Antarctica**
Area (million km^2)	0.57–0.79	1.7	12.3
Volume (km^3)[a]	0.25×10^6	2.9×10^6	24.7×10^6
SLE (m)	0.6	7.3	56.6
Total accumulation (Gt/year)[b]	720	500	1850
SLE (mm/year)	2	1.4	5.1
Mass balance	Thinning since at least 1960, at rates that more than doubled after 1990, to cause a net loss of almost 300 Gt/year since 2000.	Since approximately 1990: thickening above 2000 m, at an accelerating rate; thinning at lower elevations also accelerating to cause a net loss from the ice sheet, increasing to probably >200 Gt/year after 2005.	Since early 1990s: slow thickening in central regions and southern Antarctic Peninsula; localized thinning at accelerating rates of glaciers in Antarctic Peninsula and Amundsen Sea region. Probable near balance until 2000, shifting to losses increasing to >100 Gt/year after 2005.

[a]1 km^3 of ice is approximately 0.92 Gt.
[b]Excluding ice shelves.
Note: sea-level equivalent (SLE) includes correction for water filling regions where bedrock is below sea level.

Although little is known about ice-sheet contributions during the past 50 years, observations from satellites and aircraft since 1990 show substantial ice loss from Greenland that has doubled in the last decade, both from increased runoff and from acceleration of glaciers draining the south of the ice sheet during a period of sustained local warming (Thomas et al. 2001; Rignot and Thomas 2002; Krabill et al. 2004; Zwally et al. 2005; Rignot et al. 2008a). The Antarctic Ice Sheet is probably also losing mass, with thinning along coastal sectors of the Antarctic Peninsula, West Antarctica, and Wilkes Land in East Antarctica, and thickening further inland. Large-scale numerical models used to predict the evolution of the Greenland and Antarctic Ice Sheets in a warmer climate explain some of these changes but do not capture the rapid coastal-flow accelerations observed since the mid-1990s. If such accelerations were to be sustained in the future, these models underpredict future contributions to sea level from the ice sheets.

7.2 Mass-Balance Techniques

Mass balance refers to the balance between additions to and losses from a specified region of ice. The term is frequently used rather loosely to mean either the surface mass balance or the total mass balance. Here we shall use these more precise terms as follows: *surface mass balance* signifies the additions (usually in the form of snow) to the surface of the region of glacier, ice cap, or ice sheet under investigation, and the losses (by evaporation, snow drift, or melting and runoff) from that surface; *total mass balance* signifies the balance between all additions (snowfall, advection of ice from upstream, basal freezing, etc.) and all losses (by ice motion, surface and basal melting, etc.) from the region of ice under investigation, such as entire ice sheets, glaciers, or ice caps.

Most estimates of the surface mass balance are from repeated measurements of the length exposed of stakes planted in the snow or ice surface. Temporal changes in this length, multiplied by the density of the mass gained or lost, is the surface mass balance at the location of the stake. Various means have been devised to apply corrections for sinking of the stake bottom into the snow, densification of the snow between the surface and the stake bottom, and the refreezing of surface meltwater at depths below the stake bottom. Such measurements are time-consuming and expensive, and they are supplemented by model estimates of precipitation, sublimation, and melting. Regional climate models, calibrated by independent *in situ* measurements of temperature and pressure (e.g. Steffen and Box 2001; Box et al. 2006) provide estimates of snowfall and sublimation, and estimates of surface melting/evaporation come from energy-balance models (van de Wal and Oerlemans 1994; Hock 2005) and degree-day or temperature-index models as reviewed in Hock (2003). Within each category there is a hierarchy of models in terms of spatial and temporal resolution. Energy-balance models are physically based, require detailed input data and are more suitable for high resolution in space and time. Degree-day models are advantageous for the purposes of estimating worldwide glacier melt since the main inputs of temperature and precipitation are readily available in gridded form from atmosphere–ocean general circulation models (AOGCMs).

Techniques for measuring total mass balance include:

- the mass-balance approach, comparing total net snow accumulation with losses by total ice discharge, sublimation, and meltwater runoff;
- repeated altimetry, to estimate volume changes;
- temporal changes in gravity, to infer mass changes, except on floating ice tongues and ice shelves.

The first two provide estimates of mass balance for regions included within surveys, whereas the third gives estimates for entire regions or ice sheets. All three techniques can be applied to the big ice sheets; most studies of ice caps and glaciers apply the first, with recent studies also using repeated laser altimetry, and the third technique is applied only to large, heavily glaciated regions such as Alaska, Greenland, and Antarctica. Here, our objective is to summarize what is known about total mass balance, to assess the merits and limitations of different approaches to its measurement, and to identify possible improvements that could be made over the next few years. This assessment is based on our knowledge and understanding of the different techniques, and in some cases we estimate larger associated errors than are published, because we try to take account of errors that are neglected, assumed to be zero, or may be larger than those resulting from poorly validated models.

7.3 Ice-Sheet Mass Balance

7.3.1 Mass Balance

Snow accumulation is estimated from stake measurements, annual layering in ice cores, sometimes with interpolation using satellite microwave measurements or shallow radar sounding (Ice Sheet Mass Balance and Sea Level project (ISMASS) Committee 2004), or from atmospheric modeling (e.g. Bromwich et al. 2004). Ice discharge is the product of mean column velocity and thickness, with velocities measured *in situ* or remotely, preferably near the grounding line which is the critical juncture that separates ice that is thick enough to remain grounded from either an ice shelf or a calving front where velocity is almost depth-independent. Thickness is measured by airborne radar, seismically, or from measured surface elevations assuming hydrostatic equilibrium, for floating ice near grounding lines. Surface meltwater runoff (large on glaciers and ice caps, and near the Greenland coast and parts of the Antarctic Peninsula but small or zero elsewhere) is generally from stake measurements or from model estimates calibrated against surface observations where available (e.g. Hanna et al. 2005; Box et al. 2006). The typically small mass loss by melting beneath grounded ice is also estimated from models.

Mass-balance calculations involve the comparison of two very large numbers, and small errors in either can result in large errors in estimated total mass balance. For example, total accumulation over Antarctica, excluding ice shelves, is about 1850 Gt/year (Vaughan et al. 1999; Arthern et al. 2006; van de Berg et al. 2006) and 500 Gt/year over Greenland (Bales et al. 2001). Associated errors of ±38 to ±47 Gt/year have been assessed for Greenland by Rignot et al. (2008a), and 92 Gt/year for Antarctica for the years 1996, 2000, and 2006 by Rignot et al. (2008b).

Interferometric synthetic aperture radar (InSAR) is used to measure ice velocity from aircraft or satellite; although broad InSAR coverage and progressively improving estimates of grounding-line ice thickness have substantially improved ice-discharge estimates, incomplete data coverage and uncertainty regarding velocity/depth relationships and meltwater runoff implies errors on total discharge of approximately 5%. Consequently, assuming these errors in both snow accumulation and ice losses, current (2009) mass-balance uncertainty is approximately ±160 Gt/year for Antarctica and ±35 Gt/year for Greenland. Additional errors may result from accumulation estimates being based on data from the past few decades; at least in Greenland, we know that snowfall is increasing with time. Similarly, it is becoming clear that glacier velocities can change substantially over quite short time periods (Rignot and Kanagaratnam 2006), so these error estimates probably represent lower limits.

Repeated Altimetry

Rates of surface-elevation change (d*S*/d*t*) reveal changes in ice-sheet mass after correction for changes in depth/density profiles and bedrock elevation, or for hydrostatic equilibrium and tides if the ice is floating. Satellite radar altimetry (SRALT) has been widely used (e.g. Shepherd et al. 2002; Davis et al. 2005; Johannessen et al. 2005; Zwally et al. 2005), together with laser altimetry from airplanes (Krabill et al. 2000), and from the National Aeronautics and Space Administration's (NASA's) Ice, Cloud, and Land Elevation Satellite (ICESat) (Zwally et al. 2002a; Thomas et al. 2006). Modeled corrections for isostatic changes in bedrock elevation (e.g. Peltier 2004) are small (few millimeters per year) but with errors comparable to the correction. Those for near-surface snow-density changes (Arthern and Wingham 1998; Li and Zwally 2004) are larger (1 or 2 cm/year) and also uncertain. But of most concern is the viability of SRALT for accurate measurement of ice-sheet elevation changes.

Satellite Radar Altimetry

Available SRALT data are from altimeters with a beam width of 20 km or more, designed and demonstrated to make accurate measurements over the almost flat, horizontal ocean. Data interpretation is far more complex over sloping and undulating ice-sheet surfaces with spatially and temporally varying dielectric properties. Here, SRALT range measurements are generally off nadir, and return-waveform information is biased towards the earliest reflections (i.e. highest regions) within the large footprint, and is affected by radar penetration into near-surface snow. This probably introduces only small errors at higher elevations, where surfaces are nearly flat and there is little surface melting. But glaciers and ice streams flow in surface depressions that can be narrower than the radar footprint, so that SRALT-derived elevation changes are weighted toward slower-moving ice at the glacier sides. This is of most concern in Greenland, where other studies show big losses from outlet glaciers just a few kilometers wide (Abdalati et al. 2001). Also, there is seasonally variable melting over about half of the Greenland Ice Sheet, effectively raising and lowering the radar reflecting horizon, even at quite high elevations. SRALT estimates of d*S*/d*t* have not been validated against independent estimates except at higher elevations, where surfaces are nearly flat and horizontal and dielectric properties change but little (Thomas et al. 2001). Further, even here, different interpretations of essentially the same European Remote Sensing satellite (ERS) SRALT data from 1992–2003 (Johannessen et al. 2005; Zwally et al. 2005) give Greenland d*S*/d*t* estimates above 2000 m elevation that differ by more than quoted uncertainties, implying additional uncertainties associated with techniques used to process and interpret the data.

Errors in SRALT-derived values of d*S*/d*t* are typically determined from the internal consistency of the measurements, often after iterative removal of d*S*/d*t* values that exceed some multiple of the local value of their standard deviation.

This results in very small error estimates (e.g. Zwally et al. 2005), but neglects other potentially far larger errors, such as:

- processing/interpretation errors, discussed above, that are suggested by differing results from the same data;
- the possibility that SRALT estimates are biased by the effects of local terrain or by snow wetness. Observations by other techniques reveal extremely rapid thinning along Greenland glaciers that flow along depressions where d*S*/d*t* cannot be inferred from SRALT data, and collectively these glaciers are responsible for most of the mass loss from the ice sheet (Rignot and Kanagaratnam 2006). The zone of summer melting in Greenland increased between the early 1990s and 2005 (Box et al. 2006), probably raising the radar reflection horizon within near-surface snow. If it did, SRALT over-estimates thickening rates at higher elevations.

Both of these effects result in underestimates of total ice losses.

Aircraft and Satellite Laser Altimetry

Laser altimeters provide data that are easier to validate and interpret: footprints are small (about 1 m for airborne laser, and 60 m for ICESat), and there is negligible laser penetration into the ice. However, clouds limit data acquisition, and accuracy is affected by atmospheric conditions and particularly by laser-pointing errors. Existing laser data are sparse compared to SRALT data.

Airborne laser surveys over Greenland in 1993–4 and 1998–9 yielded elevation estimates accurate to approximately 10 cm along survey tracks (Krabill et al. 2002), but with large gaps between flight lines. ICESat orbit-track separation is also quite large, particularly in southern Greenland and the Antarctic Peninsula where rapid changes are occurring, and elevation errors along individual orbit tracks can be large (many tens of centimeters) over sloping ice. Progressive improvement in ICESat data processing is reducing these errors and, for both airborne and ICESat surveys, most errors are independent for each flight line or orbit track, so that estimates of d*S*/d*t* averaged over large areas containing many survey tracks are affected most by systematic ranging, pointing, or platform-position errors, totalling probably less than 5 cm. In Greenland, such conditions typically apply at elevations above 1500–2000 m. d*S*/d*t* errors decrease with increasing time interval between surveys. Nearer the coast there are large gaps in both ICESat and airborne coverage, requiring d*S*/d*t* values to be supplemented by degree-day estimates of anomalous melting (Krabill et al. 2000, 2004). This increases overall errors, and almost certainly underestimates total losses because it does not take full account of dynamic thinning of unsurveyed outlet glaciers.

Temporal Variations in Earth's Gravity

Since 2002, the Gravity Recovery and Climate Experiment (GRACE) satellite has measured Earth's gravity field and its temporal variability. After removing the effects of tides and atmospheric loading, high-latitude data contain information on temporal changes in the mass distribution of the ice sheets and underlying rock. Because of its high altitude, GRACE makes coarse-resolution measurements of the gravity field and its changes with time. Consequently, resulting mass-balance estimates are also at coarse resolution of several hundred kilometers. But this has the advantage of covering entire ice sheets, which is extremely difficult using other techniques. Consequently, GRACE estimates include mass changes on the many small ice caps and isolated glaciers that surround the big ice sheets. These may be quite large and these ice masses are strongly affected by changes in the coastal climate.

Error sources include measurement uncertainty, leakage of gravity signal from regions surrounding the ice sheets, and causes of gravity changes other than ice-sheet changes. Of these, the largest are the gravity changes associated with vertical bedrock motion. Velicogna and Wahr (2005) estimated a mass-balance correction of 5 ± 17 Gt/year for bedrock motion in Greenland, and a correction of 173 ± 71 Gt/year for Antarctica (Velicogna and Wahr 2006), but errors may be underestimated (Horwath and Dietrich 2006). Although other geodetic data (variations in length of day, polar wander, etc.) provide constraints on mass changes at high latitudes, unique solutions are not yet possible from these techniques.

Summary

In summary, dS/dt errors cannot be precisely quantified for either SRALT data, because of the broad radar beam and time-variable penetration, or laser data, because of sparse coverage. If the SRALT limitations discussed above are real, they are difficult if not impossible to resolve. Laser limitations result primarily from poor coverage, and can be resolved by increasing coverage, or by interpolating by, for instance, using degree-day estimates of anomalous melt losses and mass-balance estimates for individual drainage basins.

All altimetry mass-balance estimates include the following additional uncertainties.

- The first uncertainty is the density (ρ) assumed to convert thickness changes to mass changes. If changes are caused by recent changes in snowfall, the appropriate density may be as low as 300 kg/m^3; for long-term changes, it may be as high as 900 kg/m^3. This is of most concern for high-elevation regions with small dS/dt, where the safest assumption is $\rho = 600 \pm 300$ kg/m^3. For a 1-cm/year thickness change over the 1 million km^2 of Greenland above 2000 m, uncertainty would be ± 3 Gt/year.
- Also uncertain are the possible changes in near-surface snow density. Densification rates are sensitive to snow temperature and wetness, with warm conditions favoring more rapid densification (Arthern and Wingham 1998; Li and Zwally 2004). Consequently, recent Greenland warming probably caused surface lowering simply from this effect. Corrections are inferred from largely unvalidated models, and are typically less than 2 cm/year, with unknown errors. If overall uncertainty

is 0.5 cm/year, associated mass-balance errors are approximately ±8 Gt/year for Greenland and ±60 Gt/year for Antarctica.
- The final uncertainty is the rate of basal uplift. This is inferred from models and has uncertain errors. An overall uncertainty of 1 mm/year would result in mass-balance errors of about ±2 Gt/year for Greenland and ±12 Gt/year for Antarctica.

7.3.2 Mass Balance of the Greenland and Antarctic Ice Sheets

Ice locked within the Greenland and Antarctic Ice Sheets has long been considered comparatively immune to change, protected by the extreme cold of the polar regions. Most model results suggested that climate warming would result primarily in increased melting from coastal regions and an overall increase in snowfall, with net 21st-century outcomes probably being a small mass loss from Greenland and a small gain in Antarctica, and little combined impact on sea level (Church et al. 2001). Observations generally confirmed this view, although Greenland measurements during the 1990s (Krabill et al. 2000; Abdalati et al. 2001) began to suggest that there might also be a component from ice-dynamical responses, with very rapid thinning on several outlet glaciers.

Increasingly, measurements in both Greenland and Antarctica show rapid changes in the behavior of large outlet glaciers. In some cases, once-rapid glaciers have slowed to a virtual standstill, damming up the still-moving ice from further inland and causing the ice to thicken (Joughin et al. 2002; Joughin and Tulaczyk 2002). More commonly, observations reveal glacier acceleration, but in some cases changes have been very recent. In particular, velocities of tributary glaciers increased markedly very soon after ice shelves or floating ice tongues broke up (e.g. Scambos et al. 2004). This is happening along both the west and east coasts of Greenland (Joughin et al. 2004; Howat et al. 2005; Rignot and Kanagaratnam 2006) and in at least two locations in Antarctica (Joughin et al. 2003; Scambos et al. 2004; Rignot et al. 2004a). Such dynamic responses have yet to be fully explained, and are consequently not included in predictive models. What remains unclear is the response time of large ice sheets to climatic forcing. If the ice-dynamical changes observed over the last few years are sustained under global warming, the response time will be significantly shorter than earlier estimates.

Greenland

Above approximately 2000 m elevation, a state of near balance between about 1970 and 1995 shifted to a slow increase in surface mass balance (Thomas et al. 2001, 2006; Johannessen et al. 2005; Zwally et al. 2005). Nearer the coast, airborne laser altimetry surveys supplemented by modeled summer melting show a negative surface mass balance (Krabill et al. 2000, 2004), resulting in net loss from the ice sheet of 27 ± 23 Gt/year, equivalent to approximately 0.08 mm/year sea-level equivalent (SLE) between 1993–4 and 1998–9, doubling to 55 ± 23 Gt/year for 1997–2003[4].

More recently, three independent studies have also shown accelerating losses from Greenland.

- Analysis of gravity data from GRACE showed total losses of 75 ± 20 Gt/year between April 2002 and April 2004, rising to 223 ± 33 Gt/year between May 2004 and April 2006 (Velicogna and Wahr 2006; for example, see Figure 12.6). Other analyses of GRACE data show losses of 129 ± 15 Gt/year for July 2002–March 2005 (Ramillien et al. 2006), 219 ± 21 Gt/year for April 2002–November 2005 (Chen et al. 2006), and 101 ± 16 Gt/year for July 2003–July 2005. Although the large scatter in these estimates suggests that errors are larger than quoted, these results show an increasing trend in mass loss.
- Mass-balance calculations for most glacier drainage basins indicated the ice sheet was losing 110 ± 70 Gt/year in the 1960s, 30 ± 50 Gt/year or near balance in the 1970s–1980s, and 97 ± 47 Gt/year in 1996 increasing rapidly to 267 ± 38 Gt/year in 2007 (Rignot et al. 2008a) based on satellite InSAR ice velocities and aircraft radio echo sounding ice thickness.
- Comparison of 2005 ICESat data with 1998/9 airborne laser surveys showed losses over this period of 80 ± 25 Gt/year (Thomas et al. 2006), and this is probably an underestimate because of sparse coverage of regions where other investigations show large losses.

By contrast, interpretation of SRALT data from ERS-1 and -2 (Johannessen et al. 2005; Zwally et al. 2005) show thickening at high elevations, with lower-elevation thinning at far lower rates than those inferred from other approaches. The Johannessen et al. (2005) study recognized the unreliability of SRALT data at lower elevations, because of locally sloping and undulating surface topography. Zwally et al. (2005) attempted to overcome this by including dS/dt estimates for about 3% of the ice sheet derived from earlier laser-altimeter surveys, to infer a small positive mass balance of 11 ± 3 Gt/year for the entire ice sheet between April 1992 and October 2002.

The pattern of thickening/thinning over Greenland, derived from laser-altimeter data, is shown in Figure 7.1a, with the various mass-balance estimates summarized in Figure 7.1b. It is clear that the SRALT-derived estimate differs widely from the others, each of which is based on totally different methods, suggesting that the SRALT interpretations underestimate total ice loss for reasons discussed in section 2.2.1. Here we assume this to be the case, and focus on the other results shown in Figure 7.1b which strongly indicate net ice loss from Greenland at rates that increased from at least 27 Gt/year between 1993–4 and 1998–9 to about double between 1997 and 2003, to more than 80 Gt/year between 1998 and 2004, to more than 100 Gt/year soon after 2000, and to more than

[4] Note that these values differ from those in the Krabill et al. (2000, 2004) publications primarily because they take account of possible surface lowering by accelerated snow densification as air temperatures rise; moreover, they probably under-estimate total losses because the airborne laser altimetry surveys under-sample thinning coastal glaciers.

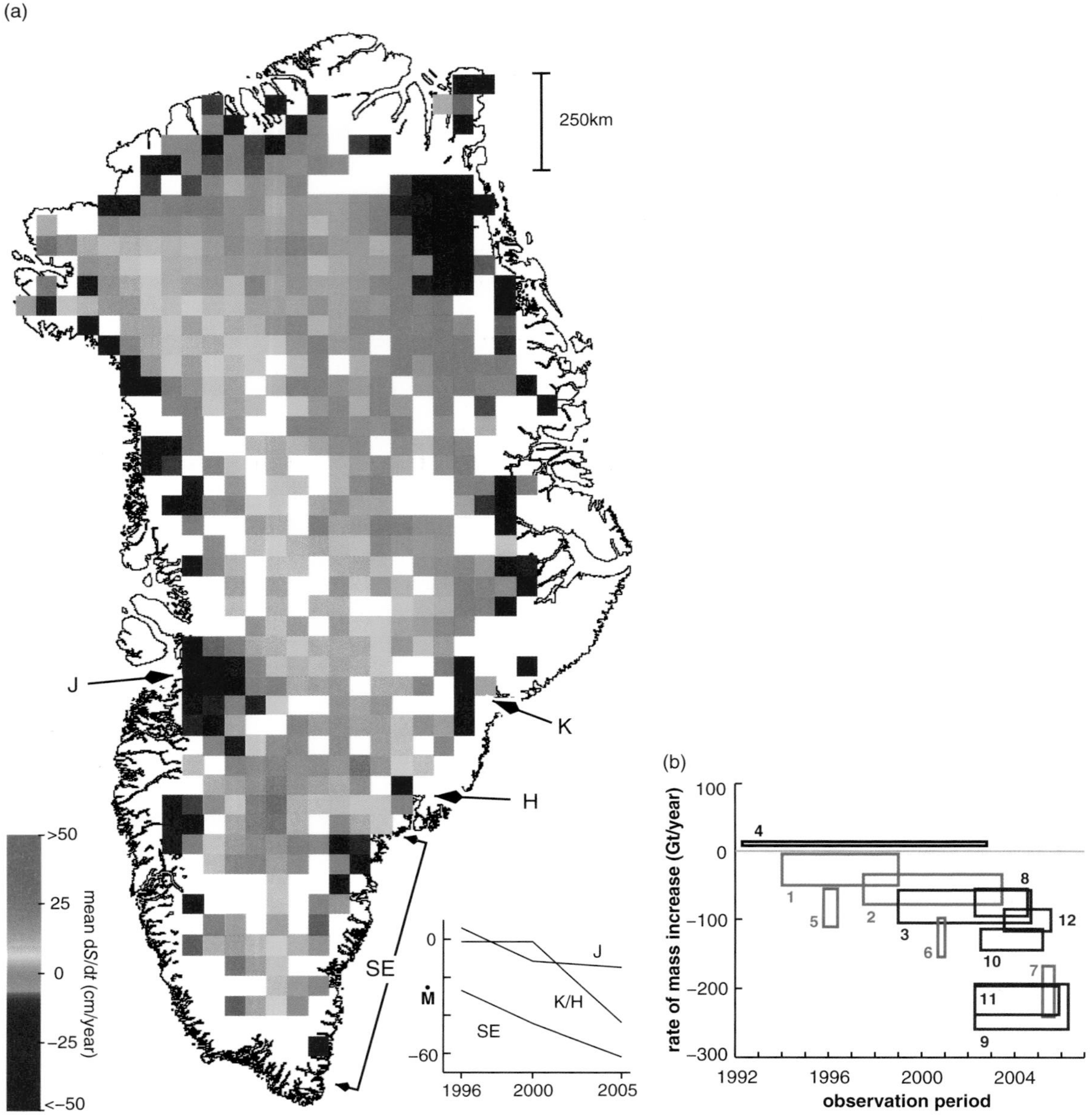

Figure 7.1 (a) Rates of elevation change (dS/dt) for Greenland derived from comparisons at more than 16 000 locations where ICESat data from October/November and May/June 2004 overlay ATM surveys in 1998–9, averaged over 50-km grid squares. Locations of rapidly thinning outlet glaciers at Jakobshavn (J), Kangerdlugssuaq (K), Helheim (H), and along the southeast coast (SE) are shown, together with plots showing their estimated mass balance ($\dot{M}$Gt/year) versus time (Rignot and Kanagaratnam 2006). (b) Mass-balance estimates for the entire ice sheet: green, Airborne Topographic Mapper (ATM; 1, 2); purple, ATM/ICESat (3, summarized in Thomas et al. 2006); black, SRALT (4, Zwally et al. 2005); red, mass balance (5–7, Rignot and Kanagaratnam 2006); blue, GRACE (8, 9, Velicogna and Wahr 2005, 2006; 10, Ramillien et al. 2006; 11, Chen et al. 2006; 12, Luthcke et al. 2006). The ATM results were supplemented by degree-day estimates of anomalous melting near the coast (Krabill et al. 2000, 2004), and probably underestimate total losses by not taking full account of dynamic thinning of outlet glaciers (Abdalati et al. 2001). SRALT results seriously underestimate rapid thinning of comparatively narrow Greenland glaciers, and may also be affected by progressively increased surface melting at higher elevations.

200 Gt/year after 2005. There are insufficient data for any assessment of total mass balance before 1990, although mass-balance calculations indicated near overall balance at elevations above 2000 m and significant thinning in the southeast (Thomas et al. 2001).

Antarctica

Determination of the mass balance of the Antarctic Ice Sheet is not as advanced as that for Greenland. Melt is not a significant factor, but uncertainties in snow accumulation are larger because fewer data have been collected, and ice thickness is poorly characterized along outlet glaciers. Instead, ice elevations, which have been improved with ICESat data, are used to calculate ice thickness from hydrostatic equilibrium at the glacier grounding line. The grounding-line position and ice velocity are inferred from Radarsat-1 and ERS-1 and -2 InSAR. For the period 1996–2000, Rignot and Thomas (2002) inferred East Antarctic growth at 20 ± 1 Gt/year, with estimated losses of 44 ± 13 Gt/year for West Antarctica, and no estimate for the Antarctic Peninsula, but the estimate for East Antarctica was based on only 60% coverage (Figure 7.2). Using improved data for 1996–2004 that provide estimates for more than 85% of Antarctica (and which were extrapolated on a basin-per-basin basis to 100% of Antarctica), Rignot et al. (2008b) found an ice

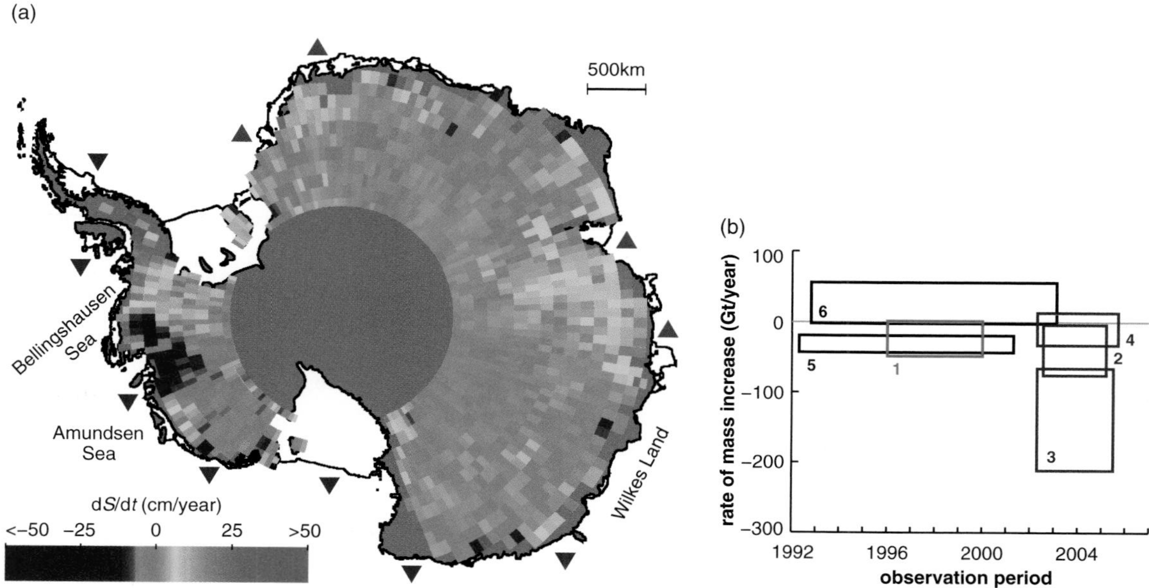

Figure 7.2 (a) Rates of elevation change (dS/dt) derived from ERS radar altimeter measurements between 1992 and 2003 over the Antarctic Ice Sheet (Davis et al. 2005). Locations of ice shelves estimated to be thickening or thinning by more than 30 cm/year (Zwally et al. 2005) are shown by purple triangles (thinning) and red triangles (thickening). (b) Mass-balance estimates for the ice sheet: red, mass balance (1, Rignot and Thomas 2002); blue, GRACE (2, Ramillien et al. 2006; 3, Velicogna and Wahr 2006; 4, Chen et al. 2006); black, ERS SRALT (5, Zwally et al. 2005; 6, Wingham et al. 2006).

loss of 106 ± 60 Gt/year for West Antarctica, 28 ± 45 Gt/year for the peninsula, and a mass gain of 4 ± 61 Gt/year for East Antarctica in 2000. In 1996, the mass loss for West Antarctica was 83 ± 59 Gt/year, but this increased to 132 ± 60 Gt/year in 2006 due to glacier acceleration. In the peninsula, the mass loss increased to 60 ± 46 Gt/year in 2006 due to the strong acceleration of glaciers in the northern peninsula following the breakup of the Larsen B ice shelf in 2002. Overall, the ice-sheet mass loss nearly doubled in 10 years, nearly entirely from West Antarctica and the northern tip of the peninsula, while little change has been found in East Antarctica. Other mass-balance analyses indicate thickening of drainage basins feeding the Filchner–Ronne Ice Shelf from portions of East and West Antarctica (Joughin and Bamber 2005) and of some ice streams draining ice from West Antarctica into the Ross Ice Shelf (Joughin and Tulaczyk 2002), but mass loss from the northern part of the Antarctic Peninsula (Rignot et al. 2005a, 2005b) and parts of West Antarctica flowing into the Amundsen Sea (Rignot et al. 2004b). In both of these latter regions, losses are increasing with time.

Although SRALT coverage extends only to within about 900 km of the poles (Figure 7.2), inferred rates of surface elevation change (dS/dt) should be more reliable than in Greenland, because most of Antarctica is too cold for surface melting (reducing effects of changing dielectric properties), and outlet glaciers are generally wider than in Greenland (reducing uncertainties associated with rough surface topography). Results show that interior parts of East Antarctica, well monitored by ERS-1 and ERS-2, thickened during the 1990s, equivalent to growth of a few tens of gigatonnes per year, depending on details of the near-surface density structure (Davis et al. 2005; Wingham et al. 2006; Zwally et al. 2005). With approximately 80% SRALT coverage of the ice sheet, and interpolating to the rest, Zwally et al. (2005) estimated a West Antarctic loss of 47 ± 4 Gt/year, East Antarctic gain of 17 ± 11 Gt/year, and overall loss of 30 ± 12 Gt/year, excluding the Antarctic Peninsula. But Wingham et al. (2006) interpret the same data to show that mass gain from snowfall, particularly in the Antarctic Peninsula and East Antarctica, exceeds dynamic losses from West Antarctica. More importantly, however, Monaghan et al. (2006) and van den Broeke et al. (2006) found very strong decadal variability in Antarctic accumulation, which suggests that it will require decades of data to separate decadal variations from long-term trends in accumulation; for instance, associated with climate warming.

The present total mass balance of Antarctica and its deglaciation history from the Last Glacial Maximum (LGM) are still poorly known. It has been shown recently that the uplift rates derived from the Global Positioning System (GPS) can be employed to discriminate between different ice-loading scenarios. There is general agreement that Antarctica was a major contributor to deglaciation at the end of the last glacial age, with the West Antarctic Ice Sheet perhaps contributing more than 15 m to rising sea level during the last 21 000 years (Clark et al. 2002). The main controversy is over whether or not the dominant Antarctic melt contribution to sea-level rise occurred during the Holocene or earlier, during the initial deglaciation phase (21 000–14 000 years ago) of northern-hemispheric ice sheets (Peltier 1998). Post-glacial rebound rates are not well constrained and are an error source for ice mass-balance assessment with GRACE satellite data. Analyses of GRACE measurements for 2002–5 show the ice sheet to be very close to balance with a gain of 3 ± 20 Gt/year (Chen et al. 2006) or net loss from the sheet ranging from 40 ± 35 Gt/year (Ramillien et al. 2006) to 137 ± 72 Gt/year (Velicogna and Wahr 2006), primarily from the West Antarctic Ice Sheet.

Using these, progressively improving data sets the mass balance has been recently updated (Rignot et al. 2008b) to show that:

- the largest losses are concentrated along the Amundsen and Bellingshausen sectors of West Antarctica and in the Indian Ocean sector of East Antarctica;
- a few glaciers in West Antarctica lose a disproportionate amount of mass. The largest mass loss is from ice flowing into Pine Island Bay. This drainage basin contains enough ice to raise sea level by 1.2 m;
- in East Antarctica, many of the glaciers are close to a state of balance.

7.3.3 Causes of Changes

In this subsection potential causes of the observed behavior of the ice sheets are discussed.

Changes in Snowfall and Surface Melting

Recent studies find no continent-wide significant trends in Antarctic accumulation over the interval 1980–2004 (van den Broeke et al. 2006; Monaghan et al. 2006), and surface melting has little effect on Antarctic surface mass balance. Modeling results indicate probable increases in both snowfall and surface melting over Greenland as temperatures increase (Hanna et al. 2008; Box et al. 2006). An update of estimated Greenland Ice Sheet runoff and surface mass balance (i.e. snow accumulation minus runoff) results presented in Hanna et al. (2005) shows significantly increased runoff losses for 1998–2003 compared with the 1961–90 climatologically "normal" period. But this was partly compensated by increased precipitation over the past few decades, so that the decline in surface mass balance between the two periods was not statistically significant. However, because there is summer melting over approximately 50% of Greenland already (Steffen et al. 2004), the ice sheet is particularly susceptible to continued warming, with probably irreversible loss of the ice sheet if local temperatures increase by more than 3°C (Gregory et al. 2004). This loss would be caused solely by imbalance between snowfall and melting, and would be accelerated by glacier acceleration of the type we are already observing.

In addition to the effects of long-term trends in accumulation/ablation rates, mass-balance estimates are also affected by interannual variability. This increases uncertainties associated with applying measured surface accumulation/ablation rates to mass-balance calculations, and it results in a lowering/raising of surface elevations measured by altimetry (e.g. van der Veen 1993). Remy et al. (2002) estimate the resulting variance in surface elevation to be around 3 m and as much as 10 m over a 30-year timescale in parts of Antarctica. This clearly has implications for the interpretation of altimeter data.

Ongoing Dynamic Ice Sheet Response to Past Forcing

The vast interior parts of an ice sheet respond only slowly to climate changes, with timescales up to 10 000 years in central East Antarctica. Consequently, current ice-sheet response probably includes a component from ongoing adjustment to past climate changes. Model results (e.g. Huybrechts 2002; Huybrechts et al. 2004) show only a small long-term change in Greenland ice-sheet volume, but Antarctic shrinkage of about 90 Gt/year, concomitant with grounding-line retreat since the LGM.

Dynamic Response to Ice-Shelf Break-Up

Recent rapid changes in marginal regions of both polar ice sheets include regions of glacier thickening and slowdown but mainly acceleration and thinning, with some glacier velocities increasing two- to sevenfold (Scambos et al. 2004). Most of these glacier accelerations closely followed reduction or loss of ice shelves. Such behavior was predicted over 30 years ago by Mercer (1978), but was discounted, as recently as the IPCC Third Assessment Report (TAR; Church et al. 2001), by most of the glaciological community based largely on results from prevailing model simulations. None of those simulations included the full stress configuration, and basal processes were also largely ignored or parameterized simplistically. Considerable effort is now underway to improve the models, but the effort is far from complete, leaving us unable to make reliable predictions of ice-sheet responses to a warming climate if such glacier accelerations were to increase in size and frequency. It should be noted that there is also a large uncertainty in current model predictions of the atmosphere and ocean temperature changes which drive the ice-sheet changes, and this uncertainty is probably at least as large as that on the marginal flow response.

Total breakup of the Jakobshavn ice tongue in Greenland was preceded by its very rapid thinning, most probably caused by a massive increase in basal melting rates (Thomas et al. 2003). Despite an increased ice supply from accelerating glaciers, thinning of more than 1 m/year, and locally more than 5 m/year, was observed between 1992 and 2001 for many small ice shelves in the Amundsen Sea and along the Antarctic Peninsula (Shepherd et al. 2003; Zwally et al. 2005). Thinning of approximately 1 m/year (Shepherd et al. 2003) preceded the fragmentation of almost all (3300 km^2) of the Larsen B ice shelf along the Antarctic Peninsula in fewer than 5 weeks in early 2002 (Scambos et al. 2003), but breakup was probably also influenced by air temperatures.

A southward-progressing loss of ice shelves along the Antarctic Peninsula is consistent with a thermal limit to ice-shelf viability (Mercer 1978; Morris and

Vaughan 2003). Cook et al. (2005) found that no ice shelves exist on the warmer side of the −5°C mean annual isotherm, whereas no ice shelves on the colder side of the −9°C isotherm have broken up. Before the 2002 breakup of Larsen B ice shelf, local air temperatures increased by more than 1.5°C over the previous 50 years (Vaughan et al. 2003), increasing summer melting and the formation of large melt ponds on the ice shelf. These may have contributed to breakup by draining into, refreezing, and wedging open surface crevasses that linked to bottom crevasses filled with sea water (Scambos et al. 2000).

Almost all ice shelves are in Antarctica, where they cover an area of approximately 1.5 million km^2 with nearly all ice streams and outlet glaciers flowing into them. The largest ones in the Weddell and Ross Sea embayments also occupy the most poleward positions and are currently still far from the viability criteria cited above. By contrast, Greenland ice shelves occupy only a few hundred square kilometers, and many are little more than floating glacier tongues. Ice shelves are nourished by ice flowing from inland and by local snow accumulation, and mass loss is primarily by iceberg calving and basal melting. Melting of up to tens of meters per year has been estimated beneath deeper ice near grounding lines (Rignot and Jacobs 2002). Significant changes are most readily caused by changes in basal melting or iceberg calving.

Ice-shelf basal melting depends on temperature and ocean circulation within the cavity beneath. Isolation from direct wind forcing means that the main drivers of sub-ice-shelf circulation are tidal and density (thermohaline) forces, but lack of knowledge of sub-ice bathymetry has hampered the use of three-dimensional models to simulate circulation beneath the thinning ice shelves.

If glacier acceleration caused by thinning ice shelves can be sustained over many centuries, sea level will rise more rapidly than currently estimated. But such dynamic responses are poorly understood and, in a warmer climate, the Greenland Ice Sheet margin would quickly retreat from the coast, limiting direct contact between outlet glaciers and the ocean. This would remove a likely trigger for the recently detected marginal acceleration. Nevertheless, although the role of outlet-glacier acceleration in the longer-term (multidecade) evolution of the ice sheet is hard to assess from current observations, it remains a distinct possibility that parts of the Greenland Ice Sheet may already be very close to their threshold of viability.

Increased Basal Lubrication

Observations on some glaciers show seasonal variations in ice velocity, with marked increases soon after periods of heavy surface melting (e.g. O'Neel et al. 2001). Similar results have also been found on parts of the Greenland Ice Sheet, where ice is moving at approximately 100 m/year (Zwally et al. 2002b). A possible cause is rapid meltwater drainage to the glacier bed, where it enhances lubrication of basal sliding. If so, there is a potential for increased melting in a warmer climate to cause an almost simultaneous increase in ice-discharge rates. However, there is little evidence for seasonal changes in the speeds of the rapid glaciers that discharge most Greenland ice. Nevertheless, this effect may result in "softening up" tributary ice that feeds the fast outlet glaciers, thus extending inland the impact of near-coastal glacier accelerations.

7.4 Mass Balance of Glaciers and Ice Caps

7.4.1 Measurements

Most measurements of the mass balance of small glaciers are obtained in one of two ways. *Direct* measurements are those in which the change in glacier surface elevation is measured with stakes and snow pits at the glacier surface (see section 7.2, Mass-Balance Techniques). The mass balance of the glacier as a whole is estimated by extrapolation from a network of such stakes. In *geodetic* measurements, the glacier surface elevation is measured at two times with reference to some external datum, usually sea level. Traditionally, the greater effort required for geodetic measurements has meant that they were limited mainly to checks on the accuracy of direct measurements (e.g. Cox and March 2004). Recent advances in remote sensing promise to alleviate this problem. Remote sensing offers much greater completeness of spatial coverage. At present, the observational database is dominated by traditional methods.

Survey of Available Material

The World Glacier Monitoring Service (WGMS; Haeberli et al. 2005a, 2005b) has collected, archived, and distributed mass-balance data based on observational results. From these and from several other new and historical sources, the annual mass-balance time series for over 300 glaciers have been constructed, quality-checked, analyzed, and presented by a number of authors (Braithwaite 2002; Cogley 2005; Cogley and Adams 1998; Dyurgerov and Meier 2005; Ohmura 2004). In the following the databases DM (Dyurgerov and Meier 2004, 2005), CA (Cogley 2005), and data published by Lemke et al. (2007) (see also Kaser et al. 2006) will be discussed in more detail. The number of single-glacier measurements of mass balance is small before 1960, and most records are short. Only about 50 extend over more than 20 years. For the most recent years no data have yet become available for some measured glaciers.

In the construction of DM, single-glacier time series were updated and were assigned to 49 larger, climatically homogeneous regions (Dyurgerov and Meier 2004, 2005). In forming the regional averages, each time series was weighted by the area of its glacier. Then the regions, weighted by their glacierized surface areas, were assigned to 13 larger regions and finally combined into six large glacier systems. Several steps of area weighting were applied to avoid suggested biases towards the use of smaller and isothermal glaciers.

The CA dataset (Cogley 2005) is obtained by applying a spatial interpolation algorithm (Cogley 2004). At each glacierized cell in a 1° × 1° grid a two-dimensional polynomial is fitted to the single-glacier observations, and the resulting estimate of specific balance is multiplied by the glacierized area of the cell. This approach offers a natural means of incorporating spatial and temporal sampling uncertainties explicitly.

In summarizing recent observations of small glaciers, Lemke et al. (2007) incorporate the analyses of CA, DM, excluding peripheral ice bodies in Greenland and Antarctica, and a similar dataset due to Ohmura (2004). The three analyses are methodologically independent treatments of very similar compilations of data and may be taken, as in Lemke et al. (2007), as a small sample of reasonable analytical outcomes. The raw measurements are correlated, but the sample average is probably the best consensus estimate of the evolution of mass balance, and its standard deviation the best estimate of uncertainty.

Annual and Pentadal Resolution

Figure 7.3 shows annual time series of total global mass balance from DM and CA, illustrating the substantial interannual variability which persists even after single-glacier measurements have been composited into a global average, with (DM) or without (CA) correction of the bias due to the non-uniform spatial distribution of the measured glaciers. The standard deviation of CA, for example, is 79 Gt/year for 1970–9 and 233 Gt/year for 1995–2004 after removal of linear trends. The variability of the spatially corrected DM series is noticeably less. Spatial bias and trends are discussed further below.

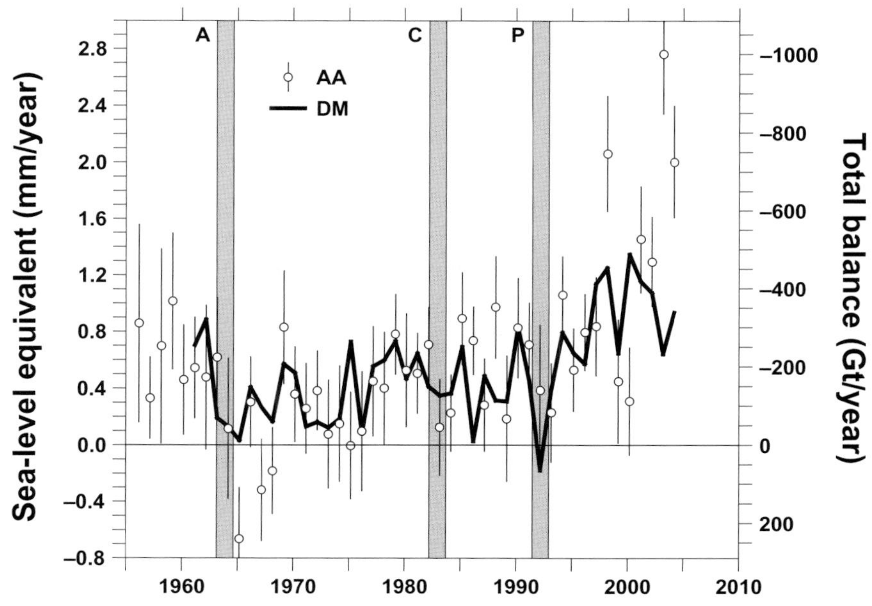

Figure 7.3 Annual global rates of small-glacier mass balance. AA, arithmetic averages of single-glacier observations from CA before spatial correction, with global area taken to be the same as in DM for ease of comparison; error bars represent twice the formal standard error. DM, averages based on area weighting as described in the text. Shaded bars are 18-month intervals following volcanic eruptions: A, Agung; C, El Chichón; P, Pinatubo.

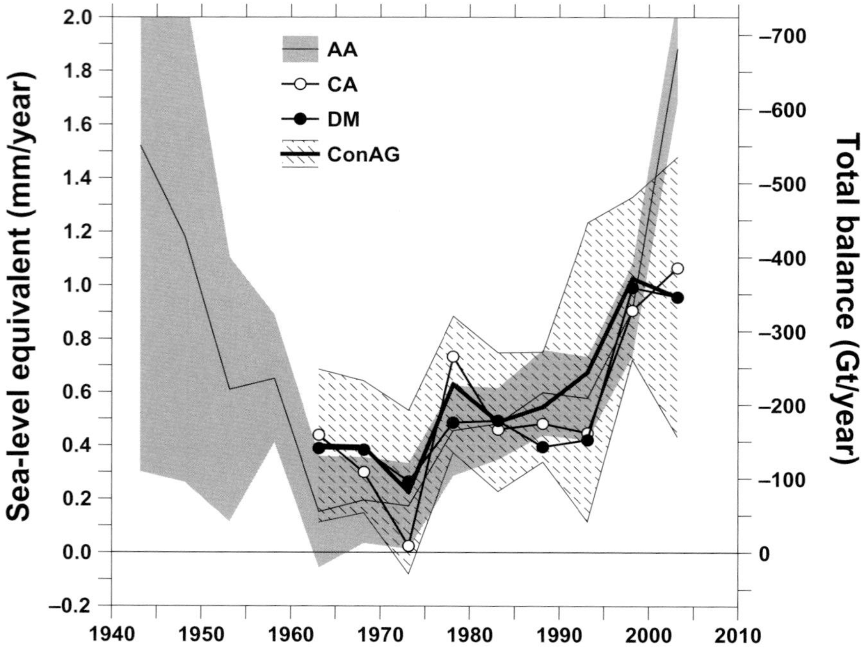

Figure 7.4 Pentadal global rates of small-glacier mass balance. For abbreviations, see Table 7.2. Confidence regions are ±2 standard errors for AA and ±2 standard deviations for ConAG. For ease of comparison, we adopt DM's total area in illustrating the series AA and CA.

Detrended annual mass-balance time series seldom or never exhibit serial correlation (Cogley and Adams 1998). That is, the autocorrelation of the series with itself is zero for lags greater than zero. Averages of the balance over n-year spans are thus less uncertain by a factor of $1/\sqrt{n}$. The raw uncertainty of mass-balance measurements is substantial, so this is one of our motives for focusing on pentadal averages below.

The strongest explosive volcanic eruptions have a recognizable impact in Figure 7.3 despite the noisiness of the series. Cooling and less-negative mass balance follow each eruption over 1–2 years, after which the glacier system returns to the previously established mode. Signals from other global climatic perturbations, such as El Niño, and regional-scale signals such as the North Atlantic Oscillation (NAO), cannot be identified.

The rate of mass loss has increased, as shown in Figure 7.4, in which annual measurements are composited into 5-year spans (see also Table 7.2). The arithmetic-average curve AA is the only curve extending before 1960 because measurements are too few at those times for area-weighting or spatial interpolation to be practicable. The early measurements suggest weakly that mass balances were negative, in circumstantial agreement with other lines of evidence such as observed patterns of retreat and the evolution of Northern Hemisphere temperatures. An earlier analysis by Meier (1984) relied on the balance amplitude, which is one half the difference between winter- and summer-season balance, to extrapolate modeled estimates and limited measurements at the global scale for 1900–61. The result, +0.46 ± 0.52 mm SLE/year, remains the only primarily

Table 7.2 Estimates of global small-glacier mass balance.

Source	Area (10^6 km^2)	Start of span	End of span	Specific balance (kg/m^2 per year)	Total balance (Gt/year)	SLE (mm SLE/year)
AA	0.572	Oct 2000	Sep 2004	−874 ± 92	−500 ± 53	+1.38 ± 0.14
CA	0.572	Oct 2000	Sep 2004	−491 ± 50	−281 ± 29	+0.78 ± 0.08
DM	0.785	Oct 2000	Sep 2004	−441 ± 93	−346 ± 73	+0.96 ± 0.20
ConAG	0.785	Oct 1960	Sep 1992	−163 ± 120	−128 ± 94	+0.35 ± 0.26
ConAG	0.785	Oct 1992	Sep 2003	−375 ± 198	−294 ± 156	+0.81 ± 0.43

SLE, sea-level equivalent. AA, Cogley (2005), arithmetic average of single-glacier balances; CA, Cogley (2005), spatial bias corrected by polynomial interpolation; DM, Dyurgerov and Meier (2005), spatial bias corrected by area-weighting of regional averages; ConAG, consensus estimate; average of estimates from CA, DM, and Ohmura (2004) including glaciers in Greenland and Antarctica.

observational estimate of the small-glacier contribution to sea-level rise for the early 20th century.

After 1960, AA tracks the other curves with fair accuracy through most of the pentades. Apparently spatial bias, while not negligible, is of only moderate significance. However the AA estimate for 2001–5 (for which only 4 years of data are available at present; Table 7.2) is starkly discordant. The discordance is due at least partly to the European heat wave of 2003. When European balances for 2003 are excluded, the annual arithmetic average increases from −1274 to −868 kg/m^2 per year. For 2001–2005, the same exclusion alters the pentadal average from −874 to −755 kg/m^2 per year. So the impact of the regional heat wave is substantial but does not explain all of the discordance. More broadly, the very negative northern mid-latitude balances for 2003 are offset by the high Arctic latitudes, where glacierization is extensive but balances are relatively few in number and only moderately negative. We emphasize that AA is not to be considered a serious estimate of global mass balance. Its confidence region is certainly too small, illustrating "the risks of sampling strategies based on studying only the places where the largest changes are taking place" (Zwally et al. 2005).

The spatially corrected pentadal series show that mass balance was close to zero in about 1970, since when it has been growing more negative. During the period approximating to that of a climatic normal (1960–92; Table 7.2), small glaciers supplied about +0.3 to +0.4 mm SLE/year. In the most recent pentade the rate is more than double this value, at about +0.8 to +0.9 mm SLE/year when Greenland and Antarctica are allowed for. The question of what to do about Greenland and Antarctic glaciers is not trivial. It is discussed in the next section.

Uncertainties are quoted as twice the standard error for AA, CA, and DM and as the standard deviation for ConAG (the consensus estimate; average of estimates from CA, DM, and Ohmura (2004) including glaciers in Greenland and Antarctica); density of ice was taken as 900 kg/m^3 and area of ocean as 362×10^6 km^2. The trend in global average pentadal mass balance is +0.018 ± 0.002 mm SLE/year for CA during 1970–2004 and is highly significant. Similar results hold for other compilations, for annual balance series, and for 1960–2004, the latter yielding slightly smaller trends, as Figure 7.4 would suggest.

The spatially corrected curves agree well. This suggests that the existing measurement network can provide a reasonable basis for global estimates when appropriate corrections are applied. Together, the independent analyses leave little or no room for reasonable doubt that glacier mass balance is negative at present and that it has become more negative since about 1970, when it was not very far from a state of equilibrium. In terms of sea-level equivalents, the new results show a larger eustatic contribution than estimated in previous calculations (Church et al. 2001). Extrapolating to the present (2008), it is probable that, the small-glacier contribution to the water balance of the ocean is nearly +1.1 mm SLE/year, and it continues to increase.

Recently, Cogley (2009a) incorporated a large number of geodetic measurements that did not contribute to the earlier analysis, as well as some of the first measurements of small-glacier mass balance by gravimetry, using data for southern Alaska from the GRACE mission (Arendt et al. 2008). The result was an increase to +1.4 mm SLE/year (averaged over 2000–5). Seemingly, much of the increase was due to better representation of calving glaciers in the newly added geodetic measurements. It has long been suspected that calving glaciers have more negative mass balances than land-terminating glaciers. Hock et al. (2009) relied on modeling of mass balance in terms of temperature and precipitation from meteorological reanalyses, and found that the mass balance of peripheral glaciers in Greenland and Antarctica was more negative than as estimated crudely in Lemke et al. (2007). It is not possible to combine the revised estimates of these two quite different studies, but they address independently two of the leading uncertainties described in section 7.4.2, and they both suggest that the current small-glacier contribution to sea-level rise is larger than estimated by Lemke et al. (2007).

Confirming the results of mass-balance analyses, Bahr et al. (2009) analyzed measurements of the accumulation-area ratio or AAR (the ratio of the area above the equilibrium-line altitude to the total area of the glacier), finding that the global average of measured AAR for 1997–2006 was 44%. A credible estimate of the average AAR for a state of balance is 57%. They suggest, relying on the method of volume-area scaling, that this discrepancy represents a commitment to continued mass loss in the future equivalent to 184 ± 33 mm SLE, simply for the glaciers to come into balance with the climate as it is today.

7.4.2 Uncertainties

Uncertainties in measured glacier changes arise from three sources: errors in the measurements, inadequate sampling in time, and inadequate sampling in space. Addressing these inadequacies in the language of conventional error analysis – that is, with error bars – is a challenging problem which, although it is not solved, is not insurmountable.

Measurement Errors

Mass-balance measurements are relatively simple, but our interest lies mainly in the interpretation of compound quantities. For example, a typical direct measurement is built up from simple raw observations with rulers and balances. These point measurements can be made with independent errors of ±50 kg/m^2 per year or better, but it is seldom possible to measure enough stakes to quantify adequately the local spatial variability. Direct measurements usually cover only the surface balance of the glacier. Internal ablation is accounted for only in reports for some Alaskan glaciers (Cox and March 2004). Its magnitude may reach a few tens of kilograms per square meter per year on temperate glaciers if the potential energy loss of draining meltwater is all converted to heat. Internal accumulation happens in the lower percolation zones of cold glaciers (those whose internal temperatures are below freezing) when surface meltwater percolates beneath the current year's accumulation of snow. It is impractical to measure, and is difficult to model with confidence in the absence of subsurface temperature or density measurements. It is treated unevenly; some measurements correct for it, but most do not. When studied in detail, its magnitude is found to be 10–100% of annual net (surface plus internal) accumulation (e.g. Bazhev 1980; Rabus and Echelmeyer 1998). There are probably many more cold glaciers than temperate glaciers, and neglect of internal accumulation is probably the largest single bias (Cogley 2009a) affecting mass-balance measurements.

The calving of icebergs is a significant source of uncertainty (e.g. Dowdeswell and Hagen 2004). Over a sufficiently long averaging period, adjacent calving and non-calving glaciers ought not to have very different balances, but calving is a process whose timescale can be quite different from the annual scale of surface mass balance and it is difficult to match the two. Most small glaciers which calve have grounded termini. The calving of floating glaciers, of course, has no immediate effect on sea level, but it may accelerate ice discharge across the grounding line. Calving glaciers are particularly numerous in Alaska, Patagonia, and at high Arctic and Antarctic latitudes, but they are underrepresented in the list of measured glaciers. Warrick et al. (1996) suggested that iceberg discharge into the sea from small glaciers totals 50 ± 30 Gt/year, or from +0.06 to +0.20 mm SLE/year, but this is an untested and uncertain result. The loss is probably greater considering that Columbia Glacier, Alaska, alone has discharged 7 Gt/year or more in recent years (O'Neel et al. 2005). The direction of the bias due to under-measurement of calving glaciers is uncertain, but probably positive (opposite to the internal-accumulation bias).

The uncertainties in geodetic measurements of mass balance are quite different. Elevations measured with respect to sea level at two different times must be collocated as accurately as possible. In modern instrumental configurations the accuracy obtainable with GPS is more than adequate. For example repeated surveys by laser altimetry can achieve meter-scale horizontal accuracy and decimeter-scale vertical accuracy (Arendt et al. 2002). However, such surveys have been possible only in the last decade, and larger errors are inevitable when comparing laser-altimeter elevations with elevations read from old topographic maps which may be uncertain by tens of meters. A further difficulty with remote-sensing measurements is that only elevations can be measured. Density must be modeled. A standard model is given by Sorge's Law (Bader 1954), which asserts that firn density beneath the surface in the accumulation zone varies only with depth, not with time. There is good evidence for this assertion in the dry snow zones of the ice sheets, but it is more problematic in the percolation zone where internal accumulation is significant and variable, and its uncertainty is a matter for conjecture when ground truth is not available.

Global mass-balance estimates suffer from uncertainty in the total area. Cogley (2003) suggests that the $1° \times 1°$ areas used in CA are uncertain by ±6–8% (of the area of the $1° \times 1°$ cell). Most of the regional areas used in DM are less uncertain, but several are based on unsatisfactory maps. Rates of area loss are not known accurately enough to be accounted for in calculations, although at least up to the present day this introduces only a small error. A further problem is deciding how to delineate the ice sheets. Different decisions yield proportionally large differences in the extent of peripheral glaciers in Greenland (0.054 million m^2 in CA against 0.076 million m^2 in DM). The definition of the Antarctic Ice Sheet is less ambiguous, but CA and DM treat the glaciers of peripheral islands differently. CA (Table 7.2) omits all of them, in effect assuming that their balance is zero. DM includes all of them, taking their area as 0.169 million m^2 (Shumskiy 1969) and assuming that their average balance is the same as that in the Canadian High Arctic. Neither of these assumptions is satisfactory, but *in situ* measurements are almost entirely lacking. Better information about the southward extension of the warming observed in the Antarctic Peninsula is needed urgently.

Temporal and Spatial Sampling Errors

The principal limitations of the existing database of mass-balance measurements, as regards temporal variability, are its lack of historical depth and its instability. The database before 1960 is so limited that very little can be said about the evolution of mass balance. This can only be addressed by turning to indirect estimates of the health of glaciers such as changes of glacier length, which require consideration of the interactions between mass balance and glacier dynamics. This problem in turn requires a modeling approach.

By "instability" we mean that measured glaciers are a shifting population. Not only does their total number fluctuate, but also the list of measured glaciers changes continually. Among the more than 300 measured glaciers the most common record length is 1 year, and many longer records have interruptions. These difficulties can be addressed by assuming that each single annual measurement is a random sample of the average balance over whatever longer period is of interest. However, the temporal variance of such a short sample is difficult to estimate satisfactorily, especially in the presence of a trend. The impact of the shifting population on sequences of annual global average mass balance is a topic which remains unexplored.

Spatial undersampling affects mass-balance measurements at two different spatial scales. On any one glacier, a small number of point measurements must represent the balance of the entire glacier. At regional and global scales, the number of measured glaciers, about 300, is tiny by comparison with the total number of glaciers, estimated at 160 000 by Meier and Bahr (1996).

On single glaciers of moderate size it is reasonable to assume that the mass balance depends only on the surface elevation. It increases from being negative (net loss) at the bottom to positive (net gain) above the equilibrium-line altitude. Variations at any one elevation are then regarded as random in character, but networks of measurement stakes are often not organized so as to capture this random variability. Trabant and March (1999), and earlier workers, have shown that a typical uncertainty for elevation-band averages of mass balance is ±200 kg/m^2 per year, consistent with measurements on radar traverses by Pinglot et al. (2001) and Pälli et al. (2002) in Svalbard. In future such measurements will improve estimates of local variability greatly, but they will not reduce the uncertainty in the whole-glacier balance by much, for Cogley (1999) showed that measurements at different elevations are highly correlated. It seems necessary to accept that measurements of glacier mass balance have substantial uncertainty due mainly to spatial variability.

Just as the point mass balance "decorrelates" slowly on single glaciers as the vertical separation between points increases, so in the horizontal direction the mass balance of any one glacier is found to be a good guide to the balance of nearby glaciers. At this scale, the distance to which single-glacier measurements yield useful information is of the order of 600 km. We can be moderately confident about estimates of regional mass balance when the region contains a moderate number of measured glaciers. Glacierized regions with few or no measured glaciers obviously pose a more serious problem. For a region without even nearby measurements there is, in a statistical sense, no better estimate than the global average with a suitably large uncertainty attached. Somewhat better estimates can plausibly be made by analogy with similar regions at different longitudes, or at the same latitude in the opposite hemisphere, but there is no reliable way to determine the error of such estimates. Among intensely glacierized regions with very few measured glaciers Patagonia, Tibet, most of the periphery of Greenland and the Russian Arctic may be singled out.

Patagonia is instructive because recent geodetic measurements covering over half of the ice in the region (Rignot et al. 2003) suggest a regional mass balance of the order of −1000 kg/m^2 per year, far more negative than the global average estimated from direct measurements alone. There is no direct way at present of knowing whether there are surprises in store in the other regions with very poor coverage.

7.4.3 Potential Improvements

The following problems stand out as being in need of attention.

- Internal accumulation: we need to seek better understanding of this bias through field work and better correction methods through climatic modeling.

- Calving glaciers: improved predictive ability is needed (e.g. van der Veen 2002), as are more routine measurements on calving glaciers.
- Spatial and temporal coverage: efforts should be made to improve the number and distribution of balance-measurement programs. Alaska, Greenland, the Russian Arctic, High Mountain Asia and Tibet, Patagonia, and Antarctica deserve special consideration. Short time series are not necessarily a poor commitment of effort and expense, but long-term programs sponsored by well-established field agencies are likely to be more productive.
- Better basic information: completion of the World Glacier Inventory (Haeberli et al. 1989; Cogley 2009b) would resolve many low-level problems, increasing the reliability of extrapolation from measured glaciers and facilitating the development of indirect (e.g. morphometric) methods of mass-balance estimation. A related problem is the distribution of glacier sizes. Most measurements are on smaller glaciers (modal size 2–4 km^2) while the actual size distribution has a mode near 256–512 km^2. There is some evidence that the smaller glaciers have more negative balances than the larger ones.
- New technology: new but already-validated measurement methods should be pursued aggressively. These methods include repeated laser altimetry (e.g. Abdalati et al. 2004; Bamber et al. 2004) and radar mapping (Rignot et al. 2003). Special attention should be paid to methods which promise large increases in the completeness of spatial coverage, such as passive-microwave monitoring of melt extent (Wang et al. 2005), and to the development of algorithms for converting such observations to mass balances.
- The error budget: apart from reducing measurement and sampling errors, more work is needed on the assimilation of direct (mostly annual) and geodetic (nearly always multiannual) balance measurements, within a uniform framework for assigning uncertainties to observations whose time spans are different.

7.5 Glacier, Ice-Cap, and Ice-Sheet Modeling

Glaciologists have struggled to find a viable method for calculating the glacier contribution to sea-level rise (Oerlemans and Fortuin 1992; Zuo and Oerlemans 1997; Oerlemans et al. 1998; Gregory and Oerlemans 1998; Raper and Braithwaite 2006). Apart from attempts with simple global heuristic models (Oerlemans 1989; Wigley and Raper 1995, 2005), the basis for regionally based projections, with one exception, has been regional mass-balance sensitivity to a 1 K warming. We explain below why this method loses accuracy as soon as the glacier areas start to change. This poses a problem for discussing uncertainties in projections of global glacier volume change, which must be derived from a concatenation of uncertainties from various steps in the calculations. However, several steps in the calculation are common to all methods, which allow some comparison of results.

Modeling the contemporary evolution of ice sheets addresses both the modeling of contemporary surface mass-balance changes and the ensuing ice-dynamic response. The latter includes the response to both contemporary atmospheric and oceanic forcing and to past changes in boundary conditions to which the ice sheets may still be responding owing to their long reaction times.

7.5.1 Glaciers

Climate input data are needed to calibrate and drive glacier and ice-cap volume-change models. For the 20th century both gridded observationally based data (New et al. 1999) and AOGCM results are available. For the future we must rely on the AOGCM projections. The problem of downscaling gridded climate data to local glacier conditions is overcome by using temperature and precipitation anomalies from the local climatology (see for example van de Wal and Wild 2001; Schneeberger et al. 2003). Uncertainties in future greenhouse gas emissions and AOGCM model response carry forward into uncertainties in the future contribution of glacier and ice-cap melt to sea-level rise. These climate-input uncertainties are not considered further here. In the following sections we discuss the effects of uncertainty in reference-period glacier volume and mass balance, as well as the models available for making projections of sea-level rise.

Uncertainty of Ice Area and Volume

A prerequisite to estimating the contribution of glacier and ice-cap melt to sea-level rise over the next century is knowledge of the ice volume available for melt. Uncertainty in the volume of ice available for melt results from incomplete glacier inventories (Braithwaite and Raper 2002). Three volume estimates in sea-level equivalents range from 0.15 to 0.37 m (Ohmura 2004; Raper and Braithwaite 2005a, 2005b; Dyurgerov and Meier 2005) for glaciers and ice caps outside Greenland and Antarctica. The glaciers and ice caps of Greenland and Antarctica (excluding the ice sheets) are estimated to contain an additional 0.34 m of sea-level equivalent (Dyurgerov and Meier 2005). These estimates represent the potential for sea-level rise from glaciers and ice caps. Most estimates of the contribution to sea-level rise under global warming exclude the Greenland and Antarctic glaciers and ice caps. One reason for this is that inventory data for these regions are poor or non-existent. But another reason is that it is not clear where the ice sheets end and the glaciers begin in, for instance, the Antarctic Peninsula and an additional problem is that ice-sheet retreat under global warming will result in ice masses becoming separated from the ice sheets and thus increasing the ice volume in the glacier and ice-cap category.

Paucity of glacier and ice-cap inventory data is the main cause for the uncertainty in the ice volumes. Outside Greenland and Antarctica, about half the ice volume may be contained in the ice caps, which have relatively large individual areas but only cover about 23% of the total area (Raper and Braithwaite 2005a). Likewise the largest proportion of glacier ice volume is contained in the glaciers with the biggest surface areas, so it is necessary to know the size distribution of glacier areas: information that is not yet available in many regions. The best way forward for reducing uncertainties in ice volume is to prioritize the addition of the largest glaciers to the glacier inventory (Meier et al. 2005; Raper and Braithwaite 2005b). Details of the size distribution of the smaller glaciers can then be estimated statistically using percolation theory (Bahr and Meier 2000) or topographic information (Raper and Braithwaite 2005a). The largest glaciers should also be a priority for ice-thickness measurements since such measurements are scarce (Bahr et al. 1997).

Uncertainties in Observed Mass Balance

Spatially distributed observed mass balances, which correspond to a reference period, are needed to calibrate or verify models of change in glacier and ice-cap volume. These reference mass balances define the degree of disequilibrium of the reference glaciers and ice caps and can be expressed as the rate of change of glacier volume or sea-level rise over the reference period. Model-based estimates of 20th-century sea-level rise from glaciers and ice caps are particularly sensitive to uncertainties in late-20th-century global mass balance (Raper and Braithwaite 2006). Their projections of ice melt were based on a 1961–90 estimate of 0.19 mm/year based on Dyurgerov and Meier (1997). Late-20th-century observation-based estimates of the rate of volume change in sea-level equivalent from glaciers and ice caps excluding Greenland and Antarctica are very uncertain, ranging from −0.05 to +0.67 mm/year for 1960–92. Clearly uncertainties in projections of volume change will reflect uncertainties in reference-period mass balance.

Modeling of Surface Mass Balance

Mass-balance modeling is concerned with the interaction of the climate with the surface of the glacier, whereas ice-dynamic modeling is concerned with the response of the ice geometry and hence changes in the ice exposure to climate. Using coupled mass-balance and ice-flow models, the response of several glaciers in different climate settings under global warming is compared in Oerlemans et al. (1998) and in Schneeberger et al. (2003) using two- and three-dimensional models respectively. The challenge for sea-level rise projections is to extend the modeling to all glaciers and ice caps as attempted by Raper and Braithwaite (2006) using a simplified representation of the mass-balance profiles and a simple geometric glacier model that treats the ice dynamics implicitly.

The challenge is to interpolate or extrapolate mass-balance model results to give estimates of the time evolution of mass balance over all glaciers and ice caps. The sensitivity of glacier mass balance to a 1°C warming is highly climate-dependent and ranges over an order of magnitude between the low sensitivity of cold/dry continental glaciers to the high sensitivity of warm/wet maritime glaciers. Based on modeled mass-balance sensitivity of 12 representative glaciers, Oerlemans and Fortuin (1992) derived a relationship between mass-balance sensitivity and annual mean precipitation. Mass-balance modeling of a further 61 glaciers confirmed this relationship (Braithwaite and Raper 2002). An extension of this approach is to use regional and seasonal mass-balance sensitivities to both changes in temperature and precipitation (Oerlemans and Reichert 2000).

Assume for a moment that glacier areas are constant over 1 year and that the mass-balance sensitivities are linear with temperature and precipitation over the changes considered. Then glacier mass-balance sensitivities multiplied by the corresponding temperature and precipitation changes and the corresponding glacier areas can correctly estimate glacier volume changes. However, this method soon becomes inaccurate for longer timescales and/or climate changes when the glacier areas over which the mass-balance sensitivities have been estimated change. This was a problem with the method of Gregory and Oerlemans (1998), which the analysis of van de Wal and Wild (2001) attempted to address by scaling down the glacier area as the volume decreased. The resulting average reduction in the glacier melt projections of about 20% by 2100 was similar to the 25% estimated in the modeling analysis of Oerlemans et al. (1998). Schneeberger et al. (2003) get a similar average reduction in 2050. However, this *ad hoc* adjustment of the area is an unsatisfactory solution from a theoretical point of view and would certainly not work for longer timescales as illustrated by the fact that the method predicts the eventual melt of the entire ice volume for any (even very small) temperature increase. The problem arises from the use of mass-balance sensitivity for projections because the mass-balance perturbation for a sustained temperature change decreases and tends to zero as a mountain glacier comes to a new equilibrium. This occurs because the melting increases preferentially at lower altitudes causing the mountain glacier to retreat up the mountain. On the other hand, for ice caps on a flat bed the mass-balance perturbation will tend to increase because melting causes them to shrink to lower elevations.

Instead of interpolating mass-balance sensitivity over all glacier areas a better approach is to interpolate the actual mass-balance profiles. The mass-balance profiles perturbed by climate changes can then be used to drive models of ice geometry. This was the approach used by Raper and Braithwaite (2006) who regress modeled grid-point mass-balance altitudinal gradients of accumulation and ablation in seven regions on annual precipitation and summer temperature from a gridded climatology and apply the resulting multiple regression equation to all grid cells with glaciers. Raper and Braithwaite (2006) find that melt projections over the next century are sensitive to uncertainty in the mass-balance profiles. For comparison with earlier studies, they increase the temperature in the model by 1 K for the summer months considered to get a globally averaged surface mass-balance sensitivity of −0.35 m/year per K, which compares with previous estimates of −0.40, −0.35, and −0.41 m/year per K (Oerlemans and Fortuin 1992; Dyurgerov and Meier 2000; Braithwaite and Raper 2002). Summer mean temperature changes in the Raper and Braithwaite (2006) analysis are simply used to perturb the equilibrium line altitude (ELA) according to the assumed lapse rate, so changes in the balance gradients with climate change are not considered. Schneeberger et al. (2003), in an analysis over several glaciers and glacierized regions, show that the use of daily compared to monthly mean input data gives similar results. However, the use of monthly mean input is clearly preferable to simply using summer mean temperature changes (as in Raper and Braithwaite 2006) since the later is not able to discriminate changes in the length of melt season.

Another limitation of the Raper and Braithwaite (2006) analysis is that they do not consider precipitation changes. Increased precipitation over the glacier areas is generally projected with global warming but this may be counteracted to some extent by less snowfall and more rainfall (Schneeberger et al. 2001). For seasonally uniform temperature rise in a variety of climate regimes, Oerlemans et al. (1998) and Braithwaite et al. (2002) find that increases in precipitation of 20–50%/K and 29–41%/K respectively are required to balance increased ablation.

Modeling of Glacier Ice Dynamics

Changes in mass balance result in an adjustment of glacier and ice cap geometry by ice flow. The study of Oerlemans et al. (1998) mainly employed two-dimensional ice-flow models based on Oerlemans and Van der Veen (1984) and the study of Schneeberger et al. (2003) employed three-dimensional models based on Blatter (1995). Only one of the ice masses in the Oerlemans et al. (1998) study was a whole ice cap (King George Island ice cap, Antarctica; Knap et al. 1996). The response of two further ice caps in Severnaya Zemlya to the same climate change scenarios was modeled by Bassford et al. (2006). These large high-latitude ice masses show generally lower rates of change relatively to lower-latitude glaciers. Glacier and ice-cap volume models that explicitly model the ice flow are impractical for use over the tens of thousands of ice masses worldwide. Therefore, the glacier and ice-cap models used by Raper and Braithwaite (2006) were based on the geometric model of Raper et al. (2000), which needs minimal input data, but requires assumptions about glacier and ice-cap hypsometry and predetermines the area-altitude distribution for any area. The ice dynamics are thus treated implicitly and following Bahr et al. (1997) area-volume scaling is used. The slower melting of ice caps, which are situated mostly at high latitudes and are estimated to contain about half the ice volume, is reported in the global analysis of Raper and Braithwaite (2006).

Summary of Uncertainties in Glacier and Icecap Modeling

One category of uncertainties in predicting glacier and ice-cap volume change is the changes in climate at the glacier surface. Contributing to the uncertainty are uncertainties in greenhouse-gas emissions, regional climate model response, and downscaling to climate change at the ice surface. There are also uncertainties associated with extrapolating scarce observations of the ice masses and their environment to all glaciers and ice caps. The sensitivity of modeled volume change to uncertainties in reference glacier and ice-cap volume, reference mass balance, and reference mass-balance gradients are summarized in Table 7.3 based on Raper and Braithwaite (2006). This analysis used simplified geometric models with implicit ice dynamics, which need further testing and verifying through

Table 7.3 Sensitivity of modeled glacier and ice cap sea-level-rise contribution to ±10% uncertainty in various reference-period inputs based on Raper and Braithwaite (2006) for a single estimate of reference-period area.

Reference (1961–90)	Sea-level rise	
	20th century	21st century
Volume ±10%	≈0%	±4%
Mass balance ±10%	±9%	±1%
Mass-balance gradient ±10%	±2%	±6%

parallel runs with more complex models, in particular on the effect of rapid melting on area-altitude distributions. In addition, it should be noted that none of the models discussed above account for the effect of basal sliding and ice calving. Also, it is always assumed that all melt finds its way directly into the oceans.

7.5.2 Ice Sheets

Modeling of Surface Mass Balance

Future mass changes of the Greenland and Antarctic Ice Sheets depend on current mass changes and on the ice-dynamic response to environmental forcing on timescales as far back as the last glacial period. Modeling contemporary surface mass-balance changes over the ice sheets is a meteorological problem. Such modeling has typically relied on atmospheric reanalysis products, mostly from the European Centre for Medium-Range Weather Forecasting (ECMWF), regional meteorological models, and on studies with atmospheric global circulation models. A major problem common to all methods, however, is the paucity of direct surface observations to validate the models. It is also important to note that surface mass balance does not include ice outflow at the margin.

Precipitation over Greenland and Antarctica is determined by the net atmospheric moisture transport. Reanalysis data have notorious problems over the ice sheets: for example, they can predict negative precipitation due to badly calibrated assimilation procedures and the limited availability of synoptic control points, especially in the interior (Bromwich et al. 2004). Studies with atmospheric global circulation models, on the other hand, have significantly refined our understanding of ice-sheet mass balance (Wild et al. 2003). However, such studies are driven by global climate scenarios which do not exactly represent the actual forcing for a given time interval, making them less useful for reconstructing detailed mass-balance changes over short periods. The best tools available today are high-resolution limited-area models driven from their lateral boundaries by reanalysis data. The latter provide the "real weather" forcing which drive the synoptic processes responsible for the formation of precipitation. Examples of such regional atmospheric models are the Dutch RACMO (Regional Atmospheric Climate Model (KNMI)) model (van Lipzig et al. 2002, 2004; van den Broeke et al. 2006), the Pennsylvania State University Polar MM5 model (Bromwich et al. 2004; Box et al. 2006), or the German HIRHAM model (Dethloff et al. 2002; HIRHAM is a regional atmospheric climate model based on a subset of the HIRLAM (high resolution limited area model) and ECHAM (an atmosphere-only version of the European Centre Hamburg climate model) models (Danish Meteorological Institute (DMI) and Max Planck Institute Hamburg)). The advantage of these is that their much higher resolution can explicitly resolve storm processes at the steep ice-sheet margins, where they are important for generating precipitation. Such models regularly also take into account the processes of sublimation and snowdrift.

In a recent investigation with RACMO version 2, van den Broeke et al. (2006) found that the Amundsen Sea sector of West Antarctica and the western Antarctic Peninsula, both data-sparse regions, receive approximately 80% more precipitation than previously assumed in continent-wide assessments (Vaughan et al. 1999). This revises downwards the imbalance of the Pine Island and Thwaites Glacier basins known for their accelerating outlet glaciers (Thomas et al. 2004a, 2004b). van den Broeke et al. (2006) did not find a significant trend in Antarctic accumulation over the period 1980–2004, contrary to the significant upward trend of +1.3 to +1.7 mm/year reported by Bromwich et al. (2004) from the dynamical retrieval method (DRM) applied to ECMWF precipitation data between 1979 and 1999. For the shorter period between 1992 and 2003, van den Broeke et al. (2006) found large interannual variability in accumulation. The absence of a significant precipitation trend over the last 50 years was recently corroborated by Monaghan et al. (2006).

In Antarctica there is hardly any surface melting at present because of the cold conditions, but runoff is an important component of the mass balance of the Greenland Ice Sheet, where summer temperatures around the margin rise well above zero. Because of a lack of direct observations, surface-melt estimates similarly have to come from models. This may involve full energy-balance calculations (Lefebre et al. 2002; Box et al. 2004; Bougamont et al. 2005) but more commonly a positive-degree approach is used. This enables calculation of the runoff on the much finer grids required to properly deal with surface topography in the narrow coastal bands where runoff typically takes place (Janssens and Huybrechts 2000). The degree-day method is an index method in which it is assumed that the total melt rate during a summer season is proportional to the total amount of positive degree days (Ohmura 2001). Although the method lacks explicit physics, its main strength is its efficiency so that one can make long runs on high-resolution grids. Its inputs are the surface temperature and the precipitation rate, the most widely available meteorological parameters.

Hanna et al. (2009) reconstructed the surface mass balance of the Greenland Ice Sheet on a 5 km × 5 km grid for the period 1958–2007. Meteorological models were forced by ERA-40 reanalysis data for 1958–2007 to retrieve annual precipitation minus evaporation and monthly surface temperature to drive the runoff/

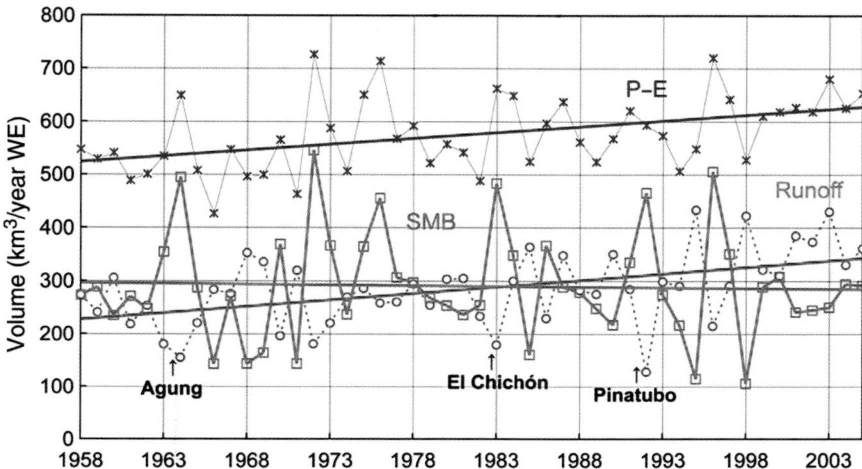

Figure 7.5 Reconstructed surface mass balance (SMB) of the Greenland Ice Sheet obtained from a positive degree-day runoff/retention model in conjunction with atmospheric re-analysis data. Units are water equivalent (WE) volume (km^3) per year. P-E, precipitation minus evaporation. (Modified from Hanna et al. 2008.)

refreezing degree-day model of Janssens and Huybrechts (2000)[5]. The surface mass balance shows a high year-to-year variability with distinct signals following major volcanic eruptions (Figure 7.5). The surface mass-balance time series reveals an upward trend of 123 km^3 in annual runoff from 1957 to 2007; moreover, six of the 10 highest-runoff years have occurred since 2001. The precipitation follows a significant increasing trend of 88.7 km^3 over this 50-year period; the additional precipitation, mainly in the form of snow accumulation, largely offsets rising Greenland runoff and the surface mass balance has an insignificant negative trend of −34.2 km^3 per year from 1958 to 2007 (Hanna et al. 2009). This recent increase in both surface melting and precipitation is also found in the studies by Box et al. (2006), using the polar MM5 regional atmosphere model for the period between 1988 and 2004, coincident with a period of marked warming over Greenland. Their study shows a coherent trend in surface mass balance with increases in the interior and decreases around the margin, an expected signature of climatic warming. However, mass-balance variability is still large and the period under investigation appears too short to establish a statistically significant trend.

A disadvantage of the comprehensive meteorological reanalyses is that they only generate data for the last few decades. Another approach, as followed by Meehl and Stocker (2007), is to scale high-resolution patterns from time-slice experiments with global circulation models with a suite of ice-sheet averaged time series superimposed on observed climatologies (Huybrechts et al. 2004; Gregory and Huybrechts 2006). Such a procedure can be generalized to calculate mass-balance changes for any given climate scenario based on the assumption that patterns of climate change are conserved in time. Figure 7.6 shows an example of resulting sea-level changes for both the Antarctic and Greenland Ice Sheets for the 20th and 21st centuries under a standard climate scenario. Over the 20th century historical forcing is applied. There is large variability due to the climate sensitivity of the various global circulation models, but for all combinations of

[5] ERA-40 is the reanalysis product provided by ECMWF (http://www.ecmwf.int/research/era/).

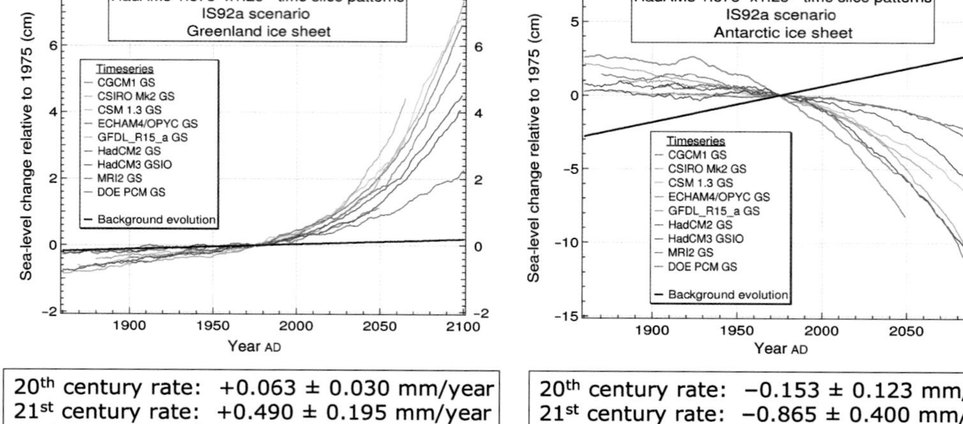

Figure 7.6 Volume changes of the Antarctic and Greenland Ice Sheets expressed in sea-level equivalent obtained from global circulation model-based experiments excluding the ice-dynamic response to contemporary mass-balance changes. The thick black line shows the long-term ice-dynamic background trend as a result of ongoing adjustment to past environmental changes as far back as the last glacial period. (Modified from Huybrechts et al. 2004.)

forcing, the Greenland Ice Sheet is projected to lose mass and the Antarctic Ice Sheet is projected to gain mass. Taking into account long-term background trends and the spread in the results, the combined effect of both ice sheets is, however, not significantly different from zero, in either the 20th or 21st century. This concerns the effect of surface mass-balance changes only, however.

Modeling of Ice Dynamics

Surface mass-balance changes obtained in the ways described above are usually fed into large-scale ice-sheet models to investigate the total ice-sheet response, thus including the residual ice-mass trend at present as well as any ice-dynamic responses to the contemporary forcing. Current state-of-the-art whole-ice-sheet models typically consider thermomechanically coupled flow, have separate treatments for grounded ice and ice shelves with some sort of stress regime coupling across grounding lines, consider isostatic adjustment of the bedrock, and have horizontal resolutions of between 10 and 40 km (e.g. Ritz et al. 2001; Huybrechts et al. 2004). Century-timescale simulations emphasize the role of surface mass-balance forcing for ice-volume changes. Unless high melting rates below ice shelves are prescribed, the ice-sheet dynamic response to contemporary forcing is usually less than 10% of the direct surface mass-balance effect. In these models the ice flow mainly reacts to perturbations in driving stress (ice thickness and surface slope) on characteristic timescales of 10^2–10^4 years dictated by the ratio of ice thickness to yearly mass turnover, physical and thermal processes at the bed and slow processes affecting ice viscosity and mantle viscosity.

Current large-scale models do not capture well the recent evidence for rapid ice-dynamical changes in several Greenland outlet glaciers and in parts of West Antarctica and the Antarctic Peninsula. This suggests that these changes are in response to processes not included in the comprehensive models such as oceanic erosion of floating ice shelves caused by ice-flow processes related to basal lubrication or inland stress transmission through longitudinal stress gradients following the loss of mechanical buttressing effects (Alley et al. 2005). Model resolution also plays a role as most of the ice is funneled in a number of fast-flowing features that transport the bulk of the ice towards the coast. A lower resolution numerically (and spuriously) widens the outlet glaciers and slows down the flow, and this affects their response. The development towards higher-resolution grids closely follows the evolution in computer power: an increase of horizontal resolution by a factor of 2 entails an increase in CPU time by a factor of 16.

A crucial issue also concerns the role of transition zones in the coupling of ice-sheet flow with ice-shelf flow. Ice shelves deform by stretching and ice sheets by shearing in horizontal planes. These contrasting flow regimes can both be modeled adequately with simplified solutions but it is unclear how they match at the grounding line and over which distance the transition takes place. In the simplest case, there is an abrupt transition between floating and grounded ice, with no stress transmission between them. In that case, grounded ice dynamics is largely unaffected by what happens in the ice shelf or at the grounding line. This view is probably valid for much of the perimeter of the East Antarctic Ice Sheet but breaks down in ice streams and the larger outlet glaciers. Early models considered transition zones in that way (Figure 7.7, left panel). Current large-scale models allow for some sort of transition zone between ice-shelf dynamics and

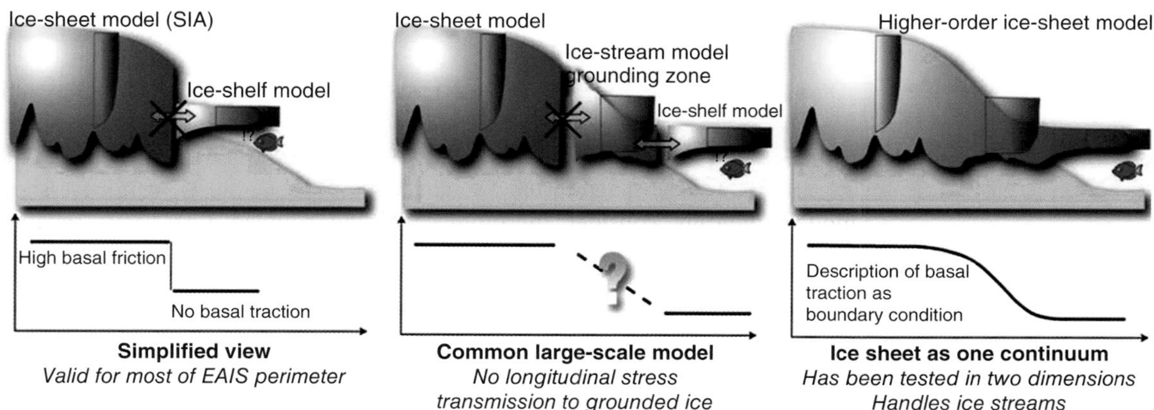

Figure 7.7 The handling of stress transfer across transition zones between floating and grounded ice and in ice streams and outlet glaciers has an important effect on marginal response timescales in numerical ice-sheet models. EAIS, East Antarctic Ice Sheet; SIA, shallow ice approximation. (After Pattyn et al. 2006.)

grounding-line dynamics but do not account for longitudinal stress transmission further inland (Figure 7.7, central panel). One solution is to consider the whole system as one continuum and solve the full stress equations over the whole model domain so that the width of the transition zone comes out as a natural manifestation of the basal boundary conditions (Figure 7.7, right panel). Such higher-order models have been tested and are able to handle stress transmission across grounding lines adequately (Pattyn et al. 2006). At this time, however, full stress tensor solutions are not computationally feasible for whole ice-sheet simulations and most progress is therefore expected from mixed or nested treatments.

A higher-order model was also applied by Payne et al. (2004) in a targeted study on Pine Island Glacier. They studied the inland transmission of perturbations at the grounding line and found two mechanisms playing. First there was an instantaneous mechanical response through longitudinal stress coupling which was felt up to 100 km inland. This was then followed by an advective-diffusive thinning wave propagating upstream on a decadal timescale with a new equilibrium being reached after about 150 years. These modeling results seem to explain many of the observations and provide a mechanism for fast adjustment to changes occurring at the margin.

Another process not linked to ice-shelf reduction is meltwater penetration to the bed, providing near-instantaneous communication between surface forcing and basal-ice dynamics. Zwally et al. (2002a) found a 10% increase in basal sliding velocity for a location in central west-Greenland that was apparently related to summertime meltwater penetration to the bed. This effect is missing from most numerical ice-sheet models. Inclusion in a model by Parizek and Alley (2004) found it to increase the sensitivity of the Greenland Ice Sheet to specified warmings by typically 10–15%, most of it after the 21st century.

Various other aspects of ice-sheet modeling are currently the subject of intense research. These are mostly addressed in schematic setups and for localized problems but can be expected to find their way into a future generation of whole ice-sheet models (Marshall 2005). Aspects include the use of more generalized flow laws incorporating anisotropy and evolutionary equations for ice crystal fabric (Pettit and Waddington 2003; Gillet-Chaulet et al. 2005), improved grounding-line schemes (Vieli and Payne 2005), a more sophisticated treatment of basal mechanics and subglacial hydrology (Tulazcyk et al. 2000; Flowers and Clarke 2002), and the continued merger of ice-sheet models with increasingly complex Earth system models (Fichefet et al. 2003; Ridley et al. 2005).

7.6 Summary and Recommendations

7.6.1 Summary

Mountain glaciers, ice caps, and the Greenland and Antarctic Ice Sheets have the potential to raise global sea level many meters. Terrestrial glaciers are shrinking all over the world. During the last decade, they have been melting at about twice the rate of the previous several decades. On the polar ice sheets, there is observational evidence of accelerating flow from outlet glaciers both in southern Greenland and in critical locations in Antarctica. Both inland snow accumulation and marginal ice melting have increased over the Greenland Ice Sheet, but there is little evidence for any significant accumulation trend over the Antarctic Ice Sheet. What are the changes occurring in glaciers and ice sheets and how are they impacting sea level?

At higher elevations, there is slow thickening on both ice sheets, at rates in Greenland that appear to be increasing with time, consistent with model predictions of increasing snowfall in a warming climate (Box et al. 2006). In some near coastal parts of both Antarctica and Greenland, thinning far exceeds that caused by increasing summer melting, indicating dynamic causes as some outlet glaciers accelerate and thin dramatically. In Greenland, thinning predominates, at rates that are increasing with time. The picture is less clear in Antarctica, but net loss appears probable, with dynamic losses also increasing with time. These results are summarized in Table 7.1. The short intervals over which balance estimates are available are clearly of concern, particularly as recent data increasingly show that conditions can change rapidly over short time intervals. Longer periods of observation will be required to improve confidence in separation of long-term trends from natural variability and in identification of causes.

It is clear from Table 7.1 that, although glaciers and ice caps contain only a very small fraction of ice on Earth, they are making the lion's share of the cryospheric contribution to sea-level rise (Figure 7.8; Meier et al. 2007). Sea-level rise contributions from all three sources (small glaciers and the two ice sheets) have probably increased significantly during the last decade.

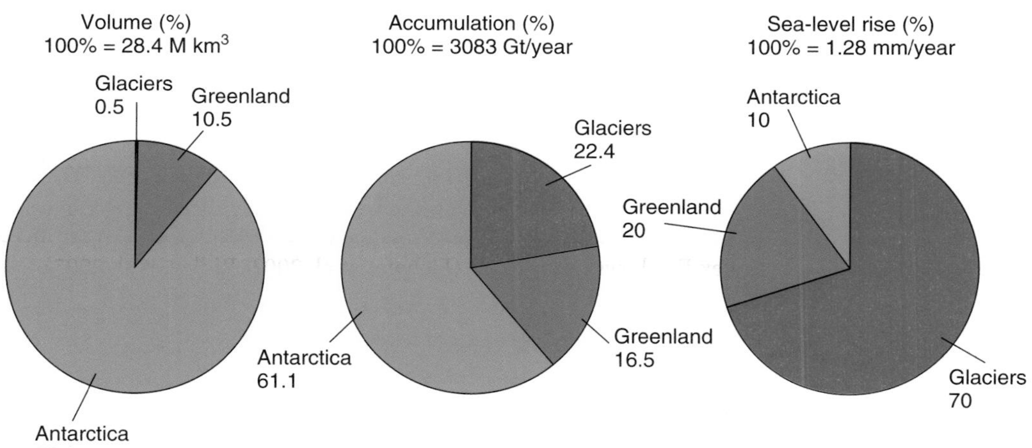

Figure 7.8 Comparison of total volume, total annual accumulation, and total contribution to sea-level rise for small glacier/ice caps and the ice sheets in Greenland and Antarctica. The total sea-level-rise contribution is based on Meier et al. (2007). M, million.

Although the precise values used in Figure 7.8 are being updated every few months, they are close enough to reality to show very clearly that sea-level rise from the ice sheets is remarkably small relative to their volumes. Certainly, we would expect low-latitude glaciers to respond most rapidly and sensitively to climate warming, with Greenland more sensitive than Antarctica because it lies on average at a lower latitude. Nevertheless, the remarkably rapid thinning of small glaciers gives an indication of just how fast glaciers can change.

Too many mountain glaciers and ice caps are losing mass more rapidly, in too many widely separated regions, to be explained by natural variability alone. And it is sobering to note that we are now observing similarly rapid changes in both Greenland and Antarctica. This is made more ominous by the fact that some of these very fast changes (doubling in speed in a few months and thinning by tens of meters per year) are caused by changes in glacier dynamics, leading to more rapid calving, and not by melting. We are seeing a transition from the slow, measured behavior long associated with polar ice sheets, to the rapid rates of change more typical of big glaciers in Alaska and Patagonia. This transition is most apparent in Greenland, where a tendency of glacier acceleration is progressing northward, leaving Greenland's southern ice dome under threat from both increased summer melting near the coasts, and increased ice discharge down glaciers that extend their influence far inland. If this continues, it is quite possible that the ice dome in southern Greenland will reach a tipping point, with accelerating positive feedback causing its ever more rapid decline and an associated sea-level rise of about 85 cm. Continued northward migration of the tendency of glacier acceleration will also make the far larger northern dome vulnerable.

At the same time, parts of Antarctica are under threat from similar changes. Ice-shelf breakup along the Antarctic Peninsula has resulted in strong acceleration of tributary glaciers, and ice shelf thinning further south in the Amundsen Sea appears also to have caused glacier acceleration. Here, the acceleration is more modest, but the glaciers are far bigger, so total ice losses are large. No one knows how far inland the zone of glacier acceleration will spread, and no one can be certain why the ice shelves are thinning and breaking up. But their thinning is probably caused by increased basal melting, implicating the ocean. And final breakup seems to be accelerated if there is sufficient surface meltwater to fill, and over-deepen, crevasses in the ice shelves, effectively wedging the ice shelf apart into fragments. Answering questions raised by these observations is of direct relevance to future sea level. At the very least, it is essential for humanity to have a reasonable knowledge of just how much change, in the climate of the ocean and atmosphere, is required to tip different types of ice mass to (effectively) irreversible collapse.

Summarizing observations of small glaciers, the rate of mass loss has not only increased, but it also appears to have accelerated, as shown in Figure 7.4 and Table 7.2. Earlier analysis by Meier (1984) showed a substantially lower rate of +0.46 ± 0.52 mm SLE/year for the period 1900–61, compared to the most recent pentade with a rate at about +0.8 to +0.9 mm SLE/year. The best estimate of the current (2008) mass balance is near to −380 to −400 Gt/year, or nearly 1.1 mm

SLE/year; this may be an underestimate if, as suspected, the inadequately measured rate of loss by calving outweighs the inadequately measured rate of gain by "internal" accumulation. Our physical understanding allows us to conclude that if the net gain of radiative energy at the Earth's surface continues to increase, then so will the acceleration of mass transfer from small glaciers to the ocean. Rates of loss observed so far are small in comparison with rates inferred for episodes of abrupt change during the last few hundred thousand years. In a warmer world the main eventual constraint on mass balance will be exhaustion of the supply of ice from glaciers, which may take place in as little as 50–100 years.

The calving of icebergs from small glaciers remains a large source of uncertainty for mass-balance estimates. Their contribution by calving is approximately 50 ± 30 Gt/year, or from +0.06 to +0.20 mm SLE/year. A major uncertainty in mass-balance estimates remains the spatial undersampling of the estimated 160 000 glaciers worldwide, given the fact that only about 300 glaciers are currently monitored.

A prerequisite to estimating the contribution of glacier and ice-cap melt to sea-level rise over the next century is knowledge of the ice volume available for melt. Uncertainty in the volume of ice available for melt results from incomplete glacier inventories. Estimates range from 0.15 to 0.37 m SLE for glaciers and ice caps outside Greenland and Antarctica.

Modeling contemporary surface mass-balance changes over the ice sheets is a meteorological problem since the models typically relied on atmospheric reanalysis products. Precipitation over Greenland and Antarctica is determined by the net atmospheric moisture transport from atmospheric models. But reanalysis data have notorious problems over the ice sheets. To overcome this problem, the use of high-resolution limited-area models driven from their lateral boundaries by reanalysis data has been quite successful.

Present large-scale models do not capture the recent evidence for rapid ice-dynamical changes in several Greenland outlet glaciers and in parts of West Antarctica and the Antarctic Peninsula. Another process not captured in present large-scale models is the meltwater penetration to the bed, providing near-instantaneous communication between surface forcing and basal-ice dynamics. These shortcomings in ice-sheet modeling are currently the subject of intense research.

7.6.2 Recommendations

Monitoring

- Reconcile estimates of ice-sheet mass balance derived using different approaches, and determine whether recent increases in mass losses are anomalous or reflect improvements in observational techniques.
- Identify causes for the apparent recent increases in mass loss to enable development of improved glacier models.

- Extend ongoing measurements of ice-thickness transects to cross each major outlet glacier in Greenland and Antarctica.
- Complete the World Glacier Inventory through sustained support for the Global Land Ice Measurements from Space (GLIMS) program.
- Extend observational coverage of terrestrial glaciers beyond traditional areas (e.g. the Alps and Alaska) to regions with poor or no coverage, thereby enabling improvements in monitoring and modeling.
- Utilize the ICESat laser and CryoSat-2 radar altimeter satellites – complemented by aircraft altimetry – to survey changes in the surface topography of the ice sheets; and based on experience gained, develop a suitable follow-on satellite.
- Utilize the GRACE satellite to infer changes in the mass of the glaciers and ice sheets.
- Seek continued access to satellite InSAR data in order to measure flow rates in glaciers and ice sheets; this will require suitable satellite missions – both existing and new – and ready access to resulting data, particularly over near-coastal regions of Greenland and Antarctica.
- Improve models to identify causes for the apparent increased mass losses from the polar ice sheets and use that as the basis for better simulations of future scenarios; particular effort is needed with respect to ocean–ice-shelf interactions, surface mass balance from climate models, and the inclusion of higher-order stress components in high-resolution ice-dynamic models.

Sustained, Systematic Observing Systems

- Sustain observations of the time-varying gravity field from GRACE and plan for an appropriate follow-on mission with finer spatial resolution to contribute to estimating changes in ice-sheet mass.
- Sustain satellite observations utilizing radar and laser altimeters, complemented by aircraft and *in situ* observations, to survey changes in the surface topography of the ice sheets.

Development of Improved Observing Systems

- Ice-sheet and glacier topography: based on experience gained with radar and laser satellite altimeters, develop and implement a suitable follow-on capability.
- Ice velocity: develop and implement an InSAR mission to observe flow rates in glaciers and ice sheets.
- Develop and implement a mission with an advanced *wide-swath* altimeter to observe the surface topography of glaciers and ice sheets.

References

Abdalati W., Krabill W., Frederick E., Manizade S., Martin C., Sonntag J. et al. (2001) Outlet glacier and margin elevation changes: near-coastal thinning of the Greenland ice sheet. *Journal of Geophysical Research* **106**(D24), 33729–41.

Abdalati W., Krabill W., Frederick E., Manizade S., Martin C., Sonntag J. et al. (2004) Elevation changes of ice caps in the Canadian Arctic Archipelago. *Journal of Geophysical Research* **109**, F04007.

Alley R.B., Clark P.U., Huybrechts P. and Joughin I. (2005) Ice-sheets and sea-level changes. *Science* **310**, 456–60.

Arendt A.A., Echelmeyer K.A., Harrison W.D., Lingle C.S. and Valentine B. (2002) Rapid wastage of Alaska glaciers and their contribution to rising sea-level. *Science* **297**, 382–6.

Arendt A.A., Luthcke S.B., Larsen C.F., Abdalati W., Krabill W.B. and Beedle M.J. (2008) Validation of high-resolution GRACE mascon estimates of glacier mass changes in the St. Elias Mountains, Alaska, USA using aircraft laser altimetry. *Journal of Glaciology* **54**(188), 778–87.

Arthern R. and Wingham D. (1998) The natural fluctuations of firn densification and their effect on the geodetic determination of ice sheet mass balance. *Climate Change* **40**, 605–24.

Arthern R., Winebrenner D. and Vaughan D. (2006) Antarctic snow accumulation mapped using polarization of 4.3-cm wavelength microwave emission. *Journal of Geophysical Research* **111**, D06107.

Bader H. (1954) Sorge's Law of densification of snow on high polar glaciers. *Journal of Glaciology* **2**, 319–23.

Bahr D.B. and Meier M. (2000) Snow patch and glacier size distribution. *Water Resources Research* **33**, 1669–72.

Bahr D.B., Meier M.F. and Peckham S.D. (1997) The physical basis of glacier volume-area scaling. *Journal of Geophysical Research* **102**, 20355–62.

Bahr D.B., Dyurgerov M.B. and Meier M.F. (2009) Sea-level rise from glaciers and ice caps: a lower bound, *Geophysical Research Letters* **36**, L03501.

Bales R., McConnell J., Mosley-Thompson E. and Csatho B. (2001) Accumulation over the Greenland ice sheet from historical and recent records. *Journal of Geophysical Research* **106**, 33813–25.

Bamber J.L., Krabill W., Raper V. and Dowdeswell J. (2004) Anomalous recent growth of part of a large Arctic ice cap: Austfonna, Svalbard. *Geophysical Research Letters* **31**, L019667.

Bassford R., Siegert M.J. and Dowdeswell J.A. (2006) Quantifying the mass balance of ice caps on Severnaya Zemlya, Russian High Arctic. III: Sensitivity of ice caps in Severnaya Zemlya to climate change. *Arctic Antarctic and Alpine Research* **38**, 21–33.

Bazhev A.B. (1980) Metody opredeleniya vnutrennego infil'tratsionnogo pitaniya lednikov. *Materialy Glyatsiologicheskikh Issledovaniy* **39**, 73–81 [in Russian; English text, Methods of determining the internal infiltration accumulation of glaciers, 147–54].

Blatter H. (1995) Velocity and stress fields in grounded glaciers: a simple algorithm for including deviatoric stress gradients. *Journal of Glaciology* **41**(138), 333–44.

Bougamont M., Bamber J.L. and Greuell W. (2005) A surface mass balance model for the Greenland Ice Sheet. *Journal of Geophysical Research* **110**, F04018.

Box J.E., Bromwich D.H. and Bai L.S. (2004) Greenland ice sheet surface mass balance 1991–2000: Application of Polar MM5 mesoscale model and in situ data. *Journal of Geophysical Research* **109**, Q16105.

Box J.E., Bromwich D.H., Veenhuis B.A., Bai L.-S., Stroeve J.C., Rogers J.C. et al. (2006) Greenland ice-sheet surface mass balance variability (1988–2004) from calibrated Polar MM5 output. *Journal of Climate* **19**, 2783–2800.

Braithwaite R.J. (2002) Glacier mass balance: the first 50 years of international monitoring. *Progress in Physical Geography* **26**, 76–95.

Braithwaite R.J. and Raper S.C.B. (2002) Glaciers and their contribution to sea level change. *Physics and Chemistry of the Earth* **27**, 1445–54.

Braithwaite R.J., Zhang Y. and Raper S.C.B. (2002) Temperature sensitivity of the mass balance of mountain glaciers and icecaps as a climatological characteristic. *Zeitschrift fur Gletscherkunde und Glazialgeologie* **38**, 35–61.

Bromwich D.H., Guo Z., Bai L. and Chen Q.-S. (2004) Modeled Antarctic precipitation. Part I: spatial and temporal variability. *Journal of Climate* **17**, 427–47.

Cazenave A. and Nerem R.S. (2004) Present-day sea level change: observations and causes. *Review of Geophysics* **42**, RG3001.

Chen J.L., Wilson C.R. and Tapley B.D. (2006) Satellite gravity measurements confirm accelerated melting of Greenland Ice Sheet. *Science* **313**, 1958–60.

Church J.A., Gregory J.M, Huybrechts P., Kuhn M., Lambeck K., Nhuan M.T., Qin D. and Woodworth P.L. (2001) Changes in Sea Level. In: *Climate Change 2001: The Scientific Basis.* Contribution of Working Group 1 to the Third Assessment Report of the Intergovernmental Panel on Climate Change (Houghton J.T., Ding Y., Griggs D.J., Noguer M., van der Linden P.J., Dai X. et al., eds), pp. 639–94. Cambridge University Press, Cambridge.

Church J.A., White N.J., Coleman R., Lambeck K. and Mitrovica, J.X. (2004) Estimates of the regional distribution of sea-level rise over the 1950 to 2000 period, *Journal of Climate* **17**, 2609–25.

Clark P.U., Mitrovica J.X., Milne G.A. and Tamisiea M.E. (2002) Sea-level fingerprinting as a direct test for the source of global meltwater pulse IA. *Science* **295**, 2438–41.

Cogley J.G. (1999) Effective sample size for glacier mass balance. *Geografiska Annaler* **81A**(4), 497–507.

Cogley J.G. (2003) *GGHYDRO: Global Hydrographic Data, Release 2.3.* Trent Technical Note 2003–1. Department of Geography, Trent University, Peterborough, Ontario.

Cogley J.G. (2004) Greenland accumulation: an error model. *Journal of Geophysical Research* **109**(D18), D18101.

Cogley J.G. (2005) Mass and energy balances of glaciers and ice sheets. In: *Encyclopaedia of Hydrological Sciences*, vol. 4 (Anderson, M.G., ed.), pp. 2555–73. Wiley, Chichester. http://www.trentu.ca/geography/glaciology/glaciology.htm.

Cogley J.G. (2009a) Geodetic and direct mass-balance measurements: comparison and joint analysis. *Annals of Glaciology* **50**(50), 96–100.

Cogley J.G. (2009b) A more complete version of the World Glacier Inventory. *Annals of Glaciology* **50**(53), 1–7.

Cogley J.G. and Adams W.P. (1998) Mass balance of glaciers other than the ice sheets. *Journal of Glaciology* **44**, 315–25.

Cook A., Fox A., Vaughan D. and Ferrigno J. (2005) Retreating glacier fronts on the Antarctic Peninsula over the past half century. *Science* **308**, 541–4.

Cox L.H. and March R.S. (2004) Comparison of geodetic and glaciological mass-balance techniques, Gulkana Glacier, Alaska, U.S.A. *Journal of Glaciology* **50**, 363–70.

Davis C.H., Li Y., McConnell J.R., Frey M.M. and Hanna E. (2005) Snowfall-driven growth in East Antarctic ice sheet mitigates recent sea-level rise. *Science* **308**, 1898–1901.

Dethloff K., Schwager M., Christensen J.H., Kiilsholm S., Rinke A., Dorn W. et al. (2002) Recent Greenland accumulation estimated from regional climate model simulations and ice core analysis. *Journal of Climate* **15**, 2821–32.

Dowdeswell J.A. and Hagen J.O. (2004) Arctic glaciers and ice caps. In: *Mass Balance of the Cryosphere* (Bamber J.L. and Payne A.J., ed.), pp. 527–57. Cambridge University Press, Cambridge.

Dyurgerov M.B. and Meier M.F. (1997) Mass balance of mountain and subpolar glaciers: a new assessment for 1961–1990. *Arctic and Alpine Research* **29**, 379–91.

Dyurgerov M.B. and Meier M.F. (2000) Twentieth century climate change: evidence from small glaciers. *Proceedings of the National Academy of Sciences USA* **97**, 1406–11.

Dyurgerov M.B. and Meier M.F. (2004) Glaciers and the study of climate and sea-level change. In: *Mass Balance of the Cryosphere* (Bamber J.L. and Payne A.J., ed.), pp. 579–622. Cambridge University Press, Cambridge.

Dyurgerov M.B. and Meier M.F. (2005) *Glaciers and the Changing Earth System: a 2004 Snapshot*. Occasional Paper No. 58. Institute of Arctic and Alpine Research, University of Colorado, Boulder, CO. http://instaar.colorado.edu/other/occ_papers.html.

Fichefet T., Poncin C., Goosse H., Huybrechts P., Janssens I. and Le Treut H. (2003) Implications of changes in freshwater flux from the Greenland ice sheet for the climate of the 21st century. *Geophysical Research Letters* **30**, 1911.

Flowers G.E. and Clarke G.K.C. (2002) A multicomponent coupled model of glacier hydrology 1. Theory and synthetic examples. *Journal of Geophysical Research* **107**(B11), 2287.

Gillet-Chaulet F., Gagliardini O., Meysonnier J., Montagnat M. and Castelnau O. (2005) A user-friendly anisotropic flow law for ice-sheet modeling. *Journal of Glaciology* **51**(172), 3–14.

Gregory J.M. and Oerlemans J. (1998) Simulated future sea-level rise due to glacier melt based on regionally and seasonally resolved temperature changes. *Nature* **391**, 474–6.

Gregory J.M. and Huybrechts P. (2006) Ice-sheet contributions to future sea-level change. *Philosophical Transactions of the Royal Society of London A* **364**, 1709–31.

Gregory J.M., Huybrechts P. and Raper S.C.B. (2004) Threatened loss of the Greenland ice-sheet. *Nature* **428**, 616.

Haeberli W., Bösch H., Scherler K., Østrem G. and Wallén C.C. (1989) *World Glacier Inventory Status 1998*. International Commission on Snow and Ice of International Association of Hydrological Sciences/UNESCO, Paris.

Haeberli W., Zemp M., Hoelzle M., Frauenfelder R., Hoelzle M. and Kääb A. (2005a) *Fluctuations of Glaciers, 1995–2000*, vol. VIII. International Commission on Snow and Ice of International Association of Hydrological Sciences/ UNESCO, Paris.

Haeberli W., Noetzli J., Zemp M., Baumann S., Frauenfelder R. and Hoelzle M. (2005b) *Glacier Mass Balance Bulletin No. 8 (2002–2003)*. International Commission on Snow and Ice of International Association of Hydrological Sciences/UNESCO, Paris.

Hanna E., Huybrechts P., Janssens I., Cappelen J., Steffen K. and Stephens A. (2005) Runoff and mass balance of the Greenland ice sheet: 1958–2003. *Journal of Geophysical Research* **110**, D13108.

Hanna E., Huybrechts P., Steffen K., Cappelen J., Huff R., Shuman C. et al. (2008) Increased runoff from melt from the Greenland Ice Sheet: a response to global warming. *Journal of Climate* **21**(2), 331–41.

Hanna E., Cappelen J., Fettweis X., Huybrechts P., Luckman A. and Ribergaard M.H. (2009) Hydrologic response of the Greenland Ice Sheet: the role of oceanographic forcing. *Hydrological Processes (Special Issue: Hydrologic Effects of a Shrinking Cryosphere)* **23**(1), 7–30.

Hock R. (2003) Temperature index melt modeling in mountain areas. *Journal of Hydrology* **282**, 104–15.

Hock R. (2005) Glacier melt: a review of processes and their modeling. *Progress in Physical Geography* **29**, 362–91.

Hock R., de Woul M., Radić V. and Dyurgerov M.B. (2009) Mountain glaciers and ice caps around Antarctica make a large sea-level rise contribution. *Geophysical Research Letters* **36**, L07501.

Holgate S.J. and Woodworth P.L. (2004) Evidence for enhanced coastal sea level rise during the 1990s. *Geophysical Research Letters* **31**, L07305.

Horwath M. and Dietrich R. (2006) Errors of regional mass variations inferred from GRACE monthly solutions. *Geophysical Research Letters* **33**, L07502.

Howat I.M., Joughin I., Tulaczyk S. and Gogineni S. (2005) Rapid retreat and acceleration of Helheim Glacier, east Greenland. *Geophysical Research Letters* **32** (L22502).

Huybrechts P. (2002) Sea-level changes at the LGM from ice-dynamic reconstructions of the Greenland and Antarctic ice sheets during the glacial cycles. *Quaternary Science Reviews* **21**, 203–31.

Huybrechts P., Gregory J., Janssens I. and Wild M. (2004) Modelling Antarctic and Greenland volume changes during the 20th and 21st centuries forced by GCM time slice integrations. *Global and Planetary Change* **42**, 83–105.

Ishii M., Kimoto M, Sakamoto K. and Iwasaki S.I. (2006) Steric sea level changes estimated from historical ocean subsurface temperature and salinity analyses. *Journal of Oceanography* **62**, 155–70.

ISMASS Committee (2004) Recommendations for the collection and synthesis of Antarctic Ice Sheet mass balance data. *Global and Planetary Change* **42**, 1–15.

Janssens I. and Huybrechts P. (2000) The treatment of meltwater retention in mass-balance parameterisations of the Greenland ice sheet. *Annals of Glaciology* **31**, 133–40.

Johannessen O., Khvorostovsky K., Miles M. and Bobylev L. (2005) Recent ice-sheet growth in the interior of Greenland. *Science* **310**, 1013–16.

Joughin I. and Tulaczyk S. (2002) Positive mass balance of the Ross Ice Streams, West Antarctica. *Science* **295**(5554), 476–80.

Joughin I. and Bamber J. (2005) Thickening of the Ice Stream Catchments Feeding the Filchner-Ronne Ice Shelf, Antarctica. *Geophysical Research Letters* **32**, L17503.

Joughin I., Tulaczyk S., Bindschadler R. and Price S.F. (2002) Changes in west Antarctic ice stream velocities: Observation and analysis. *Journal of Geophysical Research* **107**(B11), 2289.

Joughin I., Rignot E., Rosanova C., Lucchitta B. and Bohlander J. (2003) Timing of recent accelerations of Pine Island Glacier, Antarctica. *Geophysical Research Letters* **30**, 1706.

Joughin I., Abdalati W. and Fahnestock M. (2004) Large fluctuations in speed on Greenland's Jakobshavn Isbræ glacier. *Nature* **432**, 608–10.

Kaser G., Cogley J.G., Dyurgerov M.B., Meier M.F. and Ohmura A. (2006) Mass balance of glaciers and ice caps: consensus estimates for 1961–2004, *Geophysical Research Letters* **33**, L19501.

Knap W.H., Oerlemans J. and Cadée M. (1996) Climate sensitivity of the ice cap of King George Island, South Shetland Islands, Antarctica. *Annals of Glaciology* **23**, 154–9.

Krabill W., Abdalati W., Frederick E., Manizade S., Martin C., Sonntag J. et al. (2000) Greenland Ice Sheet: high-elevation balance and peripheral thinning. *Science* **289**, 428–30.

Krabill W., Abdalati W., Frederick E.B., Manizade S.S., Martin C.F., Sonntag J.G. et al. (2002) Aircraft laser altimetry measurement of elevation changes of the Greenland Ice Sheet: technique and accuracy assessment. *Journal of Geodynamics* **34**, 357–76.

Krabill W., Hanna E., Huybrechts P., Abdalati W., Cappelen J., Csatho B. et al. (2004) Greenland Ice Sheet: increased coastal thinning. *Geophysical Research Letters* **31**, L24402.

Lefebre F., H. Gallée H., van Ypersele de Strihou J.P. and Huybrechts P. (2002) Modelling of large-scale melt parameters with a regional climate model in South-Greenland during the 1991 melt season. *Annals of Glaciology* **35**, 391–7.

Lemke P., Ren J., Alley R.B., Allison I., Carrasco J., Flato G. et al. (2007) Observations: changes in snow, ice and frozen ground. In: *Climate Change 2007: The Physical Science Basis.* Contribution of Working Group I to the Fourth Assessment Report of the Intergovernmental Panel on Climate Change (Solomon S., Qin D., Manning M., Chen Z., Marquis M., Averyt K.B. et al., eds), pp. 337–83. Cambridge University Press, Cambridge.

Levitus S., Antonov J.I. and Boyer T.P. (2005a) Warming of the world ocean, 1955–2003. *Geophysical Research Letters* **32**, L02604.

Li J. and Zwally H.J. (2004) Modeling the density variation in shallow firn layer. *Annals of Glaciology* **38**, 309–13.

Lombard A., Cazenave A., Le Traon P.-Y., Guinehut S. and Cabanes C. (2006) Perspectives on present-day sea level change: a tribute to Christian le Provost. *Ocean Modelling* **56**, 445–51.

Luthcke S.B., Zwally H.J., Abdalati W., Rowlands D.D., Ray R.D., Nerem R.S. et al. (2006) Recent Greenland ice mass loss by drainage system from satellite gravity observations. *Science* **314**, 1286–9.

Marshall S.J. (2005) Recent advances in understanding ice sheet dynamics. *Earth and Planetary Science Letters* **240**, 191–204.

Meehl G.A. and Stocker T.F. (2007) Global climate projections. In: *Climate Change 2007: The Physical Science Basis.* Contribution of Working Group I to the Fourth Assessment Report of the Intergovernmental Panel on Climate Change (Solomon S., Qin D., Manning M., Chen Z., Marquis M., Averyt K.B., Tignor M. and Miller H.L. eds), pp. 747–845. Cambridge University Press, Cambridge.

Meier M.F. (1984) Contribution of small glaciers to global sea level. *Science* **226**, 1418–21.

Meier M.F. and Bahr D.B. (1996) Counting glaciers: use of scaling methods to estimate the number and size distribution of the glaciers of the world. In: *Glaciers, Ice Sheets and Volcanoes: a Tribute to Mark F. Meier* (Colbeck S.C., ed.), pp. 89–94. Special Report 96–27. US Army Corps of Engineers Cold Regions Research and Engineering Laboratory.

Meier M.F., Bahr D.B., Dyurgerov M.B. and Pfeffer W.T. (2005) Comment on 'The potential for sea level rise: New estimates from glacier and ice cap area and volume distributions'. *Geophysical Research Letters* **32**, L17501.

Meier M.F., Dyurgerov M.B., Rick U.K., O'Neel S., Pfeffer W.T., Anderson R.S. et al. (2007) Glaciers dominate eustatic sea-level rise in the 21st century. *Science* **317**, 1064–7.

Mercer J. (1978) West Antarctic ice sheet and CO2 greenhouse effect: a threat of disaster. *Nature* **271**(5643), 321–5.

Monaghan A.J., Bromwich D.H., Fogt R.L., Wang S.-H., Mayewski P.A., Dixon D.A. et al. (2006) Insignificant change in Antarctic snowfall since the International Geophysical Year. *Science* **313**, 827–30.

Morris E.M. and Vaughan D.G. (2003) Glaciological climate relationships spatial and temporal variation of surface temperature on the Antarctic Peninsula and the limit of viability of ice shelves. Antarctic peninsula climate variability: historical and paleoenvironmental perspectives. *Antarctic Research Series of the American Geophysical Union* **79**, 61–8.

New M., Hulme M. and Jones P.J. (1999) Representing twentieth century space-time climate variability. 1. Development of a 1961–1990 mean monthly terrestrial climatology. *Journal of Climate* **12**, 829–56.

O'Neel S., Echelmeyer K. and Motyka R. (2001) Short-term dynamics of a retreating tidewater glacier: LeConte Glacier, Alaska, USA. *Journal of Glaciology* **47**, 567–78.

O'Neel S., Pfeffer W.T., Krimmel R.M. and Meier M.F. (2005) Evolving force balance at Columbia Glacier, during its rapid retreat. *Journal of Geophysical Research* **110**, F03012.

Oerlemans J. (1989) A projection of future sea level. *Climatic Change* **15**, 151–74.

Oerlemans J. and Van der Veen C.J. (1984) *Ice Sheets and Climate.* Reidel, Dordrecht.

Oerlemans J. and Fortuin J.P.F. (1992) Sensitivity of glaciers and small ice caps to greenhouse warming. *Science* **258**, 115–17.

Oerlemans J. and Reichert B.K. (2000) Relating a glacier mass balance to meteorological data using a Seasonal Sensitivity Characteristic (SSC). *Journal of Glaciology* **46**, 1–2.

Oerlemans J., Anderson B., Hubbard A., Huybrechts P., Jóhannesson T., Knap W.H. et al. (1998) Modelling the response of glaciers to climate warming. *Climate Dynamics* **14**, 267–74.

Ohmura A. (2001) Physical basis for the temperature-based melt-index method. *Journal of Applied Meteorology* **40**(4), 753–61.

Ohmura A. (2004) *Cryosphere During the Twentieth Century*, pp. 239–57. Geophysical Monograph 150. American Geophysical Union, Washington DC.

Pälli A., Kohler J.C., Isaksson E., Moore J.C., Pinglot J.F., Pohjola V.A. and Samuelsson H. (2002) Spatial and temporal variability of snow accumulation using ground-penetrating radar and ice cores on a Svalbard glacier. *Journal of Glaciology* **48**(162), 417–24.

Parizek B.R. and Alley R.B. (2004) Implications of increased Greenland surface melt under global-warming scenarios: ice-sheet simulations. *Quaternary Science Reviews* **23**(9–10), 1013–27.

Pattyn F., Huyghe A., De Brabander S. and De Smedt B. (2006) Role of transition zones in marine ice sheet dynamics. *Journal of Geophysical Research* **111**, F02004.

Payne A.J., Vieli A., Shepherd A., Wingham D.J. and Rignot E.J. (2004) Recent dramatic thinning of largest West Antarctic ice stream triggered by oceans. *Geophysical Research Letters* **31**, L23401.

Peltier W.R. (1998) "Implicit ice" in the global theory of glacial isostatic adjustment. *Geophysical Research Letters* **25**, 3955–8.

Peltier W.R. (2004) Global glacial isostasy and the surface of the ice-age Earth: The ICE-5G (VM2) model and GRACE. *Annual Review of Earth and Planetary Sciences* **32**, 111–49.

Pettit E.C. and Waddington E.D. (2003) Ice flow at low deviatoric stress. *Journal of Glaciology* **49**, 359–69.

Pinglot J.F., Hagen J.O., Melvold K., Eiken T. and Vincent C. (2001) A mean net accumulation pattern derived from radioactive layers and radar soundings on Austfonna, Nordaustlandet, Svalbard. *Journal of Glaciology* **47**(159), 555–66.

Rabus B.T. and Echelmeyer K.A. (1998) The mass balance of McCall Glacier, Brooks Range, Alaska, U.S.A.; its regional relevance and implications for climate change in the Arctic. *Journal of Glaciology* **44**(147), 333–51.

Ramillien G., Lombard A., Cazenave A., Ivins E., Remy F. and Biancale R. (2006) Interannual variations of the mass balance of the Antarctic and Greenland ice sheets from GRACE. *Global and Planetary Change* **53**, 198–208.

Raper S.C.B. and Braithwaite R.J. (2005a) The potential for sea level rise: New estimates from glacier and icecap area and volume distributions. *Geophysical Research Letters* **32**, L05502.

Raper S.C.B. and Braithwaite R.J. (2005b) Reply to comment by M.F. Meier et al. on 'The potential for sea level rise: new estimates from glacier and ice cap area and volume distributions'. *Geophysical Research Letters* **32**, L17502.

Raper S.C.B. and Braithwaite R.J. (2006) Low sea level rise projections from mountain glaciers and icecaps under global warming. *Nature* **439**, 311–13.

Raper S.C.B., Brown O. and Braithwaite R.J. (2000) A geometric glacier model for sea level change calculations. *Journal of Glaciology* **46**, 357–68.

Remy F., Testut L. and Legresy B. (2002) Random fluctuations of snow accumulation over Antarctica and their relation to sea level change. *Climate Dynamics* **19**, 267–76.

Ridley J., Huybrechts P., Gregory J.M. and Lowe J. (2005) Elimination of the Greenland ice sheet in a high-CO2 climate. *Journal of Climate* **18**(17), 3409–27.

Rignot E. and Jacobs S. (2002) Rapid bottom melting widespread near Antarctic Ice Sheet grounding lines. *Science* **296**, 2020–3.

Rignot E. and Thomas R.H. (2002) Mass balance of polar ice sheets. *Science* **297**, (5586), 1502–6.

Rignot E. and Kanagaratnam P. (2006) Changes in the Velocity Structure of the Greenland Ice Sheet. *Science* **311**(5763), 986–90.

Rignot E., Rivera A. and Casassa G. (2003) Contribution of the Patagonia Icefields of South America to sea level rise. *Science* **302**(5644), 434–7.

Rignot E., Casassa G., Gogineni P., Krabill W., Rivera A. and Thomas R. (2004a) Accelerated ice discharge from the Antarctic Peninsula following the collapse of Larsen B ice shelf. *Geophysical Research Letters* **31**(18), L18401.

Rignot E., Thomas R.H., Kanagaratnam P., Casassa G., Frederick E., Gogineni S. et al. (2004b) Improved estimation of the mass balance of the glaciers draining into the Amundsen Sea sector of West Antarctica from the CECS/NASA 2002 campaign. *Annals of Glaciology* **39**, 231–7.

Rignot E., Casassa G., Gogineni, S., Kanagaratnam P., Krabill W., Pritchard H. et al. (2005a) Recent ice loss from the Fleming and other glaciers, Wordie Bay, West Antarctic Peninsula. *Geophysical Research Letters* **32**(7), 1–4.

Rignot E., Pritchard H., Thomas R., Casassa G., Krabill W., Rivera A. et al. (2005b) Mass imbalance of Fleming and other glaciers, West Antarctic Peninsula. *Geophysical Research Letters* **32**, L07502.

Rignot E., Box J.E., Burgess E. and Hanna E. (2008a) Mass balance of the Greenland ice sheet from 1958 to 2007. *Geophysical Research Letters* **35**, L20502.

Rignot E., Bamber J.L., Van den Broecke M.R., Davis C., Li Y., Van de Berg W.J. and Van Meijgaard E. (2008b) Recent Antarctic ice mass loss from radar interferometry and regional climate modelling. *Nature Geoscience* **1**, 106–10.

Ritz C., Rommelaere V. and Dumas C. (2001) Modeling the evolution of Antarctic ice sheet over the last 420000 years: implications for altitude changes in the Vostok region. *Journal of Geophysical Research* **106** (D23), 31943–64.

Scambos T., Hulbe C., Fahnestock M. and Bohlander J. (2000) The link between climate warming and break-up of ice shelves in the Antarctic Peninsula. *Journal of Glaciology* **46**, 516–30.

Scambos T., Hulbe C. and Fahnestock M. (2003) Climate-induced ice shelf disintegration in the Antarctic Peninsula. *Antarctic Research Series* **79**, 79–92.

Scambos T., Bohlander J., Shuman C. and Skvarca P. (2004) Glacier acceleration and thinning after ice shelf collapse in the Larsen B embayment, Antarctica. *Geophysical Research Letters* **31**, L18401.

Schneeberger C., Albrecht O., Blatter H., Wild M. and Hoeck R. (2001) Modelling the response of glaciers to a doubling in atmospheric CO_2: a case study of Storglaciaren, northern Sweden. *Climate Dynamics* **17**, 825–34.

Schneeberger C., Blatter H., Abe-Ouchi A. and Wild M. (2003) Modelling changes in the mass balance of glaciers of the northern hemisphere for a transient 2°XCO2 scenario. *Journal of Hydrology* **282**, 145–63.

Shepherd A., Wingham D.J. and Mansley J.A.D. (2002) Inland thinning of the Amundsen Sea sector, West Antarctica. *Geophysical Research Letters* **29**(10), 1364.

Shepherd A., Wingham D., Payne T. and Skvarca P. (2003) Larsen Ice Shelf has progressively thinned. *Science* **302**, 856–9.

Shumskiy P.A. (1969) Glaciation. In: *Atlas Antarktidy*, vol. 2 (Tolstikov E., ed.), pp. 367–400. Gidrometeoizdat, Leningrad.

Steffen K. and Box J.E. (2001) Surface climatology of the Greenland ice sheet: Greenland climate network 1995–1999. *Journal of Geophysical Research* **106**, 33951–64.

Steffen K., Nghiem S.V., Huff R. and Neumann G. (2004) The melt anomaly of 2002 on the Greenland Ice Sheet from active and passive microwave satellite observations. *Geophysical Research Letters* **31**(20), L2040210.

Thomas R., Csatho B., Davis C., Kim C., Krabill W., Manizade S. et al. (2001) Mass balance of higher-elevation parts of the Greenland ice sheet. *Journal of Geophysical Research* **106**, 33707–16.

Thomas R., Abdalati W., Frederick E., Krabill W., Manizade S. and Steffen K. (2003) Investigation of surface melting and dynamic thinning on Jakobshavn Isbrae, Greenland. *Journal of Glaciology* **49**, 231–9.

Thomas R., Rignot E., Kanagaratnam P., Krabill W. and Casassa G. (2004a) Force-perturbation analysis of Pine Island Glacier, Antarctica, suggests cause for recent acceleration. *Annals of Glaciology* **39**, 133–8.

Thomas R., Rignot E., Casassa G., Kanagaratnam P., Acuña C., Akins T. et al. (2004b) Accelerated sea-level rise from West Antarctica. *Science* **306**(5694), 255–8.

Thomas R., Frederick E., Krabill W., Manizade S. and Martin C. (2006) Progressive increase in ice loss from Greenland. *Geophysical Research Letters* **33**, L10503.

Trabant D. and March R.S. (1999) Mass-balance measurements in Alaska and suggestions for simplified observation programs. *Geografiska Annaler* **81A**(4), 777–89.

Tulazcyk S., Kamb W.B. and Engelhardt H.F. (2000) Basal mechanics of Ice Stream B, West Antarctica. II. Undrained plastic bed model. *Journal of Geophysical Research* **105**, 483–94.

van de Berg W.J., van den Broeke M.R., Reijmer C.H. and van Meijgaard E. (2006) Reassessment of the Antarctic surface mass balance using calibrated output of a regional atmospheric climate model. *Journal of Geophysical Research* **111**, D11104.

van de Wal R.S.W. and Oerlemans J. (1994) An energy balance model for the Greenland ice sheet. *Global and Planetary Change* **9**(1–2), 115–31.

van de Wal R.S.W. and Wild M. (2001) Modelling the response of glaciers to climate change, applying volume-area scaling in combination with a high resolution GCM. *Climate Dynamics* **18**, 359–66.

van den Broeke M.R., van de Berg W.J. and van Meijgaard E. (2006) Snowfall in coastal West Antarctica much greater than previously assumed. *Geophysical Research Letters* **33**, L02505.

van der Veen C.J. (1993) Interpretation of short-term ice sheet elevation changes inferred from satellite altimetry. *Climate Change* **23**, 383–405.

van der Veen C.J. (2002) Calving glaciers. *Progress in Physical Geography* **26**(1), 96–122.

van Lipzig N.P.M., van Meijgaard E. and Oerlemans J. (2002) Temperature sensitivity of the Antarctic surface mass balance in a regional atmospheric climate model. *Journal of Climate* **15**, 2758–74.

van Lipzig N.P.M., King J.C., Lachlan-Cope T.A. and van den Broeke M.R. (2004) Precipitation, sublimation, and snow drift in the Antarctic Peninsula region from a regional atmospheric model. *Journal of Geophysical Research* **109**, D24106.

Vaughan D.G., Bamber J.L., Giovinetto M., Russell J. and Cooper A.P.R. (1999) Reassessment of net surface mass balance in Antarctica. *Journal of Climate* **12**(4), 933–46.

Vaughan D.G., Marshall G.J., Connolly W.M., Parkinson C., Mulvaney R., Hodgson D.A. et al. (2003) Recent rapid regional climate warming on the Antarctic Peninsula. *Climate Change* **60**, 243–74.

Velicogna I. and Wahr J. (2005) Ice mass balance in Greenland from GRACE. *Journal of Geophysical Research* **32**, L18505.

Velicogna I. and Wahr J. (2006) Measurements of time-variable gravity show mass loss in Antarctica. *Science* **311**, 1754–6.

Vieli A. and Payne A.J. (2005) Assessing the ability of numerical ice sheet models to simulate grounding line migration. *Journal of Geophysical Research* **110**, F01003.

Wang L., Sharp M.J., Rivard B., Marshall S. and Burgess D. (2005) Melt season distribution on Canadian Arctic ice caps, 2000–2004. *Geophysical Research Letters* **32**, L19502.

Warrick R.A., Le Provost C., Meier M.F., Oerlemans J. and Woodworth P.L. (1996) Changes in sea level. In: *Climate Change 1995. The Science of Climate Change*. Contribution of Working Group 1 to the Second Assessment. Report of the Intergovernmental Panel on Climate Change (IPCC), pp. 359–405. Cambridge University Press, Cambridge.

Wigley T.M.L. and Raper S.C.B. (1995) An heuristic model for sea level rise due to the melting of small glaciers. *Geophysical Research Letters* **22**, 2749–52.

Wigley T.M.L. and Raper S.C.B. (2005) Extended scenarios for glacier melt due to anthropogenic forcing. *Geophysical Research Letters* **32**, L05704.

Wild M., Calanca P., Scherrer S.C. and Ohmura A. (2003) Effects of polar ice sheets on global sea level in high-resolution greenhouse scenarios. *Journal of Geophysical Research* **108**(D5), 4165.

Willis J.K., Roemmich D. and Cornuelle B. (2004) Interannual variability in upper ocean heat content, temperature, and thermosteric expansion on global scales. *Journal of Geophysical Research* **109**, C12036.

Wingham D., Shepherd A., Muir A. and Marshall G. (2006) Mass balance of the Antarctic ice sheet. *Philosophical Transactions of the Royal Society A* **364**, 1627–35.

Zuo Z. and Oerlemans J. (1997) Contribution of glacier melt to sea level rise since AD 1865: a regionally differentiated calculation. *Climate Dynamics* **13**, 835–45.

Zwally H.J., Abdalati W., Herring T., Larson K., Saba J. and Steffen K. (2002a) Surface melt-induced acceleration of Greenland ice-sheet flow. *Science* **297**(5579), 218–22.

Zwally H.J., Schutz R., Abdalati W., Abshire J., Bentley C., Bufton J. et al. (2002b) ICESat's laser measurements of polar ice, atmosphere, ocean, and land. *Journal of Geodynamics* **34**, 405–45.

Zwally H.J., Giovinetto M.B., Li J., Cornejo H.G., Beckley M.A., Brenner A.C. et al. (2005) Mass changes of the Greenland and Antarctic ice sheets and shelves and contributions to sea-level rise: 1992–2002. *Journal of Glaciology* **51**(175), 509–27.

8 Terrestrial Water-Storage Contributions to Sea-Level Rise and Variability

P.C.D. (Chris) Milly, Anny Cazenave, James S. Famiglietti, Vivien Gornitz, Katia Laval, Dennis P. Lettenmaier, Dork L. Sahagian, John M. Wahr, and Clark R. Wilson

8.1 Introduction

8.1.1 Purpose and Scope

A gain or loss of water by the continents generally corresponds to an equal loss or gain of water by the oceans, because water content of the global atmosphere (≈25 mm water equivalent) is tightly constrained thermodynamically. The induced change in ocean water storage, in turn, affects the global mean sea level. In this chapter, we summarize current understanding and uncertainties on contemporary continent–ocean water exchanges on timescales ranging from seasonal to centennial. We exclude from consideration the exchanges between the ocean and the ice sheets of Greenland and Antarctica, as well as the exchanges between the ocean and mountain glaciers. These exchanges are considered in other chapters of the volume. However, we do comment on exchange between the oceans and the subsurface continental cryosphere (permafrost).

8.1.2 External Constraints on the Contribution of Terrestrial Water to Present-Day Sea-Level Change

Tide-gauge measurements available since the late 19th century have indicated significant sea-level rise during the 20th century, at a rate of approximately 1.7 mm/year (Church et al. 2004; Church and White 2006; Holgate and Woodworth 2004; Holgate 2007; Jevrejeva et al. 2006). Since early 1993, sea level is accurately measured by satellite altimetry (TOPEX/Poseidon, Jason-1, and Jason-2 missions). This more than 15-year-long data set shows that, in terms of global mean, sea level is currently rising at a rate of approximately 3.4 ± 0.4 mm/year (Nerem

Understanding Sea-Level Rise and Variability, 1st edition. Edited by John A. Church, Philip L. Woodworth, Thorkild Aarup & W. Stanley Wilson.

et al. 2006; Beckley et al. 2007; Prandi et al. 2009), a value significantly higher than the mean rate recorded by tide gauges over the past decades. Note that estimates from tide gauges are similar to the altimeter estimates over the same period. On interannual to decadal timescales, the two main causes of sea-level rise are thermal expansion of the warming oceans and the net transport of freshwater mass to the oceans from melting ice sheets and mountain glaciers, and from terrestrial water reservoirs.

The Intergovernmental Panel on Climate Change (IPCC) Fourth Assessment Report (AR4) reviewed the thermal-expansion and land-ice contributions to sea-level rise for two time spans (Bindoff et al. 2007): 1961–2003 and 1993–2003. Over the decade 1993–2003, thermal expansion estimated from *in situ* hydrographic measurements accounted for about half the observed rate of sea-level rise (Willis et al. 2004; Antonov et al. 2005; Ishii et al. 2006). Being very sensitive to global warming, mountain glaciers and small ice caps have retreated worldwide in the recent decades, with significant acceleration during the 1990s. From mass-balance studies of a large number of glaciers, estimates of the glaciers' contribution to sea level have been proposed (e.g. Dyurgerov and Meier 2005; Kaser et al. 2006; Lemke et al. 2007). Glacier melting explains approximately 30% of the rate of sea-level rise over 1993–2003. Since the early 1990s, different remote-sensing observations based on airborne laser and satellite altimetry, as well as the interferometric synthetic aperture radar (InSAR) technique and space gravimetry (the Gravity Recovery and Climate Experiment (GRACE) space mission), have provided important observations of the mass balance of the ice sheets. These indicate accelerated ice-mass loss in coastal regions of Greenland and West Antarctica (see for example Lemke et al. 2007 and references therein; Rignot et al. 2008a, 2008b; Chapter 7). For 1993–2003, less than 15% of the rate of global sea-level rise was due to the mass loss of the ice sheets of Greenland and Antarctica, but this contribution has clearly increased since 2003. The sea-level budget as summarized in AR4 for the 1961–2003 and 1993–2003 time spans (Table 8.1) indicates that

Table 8.1 Sea-level budget for three time spans (1961–2003, 1993–2003, and 2003–7). Quoted errors are 1 standard deviation. The observed sea-level rate is glacial isostatic adjustment (GIA)-corrected (−0.3 mm/year removed).

Sea-level rise (mm/year)	1961–2003[a]	1993–2003[b]	2003–7[c]
1. Observed	1.8 ± 0.3	3.1 ± 0.4	2.5 ± 0.4
2. Thermal expansion	0.4 ± 0.06	1.6 ± 0.25	0.35 ± 0.2
3. Glaciers	0.5 ± 0.1	0.8 ± 0.11	1.1 ± 0.25
4. Ice sheets	0.2 ± 0.2	0.4 ± 0.2	1. ± 0.15
5. Sum of 2 + 3 + 4	1.1 ± 0.25	2.8 ± 0.35	2.45 ± 0.35

Sources: [a, b]AR4; [c]Cazenave et al. (2009), Meier et al. (2007), Rignot et al. (2008a, 2008b), Alley (2009).

for both time spans the sea-level budget is not closed. While uncertainties in sea-level observations and estimated climate components likely contribute to the difference, mass exchange with continental water stores, such as snow pack, surface water, and subsurface water, may at least partly fill the gap.

In situ hydrographic (i.e. temperature and salinity) measurements available from the newly deployed Argo system (Chapter 6) indicate that since 2003 ocean thermal expansion has increased less rapidly than during the previous decade, although sea level has continued to rise (Willis et al. 2008). From the latter observation one concludes that ocean mass increase could be the dominant factor in recent sea-level rise after 2003. Recent published studies suggest that this is indeed the case (e.g. Cazenave et al. 2009; Leuliette and Miller 2009). Accelerated glacier melting in recent years has been reported (Meier et al. 2007; Cogley 2009). Similarly, new estimates of the mass balance of the ice sheets based on InSAR and GRACE indicate accelerated ice-mass loss from these regions (Rignot et al. 2008a, 2008b; Alley 2009). While for the 1993–2003 decade total land ice was estimated to have contributed less than 50% to the rate of sea-level rise (AR4), indications are that this contribution has increased to about 80% since 2003. Moreover, thanks to GRACE space gravimetry data available since mid-2002, it has been possible to quantify the land-water contribution to sea level over the past few years, as described below.

8.1.3 Major Domains of Terrestrial Water Storage

Water is stored on land as ice sheets and glaciers (Chapter 7) and as snow pack, surface water, and subsurface water (as discussed in the present chapter). Surface water includes rivers, lakes, artificial reservoirs, the surface expression of swamps, and ephemerally inundated areas. Subsurface waters are often divided into water within a meter or two of the land surface ("soil water" or "soil moisture," which is directly accessible to plants); water in the saturated zone below the water table ("groundwater"); and the intervening "vadose zone," which can be hundreds of meters thick in arid regions and absent in humid regions.

When and where surface water is present, saturation conditions are present in the subsurface adjacent to the surface. Perennial surface water generally is indicative of a fully saturated subsurface column below the surface water. Intermittent or ephemeral surface-water bodies in arid regions may indicate subsurface saturation only near the surface and only when surface water is present.

The distinction between surface and subsurface water is sometimes useful and sometimes misleading, depending on the degree of coupling between the surface and the subsurface. From a practical standpoint, the distinction reflects ways in which observational data are collected, physics is described, and models are built. Under increasingly strong coupling and/or longer timescales, however, the distinction becomes increasingly artificial. As will be seen in this chapter, many of the uncertainties concerning variability of terrestrial water storage are the result of our ignorance of the extent of surface/subsurface coupling.

8.1.4 Major Drivers of Variations in Terrestrial Water Storage

Changes in terrestrial water storage result from climate variations, from direct human intervention in the water cycle, and from human modification of the physical characteristics of the land surface. Climate variations (which have both natural and anthropogenic causes) force changes in the surface water balance, which can increase or decrease water storage; cool and wet climatic anomalies tend to drive storage upward, and warm and dry anomalies tend to drive storage downward. Some major human activities that directly affect storage are the removal of groundwater from storage by pumping (particularly in arid regions), the creation of artificial water reservoirs by construction of dams on rivers, and irrigation of cropland. Anthropogenic changes in the physical characteristics of the land surface result from urbanization, agriculture, and forest harvesting (along with forest regrowth).

8.1.5 Overview

In this chapter we review the status of understanding of change in terrestrial water storage caused by climate variations and human activities; our main concern is with the effect of such changes on sea level. We attempt to identify the major sources of uncertainty. On the basis of the review we then provide a list of recommendations, concerning both modeling and observations, with the objective of improving this still poorly constrained contributor to sea-level change. The chapter is organized into the following sections:

- analysis tools;
- climate-driven changes of terrestrial water storage;
- direct anthropogenic changes of terrestrial water storage;
- synthesis;
- recommendations.

8.2 Analysis Tools

8.2.1 *In situ* Observations

In situ observations have their greatest utility in the evaluation of changes in surface-water storage. For any given lake, the storage variation is easily determined by monitoring of lake level and knowledge of lake-area–depth dependence. Furthermore, on sufficiently long timescales, the readily observed state of surface-water bodies might be useful as an indicator of the state of subsurface storage.

In situ gauging networks providing time series of river water levels and discharge have been installed and provide multidecadal records for many river basins, but they are distributed non-uniformly throughout the world. Gauging stations are scarce or even absent in parts of large river basins due to geographical, political, or economic limitations (Shiklomanov et al. 2002). For example, more than 20% of the freshwater discharge to the Arctic Ocean is ungauged. Portions of the North American and Siberian Arctic drainage lost more than two-thirds of their gauges between 1986 and 1999. Surface water across much of Africa is not measured.

Because of the areally extensive nature and heterogeneity of subsurface and snow stores and the inherently small spatial sampling scale of *in situ* observations, such observations alone are not of great direct utility for estimating climate-driven changes of global subsurface and snow stores. Their potential value lies more in their usefulness for evaluation and calibration of remote-sensing methods (e.g. satellite gravity and altimetry, discussed below) as well as of models that can be used to generate global storage estimates. Additionally, *in situ* measurements are of substantial benefit in the assessment of strong, localized changes in subsurface stores; such changes arise where anthropogenic disturbance is local in nature, as in the case of withdrawal of groundwater by pumping in arid regions.

8.2.2 Satellite Observations

As in the case of *in situ* observations, observations from satellite platforms serve the dual purpose of directly yielding estimates of storage and of supporting the development of models that can provide less direct estimates (e.g. Alsdorf et al. 2003, 2007; Alsdorf and Lettenmaier 2003; Milly et al. 2004). We focus here on two types of space-based systems of most direct current relevance for observation of terrestrial water storage: gravimetric and altimetric systems.

GRACE Space Gravity Data

In March 2002, a new generation of gravity missions was launched: the GRACE space mission (Tapley et al. 2004a, 2004b). GRACE provides an invaluable set of new observations allowing us to quantify the spatiotemporal change of the total terrestrial water storage (underground and surface waters, snow and ice-mass changes). In addition the GRACE data over the oceanic domain can provide information regarding the ocean-mass change (one of the two contributions to sea-level change, i.e. that resulting from water-mass addition due to land-ice melt and exchange with terrestrial storage).

GRACE allows inference of mass changes by yielding measurements of spatio-temporal variations of the gravity field with an unprecedented resolution and precision, over timescales ranging from a few days to several years. On such timescales, the mass redistribution that causes temporal gravity variations mainly occurs inside the surface fluid envelopes of the earth (oceans, atmosphere, ice caps, continental reservoirs) and is related to climate variability (both from natural and anthropogenic sources) and direct human intervention. GRACE quantifies vertically integrated water-mass changes with a precision of a few centimeters in terms of water height and a spatial resolution of approximately 300–400 km (e.g. Wahr et al. 2004; Seo et al. 2006; Ramillien et al. 2005, 2008a; Schmidt et al. 2006; Chen et al. 2004, 2005a, 2005b; Swenson and Milly 2006; Ngo-Duc et al. 2007). From these quantities and other sufficiently accurate measurements, it is also possible to estimate temporal variations of other hydrological variables, such as precipitation minus evapotranspiration, and total basin discharge (e.g. Rodell et al. 2004; Syed et al. 2005; Wahr et al. 2006; Ramillien et al. 2006a). GRACE measurements have been essential for estimates of mass balance of the ice sheets and corresponding contribution to sea level (Velicogna and Wahr 2005, 2006; Ramillien et al. 2006b; Luthcke et al. 2006; Wouters et al. 2008), ocean-mass change (Chambers et al. 2004; Lombard et al. 2007; Cazenave et al. 2009), and geographically averaged thermal expansion when combined with satellite altimetry (Chambers 2006; Lombard et al. 2007; Cazenave et al. 2009).

Temporal variations of gravity are about 1% of the magnitude of the static field. For this reason, time-variable gravity generally is expressed as anomalies with respect to the static field, and the latter is approximated by the temporal mean of a several-year series of GRACE monthly geoids. Over land, time-variable gravity anomalies mainly result from time-variable water load and can be simply expressed in terms of equivalent water height, either globally or regionally (note that GRACE also measures solid Earth processes such as glacial isostatic adjustment, GIA). The GRACE-derived equivalent water height is then usable for comparison with land-surface models (LSMs) and for other applications.

Wahr et al. (2006) estimated the accuracy of GRACE water-mass determinations. They showed that the error of individual monthly GRACE solutions depends on latitude, and is on the order of 8 mm (equivalent water height, *ewh*) near the pole and approximately 25 mm *ewh* near the equator, for a Gaussian-tapered sampling function with a 750-km radius.

Early terrestrial hydrologic applications of GRACE qualitatively confirmed the consistency of global LSM predictions with GRACE's vertically integrated water-mass change for large river basins (e.g. Tapley et al. 2004b; Wahr et al. 2004; Chen et al. 2005a, 2005b; Ramillien et al. 2005, 2008a). In some recent studies, it has been shown that GRACE is also helpful for evaluating and improving LSMs (e.g. Swenson and Milly 2006; Ngo-Duc et al. 2007; Güntner 2008) (see section 8.2.3).

Other GRACE studies have focused on sea-level change. For example, Chen et al. (2005b) have estimated the contribution of total terrestrial water change (based on GRACE) to the seasonal mean sea level. Accounting for the small water-vapor effect and correcting the altimetry-based annual mean sea level for thermal expansion, they found good agreement between GRACE-based terrestrial water storage and non-steric global mean sea level (Figure 8.1). Another study (Ramillien et al. 2008b) focused on interannual variability and trends. Analyzing GRACE data over the 27 largest river basins globally, they estimated trends in land-water storage for 2003–6 and found a net water-mass loss of approximately $-70 \pm 20\,km^3$/year, corresponding to a sea-level rise of approximately 0.2 ± 0.06 mm/year over

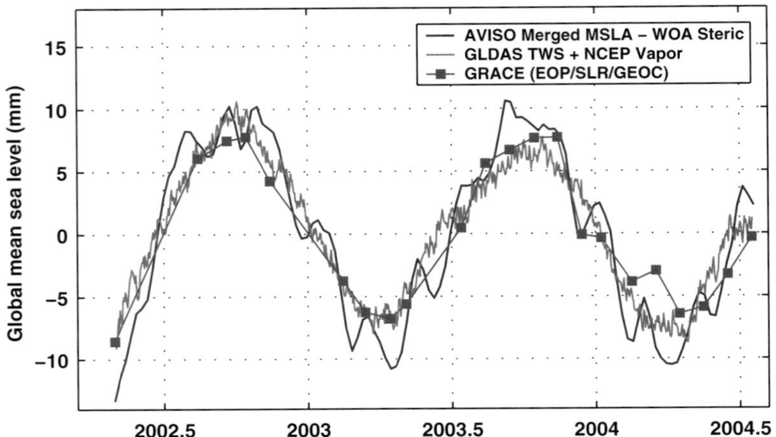

Figure 8.1 Nonsteric global mean sea level from altimeter observation (thermal expansion removed; blue), global water-mass balance (Global Land Data Assimilation System (GLDAS) LSM for total water storage on land + National Centers for Environmental Prediction (NCEP) analysis of atmospheric water storage; green), and GRACE terrestrial water storage (red). AVISO, Archivage, Validation et Interprétation de données des Satellites Océanographiques; EOP, Earth Orientation Parameter; GEOC, geocentre motion; MSLA, mean-sea-level anomaly; SLR, satellite laser ranging; TWS, terrestrial water storage; WOA, World Ocean Atlas 2001. (From Chen et al. 2005b.)

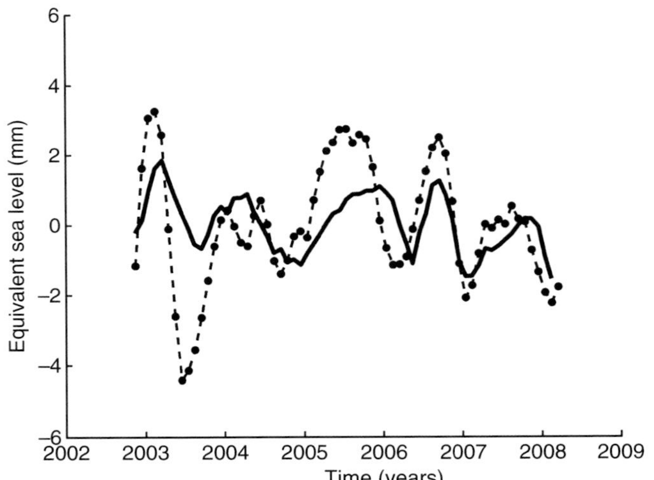

Figure 8.2 Year-to-year fluctuations in total land-water storage estimated from GRACE space gravimetry over 2003–7 (here expressed in equivalent sea level; solid curve). Detrended global mean sea level corrected for thermal expansion (annual cycle removed) is superimposed (dashed curve). (Adapted from Cazenave and Llovel 2010.)

that period (an update of this study, adding 2 years of GRACE data, gives a negative contribution to sea level, of approximately −0.2 mm/year, suggesting that over such short time spans the land-water signal is dominated by interannual variability). Interestingly, year-to-year fluctuations of GRACE-based total land-water storage correlates well with altimetry-based, detrended global mean sea level corrected for thermal expansion. This is illustrated in Figure 8.2 for 2003–8 (from Cazenave and Llovel 2010).

When averaged over the oceanic domain only, GRACE data provide an estimate of the ocean mass component to sea-level rise due to land waters and total ice-mass change. For example, Chambers et al. (2004), Chambers (2006), and Lombard et al. (2007) were able to determine directly the total water-mass contribution to seasonal sea level, in good agreement with the non-steric seasonal mean sea level. The interannual ocean mass change from GRACE was also estimated in several recent studies (e.g. Lombard et al. 2007; Willis et al. 2008; Cazenave et al. 2009; Leuliette and Miller 2009). While the results indicate unambiguous positive trend in ocean mass over the past 5 years, the exact rate of ocean mass increase is uncertain. In effect, GRACE-based ocean mass estimates need to be corrected for the GIA effect (causing negative secular change of the geoid). According to model results, this correction ranges from −1 mm/year (Paulson et al. 2007) to −2 mm/year (Peltier 2009) depending on modeling assumptions. Hence, according to the adopted GIA correction value, the ocean-mass increase since 2003 ranges between 1 and 2 mm/year.

By combining the GRACE-based ocean-mass change component with satellite altimetry-based global mean sea level, it is possible to estimate thermal expansion, without resorting to *in situ* hydrographic measurements (e.g. Chambers 2006; García et al. 2007; Lombard et al. 2007; Willis et al. 2008; Cazenave et al. 2009; Chapter 6). Finally GRACE data are increasingly used to measure the mass balance of the ice sheets and corresponding contribution to sea level (Velicogna and Wahr 2005, 2006; Chen et al. 2005, 2006; Luchtke et al. 2006; Ramillien et al. 2006b; Wouters et al. 2008).

Satellite Altimetry

During the past 15 years or more satellite radar altimetry (SRALT) has been applied to monitor water levels of inland seas, lakes, floodplains, and wetlands (e.g. Birkett 1998; Birkett et al. 2002; Mercier et al. 2002; Maheu et al. 2003; Berry et al. 2005; Frappart et al. 2005; Cretaux and Birkett 2006; Calmant and Seyler 2006; Calmant et al. 2008). Conventional nadir-viewing altimetry has limitations over land, because radar waveforms (e.g. raw radar altimetry echoes after reflection from the land surface) are more complex than their oceanic counterparts due to interfering reflections from water, vegetation canopy, and rough topography. This technique has proved, however, quite useful to measure surface elevation of extensive surface-water bodies. Water-level time series of more than 15 years in length, based on the TOPEX/Poseidon, Jason-1, European Remote Sensing satellite-1 and -2 (ERS-1/-2), and Environmental Satellite (ENVISAT) altimetry missions, are now available for several hundred continental lakes and human-made reservoirs. Internet databases include: http://www.pecad.fas.usda.gov/cropexplorer/global_reservoir for large lakes; the HYDROWEB database, http://www.legos.obs-mip.fr/soa/hydrologie/hydroweb for lakes, human-made reservoirs, rivers, and floodplains; and the River and Lakes database, http://earth.esa.int/riverandlake for large lakes and rivers.

Given the poor economic and infrastructure problems that exist for non-industrialized nations, the recent global decline in gauges, and the physics of water flow across vast lowlands, space-based measurements of surface-water elevation (and inferred discharge when possible) are of great value for a number of applications in land hydrology. Applications of direct interest for sea-level studies include LSM evaluation by altimetry-derived estimates of surface-water storage changes and possibly discharges, and direct estimates of natural and human-made surface-water-body storage change through time.

8.2.3 Models of Water Storage

The global distribution and temporal variations of continental water stores are poorly known, because comprehensive observations are not available globally. LSMs provide a link between water storage and variables that are observed or derived from data. LSMs compute the water and energy balance at the Earth's surface, yielding time variations of water storage in response to prescribed variations of near-surface atmospheric data. The required atmospheric data are the near-surface atmospheric state (temperature, humidity, and wind) and the incident water and energy fluxes from the atmosphere (precipitation and radiation). These are estimated from syntheses of observational analyses and atmospheric model "reanalyses" when the LSM is driven in "standalone" mode. Alternatively, they can be simulated by an atmospheric general circulation model when the LSM is run in "coupled" mode.

LSMs were not designed to perform calculations of water storage on land, but rather to calculate fluxes from land to atmosphere for the purpose of atmospheric modeling. This distinction is important, because a model can do very well calculating fluxes and still make large errors in computed quantities such as long-term trends in storage. Such a disparity in performance is possible because storage is a small factor in long-term average water balance. Only recently have a small number of LSMs been exercised with the problem of terrestrial water storage assessment, and it can be expected that further model developments may be needed for continued progress.

It also needs to be noted that LSMs generally do not account for changes in mass of glaciers. Instead, the presence of glaciers is prescribed, if at all, as an unchanging boundary condition. It follows that applications of LSMs to estimate changes in terrestrial water storage will not include contributions from glacier mass balance.

Global LSMs vary greatly in degree of physical realism, spatial resolution, and explicit representation of vertical and horizontal variability, and a comprehensive review is beyond the scope of this chapter. An LSM usually divides the global land mass on a regular longitude/latitude grid, with horizontal resolution anywhere from a fraction of a degree (more common in standalone applications) to 2 or 3° (in atmospheric-coupled applications). Some LSMs include subgrid heterogeneity by tracking the state of multiple subareas, or tiles, that are all assumed to experience the same atmospheric forcing. A time step on the order of an hour typically is used. For each grid cell or tile, the land is divided vertically into a vegetation layer, a snow pack, and a subsurface ("soil") domain. One or more of these, most commonly the subsurface domain, may be further discretized vertically or simply separated into a root zone and a shallow groundwater layer. Many-layer models do not explicitly distinguish "soil moisture" and "groundwater," but are nevertheless capable of generating the unsaturated and saturated zones to which these terms refer. Furthermore, most LSMs account for space-time variations in ephemeral snowpacks separately from subsurface moisture (soil moisture and groundwater).

Dynamic equations are used to describe the fluxes among the various layers. Interception (storage of water on the foliage of vegetation) is computed by balancing precipitation, throughfall, and evaporation; evaporation is limited by energy availability, which is also tracked for the various layers. Throughfall of snow forms a snowpack; sublimation and snowmelt (again, determined by energy balance) deplete the snow pack. Snowmelt and throughfall of rain infiltrate the soil surface (or run off horizontally) and moisten the surface layers of the soil. Gravity and capillary forces drive the water downward into the soil. Water is drawn from the soil by plant roots, to resupply water lost from plant tissue as a result of energy-balance-driven transpiration.

Most models have an impermeable boundary a few meters below the surface. Downward-percolating water eventually reaches this boundary and forms a saturated zone that then grows vertically. To leave the soil column, water must flow horizontally; such lateral flow to the river system generally is parameterized in such a way that it increases as the depth of the saturated zone increases. Deep storage of vadose-zone or groundwater in arid regions is tracked by almost no LSMs.

In some LSMs, when water leaves the soil column either as surface runoff or as lateral outflow from the soil column, it enters a separate model of the river system. The river model consists of a series of river channels, all of which are linked in a tree-like structure that ends at the ocean or at some point of internal drainage. Flows in the river system are usually parameterized simply in terms of a residence time of water in a link. The river model provides an important point of contact between models and observations, because streamflow is readily measured and is a sensitive indicator of the water balance of large land areas. In most models, however, the transfer of water from land to river occurs only in one direction; the reality of streamflow losses to river beds and to the atmosphere in arid regions generally is not represented.

LSMs can be tested and calibrated in various ways, but generally the available measurements of the extremely heterogeneous fields of snow pack, subsurface water, and evaporative fluxes fall far short of what is needed for exhaustive model testing (an exception is the multidecade satellite record of Northern Hemisphere snow cover extent, which has been used to evaluate the models' ability to represent interannual variability on snow cover). LSMs can be tested on a local scale at heavily instrumented sites (e.g. Henderson-Sellers et al. 1995; Chen et al. 1997). Such tests can be useful in identifying major shortcomings in model structure, but can too easily become tuning exercises in which the number of available model parameters exceeds the power of the data to falsify the model. Further, the conclusions of local tests do not easily transfer to the larger spatial scales that are relevant for sea-level assessment.

A complement to local testing of models is the use of large river basins as a control volume. Such a practice at least allows accurate determination of the areal average of the runoff flux, by means of conventional streamflow monitoring at a single site. This approach has been taken in the Global Soil Wetness Project (Dirmeyer et al. 1999). The serious shortcoming of this approach is that the basin is treated as a black box; an adequate simulation of streamflow does not ensure a realistic simulation of storage change within the basin.

The local and river-basin approaches to model evaluation mentioned above are both normally implemented in a "standalone" model. Such a framework can easily lead to incorrect conclusions if the input atmospheric forcing is not carefully evaluated and adjusted for systematic bias (Milly 1994).

GRACE is now enabling evaluation of temporal variation in continental-scale storage computed in LSMs. A number of investigators (Wahr et al. 2004; Ramillien et al. 2005; Ellett et al. 2005; Chen et al. 2005a; Seo et al. 2006; Lettenmaier and Famiglietti 2006) made preliminary comparisons of GRACE water-storage estimates with estimates from standalone LSM simulations. Swenson and Milly (2006) examined terrestrial water storage variations in several climate models that use LSMs to describe land processes. They found substantial model-specific biases in both amplitude and phase of annual storage variations, particularly in low latitudes, and suggested that these were partially associated with suboptimal descriptions of storage in the models. Ngo-Duc et al. (2007) show striking improvement in the agreement between simulated and GRACE-observed seasonal variations of water storage when a river model that has been calibrated on streamflow measurements is added to the ORCHIDEE LSM (the French global LSM) that they used in their study. Recently, Güntner (2008) investigated the potential of assimilating GRACE data for improving LSM performance.

LSMs operate on horizontal scales of tens or hundreds of kilometers, so they cannot be readily applied to some of the smaller-scale problems of anthropogenic disturbance of the hydrosphere, such as those associated with adjustments of the water table as a response to dams. Additionally, LSMs treat only the few meters nearest the land surface, so they cannot currently be applied to examine storage effects associated with groundwater mining and irrigation of arid lands. Of course, because LSMs neglect such processes, care should be exercised in the selection of river basins for LSM evaluation to ensure that anthropogenic processes do not cloud the model evaluation.

8.3 Climate-Driven Changes of Terrestrial Water Storage

8.3.1 Introduction

Climatic control of continental water storage is exerted across a range of timescales from seasonally to millions of years. We will focus mainly on the shorter end of that range (seasonal to multidecadal). These are the timescales at which climate fluctuations lead to the largest rates of change and are also those of interest for understanding present-day sea-level change.

8.3.2 Snow Pack, Soil Water, and Shallow Groundwater

The temporal variations of some of the terrestrial water stores, from seasonal to interannual and decadal timescales, have been the focus of a series of modeling studies in recent years. Such studies have made use of global LSMs that resolve snow pack, soil water, and, for some models, shallow groundwater at horizontal scales on the order of 100 km. The LSMs do not track changes in glacier mass storage, so those must be estimated by other means; cryospheric storage changes are treated in Chapter 7.

Seasonal Variation and Contribution to Sea Level

During the past decade, several studies have estimated the terrestrial water contribution to the cycle of mean sea level by use of global LSMs (Chen et al. 1998, 2005b; Minster et al. 1999; Cazenave et al. 2000; Milly et al. 2003; Ngo-Duc et al. 2005a; Chambers et al. 2004). The general approach of these studies is to estimate the annual ocean mass component from the satellite altimetry-based global mean sea level, after correcting the latter for the steric component (essentially thermal expansion) and taking into account the small annual variation of atmospheric water vapor, and then to compare the ocean-mass component to terrestrial water storage based on global LSMs or on GRACE. The annual cycle of global mean sea level has an amplitude (excursion from mean to peak or trough) of 5 mm, with a maximum in October. Because the annual cycle of steric sea level also has an amplitude of about 5 mm but is in phase opposition, once corrected for steric effects (using climatologies in general), the residual sea level displays an amplitude of 10 mm, with a maximum in September. The above studies showed that the annual cycle of sea level – corrected for ocean thermal expansion – can be satisfactorily explained by the annual variation in total terrestrial water storage simulated by LSMs, with snow pack making the largest contribution (70%).

Year-to-Year Fluctuations of the Seasonal Cycle and Contribution to Sea Level

The decade-long satellite altimetry time series provides information also on year-to-year fluctuations of the global mean annual sea-level. This change was particularly strong from 1997 to 1998, apparently because of the 1997 El Niño event.

LSMs can be used also to estimate these year-to-year fluctuations and to diagnose their causes, for example to test the hypothesis of an El Niño role in the 1997–8 difference. Ngo-Duc et al. (2005a) computed the seasonal change of global sea level by use of the ORCHIDEE LSM. They were able to simulate the drastic contrast in the annual sea level observed between 1997 and 1998. The analysis of the model results showed that the change was caused by the El Niño Southern Oscillation-driven difference in tropical precipitation over land between these two consecutive years.

Interannual to Multidecadal Variation and Contribution to Sea Level

The Land Dynamics (LaD) model of Milly and Shmakin (2002) was used by Milly et al. (2003) to quantify the contributions of time-varying storage of terrestrial waters to sea level in response to climate change on interannual to decadal timescales. A small positive sea-level trend of 0.12 mm/year was estimated for the period 1981–2000 (direct estimates of global land-water storage change since 2003 using GRACE space gravimetry indicate interannual fluctuations of ±0.2 mm/year equivalent sea level; Ramillien et al. 2008b and update of this study). The long-term trend was small, and large interannual/decadal fluctuations dominated the signal. Subsurface water was the major contributor on interannual timescales.

Ngo-Duc et al. (2005b) ran the ORCHIDEE LSM to assess the climate-driven terrestrial water change, and associated sea-level change, for the past five decades. No significant trend in sea level due to terrestrial waters was visible, but large decadal oscillations produced an overall storage range equivalent to several millimeters of sea level. A strong decreasing contribution to sea level was found during the 1970s, followed by a slow increase during the next 20 years; during the period 1975–93, the ORCHIDEE simulation showed an increase of 0.32 mm/year. During the common simulated period 1981–98, the ORCHIDEE and LaD models simulated sea-level contributions of 0.08 and 0.12 mm/year respectively. For the 1990s, however, the ORCHIDEE-implied trend in sea level was negative, at about −0.1 mm/year. As in Milly et al. (2003), the ORCHIDEE variations could be attributed to subsurface water changes caused by precipitation variations, with the largest contribution to the global mean coming from the tropics.

8.3.3 Deep Groundwater

Climate changes at millennial scales have been profound, particularly during the Pleistocene and Holocene epochs. Changes in regional precipitation can lead to large variations in water storage. In arid regions, the water table typically is deep, and net exchange of water between deep groundwater and the atmosphere occurs at a very slow rate. Consequently, the response of storage to changing climate is very slow. Arid regions such as southwestern North America may still be losing water from a groundwater system that was filled to capacity at the end of the last glaciation. A constant-rate water-table fall of 100 m (a typical current depth of the water table in arid regions) over the approximately 10 000 years of the Holocene could release water from soil having a drainable porosity of 0.3 at a rate of 3 mm/year. (Drainable porosity is the volume of water released per unit horizontal area per unit lowering of water-table height.) No estimate has been made of the fraction of global land that transitioned from humid to arid conditions following deglaciation. For a (probably overestimated) transitional area equal to 10% of the global land area, the corresponding rate of sea-level rise would be on the order of 0.1 mm/year. Because subsurface desiccation is likely to have been more heavily weighted in the earlier millennia, a substantial current sea-level signal of transient post-glacial hydrologic response appears unlikely (Walvoord et al. 2004).

8.3.4 Lakes

Lake-level time series can be constrained by paleoindices (e.g. terraced shorelines), historical records, and systematic present-day instrumental observations in some cases. On millennial and longer timescales, topographic analysis can supply estimates of upper bounds on lake storage during climatic periods of strong precipitation (Jacobs and Sahagian 1993). Millennial-scale changes in surface water may have been substantial in the past, but are unlikely to contribute significantly to the current approximately decadal–centennial rate of storage change.

During the 20th century, the Caspian Sea was a major contributor to change in global lake water storage. Although both climate variations and water-resource development contributed to 20th-century Caspian Sea level changes, climate variations appear to have played the dominant role (Golubev 1998). The level of the Caspian Sea fell about 3 m from 1900 to 1977, with a drop of about 1 m in just a few years during the 1930s. The 3-m drop generated an average sea-level rise of 0.05 mm/year for the period 1900–77. The level of the Caspian Sea rose more than 2 m over the subsequent two decades, contributing a negative trend (−0.12 mm/year) to sea level.

Figure 8.3 shows the water-volume change of the Caspian Sea for 1992–2008, measured by satellite altimetry (combining data from several satellites). For the

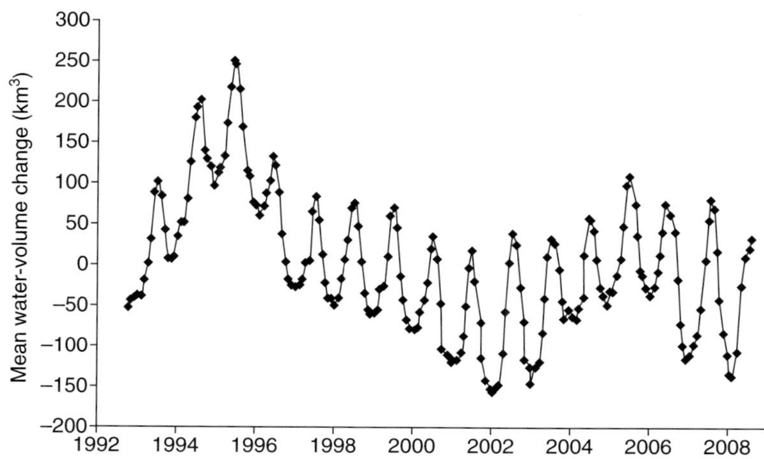

Figure 8.3 Mean water-volume change (in km^3) of the Caspian Sea measured by satellite altimetry (TOPEX/Poseidon, Jason-1, GeoSat Follow-on Satellite (GFO), and ENVISAT satellites) between 1992 and 2008. (Source: LEGOS; HYDROWEB database.)

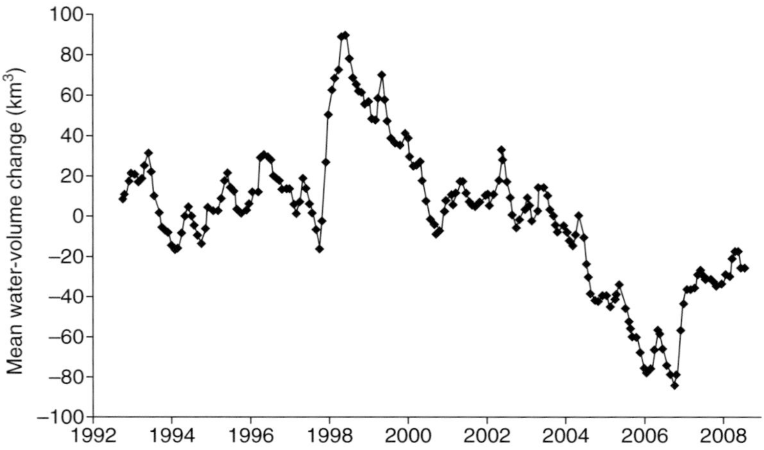

Figure 8.4 Mean water-volume change (in km^3) of Lake Victoria (East Africa) measured by satellite altimetry (TOPEX/Poseidon, Jason-1, GFO, and ENVISAT satellites) between 1992 and 2008. (Source: LEGOS; HYDROWEB database.)

period 1993–2008, the Caspian Sea volume decreased at an average rate of about $-8\,km^3$/year. During the same period, altimetry data indicate that the storage of the Aral Sea and the five Great Lakes of North America also fell. Over 1992–2008, the Aral Sea volume dropped by $-8\,km^3$/year; the Huron and Superior lake volumes also dropped by -12 and $6\,km^3$/year respectively. Storage in the major African rift-valley lakes shows high interannual variability, with a volume drop of Lake Victoria of $-4.4\,km^3$/year over 1992–2008 (see Figure 8.4 showing water-volume change of Lake Victoria). Taken together, we estimate that the aggregate storage in 15 of the largest lakes contributed about +0.11 mm/year to sea-level rise for the period 1992–2008. (the largest contributions are from Lake Huron and the Caspian and Aral Seas, the latter been strongly affected by non-climatic, anthropogenic forcing). However, it is evident that lake water storage is dominated by interannual variability over the period of altimetric records, so the trend estimated for the past approximately 15 years cannot be extrapolated back before that period.

8.3.5 Lake-Affected Groundwater

As the level of a lake rises and falls, so too does the level of the water table adjacent to the lake. Such groundwater responses have been suggested as globally significant amplifiers of both lake and reservoir storage changes (Sahagian et al. 1994; Gornitz 2001). The lateral extent of the induced groundwater storage variations can be limited by process dynamics and/or by the presence of a remote boundary of substantially lower permeability than that of the strata adjacent to the lake. A highly idealized treatment of the dynamics considers the subsurface flow to be one-dimensional and characterized by a constant transmissivity ($T = KB$, where K is saturated hydraulic conductivity and B is saturated thickness). The effective distance of lateral propagation of a water-table rise at a time t following a step rise in lake level is on the order of $(KBt/n)^{1/2}$, where n is the fillable porosity. For a 10-m layer of highly permeable material such as unconsolidated sand and gravel or well-sorted sand, one can assign typical values of $K = 0.01\,\mathrm{m^2/s}$ and $n = 0.3$; these would yield a crude upper bound on the distance of influence of the lake-level rise. The orders of magnitude of the upper-bound propagation distances after 1 and 100 years are about 3 and 30 km, respectively.

According to the calculation above, the 3-m multidecadal fall of the Caspian Sea level is not likely to have penetrated more than 30 km inland. This would imply, at most, an affected subsurface area on the order of one-sixth the area of the Caspian Sea and an induced groundwater storage volume change on the order of 5% of the lake-volume change. Despite the apparent negligibility of groundwater storage in this example, it should be noted that the potential relative contribution of induced groundwater storage to total storage associated with lake-level variations may increase as lake size decreases, because the penetration distance is independent, to first order, of the lake area. Further analysis with site-specific data for various hydrogeologic and climatic environments appears warranted.

8.3.6 *Permafrost*

In sufficiently cold regions, subsurface water deeper than about a meter remains frozen through the year. When this "permafrost" thaws as a result of a decadal to centennial climate transient, the total amount of water stored in the soil column generally decreases. Indeed, in some regions, the soil contains lenses of almost pure ice the disappearance of which explains the irregular changes observed in some landscapes following a thaw. Temperature trends in regions of permafrost generally have been positive in recent decades, and evidence suggests that large-scale thawing of permafrost is underway, perhaps with implications also for water storage (Lawrence and Slater 2005). Furthermore, as the soil column thaws and drains, the subsurface hydraulic connectivity may be enhanced, potentially leading to more free drainage of the landscape. Recently documented large-scale disappearance of lakes in the zone of discontinuous permafrost is evidence of such landscape thaw and drainage (Smith et al. 2005). Order-of-magnitude estimates suggest that this phenomenon has the potential to be an important contributor to sea-level rise in recent years. Unfortunately, such cryospheric processes are not well described in LSMs. Clearly this is an area for further research in the immediate future.

8.4 Direct Anthropogenic Changes of Terrestrial Water Storage

8.4.1 *Artificial Reservoirs*

On the basis of recent literature, Gornitz (2001) estimated that the volume impounded behind the world's largest dams grew by about 5000 km^3 during the

20th century. Other estimates are higher (Chao 1991; Vörösmarty et al. 1997; Nilsson et al. 2005; Shiklomanov and Rodda 2003), and the actual value is uncertain because of nonreporting or underreporting for some countries, and because records generally are not available for the countless reservoirs of smaller capacity (Sahagian 2000). Here we adopt a value of 7000 km^3, which is within the range of published estimates. Most reservoir water was impounded during the second half of the century, so the average rate of sea-level change associated with filling of these reservoirs was about −0.4 mm/year. A recent study by Chao et al. (2008) confirms this order of magnitude over this time span.

The temporal distribution of reservoir filling is relevant for interpreting interdecadal changes in the rate of sea-level rise. The temporal distribution of impoundment reflected in figure 8.2 of Chao (1995) (which included a large portion, but not all, of the total capacity) implies a slow deceleration in the rate of impoundment. This means that the rate of growth of reservoir storage remained positive throughout the second half of the last century, but the magnitude of the rate declined after the late 1970s. Data provided by Chao (1995) and by Shiklomanov and Rodda (2003) suggest a halving of the rate of growth of total capacity from 1950–78 to 1978–2000. Additionally, capture of sediment by reservoirs effectively reduces the overall rate of increase in global impoundment volume. For the globe, Gornitz (2001) estimates a storage-capacity decay rate of 1% per year. Taken together, these results suggest that the global effect of impoundments was greater (in absolute value) than −0.4 mm/year sea-level equivalent before 1978 and smaller than that after 1978. For a halving of the capacity growth rate in 1978, the pre-1978 rate would be about −0.5 mm/year and the post-1978 rate would be about −0.25 mm/year. We therefore adopt a rate of −0.25 mm/year to characterize recent years. The apparent deceleration in impoundment rate would have contributed in small part to the acceleration of sea-level rise that was observed late in the 20th century.

8.4.2 Dam-Affected Groundwater

When a reservoir fills behind a dam, the increase in water depth induces seepage into the subsurface. The process is similar to that discussed in section 8.3.5 in connection with climate-driven lake-level variations. Here our interest is in the response of groundwater to the initial filling of the reservoir rather than in the response to subsequent climate fluctuations. We deduce that the rate of seepage will decrease as the inverse of the square root of time, and that the cumulative amount of groundwater accumulation will grow as the square root of time. Such behavior will continue for any given reservoir until a hydraulic boundary of some kind is reached; the boundary could be either another water body or an effectively impermeable barrier. For either type of boundary, the system would equilibrate on the timescale at which the hydraulic disturbance from the dam reaches the boundary. Because water-saturated land acts as a barrier, the spatial scale of influence in humid zones will be more limited than that in arid zones.

Gornitz (2001) estimated the effect of reservoirs on global groundwater storage under the assumption that seepage losses are constant in time. Taking a seepage rate of 5% of reservoir capacity per year, Gornitz estimated a −0.7 mm/year change in sea level; that is, an effect larger than that associated with the surface-water reservoirs themselves (at 5% per year, the subsurface storage of a reservoir would be double the surface-water storage after 40 years). If instead we assume the square-root-of-time behavior and a 5% seepage loss during the first year, then the 40-year growth in groundwater storage would be about 63% of surface-water storage. In humid regions and in arid regions that have a well-defined subsurface hydraulic barrier (such as bedrock valley walls at the edge of an alluvial valley), the significance of groundwater storage would be considerably less than this.

Taking into account the considerations outlined above, the magnitudes of previous estimates of groundwater storage associated with filling of artificial reservoirs appear to have been overestimated. However, our analysis does confirm that this term might be of sufficient magnitude to warrant further quantitative assessment. Such an assessment should consider factors such as reservoir scale, climatic aridity, and hydrogeologic setting across the population of reservoirs.

8.4.3 Groundwater Mining

The artificial withdrawal of water from the ground by wells causes a reduction in storage of groundwater (Bredehoeft et al. 1982). This causes a reduction in water pressure, which induces an adjustment to natural flows. In humid regions, precipitation exceeds evapotranspiration, the voids of the earth fill almost to the land surface with water, and the ground leaks and spills excess water into the river system as runoff. Thus, the water table (the top of the saturated zone) is generally not far from the surface, and the groundwater system is tightly coupled to the other near-surface stores. As a result, groundwater storage rises and falls in response to the seasonal cycle of climate, and even to weather, and removal of water by pumping is quickly compensated by adjustments in the natural water fluxes. Relatively small adjustments in groundwater storage lead to new dynamic equilibria. Nevertheless, in major urban areas of the humid zone that rely on subsurface water supplies, large-scale "cones of depression" of water storage do develop. Relevant data are available on a piecemeal basis, but such data have not been systematically analyzed and extrapolated to global scale.

In contrast, in arid regions precipitation is much less than the potential evapotranspiration. As a consequence, the soil is dessicated by the atmosphere, and water from precipitation rarely penetrates the ground beyond the root zone of plants. Such systems can be in disequilibrium for thousands of years, as water that had been delivered to the ground during a wetter climate is gradually transported upward to the surface or laterally to topographic lows by increasingly small hydraulic gradients. In such environments, artificial withdrawal of water by pumping leads directly to a progressive decline in water storage until the withdrawal stops, for example, because the store has been depleted. The net depletion of groundwater storage that results from pumping is termed mining.

Gornitz (2001) compiled estimates of mining rates for specific countries from various sources; those explicitly reported rates totaled about 61 km^3/year (or 0.17 mm/year sea-level rise) both for recent years and for the last half-century. Gornitz extrapolated that value by assuming that the ratio of mining to total groundwater withdrawal was similar globally to what it was in the studied regions. Depending on the details of the extrapolation, this approach led to a wide range of estimates of 0.17–0.77 mm/year for the gross effect of groundwater mining on sea-level rise. However, groundwater resources are generally renewable in humid regions. Furthermore, according to Shiklomanov (1997, Figure 4.8), the major groundwater mining operations in the world are found in arid parts of the USA, Australia, and China, and in Mexico, Spain, Algeria, Tunisia, Libya, Egypt, and Saudi Arabia. This list of mining operations coincides closely, though not exactly, with the mining centers explicitly listed in the compilation by Gornitz (2001), suggesting that global extrapolation might not increase the gross effect of mining far above 0.17 mm/year. Allowing for the exclusion of some regions from Gornitz's compilation and for the likelihood that mining may be accelerating, so that past literature underestimates its magnitude, we adopt an estimated 0.2–0.3 mm/year sea-level rise for recent years, while acknowledging considerable uncertainty.

8.4.4 Irrigation

Irrigation generally causes an increase in storage in the root zone of crops. In humid regions, irrigation serves mainly to "top off" the reservoir during periods of water stress, and its global effect is likely to be small relative to the effect of arid-land irrigation. When crops are irrigated in an arid environment, part of the applied water goes into storage in the root zone, part evaporates or is transpired by the plants, and part drains vertically from the root zone to recharge the thick unsaturated zone and the deeper saturated zone.

Irrigation increases water storage both within and below the plant root zone. The storage in the root zone responds rapidly to irrigation. The leakage from the root zone to lower strata causes a transient change in storage that continues until it is balanced by increase of groundwater discharge (or other loss). Because this adjustment is much more rapid in humid regions than in arid regions (as a result of differences in water-table depth), we shall focus on arid regions in the discussion of this term below.

At the end of the 20th century, the global irrigated area was on the order of 2.5×10^6 km^2 (Shiklomanov and Rodda 2003). To obtain an upper bound on the root-zone storage of water, we assume all irrigation was put in place during the course of the last 50 years. For an average 0.1 increase in volumetric water content (irrigated state minus natural state) over a typical 1-m root zone, this amounts to an increase in water storage of 5 km^3/year and a sea-level decline of about −0.014 mm/year. The assumed 0.1 change in volumetric water content is on the order of the difference between "field capacity" (volume fraction of water in freely drained soil) and "wilting percentage" (volume fraction of water in soil when plants reach their wilting point and largely cease water uptake) for a typical soil, and the assumed rooting depth is typical for agricultural crops (Hillel 1980).

The initial rate of change in storage below the root zone in arid regions is approximately equal to the product of the rate of irrigation of arid-region croplands and the fraction of irrigation water that drains downward from the crop root zone. We estimate the global rate of irrigation as 2800 km^3/year (Shiklomanov 1997). As an approximation, we shall assume that all irrigation water is applied in arid regions, where the need for irrigation is greatest. The critical uncertain parameter is the fraction of applied irrigation water that drains below the root zone. A perfectly efficient irrigation system would allow no drainage, though it might eventually result in the accumulation of salts in the root zone. Inefficiencies in irrigation and/or intentional flushing of the root zone imply nonzero drainage. In some irrigated arid lands, downward drainage from the root zone and resultant groundwater recharge from irrigation have been of sufficient magnitude to raise the water table into the root zone, creating problems with water logging and salinization, and driving the deep-drainage fraction to zero. If, as assumed by Gornitz (2001), 5% of irrigation water goes into subsurface storage, we find a resultant sea-level decline of about −0.4 mm/year. Field studies have produced estimates of fractions of irrigation water draining vertically from the root zone that are 6–20% for high-pressure circular spray wheels (Stonestrom et al. 2004) and 40% for flood irrigation (Harill and Moore 1970). The ultimate disposition of these losses is not clear, but if all of this drainage were to enter storage (as opposed to, e.g. entering streams or being transpired by riparian native vegetation), then the associated contribution to sea-level change could be several times larger than −0.4 mm/year.

The atmosphere above irrigated land may be moistened relative to its natural state. However, increases in relative humidity are at least partially opposed by decreases in temperature and saturation humidity. Atmospheric precipitable water content typically is on the order of 20–30 mm water equivalent. Even if absolute humidity were to increase by 20% over the 200 million ha of irrigated land, the effect on sea level would be minuscule.

8.4.5 *Wetland Drainage*

In the USA, wetlands have been drained at an average rate of 2.2×10^9 m^2/year since 1780 (Mitsch and Gosselink 1993). Wetland drainage entails removal of standing water, soil moisture, and water in plants having an order of magnitude of a 1-m depth of water. From these figures, we obtain an average rate of global sea-level rise of 0.006 mm/year. Global wetland area, estimated as 8.56×10^{12} m^2 by Mitsch and Gosselink (1993), is much larger than that of the USA, but we have little information on trends in global wetland area. In Europe, about half the original wetlands have been drained for agriculture, and nearly half in the rest of the world, although inventories are very incomplete. If we assume that the fraction of global and USA wetlands drained are both 50%, and if we spread that drainage over the same 220-year period then we can infer, again very crudely, a global sea-level rise of about 0.06 mm/year. If much of the assumed global wetland drainage were additionally assumed to occur over a shorter period of time, then this estimate would be higher, but only over the shorter period of time.

8.4.6 Urbanization and Deforestation

Urbanization potentially exerts a strong impact on water balance in many ways. Replacement of vegetated areas by impermeable pavements and other structures can lead to increased surface runoff, reduced infiltration, and a lowering of the water table. On the other hand, the removal of vegetation also reduces evaporative loss, and water-delivery infrastructure can enhance recharge, leading to increase in groundwater storage. However, as in the case of other effects considered here, quantitative data allowing assessment of the global effects of urbanization are lacking.

Forests store water in living tissue. When a forest is removed, transpiration typically is reduced so that runoff is more favored in the hydrologic budget. Depending on local climate and topography, this could lead to more or less water stored in the soil. In a poorly drained environment with low slopes, the loss of a forest could cause the water table to rise as a result of decreased evapotranspiration. Alternatively, loss of a forest could cause increased surface runoff and a reduction in subsurface water fluxes and storage. As discussed elsewhere, the humid regions that are home to forests generally respond quickly to disturbance with a new equilibrium that does not require large changes in storage. Additionally, vegetation regrowth is the most common sequel to deforestation.

8.4.7 Atmospheric Water Mass

Though not formally within the scope of our review, we touch briefly here on the water content of the atmosphere. Evidence from global climate models supports a simple thermodynamic control of changes in atmospheric water content on decadal scales. Water content rises in proportion to the saturation vapor pressure of the near-surface atmosphere, which is governed by the Clausius–Clapeyron equation. Thus, a 1°C rise in global mean surface temperature translates to a 7% increase in the 25-mm water equivalent of atmospheric water content. The 0.2°C-per-decade rise in temperature typical of recent years translates to a 0.035-mm/year increase in atmospheric water content and a sea-level change of about −0.05 mm/year.

8.5 Synthesis

In section 8.3 we saw that the natural annual variation of terrestrial water storage is a dominant control of the annual cycle of global mean sea level. We also saw that climate-driven fluctuations in storage at interannual to decadal scales lead to swings in sea level on the order of a few millimeters.

Table 8.2 summarizes our understanding of land contributions to sea-level rise since the early 1990s. Those stores for which we have the most confidence contribute both positively and negatively to sea-level rise. The filling of artificial surface-water reservoirs in recent decades probably contributed about –0.25 mm/year to 1990s sea-level change; a recent global deceleration in filling of reservoirs (resulting from decreasing construction rate and sedimentation) can explain a small part of the recent (1990s) acceleration in rate of sea-level rise. Groundwater mining contributes an opposite effect of about +0.25 mm/year. The warming climate contributes about –0.05 mm/year by increasing the water content of the

Table 8.2 Estimated potential contributions of changes in terrestrial water storage to sea-level change during the decade of the 1990s. Trends assigned "medium confidence" are probably of correct sign and order of magnitude. Trends assigned "low confidence" cannot be constrained by available data to be smaller than multiple tenths of a millimeter per year in magnitude, nor are data sufficient to be sure that any of these terms is large enough to be a factor in sea-level rise. "Essentially unidirectional" trends are those whose sign and order of magnitude are probably dominated by decadal and longer timescales, as opposed to interannual variations.

	Section	1990s sea-level trend (mm/year)	Essentially unidirectional?
Medium confidence			
Reservoir filling	8.4.1	–0.25	yes
Groundwater mining	8.4.3	+0.25	Yes
Fifteen largest lakes	8.3.4	+0.1	No
Climate-driven change of snow pack, soil water, and shallow groundwater	8.3.2	–0.1	No
Atmospheric water storage	8.4.7	–0.05	Yes (under projected warming)
Low confidence, but possibly substantial magnitude			
Irrigation	8.4.4	<0	Yes
Dam-affected groundwater	8.4.2	<0	Yes
Permafrost thaw and drainage	8.3.6	>0	Yes
Lake-affected groundwater	8.3.5	?	No
Wetland drainage	8.4.5	>0	Yes
Deforestation, urbanization	8.4.6	?	No
Low confidence, probably not substantial magnitude			
Post-glacial desiccation on millennial scale	8.3.3	>0	Yes

atmosphere. Climate-driven (and anthropogenic) change in storage in 15 of the world's largest lakes from 1993 to 2006 may explain about +0.1 mm/year sea-level rise. Decadal trends on the order of a few tenths of a millimeter per year can be generated by the combination of snow-pack, soil-water, and shallow groundwater stores in response to climate variations. During the most recent decade of the 1990s, modeled climate-drive trends in these stores probably caused sea-level change of about −0.1 mm/year.

Accumulation of water below irrigated land in arid climates may contribute a substantial negative component to sea-level change, but the magnitude of this term is highly uncertain. An analysis of groundwater dynamics presented here suggests that nonequilibrium seepage to groundwater from surface-water reservoirs and lakes may be more limited than previously supposed; more detailed, site-specific analyses will be required to constrain their contributions further.

We have noted that the LSMs used for assessment of terrestrial water-storage change may not be realistic, particularly when they are applied to describe climate transients associated with melting permafrost or in deep unsaturated zones. One climate model analysis that does consider some subsurface cryospheric factors leads to an estimate on the order of 0.1 mm/year sea-level rise during recent years (Lawrence and Slater 2005). Additional cryospheric processes, neglected in that model, could make the effect even larger.

When we consider only those processes in Table 8.2 in which we place medium to high confidence, we obtain a zero net trend in sea level. This is consistent with the most likely range of ≈0 to 0.3 mm/year deduced in section 8.1.2 for the past two decades (Table 8.1).

8.6 Recommendations

8.6.1 Measurement from Space and on the Ground

Our review of the literature indicates that global, hydrologically relevant data from satellite gravimetric and altimetric missions are rapidly revolutionizing global hydrology and its ability to support sea-level analyses. The scientific payoff is great, even while space data remain limited in resolution, duration, and sampling rate. We see no evidence that available hardware and software are approaching a point of diminishing returns. At the same time, *in situ* observations are more valuable than ever, because they provide the basis for evaluation and calibration of space-based measurement platforms relevant to sea-level rise and other problems. Therefore, we recommend:

- undiminishing efforts toward the collection, archiving, and distribution of land-based measurements of groundwater and surface-water storage;
- continued vigorous development of methods for recovery of the gravity signal from GRACE measurements and for recovery of elevation signals from Jason, ENVISAT and other new altimeter satellites;

- design and execution of new space-based hydrology missions, specifically a higher-resolution gravimetry mission to follow GRACE, a wide-swath interferometric altimetry mission (e.g. SWOT, Surface Water Ocean Topography) for two-dimensional surface water elevation measurements and their derivatives with time and space, and a HYDROS/Soil Moisture and Ocean Salinity (such as the SMOS mission of the European Space Agency which was launched in November 2009)-type radiometric mission for measurement of soil moisture; and
- maintenance and enhancement of global databases and user-friendly data-delivery systems for space-based measurements of surface-water levels and volumes and column-integrated water masses.

References

Alley R. (2009) *State of Antarctic and Greenland Ice Sheet Mass Balance and Ice Sheet Modelling*. AAAS Annual Meeting, Chicago.

Alsdorf D.E. and Lettenmaier D.P. (2003) Tracking fresh water from space. *Science* **301**, 1492–4.

Alsdorf D., Lettenmaier D., Vörösmarty C. and The NASA Surfacewater Working Group (2003) The need for global, satellite-based observations of terrestrial surface waters. *EOS Transactions of the American Geophysical Union* **84**, 269–80.

Alsdorf D., Fu L.L., Mognard N., Cazenave A., Rodriguez E., Chelton D. and Lettenmaier D. (2007) Measuring global oceans and terrestrial fresh water from space. *EOS Transactions of the American Geophysical Union* **88**(n24), 253.

Antonov J., Levitus S. and Boyer T.P. (2005) Thermosteric sea level rise, 1955–2003. *Geophysical Research Letters* **32**, L12602.

Beckley B.D., Lemoine F.G., Luthcke S.B., Ray R.D. and Zelensky N.P (2007) A reassessment of global rise and regional mean sea level trends from TOPEX and Jason-1 altimetry based on revised reference frame and orbits, *Geophysical Research Letters* **34**, L14608.

Berry P.A.M., Garlick J.D., Freeman J.A. and Mathers E.L. (2005) Global inland water monitoring from multi mission altimetry. *Geophysical Research Letters* **32**, L16401.

Bindoff N., Willebrand J., Artale V., Cazenave A., Gregory J.M., Gulev S. et al. (2007) Observations: ocean climate change and sea level. In: *Climate Change 2007: The Physical Science Basis*. Contribution of Working Group 1 to the Fourth Assessment Report of the Intergovernmental Panel on Climate Change (Solomon S., Qin D., Manning M., Marquis M., Averyt K., Tignor M.M.B. et al., eds), pp. 385–432. Cambridge University Press, Cambridge.

Birkett C. (1998) Contribution of the Topex NASA radar altimeter to the global monitoring of large rivers and wetlands. *Water Resources Research* **34**, 1223–39.

Birkett C.M., Mertes L.A.K., Dunne T., Costa M.H. and Jasinski M.J. (2002) Surface water dynamics in the Amazon Basin: application of satellite radar altimetry. *Journal of Geophysical Research* **107**, 8059–80.

Bredehoeft J.D., Papadopulos S.S. and Cooper H.H. (1982) The water-budget myth. In: *Scientific Basis of Water Management*, pp. 51–7. Studies in Geophysics. US National Academy of Sciences.

Calmant S. and Seyler F. (2006) Continental surface waters from satellite altimetry. *Geosciences C.R.* **338**, 1113–22.

Calmant S., Seyler F. and Cretaux J.F. (2008) Monitoring continental surface waters by satellite altimetry. *Survey in Geophysics, Special Issue on Hydrology from Space* **29**, 247–69.

Cazenave A. and Llovel W. (2010) Contemporary sea level rise. *Annual Review of Marine Science* **2**, 145–73.

Cazenave A., Remy F., Dominh K. and Douville H. (2000) Global ocean mass variations, continental hydrology and the mass balance of Antarctica ice sheet at seasonal timescale. *Geophysical Research Letters* **27**, 3755–8.

Cazenave A., Dominh K., Guinehut S., Berthier E., Llovel W., Ramillien G. et al. (2009) Sea level budget over 2003–2008: a reeavaluation from GRACE space gravimetry, satellite altimetry and Argo. *Global Planetary Change* **65**, 83–8.

Chambers D.P. (2006) Observing seasonal steric sea level variations with GRACE and satellite altimetry. *Journal of Geophysical Research* **111**(C3), C03010.

Chambers D.P., Wahr J. and Nerem R.S. (2004) Preliminary observations of global ocean mass variations with GRACE. *Geophysical Research Letters* **31**, L13310.

Chao B.F. (1991) Man, water, and sea level. *EOS Transactions of the American Geophysical Union* **72**, 492.

Chao B.F. (1995) Anthropogenic impact on global geodynamics due to reservoir water impoundment. *Geophysical Research Letters* **22**, 3529–32.

Chao B.F., Wu Y.H. and Li Y.S. (2008) Impact of artificial reservoir water impoundment on global sea level. *Science* **320**, 212–14.

Chen J.L., Wilson C.R., Chambers D.P., Nerem R.S. and Tapley B.D. (1998) Seasonal global water mass balance and mean sea level variations. *Geophysical Research Letters* **25**, 3555–8.

Chen J.L., Wilson C.R., Tapley B.D. and Ries J.C. (2004) Low degree gravitational changes from GRACE: validation and interpretation. *Geophysical Research Letters* **31**, L22607.

Chen J.L., Rodell M., Wilson C.R. and Famiglietti J.S. (2005a) Low degree spherical harmonic influences on Gravity Recovery and Climate Experiment (GRACE) water storage estimates. *Geophysical Research Letters* **32**, L14405.

Chen J.L., Wilson C.R., Tapley B.D., Famiglietti J.S. and Rodell M. (2005b) Seasonal global mean sea level change from satellite altimeter, GRACE and geophysical models. *Journal of Geodesy* **79**, 532–9.

Chen J.L., Wilson C.R. and Tapley B.D. (2006) Satellite gravity measurements confirm accelerated melting of Greenland ice sheet. *Science* **313**, 1958–60.

Chen T.H., Henderson-Sellers A., Milly P.C.D., Pitman A.J., Beljaars A.C.M., Polcher J. et al. (1997) Cabauw experimental results from the Project for Intercomparison of Land-Surface Parameterization Schemes. *Journal of Climate* **10**, 1194–1215.

Church J.A. and White N.J. (2006) A 20th century acceleration in global sea-level rise. *Geophysical Research Letters* **33**, L01602.

Church J.A., White N.J., Coleman R., Lambeck K. and Mitrovica, J.X. (2004) Estimates of the regional distribution of sea-level rise over the 1950 to 2000 period. *Journal of Climate* **17**, 2609–25.

Cogley J.G. (2009) Geodetic and direct mass-balance measurements: comparison and joint analysis. *Annals of Glaciology* **50**(50), 96–100.

Crétaux J.-F. and Birkett C. (2006) Lake studies from satellite altimetry, *Comptes Rendus Geoscience* **338**(14–15), 1098–1112.

Dirmeyer P.A., Dolman A.J. and Sato N. (1999) The pilot phase of the Global Soil Wetness Project. *Bulletin of the American Meteorological Society* **80**, 851–78.

Dyurgerov M.B. and Meier M.F. (2005) *Glaciers and the Changing Earth System: a 2004 Snapshot*. Occasional Paper No. 58. Institute of Arctic and Alpine Research, University of Colorado, Boulder, CO. http://instaar.colorado.edu/other/occ_papers.html.

Ellett K.M., Walker J.P., Rodell M., Chen J. and Western A.W. (2005) GRACE gravity fields as a new measure for assessing large-scale hydrological models In: *MODSIM 2005 International Congress on Modelling and Simulation* (Zerger A. and Argent R.M., eds), pp. 2911–17. Modelling and Simulation Society of Australia and New Zealand, Canberra.

Frappart F., Seylerb F., Martinez J.-M., León J.G. and Cazenave A. (2005) Determination of the water volume in the Negro river sub basin by combination of satellite and in situ data. *Remote Sensing of Environment* **99**, 387–99.

García D., Ramillien G., Lombard A. and Cazenave A. (2007) Steric sea level variations inferred from combined Topex/Poseidon altimetry and GRACE gravimetry. *Pure and Applied Geophysics* **164**, 721–31.

Golubev G.N. (1998) Environmental policy-making for sustainable development of the Caspian Sea area. In: *Central Eurasian Water Crisis: Caspian, Aral and Dead Seas* (Kobori I. and Glantz M.H., eds). United Nations University.

Gornitz V. (2001) Impoundment, groundwater mining, and other hydrologic transformations: Impacts on global sea level rise. In: *Sea Level Rise, History and Consequences* (Douglas B.C., Kearney M.S. and Leatherman S.P., eds), pp. 97–119. Academic Press, San Diego.

Güntner A. (2008) Improvement of global hydrological models using GRACE data. *Survey in Geophysics* **29**, 375–97.

Harill J.R. and Moore D.O. (1970) *Effects of Ground-Water Development on the Water Regimen of Paradise Valley, Humboldt County, Nevada, 1948–68, and Hydrologic Reconnaissance of the Tributary Areas*. Bulletin 39. Nevada Division of Water Resources.

Henderson-Sellers A., Pitman A.J., Love P.K., Irannejad P. and Chen T.H. (1995) The Project for Intercomparison of Land Surface Parameterization Schemes (PILPS): Phases 2 and 3. *Bulletin of the American Meteorological Society* **76**, 489–503.

Hillel D. (1980) *Applications of Soil Physics*. Academic Press, New York.

Holgate S.J. (2007) On the decadal rates of sea level change during the twentieth century. *Geophysical Research Letters* **34**, L01602.

Holgate S.J. and Woodworth P.L. (2004) Evidence for enhanced coastal sea level rise during the 1990s. *Geophysical Research Letters* **31**, L07305.

Ishii M., Kimoto M, Sakamoto K. and Iwasaki S.I. (2006) Steric sea level changes estimated from historical ocean subsurface temperature and salinity analyses. *Journal of Oceanography* **62**, 155–70.

Jacobs D. and Sahagian D. (1993) Climate-induced fluctuations in sea level during non-glacial times, *Nature* **361**, 710–12.

Jevrejeva S., Grinsted A., Moore J.C. and Holgate S. (2006) Nonlinear trends and multiyear cycles in sea level records. *Journal of Geophysical Research* **111**, C09012.

Kaser G., Cogley J.G., Dyurgerov M.B., Meier M.F. and Ohmura A. (2006) Mass balance of glaciers and ice caps: consensus estimates for 1961–2004, *Geophysical Research Letters* **33**, L19501.

Lawrence D.M. and Slater A.G. (2005) A projection of severe near-surface permafrost degradation during the 21st century. *Geophysical Research Letters* **32**, L24401.

Lemke P., Ren J., Alley R.B., Allison I., Carrasco J., Flato G. et al. (2007) Observations: changes in snow, ice and frozen ground. In: *Climate Change 2007: The Physical Science Basis.* Contribution of Working Group I to the Fourth Assessment Report of the Intergovernmental Panel on Climate Change (Solomon S., Qin D., Manning M., Chen Z., Marquis M., Averyt K.B. et al., eds), pp. 337–83. Cambridge University Press, Cambridge.

Lettenmaier D.P. and Famiglietti J.S. (2006) Hydrology: water from on high. *Nature* **444**, 562–3.

Leuliette E. and Miller L. (2009) Closing the sea level budget with altimetry, Argo and GRACE, *Geophysical Research Letters* **36**, L04608.

Lombard A., Garcia D., Ramillien G., Cazenave A., Biancale R., Lemoine J.M. et al. (2007) Estimation of steric sea level variations from combined GRACE and satellite altimetry data. *Earth and Planetary Science Letters* **254**, 194–202.

Luthcke S.B., Zwally H.J., Abdalati W., Rowlands D.D., Ray R.D., Nerem R.S. et al. (2006) Recent Greenland ice mass loss by drainage system from satellite gravity observations. *Science* **314**, 1286–9.

Maheu C., Cazenave A. and Mechoso C.R. (2003) Water level fluctuations in the La Plata basin South America) from Topex/Poseidon altimetry. *Geophysical Research Letters* **30**(3), 1143.

Meier M.F., Dyurgerov M.B., Rick U.K., O'Neel S., Pfeffer W.T., Anderson R.S. et al. (2007) Glaciers dominate eustatic sea-level rise in the 21st century. *Science* **317**(5841), 1064–7.

Mercier F., Cazenave A. and Maheu C. (2002) Interannual lake level fluctuations in Africa from Topex-Poseidon: connections with ocean-atmosphere interactions over the Indian ocean. *Global and Planetary Change* **32**, 141–63.

Milly P.C.D. (1994) Climate, soil water storage, and the average annual water balance. *Water Resources Research* **30**, 2143–56.

Milly P.C.D. and Shmakin A.B. (2002) Global modeling of land water and energy bnalances. Part I: The Land Dynamics (LaD) model. *Journal of Hydrometeorology* **3**, 283–99.

Milly P.C.D., Cazenave A. and Gennero M.C. (2003) Contribution of climate-driven change in continental water storage to recent sea-level rise. *Proceedings of the National Academy of Sciences USA* **100**, 13158–61.

Milly P.C.D., Cazenave A., Beneveniste J., Douville H., Kosuth P. and Lettenmaier D. (2004) Space techniques used to measure change in terrestrial waters. *EOS Transactions of the Americal Geophysical Union* **85**(6), 59.

Minster J.F., Cazenave A., Serafini Y.V., Mercier F., Gennero M.C. and Rogel P. (1999) Annual cycle in mean sea level from Topex-Poseidon and ERS-1: inferences on the global hydrological cycle. *Global and Planetary Change* **20**, 57–66.

Mitsch W.J. and Gosselink J.G. (1993) *Wetlands*. Van Nostrand Reinhold, New York.

Nerem R.S, Leuliette E. and Cazenave A. (2006) Present-day sea-level change: a review. *Comptes Rendus Geosciences* **338**, 1077–83.

Ngo-Duc T., Laval K., Polcher J. and Cazenave A. (2005a) Contribution of continental water to sea level variations during the 1997–1998 El Niño–Southern Oscillation event: comparison between Atmospheric Model Intercomparison Project simulations and TOPEX/Poseidon satellite data. *Journal of Geophysical Research* **110**, D09103.

Ngo-Duc T., Laval K., Polcher J., Lombard A. and Cazenave A. (2005b) Effects of land water storage on global mean sea level over the past 50 years. *Geophysical Research Letters* **32**, L09704.

Ngo-Duc T., Laval. K., Polcher J., Ramillien G. and Cazenave A. (2007) Validation of the land water storage simulated by Organising Carbon and Hydrology in Dynamic Ecosystems (ORCHIDEE) with Gravity Recovery and Climate Experiment (GRACE) data. *Water Resources Research* **43**, W04427.

Nilsson C., Reidy C.A., Dynesius M. and Revenga C. (2005) Fragmentation and flow regulation of the world's large river systems. *Science* **308**, 405–8.

Paulson A., Zhong S. and Wahr J. (2007) Inference of mantle viscosity from GRACE and relative sea level data. *Geophyical Journal International* **171**, 497–508.

Peltier R. (2009) The closure of the budget of global sea level rise over the GRACE era: the importance and magnitudes of the required corrections for global isostatic adjustment. *Quaternary Science Reviews* **28**, 1658–74.

Prandi P., Cazenave A. and Becker M. (2009) Is coastal mean sea level rising faster than the global mean? A comparison between tide gauges and satellite altimetry over 1993-2007. *Geophysical Research Letters* **36**, L05602.

Ramillien G., Frappart F., Cazenave A. and Güntner A. (2005) Time variations of land water storage from an inversion of 2 years of GRACE geoids. *Earth and Planetary Science Letters* **235**, 283–301.

Ramillien G., Frappart F., Güntner A., Ngo-Duc T., Cazenave A. and Laval K. (2006a) Time variation of the regional evapotranspiration rate from Gravity

Recovery Climate Experiment (GRACE) satellite gravimetry. *Water Resources Research* **42**, W10403.

Ramillien G., Lombard A., Cazenave A., Ivins E., Remy F. and Biancale R. (2006b) Interannual variations of the mass balance of the Antarctic and Greenland ice sheets from GRACE. *Global and Planetary Change* **53**, 198–208.

Ramillien G., Famiglietti J. and Wahr J. (2008a) Detection of continental hydrology and glaciology signals from GRACE: a review. *Surveys in Geophysics* **29**, 361–74.

Ramillien G., Bouhours S., Lombard A., Cazenave A., Flechtner F. and Schmidt R. (2008b) Land water contributions from GRACE to sea level rise over 2002–2006. *Global and Planetary Change* **60**, 381–92.

Rignot E., Box J.E., Burgess E. and Hanna E. (2008a) Mass balance of the Greenland ice sheet from 1958 to 2007. *Geophysical Research Letters* **35**, L20502.

Rignot E., Bamber J.L., Van den Broecke M.R., Davis C., Li Y., Van de Berg W.J. and Van Meijgaard E. (2008b) Recent Antarctic ice mass loss from radar interferometry and regional climate modelling. *Nature Geoscience* **1**, 106–110.

Rodell M., Famiglietti J.S., Chen J., Seneviratne S., Viterbo P., Holl S.L. and Wilson C.R. (2004) Basin scale estimate of evapotranspiration using GRACE and other observations. *Geophysical Research Letters* **31**, L20504.

Sahagian D. (2000) Global physical effects of anthropogenic hydrological alterations: sea level and water redistribution. *Global and Planetary Change* **25**, 39–48.

Sahagian D.L., Schwartz F.W. and Jacobs D.K. (1994) Direct anthropogenic contributions to sea level rise in the twentieth century. *Nature* **367**, 54–6.

Schmidt R., Schwintzer P., Flechtner F., Reigber C., Güntner A., Döll P., Ramillien G., Cazenave A., Petrovic S., Jochmann H. and Wünsch J. (2006) GRACE observations of changes in continental water storage, *Global and Planetary Change* **50**, 112–26.

Seo K.W., Wilson C.R., Famiglietti J.S., Chen J.L. and Rodell M. (2006) Terrestrial water mass load changes from Gravity Recovery and Climate Experiment (GRACE). *Water Resources Research* **42**, W05417.

Shiklomanov I.A. (ed.) (1997) *Comprehensive Assessment of the Freshwater Resources of the World*. World Meteorological Organisation.

Shiklomanov I.A., and J.C. Rodda (eds) (2003) *World Water Resources at the Beginning of the 21st Century*. Cambridge University Press, Cambridge.

Shiklomanov A.I., Lammers R.B. and Vörösmarty C.J. (2002) Widespread decline in hydrological monitoring threatens Pan-Arctic research. *EOS Transactions of the American Geophysical Union* **83**, 13–16.

Smith L.C., Sheng Y., MacDonald G.M. and Hinzman L.D. (2005) Disappearing Arctic lakes. *Science* **308**, 1429.

Stonestrom D.A., Prudic D.E., Laczniak R.J. and Akstin K.C. (2004) Tectonic, climatic, and land-use controls on groundwater recharge in a desert environment – the southwestern United States. *American Geophysical Union, Water Science and Application* **9**, 29–47.

Swenson S.C. and Milly P.C.D. (2006) Climate model biases in seasonality of continental water storage revealed by satellite gravimetry. *Water Resources Research* **42**, W03201.

Syed T.H., Famiglietti J.S., Chen J., Rodell M., Seneviratne S.I., Viterbo P. and Wilson C.R. (2005) Total basin discharge for the Amazon and Mississippi River Basins from GRACE and a land-atmosphere water balance. *Geophysical Research Letters* **32**, L24404.

Tapley B.D., Bettadpur S., Watkins M. and Reigber C. (2004a) The Gravity Recovery and Climate Experiment: Mission overview and early results. *Geophysical Research Letters* **31**, L09607.

Tapley B.D., Bettadpur S., Ries J.C., Thompson P.F. and Watkins M.M. (2004b) GRACE measurements of mass variability in the Earth system. *Science* **305**, 503–5.

Velicogna I. and Wahr J. (2005) Ice mass balance in Greenland from GRACE. *Journal of Geophysical Research* **32**, L18505.

Velicogna I. and Wahr J. (2006). Measurements of time-variable gravity show mass loss in Antarctica. *Science* **311**, 1754–6.

Vörösmarty C.J., Sharma K.P., Fekete B.M., Copeland A.H., Holden J., Marble J. and Lough J.A. (1997) The storage and aging of continental runoff in large reservoir systems of the world. *Ambio* **26**, 210–19.

Wahr J., Swenson S., Zlotnicki V. and Velicogna I. (2004) Time-variable gravity from GRACE: first results. *Geophysical Research Letters* **31**, L11501.

Wahr J., Swenson S. and Velicogna I. (2006) Accuracy of GRACE mass estimates. *Geophysical Research Letters* **33**, L06401.

Walvoord M.A., Stonestrom D.A., Andraski B.J. and Striegl R.G. (2004) Constraining the inferred paleohydrologic evolution of a deep unsaturated zone in the Amargosa desert. *Vadose Zone Journal* **3**, 502–12.

Willis J.K., Roemmich D. and Cornuelle B. (2004) Interannual variability in upper ocean heat content, temperature, and thermosteric expansion on global scales. *Journal of Geophysical Research* **109**, C12036.

Willis J.K., Chambers D.T. and Nerem R.S. (2008) Assessing the globally averaged sea level budget on seasonal to interannual time scales. *Journal of Geophysical Research* **113**, C06015.

Wouters B., Chambers D. and Schrama E.J.O. (2008) Grace observes small scale mass loss in Greenland. *Geophysical Research Letters* **35**, L20501.

9 Geodetic Observations and Global Reference Frame Contributions to Understanding Sea-Level Rise and Variability

Geoff Blewitt, Zuheir Altamimi, James Davis, Richard Gross, Chung-Yen Kuo, Frank G. Lemoine, Angelyn W. Moore, Ruth E. Neilan, Hans-Peter Plag, Markus Rothacher, C.K. Shum, Michael G. Sideris, Tilo Schöne, Paul Tregoning, and Susanna Zerbini

9.1 Introduction

9.1.1 *Purpose and Scope*

Geodetic observations are necessary to characterize highly accurate spatial and temporal changes of the Earth system that relate to sea-level changes. Quantifying the long-term change in sea-level imposes stringent observation requirements that can only be addressed within the context of a stable, global reference system. This is absolutely necessary in order to meaningfully compare, with sub-millimeter accuracy, sea-level measurements today to measurements decades later. Geodetic observations can provide the basis for a global reference frame with sufficient accuracy. Significantly, this reference frame can be extended to all regional and local studies in order to link multidisciplinary observations and ensure long-term consistency, precision, and accuracy. The reference frame becomes the foundation to connect observations in space and time and defines the framework in which global and regional observations of sea-level change can be understood and properly interpreted. Geodetic observations from *in situ*, airborne, and spaceborne platforms measure a variety of quantities with increas-

Understanding Sea-Level Rise and Variability, 1st edition. Edited by John A. Church, Philip L. Woodworth, Thorkild Aarup & W. Stanley Wilson.

ing accuracy and resolution and address interdisciplinary science problems, including global sea-level change. In this chapter we identify critical geodetic requirements to meet the rigorous scientific demands for understanding sea-level rise and its variability, and thus contribute to improving its prediction. In particular, we stress the need for the continuity of the geodetic observational series that serve basic research, applications, and operational needs.

9.1.2 *Geodesy: Science and Technology*

Geodesy is concerned with the measurements of geometry, Earth orientation, and gravity and the geoid.

- Geometry: this refers to changes of the position of the Earth with respect to a system of quasars through time, and, in the context of sea level, changes of the surface geometry of the Earth; that is, the variations in time and space of ocean surfaces and ice covers, and of horizontal and vertical deformations of the solid Earth. Unfortunately, geodesy currently is not able to measure the vertical deformations of roughly 71% of the Earth's surface that is the ocean floor.
- Earth orientation: this is measurement of fluctuations in the orientation of our rotating planet relative to the stars, commonly divided into precession, nutation, polar motion, and changes in Earth rotation, which defines and monitors the transformation between the celestial (quasar system) and terrestrial (Earth-fixed) reference frames. Earth rotation is described by the Euler–Liouville equations describing the motion of a rather general Earth, including the solid (non-rigid) body, oceans, and atmosphere.
- Gravity and the geoid: this refers to variations in space and time of the Earth's gravity field, usually expressed as anomalies of the gravity vector, the geoid and the gravity gradient tensor. Gravity and the geoid are mathematically derived from the equations of motion of natural and artificial satellites (in post-Newtonian formulation). For more detail on these general concepts see Beutler (2005).

Each of these three "pillars of geodesy" (Rummel et al. 2005) relate to sea level, and so sea-level has always been a traditional focus of geodetic theory and practice. The surface geometry of the solid Earth defines the ocean bottom surface, which can provide a reference for measuring relative sea-level change (e.g. by tide gauges), and is essential to understanding the impacts of sea-level change (e.g. ground subsidence in Venice or Lagos or New Orleans). Sea-level changes associated with mass redistribution (e.g. melting polar ice sheets and glaciers) affect Earth rotation and polar motion. In static equilibrium, the sea surface follows the shape of the geoid. Moreover, mass redistribution associated with sea-level change also changes the shape of the geoid and the ocean bottom surface, and hence sea level (Figure 9.1). Thus geodesy is fundamental to a comprehensive understanding of sea-level variation; in fact, *sea-level variation cannot be understood outside the context of geodesy.*

Geodesists specialize in acquiring, analyzing, and interpreting space-based, ground-based, and airborne geodetic measurements that are important for

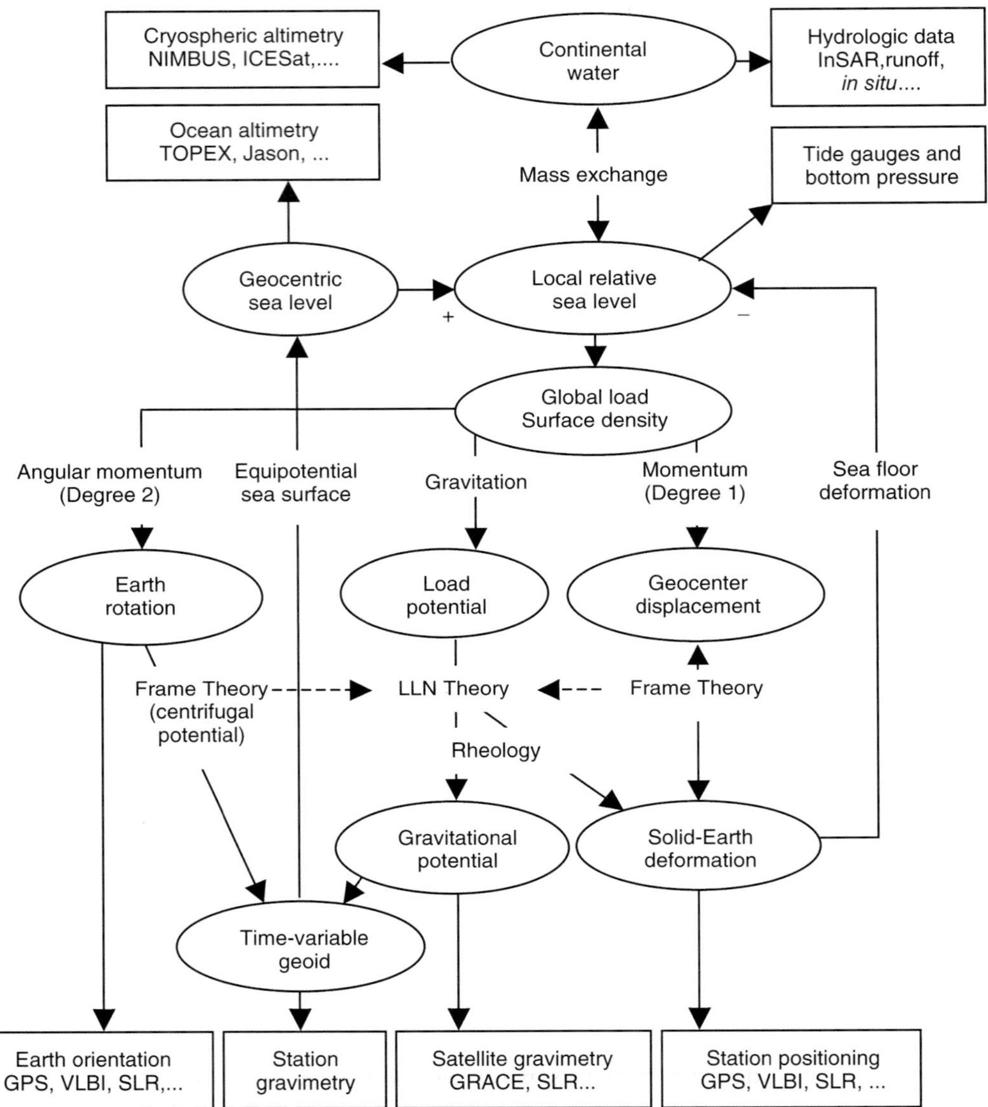

Figure 9.1 A model that incorporates self-consistency of the reference frame (Blewitt 2003), loading dynamics, passive ocean response, and Earth rotation. Closed-form inversion solutions have been demonstrated (Blewitt and Clarke 2003), thus setting the scene for data assimilation. Note that everything is a function of time, so "continental water" in its most general sense would include the entire past history of ice sheets responsible for post-glacial rebound. Arrows indicate the direction toward the computation of measurement models, phenomena are in round boxes, measurements are in rectangles, and physical principles label the arrows. GPS, Global Positioning System; GRACE, Gravity Recovery and Climate Experiment; ICESat, Ice, Cloud, and Land Elevation Satellite; InSAR, interferometric synthetic aperture radar; LLN, load Love number; SLR, satellite laser ranging; VLBI, very-long-baseline interferometry.

understanding sea-level changes. Geodetic observations and analysis contribute to an understanding of Earth processes relevant to sea-level studies. These include but are not limited to:

- hydrology and continental water storage;
- mass balance of ice sheets, ice caps, and glaciers;
- glacial isostatic adjustment (GIA);
- tides of the solid Earth and the oceans and their dissipations, and, to a lesser extent, poles and atmosphere;
- ocean circulation and ocean-bottom pressure change;
- crustal motion associated with plate tectonics, earthquakes, and volcanoes, including coastal deformations and anthropogenic subsidence;
- weather and climate, atmospheric structure, water vapor, and space weather.

Fundamental to understanding these processes is the creation and maintenance of a terrestrial reference frame (TRF) and its tie to inertial space, the celestial reference frame (CRF). The TRF and CRF provide the universal standard against which the Earth is measured; it is the foundation on which solid Earth science disciplines rest. Deficiencies in the accuracy or continuity of the TRF/CRF system limit the quality of science it can support. Observable variations in the TRF/CRF – geocenter motion and rotation irregularities – are themselves primary signals in the science of Earth change. In the context of this book, control of the reference frame over long periods of time may be a primary limiting factor for understanding sea-level change, land subsidence, crustal deformation, and ice-sheet dynamics.

9.1.3 Global Geodetic Observing System (GGOS)

Many modern geodetic techniques require a globally distributed infrastructure for collecting observations. The International Association of Geodesy (IAG) has established a variety of technique-specific scientific services since the late 1980s to facilitate global coordination and to ensure highly accurate and reliable geodetic products to support geoscientific research. And so, in the past, geodetic research concentrated on individual measurement techniques and processes, rather than on the added value that can be drawn from their integration. The Global Geodetic Observing System (GGOS) (Drewes 2005; Plag 2005) is an important new component of the IAG, and intends to give these fundamental components of geodesy a new quality and dimension in the context of Earth-system research by integrating them into a coordinated and collective observing system with utmost precision in a well-defined and reproducible global terrestrial frame. GGOS acts as an umbrella for the IAG services, and coordinates with these scientific services to ensure the development and availability of a global geodetic infrastructure and resulting science, and to identify potential gaps in services, or where new services are required to meet user needs. GGOS will aim to integrate the combination of geometric, gravimetric, and rotational data in data analysis and data assimilation, and the joint estimation or modeling of all the necessary parameters representing the difference components of the Earth system. For GGOS to meet its objectives, it must combine the greatest measurement precision (a relative precision of 0.01 parts per billion $=10^{-11}$) with utmost consistency in space, time, and applied data modeling, and with stability spanning decades. This is a key focus of GGOS.

9.1.4 Geodetic Observations as a Foundation for Assessing and Interrelating Sea-Level Measurements and Uncertainties

The rotation vector of the Earth, usually called the angular velocity vector and referred to the Earth-fixed system, is characterized by its length, the angular velocity (which in turn is directly derived from the length of day), and by two polar coordinates, which may be chosen as the angular distances of the Earth's rotation axis from the Earth's figure axis (in two orthogonal directions); these angles are called polar motion components or simply polar coordinates. This rotation vector of the solid Earth exhibits minute but complicated changes of up to several parts in 10^8 (corresponding to a variation of several milliseconds in the length of the day), and about one part in 10^6 in the orientation of the rotation axis relative to the solid Earth's figure axis (corresponding to a variation of several hundred milli-arcseconds in polar motion). The principle of conservation of angular momentum requires that changes in the rotation vector of the solid Earth must be manifestations of (1) torques acting on the solid Earth or (2) changes in the mass distribution within the solid Earth, which alter its inertia tensor. Angular-momentum transfers occur between the solid Earth and the fluid regions (the underlying liquid metallic core and the overlying hydrosphere and atmosphere) with which it is in contact; concomitant torques are due to hydrodynamic or magneto-hydrodynamic stresses acting at the fluid/solid Earth interfaces. Thus, as the angular momentum of the hydrosphere changes because of sea-level rise and ice-sheet volume variations, the angular momentum of the solid Earth will change, thereby giving rise to changes in the solid Earth's rotation vector. Similarly, the Earth's gravitational field will change as the ocean-bottom pressure changes, and, under the principle of the conservation of angular momentum, the Earth's rotation will change as the oceanic angular momentum varies due to fluctuations in the ocean-bottom pressure and velocity fields. Such variations are detectable from space-geodetic observations.

Satellite altimetry (laser and radar) measures the changes of the sea level, ice elevations, and lake/river levels with an accuracy of 1 ppb with respect to the center of mass of the Earth (Chelton et al. 2001; Schutz et al. 2005; Crétaux and Birkett 2006). Satellite altimetry measurements of the time-varying sea level, when assimilated into oceanic general circulation models along with other remotely sensed and *in situ* measurements, provide improved estimates of the three-dimensional oceanic temperature, salinity, and velocity fields. Altimetry measurements by

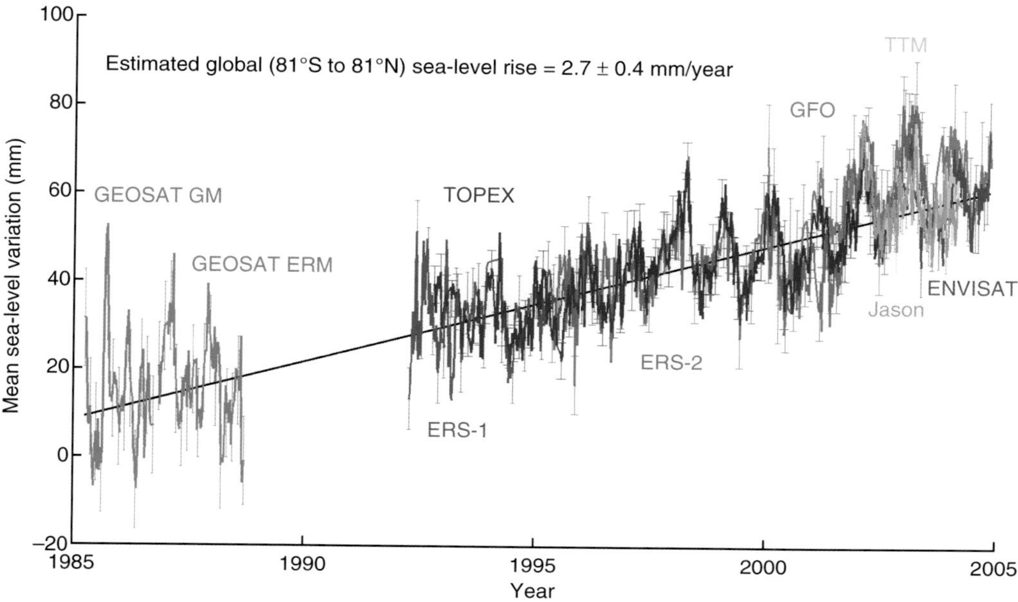

Figure 9.2 Change in global mean sea level over 1985–2004 as measured by multiple satellite altimetry missions (Kuo et al. 2006). Connecting observations from different satellite missions in a meaningful way is nontrivial and a precise, common reference frame for time periods spanning decades is essential.

TOPEX/Poseidon and its follow-ons, Jason-1 and -2, allow the determination of the sea surface height (Figure 5.5) which varies due to both thermal expansion of sea water and changes in ocean water mass arising from changes in polar ice cap, mountain glacier mass, and groundwater storage. Longer-term altimeter observations from multiple missions are clearly needed in the future and with sufficient overlap (e.g. multiyear) to permit the separate missions to be properly interconnected (e.g. Figure 9.2 from an early combination study by Kuo et al. 2006). Connecting observations from these sequential satellite missions (and often times non-overlapping missions) in a meaningful way is non-trivial, but critically essential for assessing long-term changes in sea level or other Earth process. The need to interconnect satellite mission observations in a common, precise terrestrial reference frame over decades is a driving justification for attention to the global geodetic infrastructure, and its maintenance and improvement.

Advances in the measurement of gravity with modern free-fall methods have reached accuracies of 10^{-9}g (1 µgal or 10 nm/s^2), allowing the measurement of effects of mass changes in the Earth interior or the geophysical fluids (ocean and atmosphere loading), as well as the measurement of height changes of approximately 3 mm relative to the Earth center of mass (Forsberg et al. 2005). Measurements of the temporal gravity field from space demonstrated by the Gravity Recovery and Climate Experiment (GRACE) satellite mission (Tapley et al. 2004) and *in situ* terrestrial measurements (absolute and superconducting gravimeters) provide a new global instrument for measuring mass changes of the fluid envelopes of the Earth system as well as viscoelastic response of the Earth's mantle to deglaciation, which are directly relevant to the measurement of global sea-level change. Satellite missions, such as GRACE and the European Space Agency's (ESA's) Gravity Field and Steady-State Ocean Circulation Explorer (GOCE) mission, make gravitational field observations that are sensitive to temporal and spatial variations in the Earth's mass distribution, and can be used to investigate sea-level rise and ice-sheet volume changes. The Earth's gravitational field is not sensitive to the thermal expansion of sea water; observations of the gravitational field can be used in concert with sea-level change observations to separate the change due to thermal expansion or contraction from that due to oceanic mass changes, which helps to quantify the extent to which global warming due to climate change is sequestered in the oceans (Watts and Morantine 1991).

In addition, geometrical geodesy (which today primarily implies Global Navigation Satellite Systems (GNSSs), such as the US Global Positioning System (GPS)), can directly measure displacements of the Earth's surface with a precision higher than 1 mm/year, and thus can characterize change in the Earth's shape, and the land component of relative sea-level change as measured by tide gauges. Changes in the Earth's shape can also be inverted for surface mass redistribution (Blewitt and Clarke 2003) and thus can infer time variations in the shape of the geoid that defines the sea surface in static equilibrium. Blewitt and Clarke (2003) demonstrated this technique to infer the seasonal variation in the mass component of sea-level change with no direct measurements of the ocean, independently confirming published results from TOPEX/Poseidon (corrected for steric effects), and results inferred from terrestrial hydrology (Figure 9.3). GPS is also used to position satellite altimeters in space relative to GPS stations on land, and thus the reference frame of Earth surface measurements can be made compatible with the

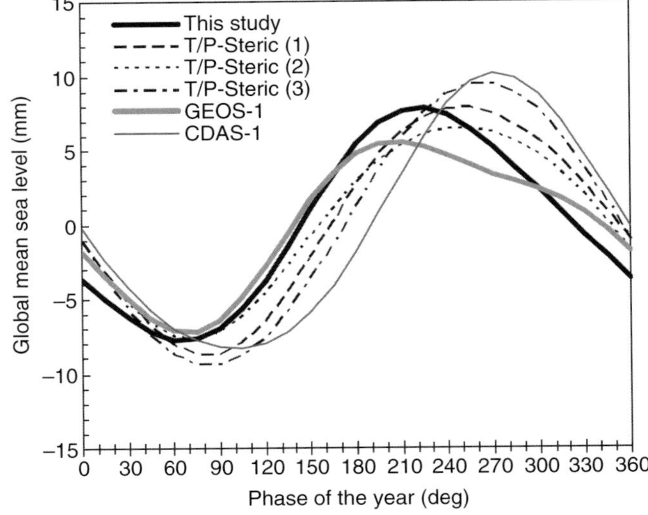

Figure 9.3 Global mean sea level determined through the theory of loading by GPS measurements of Earth's shape (Blewitt and Clarke 2003) compared with direct measurements by TOPEX/Poseidon corrected for steric effects, and terrestrial hydrology, as inferred by mass conservation.

frame of sea-surface measurements by having all observations in one reference frame.

While most geodetic observations are "point-based", interferometric synthetic aperture radar (InSAR) provides two-dimensional mapping views of Earth surface changes. The spatial resolution provided by InSAR is increasingly important for studies of deformation before, during, and after an earthquake, volcanic hazards, groundwater movements, and ice-sheet dynamics. InSAR relies on repeated imaging of a given geographic location by airborne or satellite radar platforms (such as the ESA's European Remote Sensing satellite-1 and -2 (ERS-1/-2) missions and the Canadian Radarsat 1/2 missions). With two complex radar images of the same area one can generate an interferogram as the difference in phase of the return from each pixel. The phase differences are sensitive to topography and any change in position of the imaged area. These effects can be separated using either an independent topographic data set or an additional interferogram that does not include any surface deformation; that is, with negligible temporal separation between repeat passes. A map can be constructed from the two data sets that shows the component of surface motion in the line-of-sight direction of the sensor. Interferograms can detect displacements of a few millimeters.

InSAR can measure two-dimensional glacier or ice-stream flow rates directly related to the computation of ice mass balance, which is one of the major uncertainties of sea-level change. InSAR can also contribute to the measurement and modeling of vertical land motion critical for accurate sea-level observations from coastal tide gauges.

These geodetic data sets (Earth rotation, satellite altimetry, gravity, GNSS/GPS, InSAR) provide a powerful suite of tools for investigating the causes and consequences of sea-level change.

9.2 Global and Regional Reference Systems

9.2.1 Introduction

One of the largest sources of error today in the global characterization of long-term sea-level variation is uncertainty in the TRF. For example, a 2 mm/year error in relative velocity between the mean surface of the Earth and the Earth system's center of mass can result in an error as large as 0.4 mm/year in mean global sea-level variation as determined by satellite altimetry (Kierulf and Plag 2006). The effect on local sea level can be even larger and of opposite sign (Figure 9.4). A scale-rate error of 0.1 ppb/year would map into a sea-level rate of 0.6 mm/year. These frame biases are comparable to or larger than the contributions to secular sea-level change of thermal expansion, and mass exchange with the Greenland and Antarctic Ice Sheets. It cannot be understated how important it is to make further progress in improving the TRF for studies of global change in sea level.

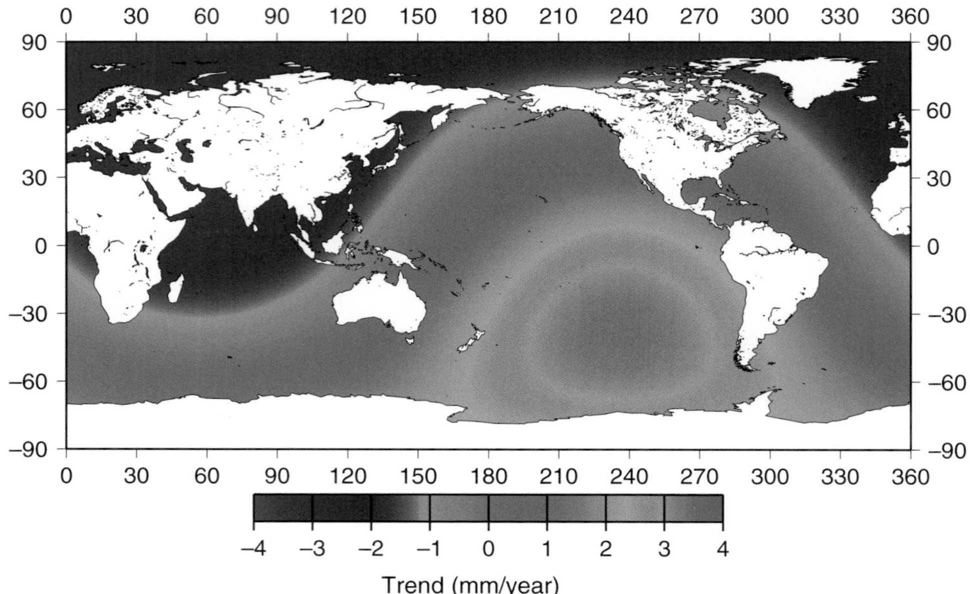

Figure 9.4 Simulated effect on sea-level error inferred by satellite altimetry caused by an error in realizing the reference frame. This example shows the effect of differences between two frames realized by the International GNSS Service (IGS): one is IGb00, which is aligned with ITRF2000; the other is the frame of the IGS precise point positioning products (Kierulf and Plag 2006). The difference in the (x, y, z) motion of the geocenter of these two frames is (−1.5, −2.2, −2.1) mm/year. The difference in sea level ranges from −3 to +3 mm/year, with a mean-sea-level error of 0.4 mm/year caused by asymmetric distribution of the ocean. (From Plag 2006.)

Yet reference frames (apart from the importance of the vertical motions of the ocean bottom) are perhaps the least understood or appreciated component of the methods that connect observations to global measures of sea-level change.

This section focuses on two main points. First, it is important to provide the scientific community a basis for understanding reference frames and their importance in the study of sea-level change. This requires careful definition of terms with this exacting application in mind. Second, there is a complex interaction through models between the reference frame and observations that can be used to assess sea-level variation. This interaction will be explored to identify where weaknesses lie. Recommendations regarding improving the reference frame for sea-level research will result. To take this discussion beyond abstract concepts, specific investigations that underscore the difficulties arising from uncertainty in the reference frame, at both the regional and global scales, will be cited.

This section should emphasize the importance of reference frames in characterizing regional and global sea-level change, and result in a better understanding of what needs to be done to reduce the level of errors introduced by complex interaction between observations, models, and reference frames.

9.2.2 *Terminology: Defining a Reference System and Frames*

Toward understanding sea-level variation, geodetic measurements are used to determine the position of the sea surface and ocean bottom in a globally consistent Terrestrial Reference System (TRS). The TRS must contain physical models (see e.g. Figure 9.1) that are gravitationally self-consistent in order to connect geometrical measurements to (1) the dynamics of the primary geodetic satellite orbits (e.g. GNSS/GPS), (2) the orbits of sensor satellites (e.g. altimeter satellites, gravity missions, InSAR), and (3) the equipotential ("level") surfaces that define sea level in static equilibrium. The connection to gravity is essential for "height variation" to be physically meaningful, allowing for interpretation within the context of a gravitationally self-consistent model. Such a system requires rigorous modeling of Earth rotation, which in turn affects the Earth's gravity field from the perspective of a terrestrial frame co-rotating with the Earth (e.g. centrifugal forces), as well as from the rotationally forced redistribution of mass in both the solid Earth response and the oceanic response (e.g. pole tides). Moreover, gravitational variation caused by mass redistribution changes the Earth's geometrical surface (including the ocean bottom) in a predictable way through surface loading models. Finally, the TRS must be connected to the "real world" through a TRF, where physical points have assigned coordinates that are consistent with the mathematical definition and physical models of the TRS.

The TRS is much more than a coordinate system. Generally accepted terms related to TRS are defined in Kovalevsky et al. (1989):

1 the ideal TRS is a mathematical, theoretical system;
2 the conventional TRS then is the sum of all conventions, parameters, constants (e.g. GM, c, etc.) that are necessary to realize the TRS;
3 a conventional TRF: any TRF is conventional by definition because of the use of (2) and in the meantime the TRF is a realization of the TRS.

The distinction must be clear between the TRS and the TRF; that is, the latter is the realization of the former, and for that conventions are required. Considering a TRS as only a coordinate system can lead to misunderstandings and potential misinterpretations. These concepts can represent a barrier to understanding the significance of geodetic observations of sea level in non-geodetic communities, which is why so much emphasis is placed on the reference system in this chapter (and observational techniques and Earth models are discussed in detail in other sections of this book). In this section the components of a TRS are defined and explained at a conceptual level with the non-geodesist in mind, avoiding the more technical and rigorous definitions that are fundamental to geodesists, for whom the definitive document today is McCarthy and Petit (2003), and references therein.

The relative positions of points in space can be completely specified by coordinates within a defined coordinate system. Coordinates actually overspecify the problem; the parameters that need to be defined are origin, orientation, and scale of the coordinate system (and their evolution in time), which are implicitly defined by assigning arbitrary coordinate values to points. For global-scale geodesy, it is convenient to choose a Cartesian coordinate system, from which transformations into other coordinate systems can be mathematically defined (for example, a conventional ellipsoidal system such as WGS-84, the World Geodetic System 1984 of the GPS). It is useful therefore when thinking about familiar systems such as WGS-84 to think of the underlying Cartesian system as being more fundamental. For example, inside a GPS receiver the GPS positioning problem is solved in a Cartesian system (x, y, z), and the position in WGS-84 (longitude, latitude, height) is calculated and displayed only as the very last step.

However, it is important to note that a TRS should not be confused with a coordinate system or a reference frame, both of which are realizations of the reference system. In general, a TRS comprises three main components, as follows.

First is a datum, which can be an ideal definition of the origin, orientation, and scale of the coordinate system, and their evolution in time. For example, the coordinates of a physical point can be defined as part of the datum definition (such as fixing the height of a benchmark for the height datum, or defining the longitude of a fiducial mark at Greenwich). Of particular relevance to sea level is the choice of vertical datum. Tide gauges measure sea level with respect to a tide-gauge datum that is only useful locally and not suitable for global studies. An attempt to measure the relative height between local tide-gauge datums does not provide an effective global solution to this problem, though it may be useful for some regional studies. For the modern TRS, the use of local fiducial marks (e.g. "fundamental stations") to define the datum is no longer used, except perhaps in some average sense over the entire reference frame (see the third item below). Thus the vertical datum is secondary, in that it must be defined in terms of the global datum parameters. Now the origin can be ideally defined as the center of mass of the Earth system, and the scale can be defined by the International System of Units (SI) meter as realized by atomic clocks together with the conventional speed of light. Observations of satellite orbits can be used to infer an origin at the center of mass of the Earth system, which connects the geometric Earth's surface to the gravity field. This is especially important for global-scale observation of sea-level change. The center of mass of the Earth system can be considered a unique, static equipotential surface (surrounding an infinitesimal point) that can be chosen as the vertical datum. However, such a choice of vertical datum, being so far away from the actual sea surface, requires an ability to position points at the Earth's surface with respect to the Earth center of mass with high accuracy. This in turn requires a stable scale, and scale plays a very practical role in this realization of the vertical datum for sea-level studies, even though it is not directly related to equipotential surfaces as such. In order to realize a practical global vertical datum it is convenient to link the gravity field to the geometry of the Earth's surface (land and sea), a natural choice of TRS origin and scale. Such a capability has only recently become possible with the advent of space geodesy, through which the origin can be realized through the dynamics of the satellite orbits by satellite laser ranging (SLR) and GNSS, and through which scale can be stabilized through observations of distant quasars using very-long-baseline interferometry (VLBI). The orientation of the modern TRS is such that the z-axis points towards the Earth's pole at some reference epoch (since the Earth's pole actually moves at a detectable level from day to day), with the x- and y-axes defining a conventional equatorial plane. The x-axis is chosen to define the meaning of Prime Meridian (zero longitude, which is no longer defined by the fiducial mark at Greenwich and actually lies approximately 200 m away), and the y-axis completes the right-handed frame, thus completing the datum definition. Among the datum parameters, the choice of orientation of the three axes is of least consequence to the problem of sea-level variation.

Second are the conventions of the reference system, which typically specify how to compute the coordinates of a point on the Earth's surface at an arbitrary time, given the epoch coordinates of that point at some initial time (the reference epoch). This transformation typically corresponds to a physical motion model (including, for example, solid Earth tides). The epoch coordinates here generally refer to parameters that are required to initialize the motion model, for example, initial position coordinates and velocity coordinates. As the motion models improve (by theory and/or experiment), so the reference system conventions might be updated from time to time. The datum can be considered part of the conventions, although it is so important that it can be useful to consider it separately.

Third is a reference frame, which is a list of epoch coordinates of a set of physical reference points (sometimes called benchmarks), derived from observations and conventions in a way that is self-consistent with the first and second components, above. This procedure is known as reference-frame realization. A reference system can have several associated reference frames derived by different realizations specific to observation systems or different spans of observations. Typically there will be a unique reference frame that is recommended as the definitive frame, and commonly such a definitive frame represents a synthesis of various observational types, using as much data as possible. Hence the definitive frame requires updating from time to time. Note that in practice, it is the reference frame that implicitly defines the origin, orientation, and scale of the reference system. If, for example, the ideal orientation cannot be realized uniquely by observations alone, one can expect different reference frame realizations to produce coordinates that might differ quite significantly. This problem can be mitigated by frame alignment of subsequent realizations to some initial frame to ensure a level of consistency. For some applications, maintaining consistency in this manner is of primary importance. For other applications, it is less desirable to maintain consistency than to achieve the highest accuracy. The latter demands progressively improving the frame's accuracy in terms of how well it is aligned with the ideal datum, such as improving the alignment of the origin with the center of mass of the Earth system. This might arise as a result of an improvement in geodetic data analysis models, or upgrades to the physical models in the TRS itself. As frames are updated in this manner, this often requires the complete reanalysis of data and its interpretation within the new system. This is the reality faced by sea-level investigations, for which the entire time series of sea level will, in general, be changed as improvements are made to the frame.

The International Terrestrial Reference System (ITRS) was developed by the geodetic community under the auspices of the International Earth Rotation and Reference Systems Service (IERS), a service of the IAG, for the most demanding scientific applications. The most accurate realizations of the ITRS are called the International Terrestrial Reference Frame (ITRF) where multitechnique geodetic solutions are rigorously combined to form the ITRF. There is no single ITRF, but rather a series of updated and improved versions. The versions are identified by the year associated with the date of last data used in the analysis, and should not be confused with the date of applicability. The most recent versions are ITRF1997, ITRF2000, and ITRF2005 (Altamimi et al. 2007). Generally, as time progresses, there is less need for frequent updates, because more time may be needed to make significant improvements through the addition of new data and improved models. However, to satisfy increasing accuracy requirements, the ITRF will continue to be updated to incorporate more advanced models for the time-dependent reference coordinates, and must be updated after large earthquakes. These successive frames provide a common reference to compare observations and results from different locations. The four main geodetic techniques used to compute accurate coordinate include: GPS, VLBI, SLR, and Doppler Orbitography Radiopositioning Integrated by Satellite (DORIS). Since the tracking network equipped with the instruments of those techniques is evolving and the period of data available increases with time, the ITRF is constantly being updated.

9.2.3 Geodetic Techniques for Realizing the ITRF

For the last few decades, continuous improvement of space geodesy techniques, in terms of technology and modeling of their observations, has drastically improved our ability to determine the terrestrial reference frame toward reaching the 1 mm accuracy level on the surface of the continents (but certainly not on the ocean bottom). The fundamental techniques through which these measurements have been acquired include GNSS satellites (GPS, Global Orbiting Navigation Satellite System (GLONASS) and future Galileo), SLR, VLBI, and DORIS. Changes in the surface geometry are measured via microwaves using radar altimetry (e.g. Fu and Chelton 2001), using lasers (e.g. ICESat, the National Aeronautics and Space Administration's (NASA's) Ice, Cloud, and Land Elevation Satellite; Schutz et al. 2005), or InSAR (Seeber 2003). The orbits of the satellites making these surface change observations must be computed as precisely as possible from precise geodetic observations using GPS, SLR, and/or DORIS, in a coherent and stable reference frame. For example, for TOPEX/Poseidon and Jason-1, orbits are computed for these radar altimeter satellites to a radial accuracy of 1–2 cm. This accuracy can be verified through intercomparison of orbits computed by independent geodetic techniques (Haines et al. 2004; Luthcke et al. 2003).

Today, the geodetic techniques that contribute to the realization of the ITRF are organized as scientific services within the IAG:

- IERS;
- International GNSS Service (IGS), formerly the International GPS Service (Dow et al. 2005);
- International VLBI Service (IVS) (Schlüter et al. 2002);
- International Laser Ranging Service (ILRS) (Pearlman et al. 2002);
- International DORIS Service (IDS) (Tavernier et al. 2006).

These scientific services, as well as the gravity field services, now within the umbrella of the International Gravity Field Service (IGFS; http://www.igfs.net/), and a possible future altimetry service, are integral components of GGOS (Rummel et al. 2005). The GGOS focus on a collective effort acknowledges that closer cooperation and understanding among the IAG services can bring significant improvements to the ITRF (http://www.ggos.org).

Each of the observational techniques has unique characteristics. VLBI connects the ITRF to the celestial reference frame and is important for realizing the scale accurately. SLR is the satellite technique that is used to locate the center of mass of the Earth system, and so defines the origin. GPS primarily contributes to the number of sites that define ITRF (densification of ITRF), and to monitoring polar motion precisely. GPS, DORIS, and SLR are used to position Earth orbiting satellites in ITRF, and GPS is used to position points and their velocities on the Earth's land and sea surfaces, such as benchmarks, tide gauges, and buoys. DORIS is the geodetic technique with the most homegenous station distribution, implementation, and operation (Fagard 2006). Connections between the techniques are enabled by collocation at a subset of ITRF sites where two or more space geodesy instruments are operated and local-site ties between monuments are measured using terrestrial high-precision surveying techniques (Figure 9.5). Conventional precise surveying techniques have been used for decades to connect different techniques, and precise leveling is still a critical method for establishing the vertical tie between tide gauges and local benchmarks.

None of the space geodesy techniques is able to provide all the necessary parameters for the TRF datum definition (origin, scale, and orientation). While satellite techniques are sensitive to the Earth center of mass (a natural TRF origin; the point around which a satellite orbits), VLBI (whose TRF origin is arbitrarily defined through some mathematical constraints) is not. The scale is dependent on the modeling of some physical parameters, and the absolute TRF orientation (unobservable by any technique) is arbitrarily or conventionally defined through specific constraints. The utility of multitechnique combinations is therefore recognized for reference-frame determination, and in particular for accurate datum definition.

Since the creation of the IERS in 1987, the implementation of the ITRF has been based on multitechnique combination, incorporating individual TRF solutions derived from space geodesy techniques as well as local ties at co-located sites. In principle, the particular strengths of one observing method can compensate

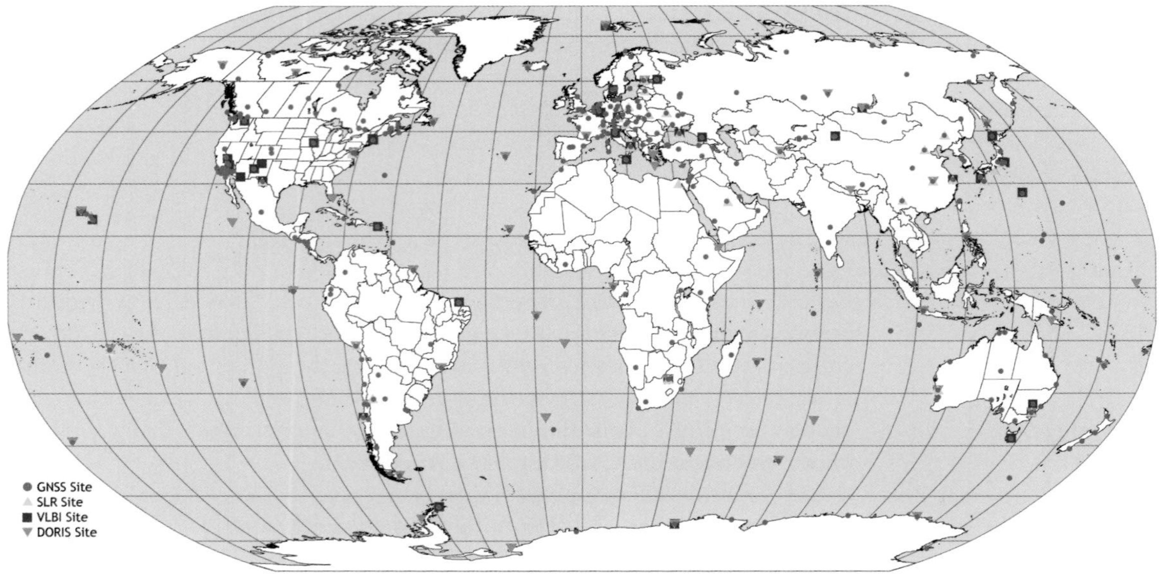

Figure 9.5 Current distribution of geodetic networks: GNSS, SLR, VLBI, and DORIS. (Courtesy of C. Noll 2008.)

for weaknesses in others if the combination is properly constructed, suitable weights are found, and accurate local ties at co-located sites are available. The ITRF quality suffers from any network degradation over time because it is heavily dependent on the network configuration. To cite only one pertinent network issue, the current configuration of co-located sites as depicted in Figure 9.6 is far from an optimal even global distribution. In particular, sites with three co-located geodetic techniques are fewer than 20 and with four techniques there are only two.

Over a decade, the stability of the ITRF2000 geocentric origin (defined by SLR) is estimated to be at the few-millimeter level and the accuracy of its absolute scale (defined by SLR and VLBI) is around 0.5 ppb (equivalent to a shift of approximately 3 mm in station heights) (Altamimi et al. 2002). While SLR currently provides the most accurate realization of the Earth's long-term center of mass for the ITRF origin, estimates of geocenter motion still need to be improved by the analysis centers of all satellite techniques. From ITRF2000 results it was found that the best scale agreement was between VLBI and SLR solutions. Geocenter stability depends on accurate dynamic modeling and observation of geodetic satellites, such as SLR and GNSS (Tregoning and van Dam 2005). Scale stability might be better ensured by minimizing source-related errors, which would imply VLBI, but it also requires accurate tropospheric delay modeling, which would imply SLR (because of the observed frequencies, the tropospheric delay effects are considerably smaller in SLR observations than in either VLBI or GNSS), so some combination of VLBI and SLR is likely to be required. The expected increase in

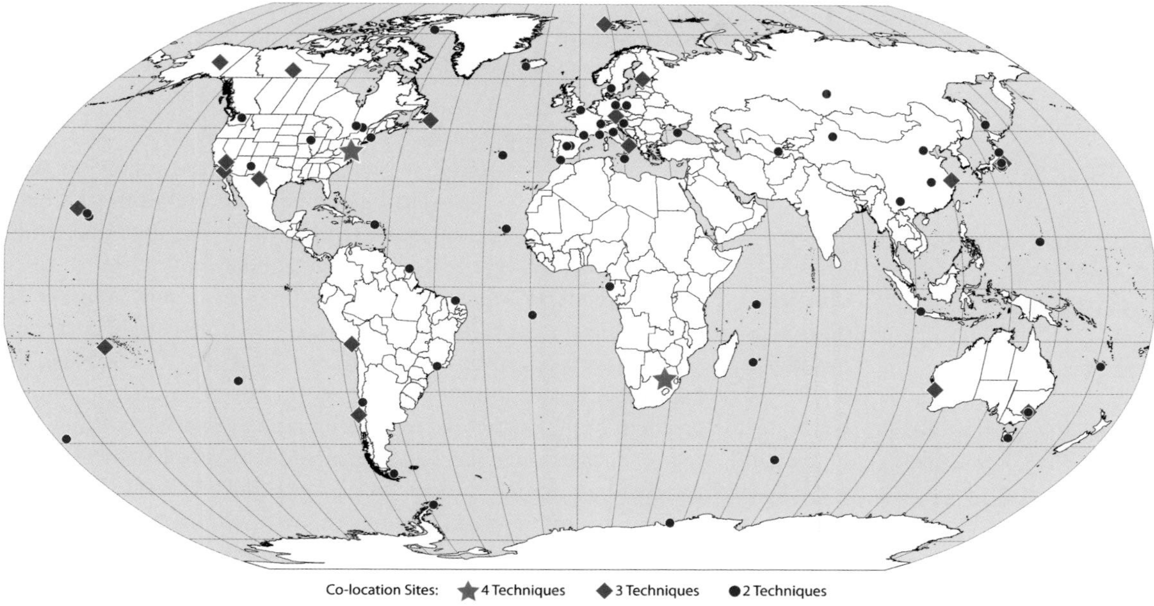

Figure 9.6 Current distribution of co-located space geodesy sites. (Courtesy of Z. Altamimi.)

numbers of GNSS satellites over the next decade to approximately 100 suggests the strong potential of GNSS to contribute significantly to both geocenter and scale stability.

Since the mid-1990s, initiated first by the IGS (Dow et al. 2005), several technique-specific analysis centers started to publish time series of daily or weekly solutions of station positions and daily Earth Orientation Parameters (EOPs). The methodology has since been extended to combine time series of results (Altamimi et al. 2005) and to extract all the benefits offered by time series combinations; for instance, detecting and monitoring non-linear station motions and other kinds of discontinuities in the time series: site instabilities, earthquake-related dislocations, seasonal loading effects, etc. Time-series combination also allows EOPs to be treated in a fully consistent way; that is, to rigorously ensure the alignment of EOPs to the combined frame. Unlike the previous ITRF solutions, ITRF2005 is based on the analysis and combinations of such time series of station positions and EOPs (Altamimi et al. 2007). The new ITRF2008 is currently under preparation and is expected to be released in 2010.

9.2.4 Errors Related to Reference Systems and their Effects

The following three aspects of reference systems lead to uncertainty in the position of a physical point that should be considered in addition to the observational error in position:

1 error in physical aspects of the conventions, in particular, the motion model;
2 error in the alignment of the frame that is used to realize the reference system; and
3 error in coordinates of the reference frame that have been used to position the points of interest.

For the challenge of understanding sea-level variation, the following issues relate to the above errors. First, the solid Earth's surface (including the ocean bottom) moves much like the sea surface itself, in the sense that it has tides (≈30 cm). Some of the motion is predictable, such as tidal motion, much of the short-period motion tends to average out in the long-term, and some of the motion is unpredictable or imprecisely known. On the timescale of a century, motion of the Earth's surface can be of the same order of magnitude as motion of the sea surface (0.1 m) and locally can exceed this by a significant amount. Naturally the problem of the impact of sea-level variations requires consideration of the land motion. A tide gauge directly measures the displacement of the sea surface relative to land at a point, and so would seem ideal. However, the use of tide-gauge data alone to infer global measures of sea-level change is fundamentally problematic due to sample bias of the relative motions of the sea surface and land over a broad range of spatial and temporal scales. Processes such as GIA have been given considerable attention and are addressed elsewhere in this book. The land beneath tide gauges may be forced to move by processes unique to (or more biased at) coastlines (versus deep ocean floors), such as coastal erosion, sedimentary loading, subsidence, atmospheric loading, anthropogenic activities, tectonic processes (e.g. strain accumulation at locked subduction zones along much of the western coasts of the Pacific rim and Sumatra), and the different ocean/land response to present-day mass redistributions such as cryospheric loading and terrestrial hydrologic loading. Other considerations include the stability of the structures to which tide gauges are attached, and the local stability of the land beneath. In general, coastlines are well known to host unique oceanic processes that can bias sea level; however, a crucial point is that this is also true for the land near the coast, in that coastlines provide a very poor sample distribution of Earth deformation processes that will not generally tend to average out on the global scale.

Second, currently, the possibility of an error in the tie of the reference frame origin to the Earth's center of mass at the 1–2 mm/year level, or an error in scale rate at the level of 0.1 mm/year, cannot be dismissed. An error in the realization of the reference frame origin at the center of mass of the solid Earth implies an error in the height of the sea surface inferred by satellite altimetric observations with very-long-wavelength (hemispheric-scale) correlated errors (Figure 9.4). If the error happened to point in the direction of large oceans such as the Pacific Ocean, this could give the erroneous impression of global sea-level change. More generally, the degree-1 terms of the spherical harmonic expansion of the ocean function imply a direct correlation of errors between realization of the origin (with time) and global mean-sea-level variation. Specifically, errors in the velocity of the origin at the level of 2 mm/year will map into global mean-sea-level errors by as much as 0.4 mm/year (Kierulf and Plag 2006). This is as large as the physical contributions of mass exchange and thermal expansion. An error in scale rate will appear as a secular change in the height of ocean altimeters, thus giving the erroneous impression of global change in sea surface height. Tide gauge measurements are immune to reference frame problems and despite the problems with sampling bias and possible land movements, they do provide a measure of ground truth that can be useful for comparison with satellite altimeter measurements.

Third, as reference systems are updated because of improved conventions or improved frames (using more recent measurements and models), the entire time series of coordinates of a monitored station will change. Even the sign of vertical velocity can change. Therefore, the concept of "measuring and archiving" the heights of tide gauges should be abandoned. The heights of stations cannot be measured absolutely, and the entire time series are always going to be subject to changes and reanalysis as reference frames continue to be updated, and certainly improved.

9.2.5 Challenges and Future Requirements

From a reference-frame perspective, the challenge for monitoring long-term variability in sea level is to define the frame origin and scale with greater accuracy than the signal to be estimated. This requires a frame stability of 0.1 mm/year, and scale stability of 0.01 ppb/year. These requirements would reduce the frame-related bias to the level of a few percentage points of the total effect of sea-level change. Current errors may be about a factor of 10 larger than this, although the level of errors are currently difficult to assess.

The most critical TRF parameters of interest to mean-sea-level studies are the origin and the scale and their long-term stability. For example, any scale bias in the TRF definition propagates directly to the height component of the stations and vice versa. As the ITRF relies on SLR to define its origin and on SLR and VLBI for its scale, the importance of these two techniques should not be underestimated for the ITRF accuracy and stability over time. Unfortunately, the distribution of the current SLR and VLBI networks and their co-locations is poor and worsening over time, threatening the long-term ITRF stability. To give a simple example, from the ITRF2005 analysis, the estimated impact of the poorly distributed SLR network and its co-locations with the other techniques induces a scale bias of about 1 ppb and 0.1 ppb/year. This is a large effect by itself and about 10 times larger than the science requirement to address sea-level change.

To meet these challenges and future requirements, the geodetic networks must be well distributed, maintained, and improved to provide the fundamental context for understanding sea-level change.

9.3 Linking GPS to Tide Gauges and Tide-Gauge Benchmarks

9.3.1 Tide Gauges and the Reference Frame

Tide gauges measure sea-level changes as variations in the relative position between the crust and the ocean surface. These measurements are difficult to interpret because they are influenced by several phenomena inducing vertical crustal movements. Vertical crustal motions at tide gauges can be measured to high accuracy independently of the sea-level reference surface by means of space geodetic techniques such as GPS and DORIS (Soudarin et al. 1999); therefore, it is possible to separate the crustal motions from geocentric sea-level variations. Tide-gauge measurements are difficult to compare because tide gauges are referred to local reference systems not yet connected on a common global datum. However, it should be pointed out that several international efforts are underway both at global (IOC 1997) and regional scales (Zerbini et al. 1996; Becker et al. 2002) which aim to overcome this challenge.

Continuous GPS is the technique of choice in vertical crustal motion determination due to the ease of use, high precision, and its direct connection to the ITRF through the products of the IGS. Simultaneous GPS measurements performed at tide gauges and at fiducial reference stations of the global reference frame can be tied in a global well-defined reference frame. The possibility to refer the tide-gauge data to the same high-precision global reference system allows comparison between the different tide-gauge data sets. This was not the case until before 1995 when tide-gauge benchmark coordinates were mostly available in the different national height systems (Wöppelmann et al. 2006).

In order to determine long-term height changes due to vertical crustal movements it is necessary to correct the GPS measurements for seasonal oscillations which can corrupt the estimate of the long-term trends up to a few millimeters per year. Loading components due to seasonal variations of the atmosphere, hydrology, and non-tidal oceanic effects (Blewitt et al. 2001; van Dam et al. 2001; Zerbini et al. 2004) have been recognized as major contributors to the observed seasonal oscillations in GPS time series. Concerning non-tidal oceanic effects, recent studies at the global and regional level (Chao et al. 2003; Zerbini et al. 2004) show that modeled bottom-pressure amplitudes taken from the Estimating the Circulation and Climate of the Ocean (ECCO) project are a factor of two smaller than those observed. Furthermore, deficiencies in the physical models can create spurious periodic effects including annual and semi-annual signals. Penna and Stewart (2003) show how mismodeled short-period tides (semidiurnal and diurnal) can alias into height time series, and Stewart et al. (2005) showed how truncation of the observation model – and even the arbitrary choice of processing data in 24-hour segments – causes similar propagation effects. Penna et al. (2006) demonstrated that mismodeled (sub-)daily periodic signals in horizontal coordinates can propagate into periodic signals in the vertical component, sometimes with an admittance greater than 100%. Boehm et al. (2006) showed that deficiencies in the Niell mapping function (used to relate the tropospheric delay in the zenith direction to the delay at any elevation angle) causes season-dependent height errors of up to 10 mm, in particular in the Southern Hemisphere. Tregoning and Herring (2006) showed that the use of actual observed atmospheric pressure values – rather than a standard sea-level atmospheric pressure model – for computing the *a priori* zenith hydrostatic delay reduces season-dependent height errors and hemisphere-dependent biases in height estimates. Watson et al. (2006) showed how improvements in the modeling of the solid Earth tides have reduced the annual signals in global GPS analysis, with the implication being that any errors in the model currently used in all space-geodetic techniques can be expected to contribute to seasonal variations that still remain in geodetic height time series.

The accuracy required by GPS to monitor tide-gauge benchmark positions on shorter timescales requires more accurate GPS positions, which in turn requires advances in network and observational configurations and geodetic data analyses. The IGS is intent on continuing to improve the accuracy of its GNSS products and, in collaboration with sister services, to strive towards meeting the demanding requirements of these longer-term studies. (http://igs.org/components/prods.html; Altamimi et al. 2002; Dow et al. 2005).

Independently from space geodetic techniques, an alternative approach to monitoring site velocities is provided by the measurement of absolute gravity at tide-gauge benchmarks. Absolute gravity does not directly estimate the vertical displacement of the crust, as gravity is affected both by mass in the ocean itself and the processes responsible for vertical crustal movement. If the processes are well understood, then it can provide independent confirmation of GPS results, particularly at inland sites (Zerbini et al. 1996; Becker et al. 2002; Teferle et al. 2006).

9.3.2 Tide-Gauge Measurements: Historical Perspective

For more than a century tide-gauge measurements in estuaries or coastal shorelines have been widely used for monitoring local sea or estuary levels, for navigation and port operations, for assimilation into tide models for scientific research and water-quality applications and for use with storm surge models for flood warning. The tide-gauge systems are or have been operated by port authorities or national maritime services with a high level of accuracy and reliability. Since 1933, the Permanent Service for Mean Sea Level (PSMSL), one of the oldest scientific services noted in other chapters in this book, has been responsible for the collection, publication, analysis, and interpretation of sea-level data from the global network of tide gauges (Woodworth and Player 2003). From the geodetic point of view, the tide-gauge system, in particular the tide-gauge zero or pole staff, is precisely leveled by conventional precise surveying techniques to a primary tide-gauge benchmark surrounded by and tied to several distant secondary benchmarks. In addition, most of the tide gauges are connected to the first-order national height system. All benchmarks should be leveled on a regular basis to ensure the long-term stability of the height reference (IOC 2006).

Recently, in the public and scientific climate discussions, data from tide gauges is increasingly important in providing long-term and reliable measures of the sea level. Here, the tide-gauge measurements are used as a primary input to study changes in local mean sea level, tidal amplitudes, surge statistics, and as boundary conditions in oceanographic circulation models. They also act as ground truth for, for example, satellite radar altimetry (SRALT). Tide gauges also help to define the global height system, and many national datums, both historically and even currently, refer their vertical measurements to "mean sea level".

An increasing number of tide-gauge benchmarks have been equipped with continuously operating GPS receivers. A small number of these stations are part of the global IGS network, and many more stations contribute to local or regional networks. There is increasing demand from the international observing systems (i.e. Global Climate Observing System (GCOS), Global Ocean Observing System (GOOS), and Global Sea Level Observing System (GLOSS)) for continuous provision of highly precise GPS height time series with an accuracy of better than 1 mm/year. An affirmed objective of the GLOSS group is to have a GPS receiver at every GLOSS Core Network station.

9.3.3 The TIGA Pilot Project

At present, about 280 tide-gauge stations are known to have continuous operating GPS stations within 10 km (Figure 9.7) (Wöppelmann et al. 2006). This number of stations is far too high for current IGS analysis centers to process, although schemes are under development for routine analysis of up to an order of magnitude more stations by 2010. In addition, the current IGS accuracy does not fully meet the requirements of the sea-level community as there is insufficient station coverage and need for improved accuracy of the vertical component.

In response to the demands by the scientific community, the IGS in 2001 initiated the GPS Tide Gauge Benchmark Monitoring Pilot Project (TIGA-PP). The pilot project includes analyzing GPS data from stations at or near public tide gauges on a continuous basis (known as cGPS@TG, for continuous GPS at tide gauges). The primary objectives of the TIGA-PP are to promote, establish, maintain, and expand a high-quality global cGPS@TG network and to compute precise daily or weekly station coordinates and velocities for this network. This goal is achieved by processing a large number of stations and also by reprocessing older data sets. TIGA-PP, as an IGS project, relies on the IGS network infrastructure, the processing capability, and expertise of the IGS community and is supported by GLOSS and PSMSL for the tide-gauge component.

The primary product is weekly sets of coordinates for analyzing vertical motions of tide gauges and tide-gauge benchmarks. All products are made publicly available to support and encourage other applications, for example sea-level studies. In particular, the products of the service facilitate the distinction between absolute

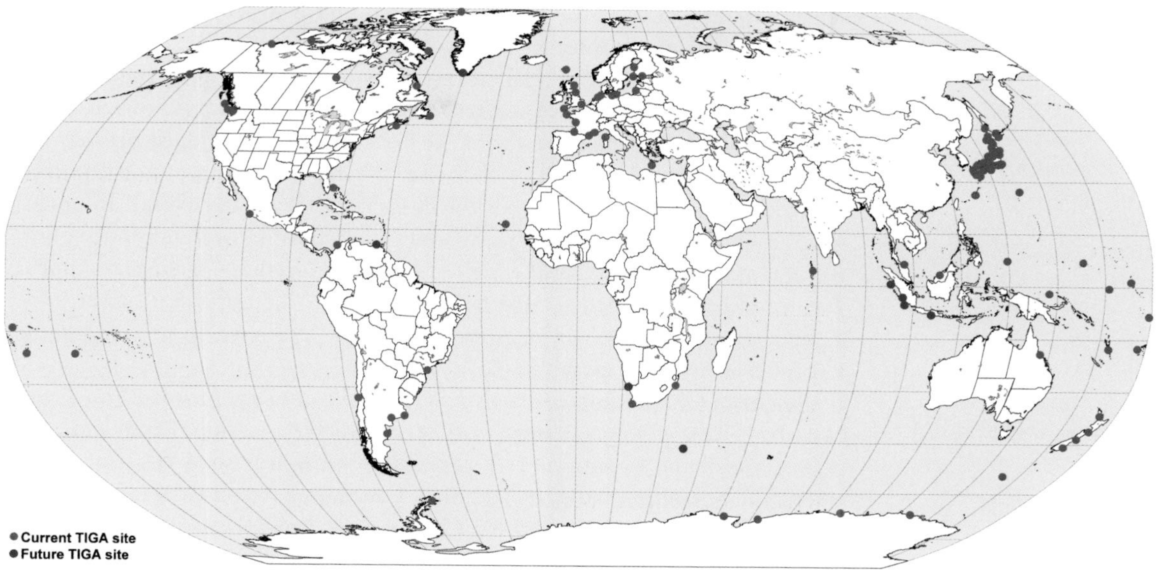

Figure 9.7 Current network of GPS stations at tide gauges contributing to the TIGA Pilot Project. (Courtesy of C. Noll 2008.)

and relative sea-level changes by accounting for the vertical uplift of the station and are, therefore, an important contribution to climate-change studies. TIGA processing may further contribute to the calibration of satellite altimeters and other oceanographic activities.

9.3.4 Current Status of TIGA

The IGS has very strict requirements on data quality and availability, and latency in data delivery of network stations. In contrast to the tide-gauge station networks, the GPS network coverage appears to be reasonably balanced geographically, with the exception of the African continent. Therefore, many of the GPS tide-gauge stations important for sea-level research are not part of the IGS network either for geographical reasons (too close to existing stations already in the IGS network) or because of latency in data provision (remote stations with poor data communications). TIGA analysis incorporates these tide-gauge GPS stations not included in IGS global analyses.

The tide-gauge stations contributing to TIGA must have a high level of reliability. One prerequisite for stations to be included in the TIGA networks is the public availability of the tide-gauge data at GLOSS data centers. Preferably, the stations are equipped with a primary GPS on or near the tide gauge and a secondary GPS station, inland (IOC 2006). Maintenance of the equipment should be repeated on a regular basis, including first-order leveling to all the available benchmarks (which is often difficult to accomplish due to increasingly limited geodetic surveying resources and capabilities).

It is imperative that the relative vertical movement between the tide gauges (and/or tide gauge benchmarks) and the nearby GPS stations is known to a higher accuracy than either the accuracy of the GPS vertical velocities or the tide-gauge estimates of relative sea-level; otherwise, the measurement of the connection between the two systems becomes the limiting factor on the overall accuracy. To observe a height tie with an accuracy of less than 0.5 mm over a distance of up to 10 km is a challenge even for the highest-precision leveling. An alternate approach that could be considered for the future is InSAR, where theoretical and experimental accuracies of velocity estimates from the permanent scatterers technique is 0.1–0.5 mm/year (Colesanti et al. 2003).

Data from GPS stations co-located at 102 tide gauges are processed on a regular basis by TIGA analysis centers (TACs). Other GPS stations complement this network to define a common reference frame. Currently six TACs process TIGA GPS on a best-effort basis. In addition, a reanalysis of past data is performed, leading to a homogeneous data set. TIGA is providing solutions with a latency of at least 460 days to permit the high-latency data from remote and manually operated stations to be included in the analysis. The TACs use almost identical processing and analytical strategies as the IGS Analysis Centers, so that the reanalysis solutions are comparable. The reprocessed past solutions are less affected, for example, by software changes, changes in the processing strategies and correction models, or station hardware failures not immediately detected. In the more recent years, additional cGPS@TGs were included in the TIGA network; thus, the reprocessing itself also needs to be repeated frequently to process newly available data.

Initial tests are being performed to combine the different solutions from the TACs. Due to TAC processing on a best-effort basis, a complete set of solutions is available only for selected weeks. In addition, complementary studies are being carried out, to identify and remove undetected jumps in station time series, or test the strategies for a TIGA dedicated combination strategy. Based on this experience, the individual TAC solutions will be improved and when necessary reprocessed. In particular, the combination of the individual solutions will be updated frequently to incorporate new stations and new solutions and to always use the newest correction models, for example from atmospheric pressure-loading models.

9.3.5 Steps for Taking TIGA Forward

There are a number of actions that can be taken to improve TIGA.

- Promote the establishment and maintenance of high-quality ties between tide-gauge and GPS stations and their benchmarks.
- Promote the establishment of dual GPS receivers (two closely located receivers) for high-quality cGPS@TG monitoring stations.

- Encourage prescise ties of tide-gauge benchmarks and nearby geodetic markers by continuous GPS, first-order leveling on a regular basis, or permanent-scatterers InSAR.
- Establish a common database of known or possible jumps in GPS and tidal time series.
- Study loading effects of shore- or island-based cGPS@TG points versus inland GPS stations.
- Study secondary ocean-loading effect caused by changes in ocean level.
- Study the effects in the height component by the combination of TIGA stations with the global geodetic reference system.
- Study the use of consistent loading corrections in TIGA processing (e.g. apply atmospheric loading corrections during the processing or during the combination of multiple analysis solutions).

9.4 Recommendations for Geodetic Observations

The ITRF must be more robust and stable over multidecadal timescales. The target accuracy is 0.1 mm/year in the realization of the center of mass of the entire Earth system ("geocenter stability"), and 0.01 ppb/year in scale stability. Geocenter stability depends on accurate dynamic modeling and observation of geodetic satellites, such as SLR and GNSS. Scale stability might be better ensured by minimizing source-related errors, which would imply VLBI, but it also requires accurate tropospheric delay modeling, which would imply SLR (because of the observed frequencies, the tropospheric delay effects are considerably smaller in SLR observations than in either VLBI or GNSS). Some combination of VLBI and SLR is likely required. The increase in GNSS satellites over the next decade to approximately 100 suggests the strong potential of GNSS to contribute significantly to both geocenter and scale stability.

The highest stability requires strong connections between the reference frames of the various geodetic systems. It is recommended to:

- implement more high-quality sites of every technique with a good (even) global distribution, and upgrade existing sites to keep pace with technological developments;
- co-locate VLBI and SLR wherever possible, and require GPS/GNSS instrumentation at every VLBI and SLR site;
- sustain and improve financial support for the technique-specific scientific services and their critical components, from infrastructure to analyses (i.e. IERS, IGS, IVS, ILRS, IDS, PSMSL, IGFS);
- place laser retroreflectors on all future GNSS satellites, and undertake research to improve laser-ranging effectiveness to Medium Earth Orbiters (MEOs);
- research biases between the various techniques, types of satellite, and instrumentation;
- improve tropospheric delay models; and
- support GGOS as the new paradigm for integrating space geodetic techniques.

To ensure long-term stability and consistency in the measurements of sea level by altimetry and space gravity missions (e.g. GRACE), it is recommended that (1) missions similar to the current altimeter and gravity missions should be continued indefinitely with sufficient overlap between missions (an overlap period of at least one year to avoid any gap between missions), and (2) a world vertical datum be established, possibly based on a precise, high-resolution global geoid model, to which all elevation-type measurements should be referred (unification of national and regional vertical datums). This will require international agreements for free exchange of gravimetric, altimetric, elevation, and other relevant data, and collaborative research work to integrate satellite, terrestrial, and airborne gravity and gradiometer data, which could be undertaken under the auspices of GGOS.

Tide gauges should be monitored for height variations using continuous GPS systems installed directly at the tide gauge. This could be accomplished by expanding efforts such as the TIGA project within IGS and by encouraging expansion of related regional activities within GLOSS. In order to have a uniform distribution of tide gauges around the globe, additional tide gauges co-located with GPS, will have to be installed in areas with poor tide-gauge coverage.

It is only through physical models that progress can be made toward understanding measurements of sea-level variability. It is recommended that research and development of comprehensive Earth models evolve in order to assimilate all geodetic data types (and including tide-gauge data) that are relevant to determining sea-level change. Such models must be self-consistent both gravitationally and with respect to the conservation of mass. At the first level, models integrating geodetic data must be developed. At the next level, the models should be integrated with terrestrial hydrological models, cryospheric models, and ocean/atmospheric circulation models.

Finally, in order to achieve the goal of determining sea-level changes at the level of 0.1 mm/year, financial support for the recommended systems and research activities must be brought up to a level capable of meeting that goal, and the support must be sustained for decades, at the very least. This requires commitments that go beyond the current typical relationships between government funding agencies and science programs. It is recommended that United Nations Educational, Scientific and Cultural Organization (UNESCO)/IOC, and elements of the former Integrated Global Observing Strategy-Partnership (IGOS-P) Coastal Zone Theme, together with the funding agencies, assess what is required by their organizations to make the funding system work for programs requiring such long-term commitment. Agreements should be sought at the international level to ensure global-scale international commitment toward solving a global-scale international problem. As a first step toward this goal, the importance of the reference frame for sea level in particular, and Earth observation in general, should be recognized by the Group on Earth Observations (GEO) as a cross-cutting activity that affects all benefit areas addressed within GEO's Global Earth Observation System of Systems (GEOSS) (http://www.earthobservations.org/index.html).

Acknowledgments

GB acknowledges support by NASA/IDS grant NNG04G099G and NASA/SENH grant NAG5-13683. Some of the research described in this chapter was carried out at the Jet Propulsion Laboratory, California Institute of Technology, under a contract with the National Aeronautics and Space Administration.

References

Altamimi Z., Sillard P. and Boucher C. (2002) ITRF2000: a new release of the International Terrestrial Reference Frame for earth science application. *Journal of Geophysical Research* **107**(B10), 2214.

Altamimi Z., Boucher C. and Willis P. (2005) Terrestrial reference frame requirements within GGOS perspective. *Journal of Geodynamics* **40**, 363–74.

Altamimi Z., Collilieux X., LeGrand J., Garyt B. and Boucher C. (2007) ITRF2005: a new release of the International Terrestrial Reference Frame based on time series of station positions and Earth Orientation Parameters. *Journal of Geophysical Research* V **112**, B09401.

Becker M., Zerbini S., Baker T., Bürki B., Galanis J., Garate J. et al. (2002) Assessment of height variations by GPS at Mediterranean and Black Sea coast tide gauges from the SELF Projects. *Global and Planetary Change* **34**, 5–35.

Beutler G. (2005) *Methods of Celestial Mechanics*, vols I and II. Springer, Berlin.

Blewitt G. (2003) Self-consistency in reference frames, geocenter definition, and surface loading of the solid Earth. *Journal of Geophysical Research* **108**(B2), 2103.

Blewitt G. and Clarke P. (2003) Inversion of Earth's changing shape to weigh sea-level in static equilibrium with surface mass redistribution. *Journal of Geophysical Research* **108**(B6), 2311.

Blewitt G., Lavallee D., Clarke P. and Nurutdinov K. (2001) A new global mode of Earth deformation: seasonal cycle detected. *Science* **294**, 2342–5.

Boehm J., Niell A., Tregoning P. and Schuh H. (2006) Global Mapping Function (GMF): a new empirical mapping function based on numerical weather model data. *Geophysical Research Letters* **33**, L07304.

Chao B., Au A.Y., Boy J.-P. and Cox C.M. (2003) Time-variable gravity signal of an anomalous redistribution of water mass in the extratropic Pacific during 1998-2002. *Geochemistry Geophysics Geosystems* **4**, 1096.

Chelton D., Ries J.C., Haines B.J., Fu L.-L. and Callahan P.S. (2001) Altimetry. In *Satellite Altimetry and Earth Sciences* (Fu L.-L. and Cazenave A., eds). Academic Press, New York.

Colesanti C., Ferretti A., Novali F., Prati, C. and Rocca F. (2003) SAR monitoring of progressive and seasonal ground deformation using the Permanent Scatterers technique. *IEEE Transactions on Geoscience and Remote Sensing* **41**, 1685–1701.

Crétaux J.-F. and Birkett C. (2006) Lake studies from satellite altimetry. *Comptes Rendus Geoscience* **338**(14–15), 1098–1112.

Dow J., Neilan R.E. and Gendt G. (2005) The International GPS Service: celebrating the 10th anniversary and looking to the next decade. *Advances in Space Research* **36**, 320–6.

Drewes H. (ed.) (2005) The Global Geodetic Observing System. *Journal of Geodynamics* **40**, 355–6.

Fagard H. (2006) Twenty years of evolution for the DORIS permanent network: from its initial deployment to its renovation. In: *DORIS Special Issue* (Willis P., ed.). *Journal of Geodesy* **80**, 429–56.

Forsberg R., Sideris M.G. and Shum C.K. (2005) The gravity field and IGGOS. *Journal of Geodynamics* **40**, 387–93.

Fu L.-L. and Chelton D.B. (2001) Large-scale ocean circulation. In: *Satellite Altimetry and Earth Sciences: a Handbook for Techniques and Applications* (Fu L.L. and Cazenave A., eds), pp. 133–69. Academic Press, San Diego.

Haines B., Bar-Sever Y., Bertiger W., Desai S. and Willis P. (2004) One-centimeter orbit determination for Jason-1: New GPS-based strategies *Marine Geodesy* **27**, 299–318.

IOC (1997) *Global Sea-level Observing System (GLOSS) Implementation Plan-1997*. Intergovernmental Oceanographic Commission Technical Series, No. 50. UNESCO, Paris.

IOC (2006) *Manual on Sea-Level Measurement and Interpretation. Volume 4 – An Update to 2006* (Aarup T., Merrifield M., Perez B., Vassie I. and Woodworth P., eds). Intergovernmental Oceanographic Commission Manuals and Guides, no. 14. IOC, Paris.

Kierulf H.P. and Plag H.-P. (2006) Precise point positioning requires consistent global products. In: *EUREF Publication No. 14 vol. BKG 35 of Mitteilungen des Bundesamtes für Kartografie und Geodäsie BKG* (Torres J.A. and Hornik H., eds), pp. 111–20. BKG, Frankfurt am Main.

Kovalevsky J., Mueller I. and Kolaczek B. (eds) (1989) *Reference Frames in Astronomy and Geophysics*. Astrophysics and Space Science Library, vol. **154**. Springer, New York.

Kuo C.Y., Shum C.K., Yi Y., Braun A., Schroeter J. and Wenzel M. (2006) Determination of 20th century global sea-level rise. *Geophysical Research Abstracts* **6**, 07741.

Luthcke S.B., Zelensky N.P., Rowlands D.D., Lemoine F.G. and Williams T.A. (2003) The 1-centimeter orbit: Jason-1 precision orbit determination using GPS, SLR, DORIS and altimeter data. *Marine Geodesy, Special Issue on Jason-1 Calibration/Validation Part 1* **26**(3–4), 399–421.

McCarthy D.D. and Petit G. (eds) (2003) *IERS Conventions*. IERS Technical Note No. 32. International Earth Rotation Service, Frankfurt am Main.

Pearlman M.R., Degnan J. and Bosworth J. (2002) The International Laser Ranging Service. *Advances in Space Research* **30**(2), 135–43.

Penna N.T. and Stewart M.P. (2003) Aliased tidal signatures in continuous GPS height time series. *Geophysical Research Letters* **30**(23), 2184.

Penna N.T., King M.A. and Steward M.P. (2006) GPS height time series: short period origins of spurious long period signals. *Journal of Geophysical Research* **112**, B02402.

Plag H.-P. (2005) The GGOS as the backbone for global observing and local monitoring: a user driven perspective. *Journal of Geodynamics* **40**, 479–86.

Plag H.-P. (2006) National geodetic infrastructure: current status and future requirements – the example of Norway. *Nevada Bureau of Mines and Geology, Bulletin* **112**.

Rummel R., Rothacher M. and Beutler G. (2005) Integrated Global Geodetic Observing System (IGGOS) – science rationale. *Journal of Geodynamics* **40**, 357–62.

Schlüter W., Himwich E., Nothnagel A., Vandenberg N. and Whitney A. (2002) IVS and its important role in the maintenance of the global reference systems. *Advances in Space Research* **30**(2), 145–50.

Schutz B.E., Zwally H.J., Shuman C.A., Hancock D. and DiMarzio J.P. (2005) Overview of the ICESat mission. *Geophysical Research Letters* **32**, L21S01.

Seeber G. (2003) *Satellite Geodesy – Foundations, Methods, and Applications*, pp. 500–5. Walter de Gruyter, Berlin.

Soudarin L., Crétaux J.F. and Cazenave A. (1999) Vertical crustal motions from the DORIS space-geodesy system. *Geophysical Research Letters* **26**(9), 1207–10.

Stewart M.P., Penna N.T. and Lichti D.D. (2005) Investigating the propagation mechanism of unmodelled systematic errors on coordinate time series estimated using least squares. *Journal of Geodesy* **79**(8), 479–89.

Tapley B.D., Bettadpur S., Ries J.C., Thompson P.F. and Watkins M.M. (2004) GRACE measurements of mass variability in the Earth system. *Science* **305**, 503–5.

Tavernier G., Fagard H., Feissel-Vernier M., Le Bail K., Lemoine F., Noll C. et al. (2006) The International DORIS Service: genesis and early achievements. *Journal of Geodesy* **80**, 403–17.

Teferle F.N., Bingley R.M., Williams S.D.P., Baker T.F. and Dobson A.H. (2006) Using continuous GPS and absolute gravity to separate vertical land movements and changes in sea level at tide gauges in the UK. *Philosophical Transactions of the Royal Society of London A* **364**, 917–30.

Tregoning P. and van Dam T.M. (2005) Effects of atmospheric pressure loading and seven-parameter transformations on estimates of geocenter motion and station heights from space geodetic observations. *Journal of Geophysical Research* **110**, B03408.

Tregoning P. and Herring T.A. (2006) Impact of a priori zenith hydrostatic delay errors on GPS estimates of station heights and zenith total delays. *Geophysical Research Letters* **33**, L23303.

van Dam T.M., Wahr J., Milly P.C.D., Shmakin A., Blewitt G., Lavallée D. and Larson K. (2001) Crustal displacements due to continental water loading. *Geophysical Research Letters* **28**(4), 651–4.

Watson C., Tregoning P. and Coleman R. (2006) The impact of solid Earth tide models on GPS coordinate and tropospheric time series. *Geophysical Research Letters* **33**, L08306.

Watts R.C. and Morantine M.C. (1991) Is the greenhouse gas-climate signal hiding in the deep ocean? *Climatic Change* **18**(4), iii–vi.

Woodworth P.L. and Player R. (2003) The Permanent Service for Mean Sea Level: an update to the 21st century. *Journal of Coastal Research* **19**, 287–95.

Wöppelmann G., Zerbini S. and Marcos M. (2006) Tide gauges and geodesy: a secular history of interactions and synergies. *Comptes Rendus Geosciences* **338**(14–15), 980–91.

Zerbini S., Plag H.-P., Baker T., Becker M., Billiris H., Bürki B. et al. (1996) Sea-level in the Mediterranean: a first step towards separating crustal movements and absolute sea-level variations. *Global and Planetary Change* **14**(1–2), 1–48.

Zerbini S., Matonti F., Raicich F., Richter B. and van Dam T. (2004) Observing and assessing non-tidal ocean loading using ocean, continuous GPS and gravity data in the Adriatic area. *Geophysical Research Letters* **31**, L23609.

10 Surface Mass Loading on a Dynamic Earth: Complexity and Contamination in the Geodetic Analysis of Global Sea-Level Trends

Jerry X. Mitrovica, Mark E. Tamisiea, Erik R. Ivins, L.L.A. (Bert) Vermeersen, Glenn A. Milne, and Kurt Lambeck

10.1 Introduction

Constraints on sea-level variations, whether derived from the long-standing tide-gauge record of the Permanent Service for Mean Sea Level (PSMSL) (e.g. Woodworth and Player 2003; www.psmsl.org) or more recent measurements from satellite altimetry and gravity missions, provide an important lens for observing the evolution of the Earth system. The resolving power of this lens is greatly enhanced by consideration of the interplay of geodynamics with sea level. While ocean scientists seek an understanding of sea-level variability through the internal dynamics and thermal energetics of the ocean, geodynamicists seek an understanding of the roles played by solid Earth movements, changes in gravity, and fundamental variability in Earth rotation. Without the perspective of geodynamics it may be difficult to decipher modern global change signals such as cryospheric mass flux and ocean thermal expansion. These geodynamical processes, for example, include glacial isostatic adjustment (GIA). This chapter summarizes research in solid Earth geophysics, ranging from the relatively straightforward to the conceptually complex, that has bearing on the problem of global sea-level rise. The chapter begins with a discussion of recent efforts in

Understanding Sea-Level Rise and Variability, 1st edition. Edited by John A. Church, Philip L. Woodworth, Thorkild Aarup & W. Stanley Wilson.

regard to the prediction of the GIA process. These include studies that have improved constraints on our understanding of Late Pleistocene ice histories and Earth structure, and that have advanced the theoretical treatment of sea-level change, viscoelastic deformation and rotational dynamics. Next, the notion of the fingerprinting of GIA-corrected sea-level trends by highlighting the spatially variable nature of the sea-level response to rapid melting of ice sheets and glaciers is discussed. In particular, recent efforts to apply this physics to extract source information from complex sea-level trends, and thus to move beyond simple global averages of sea-level change, are summarized.

Modern analyses of sea-level variability are generally concerned with the direct contributions from continent-derived freshwater flux and thermosteric and halosteric ocean changes. However, the response of the solid Earth to a suite of geophysical processes also has a direct contribution to ongoing sea-level change. Moreover, tide gauges, satellite altimetry, and gravity missions all observe different components of sea-level change, and tying these observations together requires an understanding of the solid Earth deformation field. These deformations thus play a critical role for correctly interpreting disparate sea-level observations.

The redistribution of water, for example meltwater from glaciers, damming of rivers, and movement of groundwater, will deform the Earth's crust and perturb the geopotential. In addition, the ocean basins and geoid are continuing to respond to past loading events, most notably the Late Pleistocene glacial cycles but also to tectonic processes. Seismic deformation also induces contributions to the present-day sea-level variations. The connection between sea level and the solid Earth may be manifest in even more subtle effects, including changes in the Earth's rotation axis; that is, polar-motion and length-of-day variations. This chapter provides a broad overview of the role that solid Earth studies are playing in efforts to constrain global sea-level variability. This includes both the key results emerging during the past decade or more, and an outline of some exciting new developments that hold special promise for future research.

10.1.1 The Tide-Gauge Record

The three most common data sets used to study modern sea-level change are, from longest to shortest time span, tide gauges, satellite altimetry, and spaceborne gravity observations. Tide gauges, which measure the local vertical position of the surface of the ocean with respect to the land surface (or crust), particularly highlight the impact of the solid Earth on sea-level estimation. The complex spatial variation evident in tide-gauge estimates of 20th-century sea-level trends has been a major source of uncertainty from the very earliest efforts to constrain the amplitude of globally averaged sea-level rise (for a history, see Douglas 1991). Indeed, this geographic variation partly motivated the careful culling of the data set to remove sites that were subject to potential contamination from tectonic signals and urban groundwater pumping, or that were of insufficient time span to robustly separate secular trends from decadal fluctuations. As an example, an estimate of global sea-level rise of 1.8 ± 0.1 mm/year by Douglas (1991) was based on a subset of 23 widely distributed sites from the approximately 2000 records within the PSMSL database. Peltier and Tushingham (1989, 1991) used data from several hundred sites in their analysis, and the increased uncertainty in their estimate (2.4 ± 0.7 mm/year) reflects, in part, their more permissive selection criteria.

Since the late 1980s, tide-gauge analyses have accounted for the ongoing sea-level signal due to the Late Pleistocene glacial cycles, or GIA. Most commonly, this effort has involved global GIA corrections based on standard numerical techniques applied to one-dimensional Earth models (e.g. Peltier and Tushingham 1989, 1991; Trupin and Wahr 1990; Douglas 1991). Site selection has also been used to minimize the GIA correction (and its uncertainty). For example, Lambeck et al. (1998a) used about 25 tide-gauge records from a region of the Baltic Sea that separates areas of uplift and submergence; their estimate of a 100-year-long average sea-level rise of 1.1 ± 0.2 mm/year is thus relatively insensitive to errors in the ice and Earth models adopted for the numerical GIA correction. There have also been efforts to correct sea-level records for the influence of GIA using observations. Milne et al. (2001) used vertical crustal deformations estimated from the Fennoscandian BIFROST GPS network (and models of the associated geoid warping) to remove the GIA signal from nearby tide-gauge records.

There has generally been an expectation that an accurate GIA correction will reduce the geographic variation in the sea-level trends estimated from globally distributed sites. However, removing rates from a single global model raises the potential for bias in the estimates of sea-level rise, particularly when analyzing regional networks. For example, Davis and Mitrovica (1996) showed that a moderate increase in the deep-mantle viscosity model adopted by the previous studies significantly reduced GIA-corrected tide-gauge trends along the US east coast, and they inferred a regional sea-level rise of 1.5 ± 0.3 mm/year. Moreover, plausible lateral variations in mantle viscosity and lithospheric thickness could alter the mean GIA correction by 0.5 mm/year at tide-gauge sites located well away from the Pleistocene ice centers (Mitrovica and Davis 1995). It is essential to recognize the global nature of such GIA corrections, as, for example, they may even have to be accounted for in studies of interseismic (tectonic) plate-motion studies (e.g. Savage and Thatcher 1992).

More recent tide-gauge estimates of globally averaged sea-level rise fall in the range of 1.5–2.0 mm/year (Douglas 1997; Holgate and Woodworth 2004). Additionally, estimates have come from analyses of shorter time-series satellite-altimetric data sets (Cazenave et al. 1998; Nerem and Mitchum 2001; Leuliette et al. 2004) and combinations of tide-gauge and altimetric data (Church et al. 2004). This last, joint inference (1.8 ± 0.3 mm/year) is consistent with the tide-gauge-only results for the second half of the 20th century; however, estimates from altimetry data alone from 1993 (and also from tide-gauge data) are approximately 1 mm/year larger than the estimates for earlier decades (e.g. Leuliette et al. 2004). GIA corrections to altimetry data are significant. In particular, Peltier (2001) has argued that this correction would increase altimetry-based estimates of global sea-level rise by 0.3 mm/year. There have also been indications of a positive sampling bias in the tide-gauge estimates, which would further exacerbate the 1 mm/year discrepancy. Specifically, Holgate and Woodworth (2004) argued, using altimetry data, that the sea-level rise within coastal zones (where tide-gauge sites are generally located) may be accentuated relative to the open ocean. Thus, the tide-gauge-derived estimates cited above might represent upper bounds on the 20th-century trend. Adding to this complexity is a recent suggestion that the coast-versus-open-ocean discrepancy was reversed during the late 1970s and late 1980s, and that an oscillation between ocean states over that past 50 years may mean that there is, essentially, minimal difference between longer-term rates of coastal and open-ocean sea-level rise (White et al. 2005).

What are the amplitude and the origin of the trend of sea-level rise? What is the partitioning of the signal into contributions from steric effects (the strongest of which is thermal expansion) and recent mass flux from polar ice sheets (Greenland, Antarctic) and small glaciers? How rapidly might they vary in time? The nature of the partitioning has remained contentious (Cazenave and Nerem 2004) despite improvements in the direct mapping of three-dimensional ocean heat content (Levitus et al. 2000; Willis et al. 2004; Antonov et al. 2005) and changes in the volume of ice reservoirs, particularly mountain glaciers (e.g. Dyurgerov and Meier 2005).

The recent literature has provided several avenues by which this uncertainty may be improved. The first is the notion of the fingerprinting of global sea-level rise (Mitrovica et al. 2001a; Plag and Jüttner 2001; Tamisiea et al. 2003; Plag 2006). It has long been known that the rapid melting of an ice sheet will produce a unique, spatially nonuniform change in sea level (Woodward 1888; Farrell and Clark 1976; Clark and Lingle 1977; Clark and Primus 1987; Nakiboglu and Lambeck 1991; Conrad and Hager 1997). Thus, the geographic variation in GIA-corrected tide-gauge trends may permit one to estimate not only the amplitude of global sea-level rise, but also, for sufficiently accurate and widely distributed sites, the individual sources for this rise. The technique is not limited to relative sea-level variations measured by tide gauges (i.e. changes in sea-surface height relative to the ocean bottom); rather, it is also applicable to GIA-corrected patterns of sea-surface changes measured by altimeters and geoid changes constrained by satellite gravity observations (Tamisiea et al. 2001). More generally, the concept of sea-level fingerprinting has served to counter the misconception that residual (GIA-corrected) tide-gauge trends should ideally show little geographic variation.

10.1.2 Constraints from Gravity Field Variability and Long-Term Polar Motion (True Polar Wander)

Secular perturbations in the Earth's gravitational field and rotation vector have also, after correction or simultaneous inversion for the GIA signal, provided constraints on individual meltwater contributions to global sea-level rise. These efforts have included analyses of trends in either regional or long-wavelength components of the gravitational field. For regional studies, many have used satellite gravity data obtained from the Gravity Recovery and Climate Experiment (GRACE) to estimate the mass balance of glaciers and ice sheets. As a specific example, Tamisiea et al. (2005) used GRACE data from 2002–4 to estimate a trend in the melting of Alaskan ice complexes of 0.31 ± 0.09 mm/year (equivalent sea level), in agreement with aerial altimetry mapping (Arendt et al. 2002) from an earlier time period (see also Chen et al. 2006a). Estimates have also been derived for the mass balance in Antarctica (e.g. Velicogna and Wahr 2006a), and Greenland (e.g.Velicogna and Wahr 2005, 2006b; Luthcke et al. 2006; Chen et al. 2006b). While the time series is still short, the monthly GRACE fields allow for good sampling of interannual variations. Horwath and Dietrich (2006) suggest a potential underestimation of correlated errors in the GRACE data that reach a maximum level at high latitude. The impact of this issue on polar mass flux may be significant.

Numerous studies have focused on satellite-derived secular trends in the long-wavelength coefficients of the Earth's geopotential, although the most robust of these coefficients, the so-called secular oblateness (dJ_2/dt) observation (Yoder et al. 1983; Cheng et al. 1997; Nerem and Klosko 1996) is insufficient to separate individual contributions from the polar ice reservoirs (Mitrovica and Peltier 1993; Ivins et al. 1993). To overcome this obstacle, efforts have been made to incorporate into the analysis other low-order zonal harmonics (James and Ivins 1997; Tosi et al. 2005) and the secular reorientation of the rotation pole relative to the surface geography, or true polar wander (TPW; James and Ivins 1997; Johnston and Lambeck 1999).

In regard to this class of studies, Munk's (2002) analysis of the dJ_2/dt observable (or equivalently the secular change in the length of day, LOD; Wu and Peltier 1984), TPW, and integrated changes in LOD over the last three millennia from observations of ancient eclipses (Stephenson and Morrison 1995) has come to define an "enigma" in the interpretation of global sea-level rise. Specifically, Munk (2002) argued that predictions generated from GIA models simultaneously fit all three data sets (last three millennia LOD, nearly three-decade-long satellite-based oblateness change record, and TPW from astronomical and astrometric records), thus leaving no residual room for signals associated with recent mass flux from global ice reservoirs. While this would suggest that thermal expansion dominates the trends observed from tide-gauge data (since they have no impact on oblateness variation nor on TPW), Munk (2002) also argued that the ocean heat budget is insufficient for this to be the case. Hence, the enigma. Mitrovica et al. (2006), however, showed that the three data sets can be reconciled with a contribution from recent melting, provided one applies a new, more accurate theory of load-induced TPW (Mitrovica et al. 2005) and, more importantly, accounts for the possibility of a greater uncertainty in the dJ_2/dt constraint due to a previously underappreciated dispersion and phase discrepancy in the solid Earth 18.6-year tide (Benjamin et al. 2006; also see Lambeck and Nakiboglu 1983 and Ivins and Sammis 1995). In this case, when uncertainties in the signal from the 18.6-year tide are accounted for, the GIA-corrected residuals do permit approximately 1 mm/year of meltwater flux.

Our discussion of solid Earth dynamics will have two underlying themes. These relate to the impact of solid Earth deformations on: (1) direct observations of sea-level change (e.g. tide gauges, altimetry); and (2) observations that can provide constraints on the magnitude and source of the mass contributions to sea-level change (e.g. TPW). The accuracy with which the community can predict this motion governs the extent to which the climate-sensitive component of the observed signal can be isolated and appropriately interpreted. Furthermore, the question of accuracy requires, in turn, a discussion of ongoing improvements in the theoretical framework and model parameterizations that define the state of the art.

10.2 Glacial Isostatic Adjustment

As indicated in the Introduction, corrections for the GIA process are a key element in the ultimate analysis of observations related to global sea level. In recent years there have been significant improvements in numerical techniques and underlying theory associated with such predictions, and further refinements in the input fields required for the simulations. The latter include both the space-time history of Late Pleistocene ice cover and parameters governing the viscoelastic structure of the planet. Each area of improvement is considered, in turn, below.

The vast majority of GIA studies, over the last quarter century, have ultimately been based on the viscoelastic Love number theory developed by Peltier (1974) and Wu (1978) with later augmentation by Wu and Peltier (1982) and Peltier (1985). The Love numbers describe the response of a spherically symmetric, self-gravitating, viscoelastic planet to the application of an impulse point load, and they can be suitably combined to treat any geophysical observable of interest. One of the most important developments in GIA research over the last decade has been the effort, by several independent groups, to describe and implement numerical methods for treating more realistic Earth models in which viscoelastic parameters vary in three dimensions. In their current stage of development, these include spherical codes based on spectral (Martinec 2000), finite-element (Wu and van der Wal 2003; Zhong et al. 2003), and finite-volume (Latychev et al. 2005a) methodologies, as well as a complete normal-mode perturbation theory (Tromp and Mitrovica 1999). This effort has been motivated, in part, by the recognition that best-fit model predictions to different regional geological sea-level data prefer distinct mantle and lithospheric properties. For example, the GIA response along the coastline of Australia requires a different upper-mantle viscosity value than does data from Fennoscandia (Nakada and Lambeck 1991). Papers that use the new codes to investigate the impact of lithospheric thickness variations, plate boundaries, and upper- and lower-mantle viscosity heterogeneities on a suite of GIA-related observables are now emerging (e.g. Latychev et al. 2004, 2005a, 2005b; Kaufmann et al. 2005; Martinec and Wolf 2005; Paulson et al. 2005; Wu et al. 2005; Wang and Wu 2006). However, while these early efforts have provided significant insight into the potential limitations of a spherically symmetric GIA theory, they have yet to be brought fully to bear on problems related to global sea-level rise. An exception is the study by Kendall et al. (2006), which demonstrated that plausible three-dimensional variations in Earth structure have a significant impact on GIA corrections to the global tide-gauge record.

One simple method of treating the problem of lateral heterogeneity in mantle viscosity is to formulate regional calculations, especially when clearly necessitated by the data themselves and substantiated by basic inferences from seismic velocity studies. The first example of this type of study was an analysis of the Australian region by Nakada and Lambeck (1991). This approach assumes that regional calculations can be accurately performed using one-dimensional (depth-varying) Earth models tuned to local, underlying Earth structure. Some studies (Wu 2006) have argued that such an approach is sound, while others have highlighted the limitations of this simple procedure, particularly in accounting for upper-mantle variability or in treating horizontal crustal motions (Whitehouse et al. 2005; Kendall et al. 2006).

The analyses described above have been based on Earth models with linear, Maxwell viscoelastic rheologies. A small set of studies have considered the impact, on various GIA predictions, of incorporating nonlinear rheologies (e.g. Gasperini et al. 1994; Giunchi and Spada 2000; Wu 2002).

10.2.1 Recent Improvements in Sea-Level Theory

A second major area of improvement in long-standing GIA theory has centered on calculations of post-glacial relative sea-level changes. The prediction of relative sea-level variations is complicated. The ocean redistribution is governed (in an equilibrium theory) by the gravitational field of the planet and by deformations of the solid surface, yet the gravitational field is, in turn, perturbed by the direct gravitational effect of the ocean (and ice) redistribution and the solid Earth deformation it produces. This circularity gives rise to an integral equation that has come to be known as the sea-level equation and which was first derived for a deforming planet in the seminal study of Farrell and Clark (1976). Farrell and Clark's (1976) version of the sea-level equation was based on a nonrotating Earth and a very specific limitation on the ocean basin geometry shown schematically in Figure 10.1a. Specifically, Farrell and Clark (1976) assumed that there was no marine-based ice and that the edges of the ocean were steep vertical cliffs, such that changes in ice volume would not lead to a transgression or regression. The first global applications of the Farrell and Clark (1976) theory were based on a disk load discretization of the ice and ocean mass loads (e.g. Wu and Peltier 1983). This approach was later replaced by the pseudo-spectral algorithm developed by Mitrovica and Peltier (1991), which now is the standard approach for spherically symmetric calculations.

Over the last decade there has been a series of efforts to overcome the fixed-shoreline assumption in Figure 10.1a and to treat the more realistic scenarios

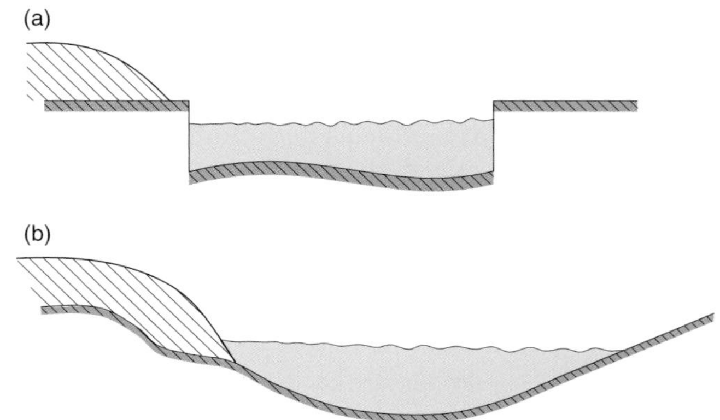

Figure 10.1 Cartoon showing the shoreline geometry treated in (a) the traditional sea-level equation of Farrell and Clark (1976), or (b) the more realistic case where shoreline migration will result from either the growth or ablation of grounded, marine-based ice or a local rise or fall in sea level.

shown in Figure 10.1b. Johnston (1993), for example, was the first to incorporate the onlap and offlap of water (and the associated change in the water load) at shorelines subject to a local rise or fall of sea level, respectively. His analysis was followed by Peltier (1994) and Milne (1998) (see also Milne et al. 1999), who each adopted a distinct revision to the Farrell and Clark (1976) theory. More recently, Peltier and Drummond (2002) have further revised the Peltier (1994) version of the sea-level theory to include a so-called broad-shelf effect. In an effort to assess the relative accuracies of these approaches, Mitrovica and Milne (2003) derived a generalized sea-level equation. They argued that the Peltier (1994) methodology was less accurate than the approaches advocated by Johnston (1993) and Milne (1998), and this was confirmed by quantitative analysis in Kendall et al. (2005). Moreover, Mitrovica (2003) demonstrated that the broad-shelf effect described by Peltier and Drummond (2002) is included in the Johnston (1993) and Milne (1998) theories.

The next extension to the Farrell and Clark (1976) theory involved the inundation of water into regions vacated by melting (grounded) marine-based ice, or movement of water out of areas of advancing marine-based ice. The former was denoted "water dumping" by Milne (1998) and Milne et al. (1999), who extended their sea-level theory to incorporate both the impact of the accommodation space created by the retreating ice on mass conservation (and thus the eustatic sea-level curve) and also the loading effect of the water inundation. Peltier (1998a) independently treated the same effect, which he associated with the term "implicit ice". His procedure computes, *a posteriori*, the volume of water within regions once covered by marine ice; the ice-mass equivalent to this volume is the primary contributor to the implicit-ice component of the load. This procedure takes into account the influence of the inundation on the eustatic curve, but it does not appear to account for the loading effect of this inundation (see Mitrovica 2003). Calculations by Kendall et al. (2005), based on the generalized sea-level equation of Mitrovica and Milne (2003), have confirmed the accuracy of the Milne et al. (1999) extensions to the Farrell and Clark (1976) theory.

Sea-level calculations by the group at the Australian National University (ANU) have also incorporated the tandem processes for shoreline migration shown in Figure 10.1, as described in detail by Lambeck et al. (2003) and references cited therein. Peltier (2002) has questioned the accuracy of these calculations via a criticism of Yokoyama et al. (2000) (for a rebuttal of these criticisms, see Lambeck et al. 2002b). However, Mitrovica (2003), Mitrovica and Milne (2003), and Kendall et al. (2005), have demonstrated, using generalized theoretical descriptions supported by numerical calculation, that the ANU theory is technically sound. The final extension to the sea-level theory accounts for rotational effects.

Specifically, the redistribution of ice and water loads perturbs the rotation vector of the planet, and this perturbation acts, in turn, to load the planet via a perturbation in the centrifugal potential (the so-called rotational driving potential). The loading is manifest in the complete suite of GIA-related observables (e.g. three-dimensional crustal motions, sea-level variations, geoid anomalies; e.g. Milne and Mitrovica 1996; Mitrovica et al. 2001b) and this signal has come to be known as rotational feedback (Peltier 1998b). The first calculation of this effect was by Han and Wahr (1989), while a complete, gravitationally self-consistent extension to the sea-level equation to include the feedback mechanism was first derived by Milne and Mitrovica (1996). This effort has been followed by others (e.g. Milne and Mitrovica 1998; Peltier 1998b; see also section 10.2.4).

The generalized sea-level equation derived by Mitrovica and Milne (2003) is based on an exact expression for the change in sea level (and the ocean load) in response to the growth or ablation of marine-based ice or local sea-level rise or fall. The theory is also valid for an Earth model of arbitrary complexity. Kendall et al. (2005) have provided algorithms for the solution of this generalized sea-level equation; in the special case of rotating, spherically symmetric Earth models their algorithm defines an extended pseudo-spectral formulation. The reader is referred to this paper for a detailed examination of the (site-dependent) importance of the two shoreline migration mechanisms (Figure 10.1) in predictions of GIA-induced relative sea-level histories. An analogous assessment of the impact of rotational feedback can be found, for example, in Milne and Mitrovica (1996, 1998) and Peltier (2004). Finally, note that there have been efforts to apply these extensions to the traditional sea-level theory in the case of three-dimensional viscoelastic Earth models. Paulson et al. (2005), for example, have coupled a sea-level calculator for a rotating Earth with retreat of marine-based ice onto their finite-element response code. A similar coupling is accomplished in a formulation by Martinec and Hagedoorn (2005), where the sea-level equation is solved in the time domain.

The move toward three-dimensional Earth models and improvements in the sea-level theory represent technical advances in the study of GIA. Over the last few years there has also been a continuation of long-standing efforts to improve constraints on the two key inputs to such models; namely, the radial Earth model (defined by the thickness of an elastic lithosphere and the profile of mantle viscosity) and the Late Pleistocene ice history.

10.2.2 Inversions for Mantle Viscosity

Inferences of mantle viscosity based on GIA data are sensitive to uncertainties in the Late Pleistocene ice-sheet history, and various data parameterizations have been introduced in an effort to reduce this nonuniqueness. As an example, Nakada and Lambeck (1989) advocated the use of differential Late Holocene highstands between pairs of far-field (e.g. Australia) sites. Forward model predictions indicate that far-field sea-level histories are relatively insensitive to details of the ice history, and using the difference between highstands at nearby sites removes any signal associated with residual melting since the end of the main deglaciaton phase. Others have argued that decay times inferred from post-glacial sea-level curves at sites once covered by major ancient ice sheets (e.g. Laurentia, Fennoscandia) also provide constraints on viscosity that are relatively uncontaminated by uncertainties in the Pleistocene ice geometry (e.g. Mitrovica 1996; Mitrovica and Forte 1997, 2004; Peltier 1998b).

Formal inversions of GIA data sets have yielded various inferences of mantle viscosity. Mitrovica and Forte (2004), following earlier work (Mitrovica and Forte 1997), simultaneously inverted data related to mantle convection and GIA, the latter including decay times from Hudson Bay and Fennoscandia as well as the Fennoscandian relaxation spectrum (FRS; Wieczerkowski et al. 1999). Their resulting profiles were characterized by an increase of three orders of magnitude in viscosity from the upper mantle (with a mean value of 4×10^{20} Pa s) to a high-viscosity zone near 2000 km depth. Kaufmann and Lambeck (2000) also preferred a large increase of viscosity with depth on the basis of an inversion of a widely distributed set of relative sea-level histories, rotation perturbations, and long-wavelength rates of change in the gravity field, though their preferred profile shows a stiffer region at the top of the lower mantle (i.e. just below the seismic discontinuity at 670 km depth). The increase by an order of magnitude in the deeper lower mantle has been a long-standing position advocated by those examining the pressure dependence of all microphysical, defect-related, creep mechanics in the mantle (Ivins et al. 1993; Ranalli 2001; Monnereau and Yuen 2002). Peltier's analyses of decay times and the FRS (e.g. Peltier 1998b) find a viscosity profile (VM2) that increases only moderately with depth (Peltier 2004). Thus, the long-standing effort to establish a "reference" radial profile of mantle viscosity continues to be mired in some controversy. A reasonable bound for viscosity profiles may be that offered by Mitrovica and Forte (2004) as advocated recently by Tosi et al. (2005) and Ivins and James (2005). A new analysis of GIA data near Churchill on the coast of Hudson Bay is consistent with this reasonable bound (Wolf et al. 2006).

To complicate this issue, there has also been a growing appreciation of the potential impact of fine-scale radial structure of the lithosphere on GIA observables (e.g. Klemann and Wolf 1999; Di Donato et al. 2000; Kendall et al. 2003) and the potential bias introduced in one-dimensional inferences by the presence of lateral variations in structure (Paulson et al. 2005). However, key GIA data sets have been improved, including the FRS (Klemann and Wolf 2005), and these have helped to rectify inconsistencies noted in previous studies (Mitrovica and Forte 2004).

10.2.3 Reconstructions of Late Pleistocene Ice Histories

A number of Late Pleistocene ice-sheet reconstructions at both regional and global scales have been published in the past few decades. The most widely used global-scale ice models have been those developed by Peltier and colleagues (e.g. Wu and Peltier 1983; Tushingham and Peltier 1991), including ICE-5G (Peltier 2004). However, a global model has also been iteratively developed and employed by Lambeck and colleagues at the ANU (Lambeck et al. 2002a). A substantially larger number of regional ice models have also been developed, including for the British Isles and Ireland (e.g. Lambeck 1993; Shennan et al. 2002), Fennoscandia (e.g. Lambeck et al. 1998b; 2010), Greenland (e.g. Huybrechts et al. 2004; Tarasov and Peltier 2002; Fleming and Lambeck 2004), and Antarctica (e.g. James and Ivins 1998; Nakada et al. 2000; Huybrechts et al. 2004; Ivins and James 2005). Progress has also been made in establishing ice-load histories for smaller regional ice complexes, such as those in southern Patagonia (Sugden et al. 2002), coastal Alaska (Motyka 2003; Barclay et al. 2003), the Canadian Codillera (Clague and James 2002), Svalbard ice cap (Lonne and Lysa 2005), and the European Alps (Florineth and Schlüchter 2000; Stocchi et al. 2005).

The veracity of a particular ice model is dependent on the quality of the observational evidence and the level of glaciological complexity considered in its construction. In general terms, the observational evidence can be classed as one of two types: (1) those obtained from geological field evidence which provides information on, for example, the spatial extent of the ice sheet and ice-flow directions and (2) those that relate to the response of the solid Earth and gravity field to the ice sheet, such as relative sea-level changes, crustal deformation, and changes in gravity and Earth rotation (the GIA observables). Most previous studies have been limited in that either a subset of the available observational data were employed or the degree of glaciological complexity was limited. For example, a number of glaciologically realistic models have been developed that are broadly consistent with geological field data on lateral ice extent and ice-flow patterns but have not been calibrated to fit GIA observables (e.g. Huybrechts et al. 2004). In comparison, models that have been calibrated to fit both GIA and geological field constraints are, commonly, glaciologically simplistic (e.g. Lambeck et al. 1998b; Shennan et al. 2002). Only a small number of studies have attempted to calibrate complex glaciological models using both GIA and geological data (e.g. Tarasov and Peltier 2002).

The quality and quantity of field evidence has significantly improved over the past decade due largely to the successful application of modern chronological dating methods to moraine debris and other organic markers. The improved field constraints have had a significant impact on ice-model reconstructions employed in GIA sea-level modeling. This is particularly true in the Eurasian Arctic and the Antarctic. Two of the greatest uncertainties in ice-sheet reconstruction at the time ICE-3G was published by Tushingham and Peltier (1991) owe their origin to the inaccessibility of data in the Arctic coastal waters of the former Soviet Union (Mangerud et al. 2002) and to the lack of field constraints on the size and timing of the collapse of the Antarctic Ice Sheet at the Last Glacial Maximum (LGM).

During the past 15 years a large degree of this uncertainty has been significantly reduced due to improved field observations. First, there is now a wealth of coastal sediment core data that has been combined with continental and Arctic island data (Svendsen et al. 2004) and these clearly rule out, for example, the hypothesis of a large marine-based East Siberian Ice Sheet (which is included in ICE-3G). (See Lambeck 1995, who earlier reached the same conclusion from the inversion of sea-level data from Arctic Europe.) This has led to a significant improvement in constraining the volume and deglaciation history of the Kara Sea Ice Sheet (Svendsen et al. 2004). This revision is accounted for in the reconstructions of both Lambeck et al. (1998b) and ICE-5G (Peltier 2004).

Reconstructions of Antarctic Ice Sheet evolution following the LGM have also been significantly revised in recent years due to the improved field constraints now available. Early models were heavily influenced by the series of maps published through the CLIMAP project (Denton and Hughes 1981). In addition, studies that considered far-field sea-level observations concluded that a relatively large volume of post-LGM melt (totaling 37 m of eustatic sea-level rise) was required from the Antarctic Ice Sheet to fit these observations (Nakada and Lambeck 1988). Other early ice models, such as ICE-3G (Tushingham and Peltier 1991), also contained a relatively large LGM Antarctic Ice Sheet (equivalent to 26 m of eustatic rise). However, neither of these models were tested against observations from the Antarctic continent.

A wealth of marine, ice-proximal, and rock-outcrop-based glacio-geomorphological data collected over the past decade appears to be inconsistent with post-LGM melt from Antarctica in excess of approximately 20 m. A summary compilation by Bentley (1999) revises downward the total meltwater contribution to 8–12 m, and Denton and Hughes (2002) have bounded an assessment at 3–17 m. Using the Denton and Hughes (2002) maximum estimate as a starting point, Ivins and James (2005) assembled a load history ("IJ05") with additional constraints available since the early 2000s. The additional information constraining "IJ05" included new maximum ice thickness estimates in the central part of west Antarctica (e.g. Licht 2004; Waddington et al. 2005), the mountain ranges near the eastern coast of the Weddell Sea (Fogwill et al. 2004), and additional information across the entire continent and its margins. The model retains one important feature that was introduced into the global models by Nakada and Lambeck (1988), the necessity of a late meltwater component from Antarctica, but that is forced to obey the known constraints on the volume of freshwater input to the oceans in the far field over the past two to three millennia (e.g. Lambeck et al. 2004).

While ice-sheet reconstructions have improved dramatically in the past decade, it is also evident that there remain first-order uncertainties to resolve. A good example of this is the evolution in global ice cover during the occurrence of meltwater pulse IA (15 000–13 000 calibrated years before present). A number of publications have challenged the conventional view (Peltier 1994) that the meltwater was sourced solely from the Laurentide Ice Sheet, and in particular its southern sector, by demonstrating that this scenario is inconsistent with far-field sea-level records (Clark et al. 2002; Bassett et al. 2005). Improved collaboration between the modeling (GIA and ice) and field communities will certainly aid in resolving these outstanding issues and will lead to improved accuracy in GIA model predictions of recent sea-level change.

10.2.4 Modeling of TPW

GIA-induced perturbations in the rotation vector have two important connections to the study of 20th-century sea-level changes. First, rotation data, after correction for the GIA process, provide a constraint on recent mass balance of ice reservoirs (Munk 2002). Second, TPW driven by GIA will perturb relative sea level and the position of the sea surface, and this rotational feedback will form part of the ice-age correction to these observations.

Changes in the rotation vector for a viscoelastically deforming planet subject to changing surface mass loads are computed by solving the standard continuum mechanical system coupled to the equations governing conservation of angular momentum (Euler's equations). Consider an idealized planet subject to a localized mass load. The rotation vector of the planet will move away from the load, or, alternatively, the load will move toward the equator. However, this tendency will be opposed by the background rotational bulge (supported, in part, by mantle convection) and will act to resist polar wander. A complication in this simple picture is that the effective mass load will slowly diminish as it adjusts toward isostatic equilibrium (i.e. as the solid Earth below it subsides) and the bulge resistance will weaken as a large component of the rotational bulge adjusts toward the new orientation of the rotation vector. This adjustment occurs as the bulge responds to the change in centrifugal potential. Any rotation theory must provide both a formalism for computing the viscoelastic response to surface mass and potential loads, and an expression for the background bulge (or ellipticity) upon which these loads are applied. In standard GIA theory of Wu and Peltier (1984) the former is derived in terms of load and tidal Love numbers, and, for spherically symmetric Earth models, these expressions remain valid. However, Mitrovica et al. (2005) have shown that the treatment of the background ellipticity, or complete degree-2 shape of the solid Earth, that has been adopted in the standard theory introduces substantial errors into predictions of the rotational response.

In the Wu and Peltier (1984) theory, the background ellipticity is replaced by the equilibrium rotational form of the same Earth model adopted in the loading calculation. That is, the background flattening is assumed to be the same as the form one would have by spinning this Earth model at the current rotation rate and waiting an infinite time. The Earth model adopted in GIA

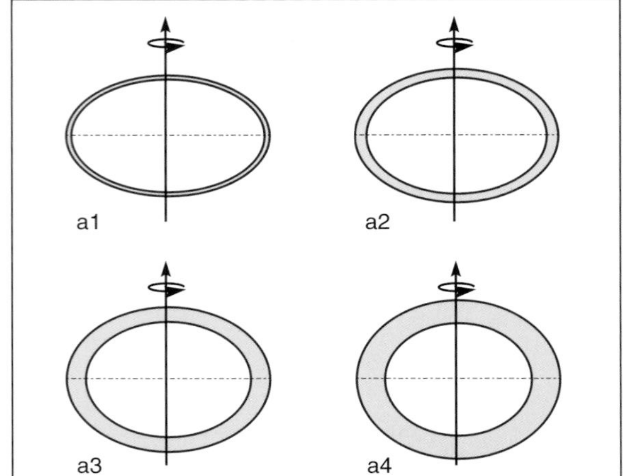

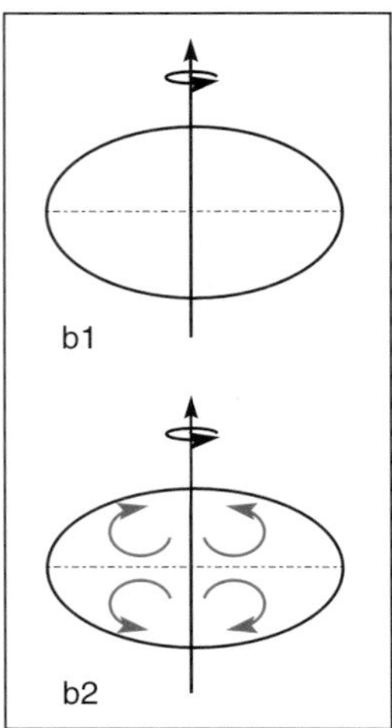

Figure 10.2 Schematic illustration of the background ellipticity adopted in the old (Wu and Peltier 1984) and new (Mitrovica et al. 2005) theories for load-induced perturbations in Earth rotation. (a) The old theory assumes that the background ellipticity may be replaced by the form established by rotating a planet with an elastic lithosphere of prescribed thickness (blue region in each plot) and waiting an infinite time. This approach yields an ellipticity that is a function of the adopted elastic thickness, as illustrated by the different plots on each frame. (b) The new theory ties the background form to the observed ellipticity using two steps. First, one considers the ellipticity established by a rotating planet with no lithosphere (the hydrostatic form; b1). This form is then augmented to include the nonhydrostatic ellipticity of the planet (b2), which is thought to be connected to convective flow in the Earth's mantle (hence the red flow lines).

loading calculations is defined by an elastic lithosphere and a mantle viscosity field. At infinite time, the viscous stresses in the model would relax completely in response to a constant rotation, but the elastic lithosphere would not. Thus, the equilibrium rotational form of the model would be a function of the adopted thickness of the elastic lithosphere, as illustrated schematically in Figure 10.2a.

The background form adopted in the rotation theory should, in fact, be tied to the observed ellipticity of the real Earth since the ice-age loading is superimposed on this form. In this regard, the standard theory errs by introducing a model dependence in the background form (the higher the adopted lithospheric thickness, the less the background ellipticity (Figure 10.2a) and the more unstable the rotation pole). A more accurate approach would be to compute the background form using a model which has no elastic lithosphere (i.e. no infinite time strength; Figure 10.2b1). This would represent the so-called hydrostatic form, and calculations show that it is closer to the Earth's background ellipticity than predictions generated using a model with an elastic lithosphere (Mitrovica et al. 2005). However, this approach has a shortcoming that highlights a second problem with the standard theory. Namely, the theory neglects nonhydrostatic contributions to the Earth's ellipticity. Nakiboglu (1982) identified a significant, nonhydrostatic "excess ellipticity" that is widely considered to be due to convective flow in the Earth's mantle. Thus, a second important improvement to the rotation theory would be to augment the hydrostatic form with Nakiboglu's (1982) nonhydrostatic contribution (Figure 10.2b2), as advocated by Mitrovica et al. (2005). The net effect of this augmentation is a further stabilization of the rotation vector.

Figure 10.3 summarizes, following Mitrovica et al. (2006), the sea-level enigma cited by Munk (2002). In Figures 10.3a and b the solid black lines are predictions of the GIA-induced secular trend in J_2 (which is proportional to changes in the rotation rate) and polar motion, respectively, where the latter is computed using the theory of Wu and Peltier (1984), as a function of the lower-mantle viscosity

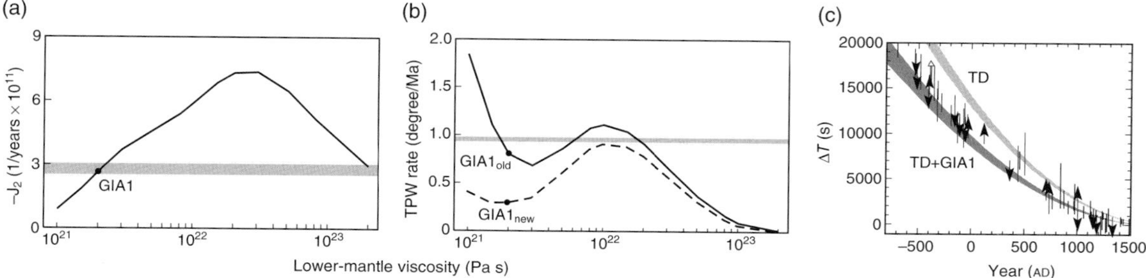

Figure 10.3 (a) Prediction of the present-day variation in the J_2 coefficient of the geopotential due to GIA as a function of the adopted lower-mantle viscosity. The shaded region represents a typical observational constraint (Nerem and Klosko 1996). The result GIA1 refers to the specific lower mantle viscosity used by Munk (2002) (2×10^{21} Pa s). (b) As in (a), except for predictions of the rate of TPW based on the old (Wu and Peltier 1984; black solid line) and new (Mitrovica et al. 2005; dashed line) rotation theories, respectively. The shaded region is the range in observed values from McCarthy and Luzum (1996) and Gross and Vondrák (1999). (c) Ancient eclipse data compiled by Stephenson and Morrison (1995) showing the discrepancy in time between the occurrence of specific eclipses and the timing expected on the basis of the current rotation rate. These data indicate an integrated slowing of the Earth's rotational clock by 4–5 h over the last 2.5 millennia. The shaded region labeled TD is the slowing expected on the basis of tidal dissipation, while the region TD+GIA1 is the total shift when the tidal dissipation curve is augmented by a GIA-induced nontidal acceleration predicted on the basis of model GIA1.

of the Earth model. The predictions were generated using an ice history based on the ICE-5G model (Peltier 2004). The light-shaded regions on each frame represent typical observational constraints cited for these data (Nerem and Klosko 1996; McCarthy and Luzum 1996; Gross and Vondrák 1999). The results labeled GIA1 were based on a model adopted by Munk (2002) (lower mantle viscosity of 2×10^{21} Pa s), following Peltier (1998b). This model provides a reasonably good fit to both data sets. The final frame, Figure 10.3c, shows individual eclipse records compiled by Stephenson and Morrison (1995), which yield a time shift (ΔT) between the occurrence of a given eclipse measured by ancient astronomers and the timing predicted on the basis of the Earth's present rotation rate. The light shaded region labeled TD represents the integrated slowing of the Earth's rotational clock predicted on the basis of tidal dissipation (Stephenson and Morrison 1995). Finally, the region TD+GIA1 is the slowing predicted when the curve TD is augmented by a GIA-induced nontidal acceleration associated with the model GIA1.

According to Munk (2002), the simultaneous reconciliation of the dJ_2/dt, TPW, and ancient eclipse data using the model GIA1 leaves no room for a signal from present-day melting of ice reservoirs as required by 20th-century sea-level observations. One might choose a different lower-mantle viscosity in order that a residual arises between the dJ_2/dt datum and the GIA prediction sufficient to permit melting from ice reservoirs equivalent to 1 mm/year (for example) of sea-level rise. However, such a choice would also yield a residual with the eclipse observations, and so the inferred melting would have to extend through the last few millennia, not just the 20th century, to fit these observations.

How to resolve this enigma? A number of studies have noted the trade-off between deep-mantle viscosity and meltwater flux on the basis of an analysis of dJ_2/dt and TPW observations (James and Ivins 1997; Klosko and Chao 1998; Johnston and Lambeck 1999; Tosi et al. 2005) and have ignored the eclipse observations. In this regard, the eclipse data may be subject to larger errors than considered by Stephenson and Morrison (1995) (see, for example, Steele 2005).

Mitrovica et al. (2006) have demonstrated that it is possible to simultaneously reconcile all three data sets in Figure 10.3 while permitting approximately 1 mm/year of anomalous 20th-century melting. The dashed line in Figure 10.3b is the prediction of GIA-induced TPW generated using the new rotation theory. These revised rates are smaller than those generated using the old TPW theory, reflecting the increased rotational stability when a more accurate treatment for the background ellipticity is employed. The difference is not small; indeed, the prediction based on model GIA1 is reduced by approximately 65% when the new theory is applied, and the result is a prediction which is far too small to fit the observational constraint. The accord cited by Munk (2002) disappears with the application of the new, more accurate and physically based, rotation theory. In addition, Benjamin et al. (2006) and Mitrovica et al. (2006) showed, following Cox et al. (2003), that the oft-cited error level in the dJ_2/dt observation significantly underestimates the uncertainty associated with the 18.6-year tide signal. A revised estimate of this error introduces sufficient inconsistency between the dJ_2/dt and eclipse observations for significant melting onset in the 20th century.

The ongoing development of software for treating GIA on three-dimensional Earth models will set the stage for incorporating a fully triaxial Earth in studies of TPW (Nakada, 2009; Matsuyama et al., 2010) and these may uncover additional sensitivities. The diminished GIA signal in TPW predicted by the new theory of Mitrovica et al. (2005) has pointed toward an anomalous, late-20th-century melting of the Earth's land-based cryosphere, but it also opens the possibility for other effects to have greater poignancy, including mantle processes and plate tectonics (e.g. Vermeersen et al. 1994; Steinberger and O'Connell 1997).

10.3 Sea Level, Sea Surface, and the Geoid

Predictions of the GIA contribution to current changes in the sea-surface and the geoid are becoming increasingly important due to the increase in length of the altimetric time series, the new time-variable gravity data sets (Tapley et al. 2004), and high-resolution gravity and geoid solutions expected from the Gravity Field and Steady-State Ocean Circulation Explorer (GOCE) satellite launched in March 2009. While the ongoing contribution due to GIA has been accounted for in tide-gauge studies since the late 1980s, a GIA correction is frequently neglected in sea-surface/geoid analyses (for an early exception, see Nakiboglu and Lambeck 1981).

It is important to clarify the meaning of "sea surface" and "geoid" because the terminology adopted within some of the GIA literature diverges from the usual geodetic usage. Solutions of the sea-level equation associated with GIA yield predictions of the time-varying (through the glacial cycles and into the current interglacial) evolution of the sea-surface and the solid surface; that is, the two bounding surfaces defining relative sea level. Within these solutions, mass is conserved by ensuring that changes in volume bounded between these two surfaces are consistent with changes in the mass of the ice sheets, and with the continental hydrological system in general. While tide gauges measure relative sea-level changes, sea-surface variations are observed through altimetry. The equilibrium, or static, theory adopted in GIA studies (see Farrell and Clark 1976) assumes that the sea surface always remains an equipotential. The equipotential that defines the sea surface may change over time both because of ongoing mass redistribution in the Earth system due to GIA (post-glacial rebound, peripheral subsidence, etc.) and because of hydrological flux into or out of the ocean. There may also be changes caused by large-scale tectonic processes operating at 10^6-year timescales and longer, as discussed recently by Barletta and Sabadini (2006), but these tend to be smaller than the GIA influences discussed here. GIA studies that solve the sea-level equation combine geoid variations (in the sense used by geodesists) with changes from one equipotential to another into a single prediction that is called the "geoid" but is actually the sea surface. This distinction is important to remember when comparing GIA predictions to satellite-derived time-variable gravity changes.

As we are presently in an interglacial, GIA predictions will not involve an ongoing exchange of water between the continents and oceans. However, this does not mean that GIA will not produce an ongoing change in the globally averaged sea surface. As is clear from the preceding paragraph, the mean sea surface will change because of ongoing GIA-induced deformations (which change the shape of the ocean basin). As an example, the collapsing forebulge caused by the last glaciation would extend into the ocean off the North American east coast. Globally averaged over the oceans, this effect has a nonzero contribution to the sea surface. Overall, the changing shape of the ocean basins due to GIA causes the mean sea surface to decrease by approximately 0.3 mm/year (Peltier 2001). Thus, estimates of sea-level rise from altimetry will increase by roughly this amount once the GIA contribution is removed (the uncertainty derives from continued uncertainties in the ice history and Earth model, as discussed above).

The quasi-static low-degree geoid harmonics contain limited information on GIA since these harmonics contain signals from multiple geophysical process (e.g. mantle convection). However, the importance of GIA will increase at intermediate wavelengths in which the spectral power of the Late Pleistocene ice and water loading peaked (Mitrovica and Peltier 1989). For example, Van der Wal et al. (2004) and Schotman and Vermeersen (2005) have shown that if shallow low-viscosity (crustal, asthenospheric) zones exist, these can produce elongated ribbon-like signatures in the geoid that should be well above the detection level of GOCE.

Ultimately, a more robust framework for separating the GIA signal from other geophysical processes lies in the time rather than spatial scale of the signal. Tamisiea et al. (2007) have generated the first map of the free-air gravity trend over Laurentia using 4 years of GRACE satellite gravity measurements (see also Paulson et al. 2007). These trends, in contrast to the static field, will have negligible contributions from slowly varying processes such as mantle convection or the emplacement of crustal density heterogeneities; moreover, the GRACE time window is sufficient to separate the GIA signal from annual and interannual hydrological mass flux. The Tamisiea et al. (2007) maps indicate that the sub-Arctic Laurentian ice complex was composed of two major domes, one to the east and the other to the west of Hudson Bay. Furthermore, their analysis indicates that GIA contributed about 25–45% of the static gravity field over the same region.

One area of research that is in need of development is the prediction of the surface load response to sediment transport during recent and Holocene times. One impediment to this effort has been the lack of a coordinated effort to characterize the global responses to the world's great sedimentary transport systems during and after glacial – interglacial transitions (Blum and Törnqvist 2000). A study of coastal sea level change by Gehrels et al. (2004) refers to sediment loading as one of the "untested factors" that could have a substantial local effect. In this regard, it may be particularly important to model the load response near the mouths of the world's major rivers, especially where they are susceptible to large changes in supply during the Late Holocene (Ivins et al. 2007). As an example, an effort to incorporate sediment loading within the Gulf of Mexico into local predictions of post-glacial sea-level change has been described by Simms et al. (2007). Moreover, it is also important to recognize the coupling between post-glacial sea-level change and the pattern of sediment transport (Whitehouse et al. 2007). Finally, research in this area should include anthropogenically induced sediment imbalances, such as in the Yangtze Delta where sediment deficits are very large compared to the regional geological record due to the construction of the Three Gorges Dam (Yang et al. 2003).

10.4 Rapid Melting and Sea-Level Fingerprints

There has been a commonly held view that recent melting of global ice reservoirs should produce a more or less uniform sea-level change. In fact, a long history of studies, dating back to Woodward (1888), have demonstrated the significant geographic variability in sea level that would follow the rapid melting of an ice sheet (e.g. Farrell and Clark 1976; Clark and Lingle 1977; Clark and Primus 1987; Nakiboglu and Lambeck 1991; Conrad and Hager 1997; Mitrovica et al. 2001a; Plag and Jüttner 2001; Tamisiea et al. 2001, 2003; Plag 2006). The physics governing sea-level change in this case is shown, schematically, in Figure 10.4 (from Tamisiea et al. 2003). An ice sheet exerts a gravitational pull, or tide, on

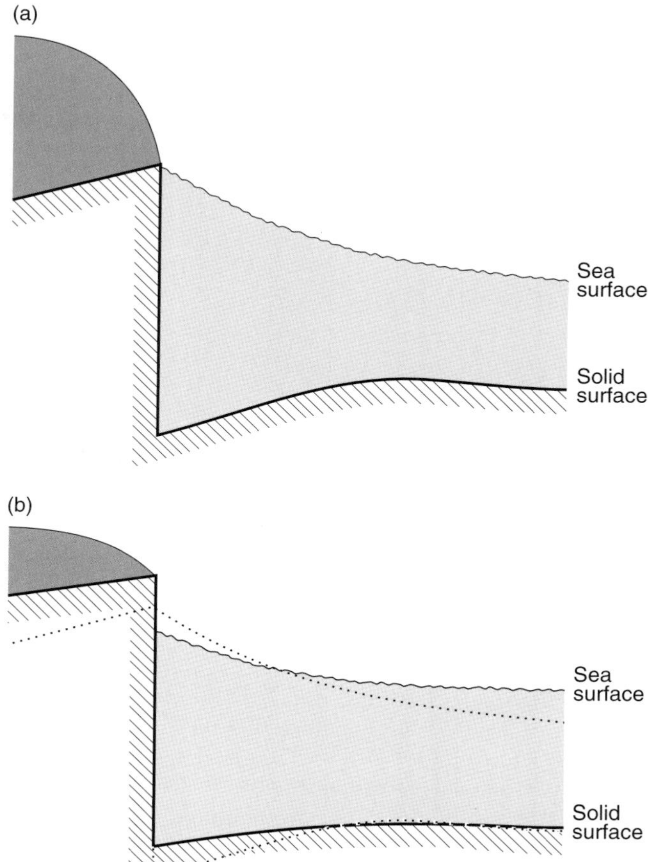

Figure 10.4 Schematic illustration of the physics of sea-level change in response to the rapid melting of an ice sheet (after Tamisiea et al. 2003). (a) The sea surface and solid surface warping in the presense of an ice sheet. (b) Sea and solid surfaces after a melting event (original surfaces are given by the dotted lines). Relative sea level in each frame is given by the thickness of the shaded region; that is, the height of the sea surface relative to the sea bottom.

the surrounding ocean, and this piling up of water is also responsible, together with the ice load, for a deformation of the solid surface (Figure 10.4a). If the ice sheet melts, then the tide relaxes and water moves from the near field to the far field (Figure 10.4b). This redistribution is accompanied by a deformational response of the solid surface: uplift in the near field and subsidence in the far field. In addition, of course, meltwater enters the ocean, and thus the integrated change in relative sea level (that is, the change in volume bounded by the sea surface and ocean bottom) must increase to conserve mass. The net effect of these processes is that relative sea-level falls in the vicinity of the ablating ice mass, and it rises by increasing amounts as one moves away from the ice.

Mitrovica et al. (2001a) and Plag and Jüttner (2001) have used the terms "patterns" and "fingerprints" to describe the expected sea-level variation. This terminology is motivated by the fact that changes in mass of each ice sheet or glacier would have a distinct sea-level geometry. As a consequence, the traditional approach of taking the mean of GIA corrected sea-level data (e.g. tide-gauge rates)

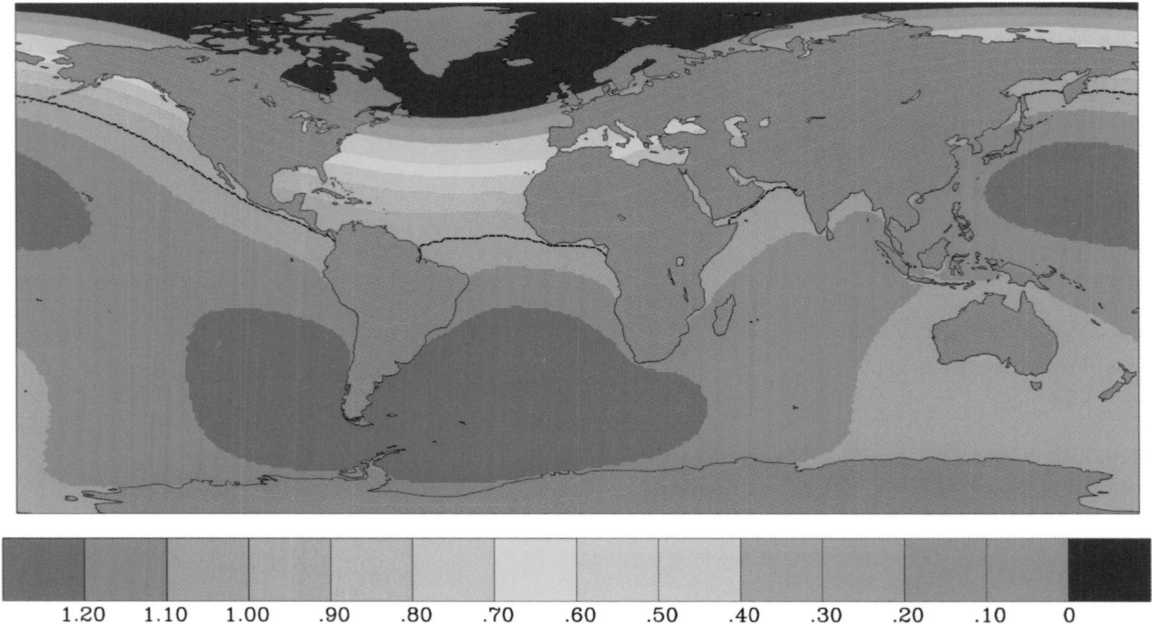

Figure 10.5 Sea-level fingerprint of present-day melting from the Greenland ice sheet (after Mitrovica et al. 2001a). The calculation assumes that mass loss is uniform over the ice sheet and equivalent to 1 mm/year of eustatic sea-level rise. The prediction is based on a gravitationally self-consistent sea-level theory which incorporates rotational feedback. The dashed line marks the 1 mm/year contour. The dark blue zone represents the near-field region that would experience a sea-level fall; note that this region includes locations as distant as Norway, Scotland, and Newfoundland.

ignores the potentially powerful constraint embedded within these spatially varying trends. In particular, these patterns provide a method for estimating not only the globally averaged sea-level trend, but also, for sufficiently distributed data sets, the sources of the meltwater.

As an illustrative example of this approach, Figure 10.5 (from Mitrovica et al. 2001a) shows a prediction of the relative sea-level change resulting from melting from Greenland equal to a eustatic (i.e. geographically uniform) sea-level rise of 1 mm/year. The calculations include the effect of rotational feedback. The dark blue zone represents the area in which sea-level falls, while the dashed line is the 1 mm/year contour; that is, areas to the north of this line experience a sea-level rise (or indeed fall) below the eustatic value of the melting, while areas to the south experience a sea-level rise above the eustatic value. The significant departures from this eustatic value reflect the rather dramatic spatial variability in sea-level change that accompanies the rapid melting of an ice complex. Similar calculations were performed by Mitrovica et al. (2001a) for melting from the Antarctic Ice Sheet and a tabulation of mountain glacier flux from Meier (1984).

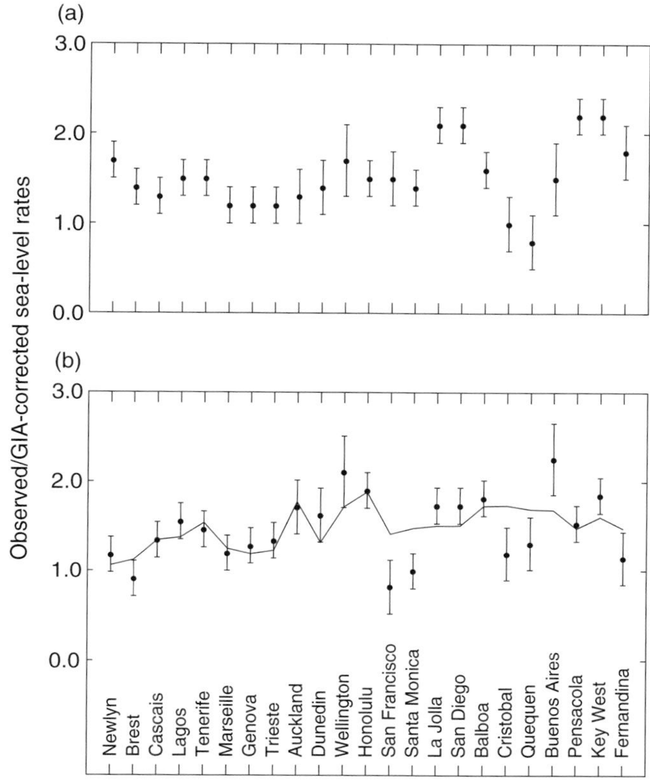

Figure 10.6 Tide-gauge fingerprint analysis (adapted from Mitrovica et al. 2001a). (a) Long-term relative sea-level trends (in mm/year) determined from the PSMSL tide-gauge database (http://www.pol.ac.uk/psmsl) for 23 sites included in the analysis of Douglas (1997). (b) Same data after correction for GIA using a prediction based on the ICE-3G deglaciation history (Tushingham and Peltier 1991) and an Earth model characterized by an elastic lithospheric thickness of 120 km, an upper-mantle viscosity of 10^{21} Pa s and a lower-mantle viscosity of 5×10^{21} Pa s. The solid line on the figure represents the best least-squares fit to the GIA-corrected data based on a weighted sum of fingerprints associated with melting from Greenland, the Antarctic, and mountain glaciers, plus a geographically uniform signal.

Figure 10.6a shows secular sea-level trends determined at a set of 23 sites using the PSMSL database (Woodworth and Player 2003). The sites were adopted in the analysis of Douglas (1997). In Figure 10.6b these rates have been corrected for GIA using a spherically symmetric numerical calculation based on the ICE-3G deglaciation history (Tushingham and Peltier 1991). This GIA correction does not reduce the geographic variation evident in the raw trends (the same is true for a GIA calculation in which the lower-mantle viscosity is reduced to a value of 2×10^{21} Pa s, as adopted by Douglas 1997). The mean GIA-corrected rate is 1.5 ± 0.1 mm/year. The solid line on Figure 10.6b represents the result of a fingerprint analysis. The geographic variation in the GIA-corrected trends was fit using a combination of the fingerprint from the set of mountain glaciers tabulated by Meier (1984) plus a weighted sum of fingerprints associated with Greenland (i.e. Figure 10.5) and Antarctic mass flux and "other" sources. In principle, these other sources would represent the signal from thermal expansion, groundwater fluxes, etc.; this simple exercise assumes that the fingerprint for them is globally uniform. The least-squares procedure yielded the weighting for the various fingerprints. Since most of the tide-gauge records are in the Northern Hemisphere, and since the Antarctic fingerprint is relatively uniform in the north, the fitting procedure could not independently determine the weighting for Antarctic flux and the "other" contributions, and thus the sum of these latter weightings was estimated. The fit of the solid line to the data is statistically significantly better (at 90% confidence) than a simple mean value through these rates. The former clearly captures the variation across most of the sites and, in particular, it reconciles the somewhat lower (by ≈ 0.5 mm/year) rates in Europe relative to the remaining sites. The lower trends in Europe have represented a long-standing enigma within the literature (Douglas 1991; Woodworth et al. 1999). The fingerprint analysis is able to reconcile them by incorporating significant melting from Greenland (0.6 mm/year of equivalent eustatic sea-level rise); a contribution that will cause the European sea-level rates to be greatly reduced from the global average (Figures 10.5 and 10.6).

A powerful aspect of the fingerprint approach is that it allows for the possibility that sea-level rates in the vicinity of the tide gauges may not be representative of the global average; that is, an average of a set of tide-gauge samples of the predicted field in Figure 10.5 may not be close to the global average used in the calculation (1 mm/year), but the geometries evident in these rates would allow one to nevertheless infer the global average. This sampling issue was highlighted by Plag (2006), whose fingerprint analysis was applied to the approximately 1000 tide-gauge sites in the Revised Local Reference (RLR) subset of the PSMSL database with data in the time window 1950–98. Plag's (2006) analysis included a global fingerprint for thermosteric sea-level change taken from either Ishii et al. (2003) or Levitus et al. (2000), fingerprints for uniform melting of Antarctic and Greenland ice (as in Figure 10.5), a eustatic "other term", and a GIA correction with a free multiplicative scaling. No fingerprint for mountain glacier melting was included. The tide-gauge rates were, furthermore, binned into "regression grids" of various size (ranging from 1 to 10°), with a weighting in each grid based on the time window of the individual records. While Plag (2006) found that the fingerprints explained only about 15% of the data variance, a significant correlation (coefficient 0.38 ± 0.07) existed between the model and observations. Plag (2006) inferred a contribution to global sea-level rise of 0.39 ± 0.11 and 0.10 ± 0.05 mm/year from the Antarctic and Greenland ice complexes, respectively, and a steric effect of 0.35 mm/year. His estimate of global sea-level rise was 1.05 ± 0.75 mm/year. The least-squares fits showed a suite of correlations; for example, the estimate of Antarctic melting was particularly sensitive to the choice of GIA model, and the estimate was, as in Mitrovica et al. (2001a), correlated to the magnitude of the "other" (eustatic) signal.

The Mitrovica et al. (2001a) (see also Tamisiea et al. 2003; Plag 2006) fingerprint studies define approaches that mirror end-member philosophies adopted in earlier tide-gauge analyses. Should fingerprint analyses of tide-gauge trends be based on a small, carefully selected set of records culled for the contaminating influence of tectonics, groundwater pumping, etc? Or do the benefits of a larger data set outweigh the potential problems with data quality in individual records?

The answer to this question, and the robustness of a fingerprinting exercise, will ultimately be connected to one's ability to quantify the impact of systematic noise sources and to accurately model or constrain the geometry associated with the primary contributors to the sea-level variability. For example, ice and Earth model uncertainties in the GIA models have been discussed in this chapter, and these indicate that a simple scaling factor is unlikely to capture the inherent uncertainty in this correction. Moreover, the sea-level fingerprints of present-day melting events can be more complex, at least locally, than predictions based on uniform mass-loss events. As an example of the latter, consider the fingerprint of recent melting from the Antarctic Ice Sheet. Using an *a priori* glaciological estimate, Ivins et al. (2005) computed the elastic-gravitational response to the combined ice loss predicted by Rignot and Thomas (2002), Rignot et al. (2003), and Rignot et al. (2004) for individual drainage basins of Antarctica, the Patagonian ice fields, and the Antarctic Peninsula region. The ice mass loss is equivalent to a eustatic sea-level rise of 0.32 mm/year and 0.105 mm/year, respectively, for Antarctica (with the Antarctic Peninsula) and southernmost South America. The prediction of the time-varying geoid for this history is shown in Figure 10.7. While the effects of load self-gravitation were not included in this calculation, the

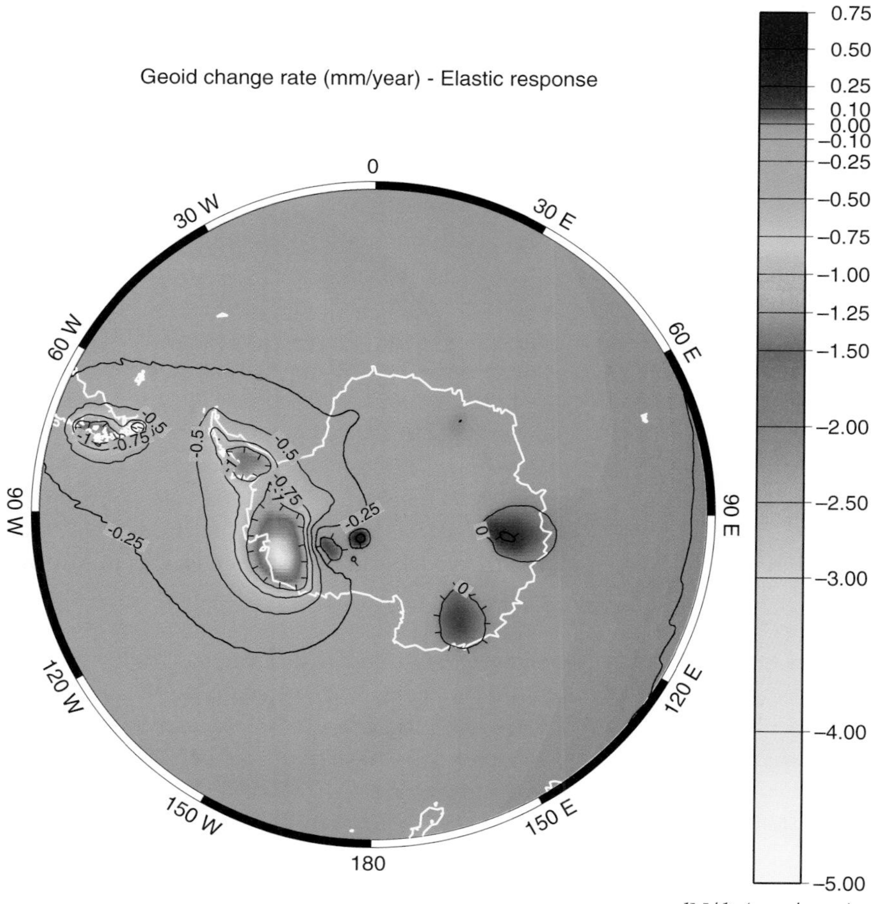

Figure 10.7 Computed secular rate of geoid change due to late-20th century ice-mass flux estimated by Rignot and Thomas (2002) for Antarctica and of Rignot et al. (2003) for Patagonia. The calculation was purely elastic (load self-gravitation is not included) with Love numbers for a PREM Earth model (from Ivins et al. 2005).

difference between Figure 10.7 and zonally distributed circum-Antarctic fingerprints for the geoid (Tamisiea et al. 2001) and relative sea level (Mitrovica et al. 2001a; Plag and Jüttner 2001; Plag 2006) computed assuming a uniform mass loss are significant.

These issues also arise in a second avenue of application for the fingerprinting approach; namely, studies of sea-level change based on satellite altimetry and gravity data sets (e.g. Tamisiea et al. 2001). In this case, a generally improved spatial coverage relative to the tide-gauge database is accompanied by a diminished time window of observation. Tamisiea et al. (2003) have shown that the sea-surface fingerprint of recent melting from the Alaskan mountain glacier system could provide a robust constraint on the local mass flux and should be a target for altimetric studies. In this case, vertical uplift and tide-gauge measurements would be complicated both by a multistage load history and the active continental tectonic environment within the region (Larsen et al. 2005). Klemann et al. (2007), for example, have demonstrated that slab geometry and rheological complexities, such as a water-weakened mantle wedge, may make a model-based removal of the local vertical crustal motion difficult in such areas. Whereas their focus was the impact of Little Ice Age mass loss on the present-day deformation in southernmost Patagonia, the complications they infer probably apply to any coastal subduction environment with surface ice-load loss. Finally, GRACE satellite data have recently inferred mass loss over the polar ice sheets on the basis of the regional trends in gravity (e.g. Velicogna and Wahr 2006a, 2006b; Ramillien et al. 2006; Velicogna, 2009). Both studies have commented on the potentially significant impact of uncertainties in the GIA correction.

10.5 Great Earthquakes

Subocean earthquakes are capable of triggering tsunamis and thereby constitute an extreme category of short-timescale, cataclysmic, solid Earth-induced sea-level variations. This section does not deal with tsunamis (which can also be triggered by landslides), but instead concentrates on the non-inertially forced sea-level variations due to earthquakes.

The influence of great earthquakes on regional sea level may be large and can extend over thousands of kilometers from the epicenter. Indeed, historical records show meters of sea-level change recorded from co-seismic and post-seismic motions. Figure 10.8 shows the co-seismic sea-level change associated with the December 26, 2004, Aceh-Andaman earthquake in Sumatra, Indonesia, within 100 km of the rupture. In the before-and-after pictures taken from an imaging instrument aboard the TERRA satellite, the emergence and submergence (a pivot line is crossed between frames a, b and c, d of Figure 10.8) magnitudes are of the order 1–2 m. Recent work by Han et al. (2006) shows that the gravitational effects of the earthquake were detected by the GRACE satellite pair and have a surface gravity change of 10–15 μgal over many hundreds of kilometers of open ocean.

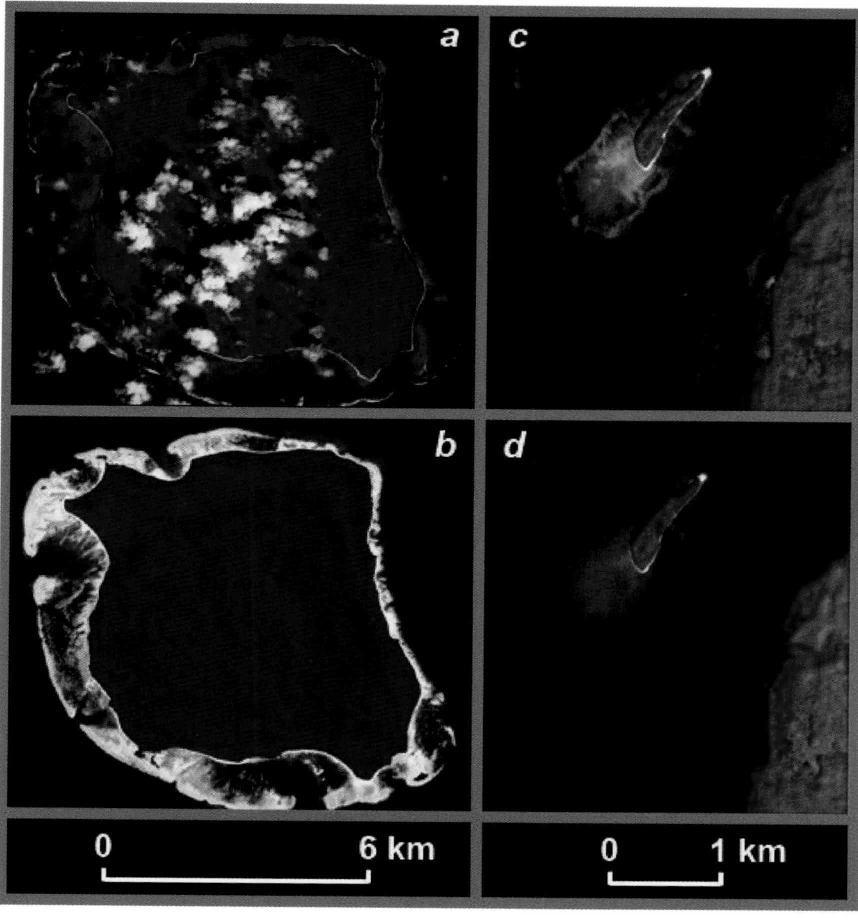

Figure 10.8 (a) Pre-earthquake and (b) post-earthquake Advanced Spaceborne Thermal Emission and Reflection Radiometer (ASTER) images of North Sentinel Island. The image shows sea-level lowering caused by the great Aceh-Andaman earthquake of 2004 ($M_w \approx 9.2$). Here the emergence of the coral reef surrounding the island is very clear in the image. Co-seismic, tidally corrected, uplift is 1–2 m. (c) Pre-earthquake and (d) post-earthquake ASTER images of a small island off the northwest coast of Rutland Island, 38 km east of North Sentinel Island, showing submergence of the coral reef surrounding the island. Images (a) and (b) are about 50 km from the rupture on the underthrusting Sunda Trench. (Adapted from Meltzner et al. 2006.)

Additionally, co-seismic strain from this magnitude 9+ event will remain at the millimeter level (or more) at a distance of approximately 5000 km (Vigny et al. 2005) and the global impact of this earthquake is at the level of 1 mm permanent co-seismic displacement (Kreemer et al. 2006).

Over the last decade there have been efforts to combine numerical models of solid Earth deformation with large catalogs of seismic events to estimate the cumulative impacts of this seismicity on global sea level. In a recent study, Melini and Piersanti (2006) estimated a mean-sea-level signal at PSMSL tide-gauge stations of as much as 0.25 mm/year. The large majority of this signal originates from the very largest thrust events (1960 Chile 1964 Alaska). Thus, the history of seismicity, and future events, may contribute nonnegligibly to observed sea-level trends.

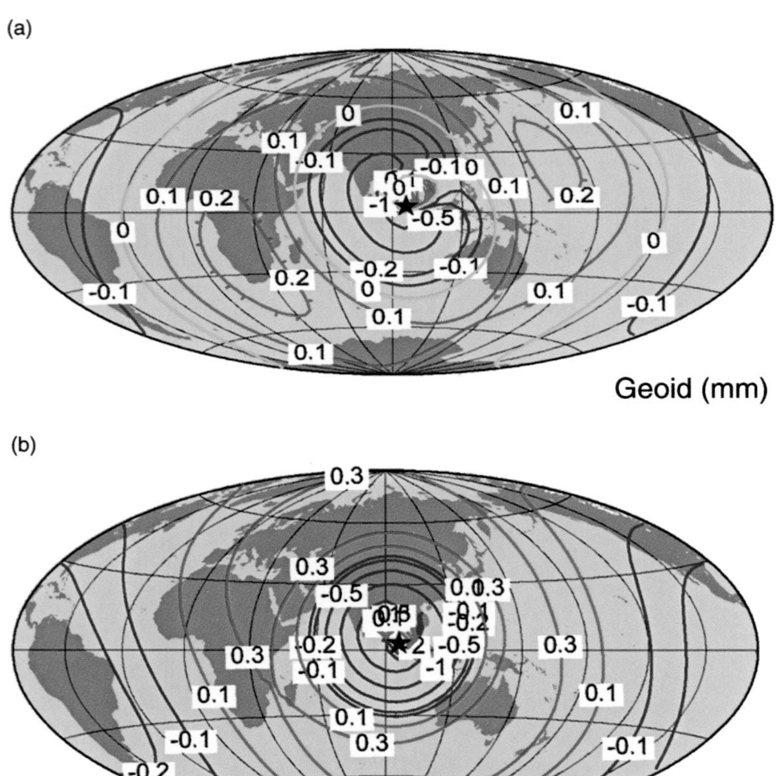

Figure 10.9 (a) Theoretical co-seismic geoid changes and (b) vertical bedrock surface motion caused by the Aceh-Andaman earthquake of December 26, 2004. The black star shows the epicenter. An asymptotic spherical Love number approach by Sun (2004) and colleagues also provided reasonable fits to the near and intermediate distance vertical displacement field (Figure 10.8) using static dislocations. (This figure is provided courtesy of W. Sun, and was presented at the May 2006, Joint Spring AGU Meeting; see Sun and Okubo 2006.)

Finally, it is clear that mantle flow and the associated evolution of plate boundaries also leads to regional sea-level changes (Menard 1978; Mitrovica et al. 1989). However, models of the global consequences of mantle flow suggest that the associated sea-level changes are relatively small (e.g. Barletta and Sabadini 2006). More intriguing is the question of the co-seismic impact of great thrust events on the sea-level record. Calculations of the predicted global static geoid and vertical displacement changes due to the great Aceh-Andaman earthquake ($M_w = 9.1$–9.3) are shown in Figures 10.9a and b, respectively. Using a dislocation theory in an otherwise classical formulation for a self-gravitating, spherical, nonrotating, elastic Earth, Sun and Okubo (2006) show that the predicted sea-level change, in parts of the globe as far away as northern Canada and Corsica (compared with southern Iran, for example), are about 0.6–0.8 mm, peak to peak. That such large global static deformations of the solid Earth affect sea level at this level was previously unrecognized.

10.6 Final Remarks

This chapter has provided a survey of recent advances in understanding of solid Earth processes that have bearing on the analysis of global sea-level trends. Some of the advances in GIA predictions, for example improvements in models for Late Pleistocene ice history and viscoelastic Earth structure, have emerged from a continuous and ongoing process of refinement dating back many decades. Others, for example the revised theory for deglaciation-induced TPW, have uncovered important issues that have been hidden in the complexity of traditional theories. In either case, the advances have led to significant improvements in the accuracy of GIA predictions and have had direct and immediate impact on analyses of global sea-level trends. In the next decade, these improvements will continue to emerge, notably the effort, now in its infancy, to model the ice-age response of complex, three-dimensional Earth models.

The chapter has also highlighted the notion of sea-level fingerprinting and has advocated the procedure as a framework for extracting information regarding source contributions from the complexity of observed sea-level trends. The physics underlying the philosophy of fingerprinting is straightforward, and the challenges to future applications of the procedure are equally clear. How can we improve constraints on the fingerprints associated with thermosteric effects and recent ice mass flux? In particular, what other data sets can be invoked to limit the range of possible melting geometries, or rather to expand the range from the simple scenarios considered to date? How can the uncertainty in the GIA signal, which is still significant, be best parameterized within the fingerprint procedure? What are the important sources of noise, and which processes (e.g. terrestrial hydrological redistributions) need to be included in an expanded regression analysis? In regard to tide-gauge records, what is the appropriate balance to strike between retaining information and culling the data set to avoid systematic sources of error, and how might this data set be combined with others (e.g. GRACE gravity) in an extended fingerprint analysis? Once again, these are research questions that will be targets for study over the next decade.

The *essential conclusions* regarding surface mass loadings may be summarized as follows:

- uncertainties in mantle viscosity and Late Pleistocene ice mass histories have a significant impact on estimates of global sea-level rise from tide-gauge observations;
- differences between current models for both Antarctic and Greenland deglaciation histories are considerable and will influence interpretations of altimetry and time-variable gravity data;
- fingerprinting of sea-level data (e.g. for polar-ice melt signatures) offers promise of understanding mass balances given enhanced relative sea-level data sets.

Improved constraints on relative sea-level variations should be sought in all parts of the world. These relative sea-level data provide important feedback into all models of ice-mass change and sea-level mass change.

The main *forefront questions* for research during the coming decade should be aimed at:

- improving the constraints on the fingerprints associated with thermosteric effects and recent ice-mass flux;
- determining what other data sets can be invoked to limit the range of possible melting geometries, or to expand the range from the simple scenarios considered to date;
- how can the uncertainty in the GIA signal, which is still significant, be best parameterized within the fingerprint procedure?
- what are the important sources of noise, and which processes (e.g. terrestrial hydrological and sedimentary load redistributions) need to be included in an expanded regression analysis?
- in regard to tide-gauge records, what is the appropriate balance to strike between retaining information and culling the data set to avoid systematic sources of error, and how might this data set be combined with others (e.g. GRACE and GOCE gravity, altimetry) in an extended fingerprint analysis?

The most fruitful avenues for future *modeling* will include the following:

- truly three-dimensional Earth modeling that incorporates more realistic lateral variations in (possibly nonlinear) rheologies and lithosphere for regional applications when there is a clear necessity for it (Occam's razor);
- continued research into GIA-induced TPW;
- ice-sheet models that are internally dynamically consistent and that are consistent with all known geological, ice-core, sedimentary-core, trimline, cosmogenic rock exposure dates, and other data;
- non-ice-sheet-related surface forcings (tectonic, co- and postseismic, sedimentary, hydrological, changes in ocean currents, etc.);
- standardization of numerical models with clear instructions on how to use them (e.g. valid for global applications but not for regional ones, etc.).

One of the principal surface load responses that must be integrated into the interpretation of the sources of sea-level variability comes from GIA, and to this end, we may recommend the following emphasis on *data collection* and analysis:

- continuous GPS networks with sufficient density to determine uplift patterns as in the BIFROST project;
- follow-on GRACE gravity mission for continuity of the gravity data sets and for observing temporal variations at higher resolution;
- follow-on Jason mission *with higher-latitude coverage* for ocean surfaces, notably for fingerprinting, and a dedicated altimetry mission (e.g. the recently launched CryoSat-2 and an ICESat follow-on) for ice-mass surfaces;

- continuity of tide gauges and network densification in the Southern Hemisphere;
- expansion of the relative sea-level data sets from geological and archaeological indicators;
- in regions where rapid ongoing ice retreat occurs, the crustal motion response should also be examined using advanced interferometric synthetic aperture radar (InSAR) techniques.

Acknowledgments

Funding for this work has been provided by the Natural Sciences and Engineering Research Council of Canada, the Natural Environment Research Council of the United Kingdom, National Aeronautics and Space Administration (NASA; Earth Sciences, Solid Earth and Surface Processes Focus Area; Interdisciplinary Science Program, NNG04GL69G; and GRACE Science Team, NNG04GF09G), Delft University, The Netherlands, and the Antarctic and Climate Research Ecosystems Cooperative Research Centre, Hobart, Australia. Part of this work was performed at the Jet Propulsion Laboratory, California Institute of Technology.

References

Antonov J., Levitus S. and Boyer T.P. (2005) Thermosteric sea level rise, 1955–2003. *Geophysical Research Letters* **32**, L12602.

Arendt A.A., Echelmeyer K.A., Harrison W.D., Lingle C.S. and Valentine B. (2002) Rapid wastage of Alaska glaciers and their contribution to rising sea-level. *Science* **297**, 382–6.

Barclay D.J., Gregory C.W. and Calkin P.E. (2003) An 850 year record of climate and fluctuations of the iceberg-calving Nellie Juan Glacier, south central Alaska, U.S.A.. *Annals of Glaciology* **36**, 51–6.

Barletta V.R. and Sabadini R. (2006) Investigating superswells and sea-level changes caused by superplumes via normal mode relaxation theory. *Journal of Geophysical Research* **111**, B04404.

Benjamin D., Wahr J., Ray R., Egbert G. and Desai S.D. (2006) Constraints on mantle anelasticity from geodetic observations, and implications for the J2 anomaly. *Geophysical Journal International* **165**, 3–16.

Bentley M.J. (1999) Volume of Antarctic ice at the Last Glacial Maximum, and its impact on global sea-level change. *Quaternary Science Reviews* **18**, 1569–95.

Blum M.D. and Törnqvist T.E. (2000) Fluvial responses to climate and sea-level change: a review and look forward. *Sedimentology* **47**, 2–48.

Cazenave A. and Nerem R.S. (2004) Present-day sea level change: observations and causes. *Review of Geophysics* **42**, RG3001.

Cazenave A., Dominh K., Gennero M.C. and Ferret B. (1998) Global mean sea-level changes observed by TOPEX-Poseidon and ERS-1. *Physics and Chemistry of the Earth* **23**, 1069–75.

Chen J.L., Tapley B.D. and Wilson C.R. (2006a) Alaskan mountain glacial melting observed by satellite gravimetry. *Earth and Planetary Science Letters* **248**, 368–78.

Chen J.L., Wilson C.R. and Tapley B.D. (2006b) Satellite gravity measurements confirm accelerated melting of Greenland Ice Sheet. *Science* **313**, 1958–60.

Cheng M.K., Shum C.K. and Tapley B.D. (1997) Determination of long-term changes in the Earth's gravity field from satellite laser ranging observations. *Journal of Geophysical Research* **102**, 22377–90.

Church J.A., White N.J., Coleman R., Lambeck K. and Mitrovica, J.X. (2004) Estimates of the regional distribution of sea-level rise over the 1950 to 2000 period, *Journal of Climate* **17**, 2609–25.

Clague J.J. and James T.S. (2002) History and isostatic effects of the last ice sheet in southern British Columbia. *Quaternary Science Reviews* **21**, 71–87.

Clark J.A. and Lingle C.S. (1977) Future sea-level changes due to West Antarctic ice-sheet fluctuations. *Nature* **269**, 206–9.

Clark J.A. and Primus J.A. (1987) Sea-level changes resulting from future retreat of ice sheets: an effect of CO_2 warming of the climate. In: *Sea-Level Changes* (Tooley M.J. and Shennan I., eds), pp. 356–70. Institute of British Geographers, London.

Clark P.U., Mitrovica J.X., Milne G.A. and Tamisiea M.E. (2002) Sea-level fingerprinting as a direct test for the source of global meltwater pulse IA. *Science* **295**, 2438–41.

Conrad C. and Hager B.H. (1997) Spatial variations in the rate of sea-level rise caused by present-day melting of glaciers and ice sheets. *Geophysical Research Letters* **24**, 1503–6.

Cox C.M., Boy A. and Chao B.F. (2003) Time-variable gravity: using satellite-laser-ranging as a tool for observing long-term changes in the earth system. *Proceedings of the 13th International Workshop on Laser Ranging* (Noomen R., Klosko S., Noll C. and Pearlman M., eds), pp. 9–19.

Davis J.L. and Mitrovica J.X. (1996) Glacial isostatic adjustment and the anomalous tide gauge record of eastern North America, *Nature* **379**, 331–3.

Denton G.H. and Hughes T. (1981) *The Last Great Ice Sheets*. Wiley-Interscience, New York.

Denton G.H. and Hughes T. (2002) Reconstructing the Antarctic Ice Sheet at the Last Glacial Maximum. *Quaternary Science Reviews* **21**, 193–202.

Di Donato G., Mitrovica J.X., Sabadini R. and Vermeersen L.L.A. (2000) The influence of a ductile crustal zone on glacial isostatic adjustment: geodetic observables along the U.S. East Coast. *Geophysical Research Letters* **27**, 3017–20.

Douglas B.C. (1991) Global sea level rise. *Journal of Geophysical Research* **96**, 6981–92.

Douglas B.C. (1997) Global sea-rise: a redetermination. *Surveys in Geophysics* **18**, 279–92.

Dyurgerov M.B. and Meier M.F. (2005) *Glaciers and the Changing Earth System: a 2004 Snapshot.* Occasional Paper No. 58. Institute of Arctic and Alpine Research., University of Colorado, Boulder, CO. http://instaar.colorado.edu/other/occ_papers.html.

Farrell W.E. and Clark J.A. (1976) On postglacial sea level. *Geophysical Journal of the Royal Astronomical Society* **46**, 647–67.

Fleming K. and Lambeck K. (2004) Constraints on the Greenland Ice Sheet since the Last Glacial Maximum from sea-level observations and glacial-rebound models. *Quaternary Science Reviews* **23**, 1053–77.

Florineth D. and Schlüchter C. (2000) Alpine evidence for atmospheric circulation patterns in Europe during the Last Glacial Maximum. *Quaternary Research* **54**, 295–308.

Fogwill C.J., Bentley M.J., Sugden D.E., Kerr A.R. and Kubik P.W. (2004) Cosmogenic nuclides ^{10}Be and ^{26}Al imply limited Antarctic ice sheet thickening and low erosion in the Shackleton Range for >1 m.y. *Geology* **32**, 265–8.

Gasperini P., Da Forno G. and Boschi E. (1994) Linear or non-linear rheology in the Earth's mantle: the prevalence of power-law creep in the postglacial isostatic readjustment of Laurentia. *Geophysical Journal International* **157**, 1297–1302.

Gehrels W.R., Milne G.A., Kirby J.R., Patterson J.R. and Belknap D.F. (2004) Late Holocene sea-level changes and isostatic crustal movements in Atlantic Canada. *Quaternary International* **120**, 79–89.

Giunchi C. and Spada G. (2000) Postglacial rebound in a non-newtonian, spherical Earth. *Geophysical Research Letters* **27**, 2065–8.

Gross R.S. and Vondrák J. (1999) Astrometric and space-geodetic observations of polar wander. *Geophysical Research Letters* **26**, 2085–8.

Han D. and Wahr J. (1989) Post-glacial rebound analysis for a rotating Earth. In: *Slow Deformations and Transmission of Stress in the Earth.* (Cohen S. and Vanicek P., eds), pp. 1–6. AGU Monograph Series No. 49. American Geophysical Union, Washington DC.

Han S-C., Shum C.K., Bevis M., Ji C. and Kuo C-Y. (2006) Crustal dilatation observed by GRACE after the 2004 Sumatra-Andaman earthquake. *Science* **313**, 658–62.

Holgate S.J. and Woodworth P.L. (2004) Evidence for enhanced coastal sea level rise during the 1990s. *Geophysical Research Letters* **31**, L07305.

Horwath M. and Dietrich R. (2006) Errors of regional mass variations inferred from GRACE monthly solutions. *Geophysical Research Letters* **33**, L07502.

Huybrechts P., Gregory J., Janssens I. and Wild M. (2004) Modelling Antarctic and Greenland volume changes during the 20th and 21st centuries forced by GCM time slice integrations. *Global and Planetary Change* **42**, 83–105.

Ishii M., Kimoto M. and Kachi M. (2003) Historical ocean subsurface temperature analysis with error estimates. *Monthly Weather Review* **131**, 51–73.

Ivins E.R. and Sammis C.G. (1995) On lateral viscosity constrast in the mantle and the rheology of low-frequency geodynamics. *Geophysical Journal International* **123**, 305–22.

Ivins E.R. and James T.S. (2005) Antarctic glacial isostatic adjustment: a new assessment. *Antarctic Science* **17**, 541–53.

Ivins E.R., Sammis C.G. and Yoder C.F. (1993) Deep mantle viscous structure with prior estimate and satellite constraint. *Journal of Geophysical Research* **98**, 4579–4609.

Ivins E.R., Rignot E., Wu X-P., James T.S. and Casassa G. (2005) Ice mass balance and Antarctic gravity change: Satellite and terrestrial perspectives, In: *Earth Observation with CHAMP: Results with Three Years in Orbit* (Reigber C., Lühr H., Schwintzer P. and Wickert J., eds), pp. 3–12. Springer-Verlag, Heidelberg.

Ivins E.R., Dokka R.K. and Blom R.G. (2007) Post-glacial sediment load and subsidence in coastal Louisiana. *Geophysical Research Letters* **34**, L16303.

James T.S. and Ivins E.R. (1997) Global geodetic signatures of the Antarctic ice sheet. *Journal of Geophysical Research* **102**, 605–33.

James T.S. and Ivins E.R. (1998) Predictions of Antarctic crustal motions driven by present-day ice sheet evolution and by isostatic memory of the Last Glacial Maximum. *Journal of Geophysical Research* **103**, 4993–5017.

Johnston P. (1993) The effect of spatially non-uniform water loads on predictions of sea level change. *Geophysical Journal International* **114**, 615–34.

Johnston P. and Lambeck K. (1999) Postglacial rebound and sea level contributions to changes in the geoid and the Earth's rotation axis. *Geophysical Journal International* **136**, 537–58.

Kaufmann G. and Lambeck K. (2000) Mantle dynamics, postglacial rebound and the radial viscosity profile. *Physics of the Earth and Planetary Interiors* **121**, 301–24.

Kaufmann G., Wu P. and Ivins E.R. (2005) Lateral viscosity variations beneath Antarctic and their implications on regional rebound motions and seismotectonics. *Journal of Geodynamics* **39**, 165–81.

Kendall R.A., Mitrovica J.X. and Sabadini R. (2003) Lithospheric thickness inferred from Australian post-glacial sea-level change: the influence of a ductile crustal zone. *Geophysical Research Letters* **30**, 1461–4.

Kendall R.A, Mitrovica J.X. and Milne G.A. (2005) On post-glacial sea level: II. Numerical formulation and comparative results on spherically symmetric models. *Geophysical Journal International* **161**, 679–706.

Kendall R.A., Latychev K., Mitrovica J.X., Tamisiea M.E. and Davis J. (2006) Decontaminating tide gauge sea-level records for the influence of glacial isostatic adjustment: the potential impact of 3-D Earth structure. *Geophysical Research Letters* **33**, L01029.

Klemann V. and Wolf D. (1999) Implications of a ductile crustal layer for the deformation caused by the Fennoscandian ice sheet. *Geophysical Journal International* **139**, 216–26.

Klemann V. and Wolf D. (2005) The eustatic reduction of shoreline diagrams: implications for the inference of relaxation-rate spectra and the viscosity stratification below Fennoscandia, *Geophysical Journal International* **162**, 249–56.

Klemann V., Ivins E.R., Martinec Z. and Wolf D. (2007) Models of active glacial isostasy roofing warm subduction: case of the South Patagonia Ice Field. *Journal of Geophysical Research* **112**, B09405.

Klosko S. and Chao B. (1998) Secular variations of the zonal gravity field, global sea-level and polar motion as geophysical constraints. *Physics and Chemistry of the Earth* **23**, 1091–1102.

Kreemer C., Blewitt G., Hammond W.C. and Plag H-P. (2006) Global deformation from the great 2004 Sumatra-Andaman Earthquake observed by GPS: Implications for rupture process and global reference frame. *Earth Planets and Space* **58**, 141–8.

Lambeck K. (1993) Glacial rebound of the British Isles-II. A high resolution, high-precision model. *Geophysical Journal International* **115**, 960–90.

Lambeck K. (1995) Constraints on the Late Weichselian Ice Sheet over the Barents Sea from observations of raised shorelines. *Quatenary Science Reviews* **14**, 1–16.

Lambeck K. and Nakiboglu S.M. (1983) Long-period Love numbers and their frequency-dependence due to dispersion effects. *Geophysical Research Letters* **10**, 857–60.

Lambeck K., Smither C. and Ekman M. (1998a) Tests of glacial rebound models for Fennoscandinavia based on instrumented sea- and lake-level records. *Geophysical Journal International* **135**, 375–87.

Lambeck K., Smither C. and Johnston P. (1998b) Sea-level change, glacial rebound and mantle viscosity for northern Europe. *Geophysical Journal International* **134**, 102–44.

Lambeck K., Yokoyama Y. and Purcell A. (2002a) Into and out of the Last glacial Maximum Sea Level change during Oxygen Isotope Stages 3-2. *Quaternary Science Reviews* **21**, 343–60.

Lambeck K., Yokoyama Y., Purcell A. and Johnston P. (2002b) Reply to comment by W.R. Peltier, *Quaternary Science Reviews* **21**, 415–18.

Lambeck K., Purcell A., Johnston P., Nakada M. and Yokoyama Y. (2003) Water-load definition in the glacio-hydro-isostatic sea-level equation. *Quaternary Science Reviews* **22**, 309–18.

Lambeck K., Anzidei M., Antonioli F., Benini A. and Esposito E. (2004) Sea level in Roman time in the central Mediterranean and implications for modern sea level rise. *Earth and Planetary Science Letters* **224**, 563–75.

Lambeck K., Purcell A., Zhao J. and Svensson, N-O. (2010) The Scandinavian Ice Sheet: from MIS 4 to the end of the Last Glacial Maximum. *Boreas* **39**(2), 410–35.

Larsen C.F., Motyka R.J., Freymueller J.T., Echelmeyer K.A. and Ivins E.R. (2005) Rapid viscoelastic uplift in southeast Alaska caused by post-Little Ice Age glacial retreat. *Earth and Planetary Science Letters* **237**, 548–60.

Latychev K., Mitrovica J.X., Tamisiea M.E., Tromp J. and Moucha R. (2004) Influence of lithospheric thickness variations on 3-D crustal velocities due to glacial isostatic adjustment. *Geophysical Research Letters* **31**, L01304.

Latychev K., Mitrovica J.X., Tromp J., Tamisiea M.E., Komatitsch D. and Christara C.C. (2005a). Glacial isostatic adjustment on 3-D Earth models: a finite volume formulation. *Geophysical Journal International* **161**, 421–44.

Latychev K., Mitrovica J.X., Tamisiea M.E., Tromp J., Christara C. and Moucha R. (2005b). GIA-induced secular variations in the Earth's long wavelength gravity field: Influence of 3-D viscosity variations. *Earth and Planetary Science Letters* **240**, 322–7.

Leuliette E.W, Nerem R.S. and Mitchum G.T. (2004) Calibration of TOPEX/Poseidon and Jason altimeter data to construct a continuous record of mean sea level change. *Marine Geodesy* **27**, 79–94.

Levitus S., Antonov J.I., Boyer T.P. and Stephens C. (2000) Warming of the world ocean. *Science* **287**, 2225–9.

Licht K.J. (2004) The Ross Sea's contribution to eustatic sea-level during meltwater pulse 1A. *Sedimentary Geology* **165**, 343–53.

Lonne I. and Lysa A. (2005) Deglaciation dynamics following the Little Ice Age on Svalbard: Implications for shaping of landscapes at high latitudes. *Geomorphology* **72**, 300–19.

Luthcke S.B., Zwally H.J., Abdalati W., Rowlands D.D., Ray R.D., Nerem R.S., Lemoine F.G., McCarthy J.J. and Chinn D.S. (2006) Recent Greenland ice mass loss by drainage system from satellite gravity observations. *Science* **314**, 1286–9.

Mangerud J., Astakhov V. and Svendsen J.I. (2002) The extent of the Barents-Kara Ice Sheet during the Last Glacial Maximum. *Quaternary Science Reviews* **21**, 111–19.

Martinec Z. (2000) Spectral-finite element approach to three-dimensional viscoelastic relaxation in a spherical earth. *Geophysical Journal International* **142**, 117–41.

Martinec Z. and Hagedoorn J. (2005) Time-domain approach to linearized rotational response of a three-dimensional viscoelastic earth model induced by glacial-isostatic adjustment: I. Inertia-tensor perturbations. *Geophysical Journal International* **163**, 443–65.

Martinec Z. and Wolf D. (2005) Inverting the Fennoscandian relaxation-time spectrum in terms of an axisymmetric viscosity distribution with a lithospheric root. *Journal of Geodynamics* **39**, 143–63.

Matsuyama I., Mitrovica J.X., Daradich A., and Gomez N. (2010) The rotational stability of a triaxial ice-age Earth. *Journal of Geophysical Research* **115**, B05401.

McCarthy D.D. and Luzum B.J. (1996) Path of the mean rotation pole from 1899 to 1994. *Geophysical Journal International* **125**, 623–9.

Meier M.F. (1984) Contribution of small glaciers to global sea level. *Science* **226**, 1418–21.

Melini D. and Piersanti A. (2006) Impact of global seismicity on sea level change assessment. *Journal of Geophysical Research* **111**, B03406.

Meltzner A.J., Sieh K., Abrams M., Agnew D.C., Hudnut K.W., Avouac J.P. and Natawidjaja D.H. (2006) Uplift and subsidence associated with the great Aceh-Andaman earthquake of 2004. *Journal of Geophysical Research* **111**, B02407.

Menard H.W. (1978) Fragmentation of the Farallon plate by pivoting subduction. *Journal of Geololgy* **86**, 99–110.

Milne G.A. (1998) *Refining Models of the Glacial Isostatic Adjustment Process*. PhD Thesis, University of Toronto, Toronto.

Milne G.A. and Mitrovica J.X. (1996) Postglacial sea-level change on a rotating Earth: first results from a gravitationally self-consistent sea-level equation. *Geophysical Journal International* **126**, F13–20.

Milne G.A. and Mitrovica J.X. (1998) Postglacial sea-level change on a rotating *Earth. Geophysical Journal International* **133**, 1–10.

Milne G.A., Mitrovica J.X. and Davis J.L. (1999) Near-field hydro-isostasy: the implementation of a revised sea-level equation. *Geophysical Journal International* **139**, 464–82.

Milne G.A., Davis J.L., Mitrovica J.X., Scherneck H.-G., Johansson J.M., Vermeer M. and Koivula H. (2001) Space-geodetic constraints on glacial isostatic adjustment in Fennoscandia. *Science* **291**, 2381–5.

Mitrovica, J.X. (1996) Haskell [1935] revisited. *Journal of Geophysical Research* **101**, 555–69.

Mitrovica J.X. (2003) Recent controversies in predicting post-glacial sea-level change. *Quaternary Science Reviews* **22**, 127–33.

Mitrovica J.X. and Peltier W.R. (1989) Pleistocene deglaciation and the global gravity field. *Journal of Geophysical Research* **96**, 13651–71.

Mitrovica J.X. and Peltier W.R. (1991) On post-glacial geoid subsidence over the equatorial oceans. *Journal of Geophysical Research* **96**, 20053–71.

Mitrovica J.X. and Peltier W.R. (1993) Present-day secular variations in the zonal harmonics of the Earth's geopotential. *Journal of Geophysical Research* **98**, 4509–26.

Mitrovica J.X. and Davis J.L. (1995) Present-day post-glacial sea level change far from the Late Pleistocene ice sheets: Implications for recent analyses of tide gauge records. *Geophysical Research Letters* **22**, 2529–32.

Mitrovica J.X. and Forte A.M. (1997) Radial profile of mantle viscosity: Results from the joint inversion of convection and postglacial rebound observables. *Journal of Geophysical Research* **102**, 2751–69.

Mitrovica J.X. and Milne G.A. (2003) On post-glacial sea level: I. General theory, *Geophysical Journal International* **154**, 253–67.

Mitrovica J.X. and Forte A.M. (2004) A new inference of mantle viscosity based upon joint inversion of convection and glacial isostatic adjustment data. *Earth and Planetary Science Letters* **225**, 177–89.

Mitrovica J.X., Beaumont C. and Jarvis G.T. (1989) Tilting of continental interiors by the dynamical effects of subduction. *Tectonics* **8**, 1079–94.

Mitrovica J.X., Tamisiea M.E., Milne G.A. and Davis J.L. (2001a) Recent mass balance of polar ice sheets inferred from patterns of global sea-level change. *Nature* **409**, 1026–9.

Mitrovica J.X., Milne G.A. and Davis J.L. (2001b) Glacial isostatic adjustment on a rotating Earth. *Geophysical Journal International* **147**, 562–79.

Mitrovica J.X., Wahr J., Matsuyama I. and Paulson A. (2005) The rotational stability of an ice-age Earth. *Geophysical Journal International* **161**, 491–506.

Mitrovica J.X., Wahr J., Matsuyama I., Paulson A. and Tamisiea M.E. (2006) Reanalysis of ancient eclipse, astronomic and geodetic data: a possible route to resolving the enigma of global sea-level rise. *Earth and Planetary Science Letters* **243**, 390–9.

Monnereau M. and Yuen D.A. (2002) How flat is the lower-mantle temperature gradient? *Earth and Planetary Science Letters* **202**, 171–83.

Motyka R.J. (2003) Little Ice Age subsidence and Post Little Ice Age uplift at Juneau, Alaska inferred from dendrochronology and geomorphology. *Quaternary Research* **59**, 300–9.

Munk W. (2002) Twentieth century sea level: an enigma. *Proceedings of the National Academy of Sciences USA* **99**, 6550–5.

Nakada M. (2009) Polar wander of the Earth associated with the Quaternary glacial cycles on a convecting mantle. *Geophysical Journal International* **179**, 569–78.

Nakada M. and Lambeck K. (1988) The melting history of the late Pleistocene Antarctic ice sheet. *Nature* **333**, 36–40.

Nakada M. and Lambeck K. (1989) Late Pleistocene and Holocene sea-level change in the Australian region and mantle rheology. *Geophysical Journal of the Royal Astronomical Society* **96**, 497–517.

Nakada M. and Lambeck K. (1991) Late Pleistocene and Holocene sea-level change: evidence for lateral mantle viscosity structure? In: *Glacial Isostasy, Sea-Level and Mantle Rheology* (Sabadini R., Lambeck K. and Boschi E., eds), pp. 79–94. Kluwer, Dordrecht.

Nakada M., Kimuru R., Okuno J., Moriwaki K., Miura H. and Maemoku H. (2000) Late Pleistocene and Holocene melting history of the Antarctic ice sheet derived from sea-level variations. *Marine Geology* **167**, 85–103.

Nakiboglu S.M. (1982) Hydrostatic theory of the Earth and its mechanical implications. *Physics of the Earth and Planetary Interiors* **28**, 302–11.

Nakiboglu S.M. and Lambeck K. (1981) Deglaciation related features of the Earth's gravity field. *Tectonophysics* **72**, 289–303.

Nakiboglu S.M. and Lambeck K. (1991) Secular sea level change. In: *Glacial Isostasy, Sea Level and Mantle Rheology* (Sabadini R., Lambeck K. and Boschi E., eds), pp. 237–58. Kluwer Academic Publishers, Dordrecht.

Nerem R.S. and Klosko S.M. (1996) Secular variations of the zonal harmonics and polar motion as geophysical constraints. In: *Global Gravity Field and its Variations, International Association of Geodesy Symposium 116* (Rapp R., Cazenave A. and Nerem R.S., eds), pp. 152–63. Springer Verlag, Berlin.

Nerem R.S. and Mitchum G.T. (2001) Observations of sea level change from satellite altimetry. In: *Sea Level Rise: History and Consequences* (Douglas B.C., Kearney M.S. and Leatherman S.P., eds), pp. 121–63. Academic Press, London.

Paulson A., Zhong S. and Wahr J. (2005) Modelling post-glacial rebound with lateral viscosity variations. *Geophysical Journal International* **163**, 357–71.

Paulson A., Zhong, S. and Wahr J. (2007) Inference of mantle viscosity from GRACE and relative sea level data. *Geophyical Journal International* **171**, 497–508.

Peltier W.R. (1974) The impulse response of a Maxwell Earth. *Reviews of Geophysics* **12**, 649–69.

Peltier W.R. (1985) The LAGEOS constraint on deep mantle viscosity: Results from a new normal mode method for the inversion of visco-elastic relaxation spectra. *Journal of Geophysical Research* **90**, 9411–21.

Peltier W.R. (1994) Ice age paleotopography, *Science* **265**, 195–201.

Peltier W.R. (1998a) "Implicit ice" in the global theory of glacial isostatic adjustment. *Geophysical Research Letters* **25**, 3955–8.

Peltier W.R. (1998b) Postglacial variations in the level of the sea: Implications for climate dynamics and solid-Earth geophysics. *Reviews of Geophysics* **36**, 603–89.

Peltier W.R. (2001) Global glacial isostatic adjustment and modern instrumental records of relative sea level history. In: *Sea Level Rise: History and Consequences* (Douglas B.C., Kearney M.S. and Leatherman S.P., eds), pp. 61–95. Academic Press, London.

Peltier W.R. (2002) Comments on the paper of Yokoyama et al. (2000), entitled "Timing of the Last Glacial Maximum from observed sea level minima". *Quaternary Science Reviews* **21**, 409–14.

Peltier W.R. (2004) Global glacial isostasy and the surface of the ice-age Earth: The ICE-5G (VM2) model and GRACE. *Annual Review of Earth and Planetary Sciences* **32**, 111–49.

Peltier W.R. and Tushingham A.M. (1989) Global sea-level rise and the greenhouse effect: Might they be connected? *Science* **244**, 806–10.

Peltier W.R. and Tushingham A.M. (1991) Influence of glacial isostatic adjustment on tide gauge measurements of secular sea level change. *Journal of Geophysical Research* **96**, 6779–96.

Peltier W.R. and Drummond R. (2002) A "broad-shelf effect" upon postglacial relative sea level history. *Geophysical Research Letters* **29**, 1169.

Plag H.-P. (2006) Recent relative sea-level trends: an attempt to quantify the forcing factors. *Philosophical Transactions of the Royal Society of London A* **364**, 821–44.

Plag H.-P. and Jüttner H.-U. (2001) Inversion of global tide gauge data for present-day ice load changes. In: *Proceedings of the Second International Symposium on Environmental Research in the Arctic and Fifth Ny-Alesund Scientific Seminar* (Yamanouchi T., ed.), pp. 301–17. Memoirs of the National Institute of Polar Research, Special Issue 54.

Ramillien G., Lombard A., Cazenave A., Ivins E., Remy F. and Biancale R. (2006) Interannual variations of the mass balance of the Antarctic and Greenland ice sheets from GRACE. *Global and Planetary Change* **53**, 198–208.

Ranalli G. (2001) Mantle rheology: radial and lateral viscosity variations inferred from microphysical creep laws. *Journal of Geodynamics* **32**(4–5), 425–44.

Rignot E. and Thomas R.H. (2002) Mass balance of polar ice sheets. *Science* **297**(5586), 1502–6.

Rignot E., Rivera A. and Casassa G. (2003) Contribution of the Patagonia Icefields of South America to sea level rise. *Science* **302**(5644), 434–7.

Rignot E., Casassa G., Gogineni P., Krabill W., Rivera A. and Thomas R. (2004) Accelerated ice discharge from the Antarctic Peninsula following the collapse of Larsen B ice shelf. *Geophysical Research Letters* **31**(18), L18401.

Savage J.C. and Thatcher W. (1992) Interseismic deformation at the Nankai Trough, Japan, Subduction Zone. *Journal of Geophysical Research* **97**, 11117–35.

Schotman H.H.A. and Vermeersen L.L.A. (2005) Sensitivity of glacial isostatic adjustment models with shallow low-viscosity earth layers to the ice-load history in relation to the performance of GOCE and GRACE. *Earth and Planetary Science Letters* **236**, 828–44.

Shennan I., Peltier W.R., Drummond R. and Horton B.P. (2002) Global to local scale parameters determining relative sea-level changes and the post-glacial isostatic adjustment of Great Britain. *Quaternary Science Reviews* **21**, 397–408.

Simms A.R., Lambeck K., Purcell A., Anderson J.B. and Rodriguez A.B. (2007) Sea-Level history of the Gulf of Mexico since the Last Glacial Maximum with implications for the melting history of the Laurentide Ice Sheet. *Quaternary Science Reviews* **26**, 920–40.

Steele J.M. (2005) Ptolomy, Babylon and the rotation of the Earth. *Astronomy and Geophysics* **46**, 5.11–5.15.

Steinberger B. and O'Connell R.J. (1997) Changes of the Earth's rotation axis owing to advection of mantle density heterogeneities. *Nature* **387**, 169–73.

Stephenson F.R. and Morrison L.V. (1995) Long-term fluctuations in the Earth's rotation: 700 BC to AD 1990. *Philosophical Transactions of the Royal Society of London A* **351**, 165–202.

Stocchi P., Spada G. and Cianetti S. (2005) Isostatic rebound following the Alpine deglaciation: impact on the sea level variations and vertical movements in the Mediterranean region. *Geophysical Journal International* **162**, 137–47.

Sugden D.E., Hulton N.R.J. and Purves R.S. (2002) Modelling the inception of the Patagonian ice sheet. *Quaternary International* **95–96**, 55–64.

Sun W.-K. (2004) Short note: asymptotic theory for calculating deformations caused by dislocations buried in a spherical earth – gravity change. *Journal of Geodesy* **78**, 76–81.

Sun W.-K. and Okubo S. (2006) Temporal gravity changes due to earthquakes and volcanoes – recent advances in theoretical and observational studies in Japan. *EOS Transactions of the American Geophysical Union* **87**(36), Joint Assembly Supplement, abstract G44A-03.

Svendsen J., Alexanderson H., Astakhov V., Demidov I., Dowdeswell J., Funder S. et al. (2004) Late Quaternary ice sheet history of Northern Eurasia. *Quaternary Science Reviews* **23**, 1229–71.

Tamisiea M.E., Mitrovica J.X., Milne G.A. and Davis J.L. (2001) Global geoid and sea level changes due to present-day ice mass fluctuations. *Journal of Geophysical Research* **106**, 30849–63.

Tamisiea M.E., Mitrovica J.X. and Davis J.L. (2003) A method for detecting rapid mass flux of small glaciers using local sea level variations. *Earth and Planetary Science Letters* **213**, 477–85.

Tamisiea M.E., Leuliette E., Davis J.L. and Mitrovica J.X. (2005) Constraining hydrological and cryospheric mass flux in southeastern Alaska using space-based gravity measurements. *Geophysical Research Letters* **32**, L20501.

Tamisiea M.E., Mitrovica J.X. and Davis J.L. (2007) GRACE gravity data constrain ancient ice geometries and continental dynamics over Laurentia. *Science* **316**(5826), 881–3.

Tapley B.D., Bettadpur S., Ries J.C., Thompson P.F. and Watkins M.M. (2004) GRACE measurements of mass variability in the Earth system. *Science* **305**, 503–5.

Tarasov L. and Peltier W.R. (2002) Greenland glacial history and local geodynamic consequences. *Geophysical Journal International* **150**, 198–229.

Tosi N., Sabadini R., Marotta A.M. and Vermeersen L.L.A. (2005) Simultaneous inversion for the Earth's mantle viscosity and ice mass imbalance in Antarctica and Greenland. *Journal of Geophysical Research* **110**, B07402.

Tromp J. and Mitrovica J.X. (1999) Surface loading of a viscoelastic planet-I. General theory. *Geophysical Journal International* **137**, 847–55.

Trupin A.S. and Wahr J.M. (1990) Spectroscopic analysis of global tide gauge sea level data. *Geophysical Journal International* **100**, 441–53.

Tushingham A.M. and Peltier W.R. (1991) Ice-3G: A new global model of late Pleistocene deglaciation based upon geophysical predictions of postglacial relative sea level. *Journal of Geophysical Research* **96**, 4497–4523.

Van der Wal W., Schotman H.H.A. and Vermeersen L.L.A. (2004) Geoid heights due to a crustal low viscosity zone in glacial isostatic adjustment modeling: a sensitivity analysis for GOCE. *Geophysical Research Letters* **31**, L05608.

Velicogna, I. (2009) Increasing rates of ice mass loss from the Greenland and Antarctic ice sheets revealed by GRACE. *Geophysical Research Letters* **36**, L19503.

Velicogna I. and Wahr J. (2005) Ice mass balance in Greenland from GRACE. *Journal of Geophysical Research* **32**, L18505.

Velicogna I. and Wahr J. (2006a). Measurements of time-variable gravity show mass loss in Antarctica. *Science* **311**, 1754–6.

Velicogna I. and Wahr J. (2006b) Acceleration of Greenland Ice Mass Loss in Spring 2004. *Nature* **443**, 329–31.

Vermeersen L.L.A., Sabadini R., Spada G. and Vlaar N.J. (1994) Mountain building and Earth rotation. *Geophysical Journal International* **117**, 610–24.

Vigny C., Simons W.J.F., Abu S., Bamphenyu R., Satirapod C., Choosakul N. et al. (2005) Insight into the 2004 Sumatra-Andaman earthquake from GPS measurements in southeast Asia. *Nature* **436**, 201–6.

Waddington E.D., Conway H, Steig E.J., Alley R.B., Brook E.J., Taylor K. and White J.W.C. (2005) Decoding the dipstick: thickness of Siple Dome, West Antarctica, at the Last Glacial Maximum. *Geology* **33**, 281–4.

Wang H.S. and Wu P. (2006) Effects of lateral variations in lithospheric thickness and mantle viscosity on glacially induced surface motion on a spherical, self-gravitating Maxwell Earth. *Earth and Planetary Science Letters* **244**, 576–89.

White N.J., Church J.A. and Gregory J.M. (2005) Coastal and global averaged sea level rise for 1950 to 2000. *Geophysical Research Letters* **32**, L01061.

Whitehouse P.L., Latychev K., Milne G.A., Mitrovica J.X. and Kendall R. (2005) The impact of 3-D Earth structure on Fennoscandian glacial isostatic adjustment: Implications for space-geodetic estimates of present-day crustal deformations. *Geophysical Research Letters* **33**, L01029.

Whitehouse P.L., Allen M.B. and Milne G.A. (2007) Glacial isostatic adjustment as a control on coastal processes: an example from the Siberian Arctic. *Geology* **35**, 747–50.

Wieczerkowski K, Mitrovica J.X. and Wolf D. (1999) A revised relaxation time spectrum for Fennoscandia. *Geophysical Journal International* **139**, 69–86.

Willis J.K., Roemmich D. and Cornuelle B. (2004) Interannual variability in upper ocean heat content, temperature, and thermosteric expansion on global scales. *Journal of Geophysical Research* **109**, C12036.

Wolf D., Klemann V., Wünsch J. and Zhang F-P. (2006) A reanalysis and reinterpretation of geodetic and geologic evidence of glacial-isostatic adjustment in the Churchill region, Hudson Bay. *Surveys in Geophysics* **27**, 19–61.

Woodward R.S. (1888) On the form and position of mean sea level. *US Geological Survey Bulletin* **48**, 87–170.

Woodworth P.L. and Player R. (2003) The Permanent Service for Mean Sea Level: an update to the 21st century. *Journal of Coastal Research* **19**, 287–95.

Woodworth P.L., Tsimplis M.N., Flather R.A. and Shennan I. (1999) A review of the trends observed in British Isles mean sea level data measured by tide gauges. *Geophysical Journal International* **136**, 651–70.

Wu P. (1978) *The Response of a Maxwell Earth to Applied Surface Mass Loads: Glacial Isostatic Adjustment.* MSc Thesis, University of Toronto, Toronto.

Wu P. (2002) Effects of nonlinear rheology on degree 2 harmonic deformation in a spherical self-gravitating earth. *Geophysical Research Letters* **29**, 1198.

Wu P. (2006) Sensitivity of relative sea levels and crustal velocities in Laurentide to radial and lateral viscosity variations in the mantle. *Geophysical Journal International* **165**, 401–13.

Wu P. and Peltier W.R. (1982) Viscous gravitational relaxation. *Geophysical Journal of the Royal Astronomical Society* **70**, 435–85.

Wu P. and Peltier W.R. (1983) Glacial isostatic adjustment and the free air gravity anomaly as a constraint on deep mantle viscosity. *Geophysical Journal of the Royal Astronomical Society* **74**, 377–449.

Wu P. and Peltier W.R. (1984) Pleistocene deglaciation and the Earth's rotation: A new analysis. *Geophysical Journal of the Royal Astronomical Society* **76**, 753–91.

Wu P. and van der Wal W. (2003) Postglacial sea levels on a spherical, self-gravitating viscoelastic Earth: effects of lateral viscosity variations in the upper mantle on the inference of viscosity contrasts in the lower mantle. *Earth and Planetary Science Letters* **211**, 57–68.

Wu P., Wang H. and Schotman H. (2005) Postglacial induced surface motions, sea-levels and geoid rates on a spherical, self-gravitating laterally homogeneous earth. *Journal of Geodynamics* **39**, 127–42.

Yang S.L., Belkin I.M., Belkina A.I., Zhao Q.Y., Zhu J. and Ding P.X. (2003) Delta response to decline in sediment supply from the Yangtze River: evidence of the recent four decades and expectations for the next half-century. *Estuarine, Coastal and Shelf Science* **57**, 689–99.

Yoder C.F., Williams J.G., Dickey J.O., Schutz B.E., Eanes R.J. and Tapley B.D. (1983) Secular variation of Earth's gravitational harmonic J2 coefficient from Lageos and nontidal acceleration of Earth rotation. *Nature* **303**, 757–62.

Yokoyama Y., Lambeck K., De Deckker P., Johnston P. and Fifield K. (2000) Timing of the Last Glacial Maximum from observed sea-level minima. *Nature* **406**, 713–16.

Zhong S., Paulson A. and Wahr J. (2003) Three-dimensional finite element modeling of Earth's viscoelastic deformation: effects of lateral variations in lithospheric thickness. *Geophysical Journal International* **155**, 679–95.

11 Past and Future Changes in Extreme Sea Levels and Waves

Jason A. Lowe, Philip L. Woodworth, Tom Knutson, Ruth E. McDonald, Kathleen L. McInnes, Katja Woth, Hans von Storch, Judith Wolf, Val Swail, Natacha B. Bernier, Sergey Gulev, Kevin J. Horsburgh, Alakkat S. Unnikrishnan, John R. Hunter, and Ralf Weisse

11.1 Introduction

Coastal impacts of sea-level change can result from individual extreme sea-level and wave events, or long-term fluctuations in mean sea level, or most likely from a combination of processes. An example of a combined impact is the damage caused by Hurricane Katrina at New Orleans, which resulted in unprecedented storm-surge levels and failure of coastal defenses. This was compounded by the rate of local mean sea-level rise relative to the land level of the Mississippi Delta of several times the global average, as occurs naturally in all major deltas, together with anthropogenic changes to the delta wetlands. On much longer timescales, extremes and mean-sea-level change are both major factors in determining coastal evolution including the development of coastal ecosystems.

It will be seen below that, although it is difficult to determine how mean sea level has changed in the past and will change in the future and to determine the reasons for change (the main topics of this volume), the very nature of extreme events makes estimation of future extreme levels a more difficult task. However, for many practical purposes, the study of extremes is far more important than that of mean sea level alone. Extremes often result in loss of life and great damage to infrastructure and the environment, and knowledge of their historical, and potential future, amplitudes and frequencies determines the scale of resources required for adaptation and coastal protection (see Figure 11.1).

Understanding Sea-Level Rise and Variability, 1st edition. Edited by John A. Church, Philip L. Woodworth, Thorkild Aarup & W. Stanley Wilson.

Figure 11.1 The flooding at Sea Palling, Norfolk, UK, due to the storm surge of January 31–February 1, 1953. This storm surge led to major investment in coastal protection along the east coast of the UK and in the Netherlands. (Picture courtesy of Eastern Daily Press.)

This chapter discusses changes in extreme sea levels and waves and is divided into four parts. First, we review changes in extreme sea levels and waves in the recent past. Then we discuss changes in the atmospheric storm events that drive extreme sea-level changes. There follows a review of recent advances in the modeling of future extreme events. (The reader is referred to the list of abbreviations and acronyms at the front of the book for models mentioned in the text.) The European shelf, Bay of Bengal and Australian regions have been investigated in greater detail than most other areas, and are selected for this section as special case studies of future change. Finally, we highlight issues that we believe need to be addressed in order to further understand the changes of the past and better predict those of the future.

11.2 Evidence for Changes in Extreme Sea Levels and Waves in the Recent Past

11.2.1 Past Changes in Extreme Sea Levels

As extreme events often result in flooding and loss of life, an important question is whether their amplitudes and frequencies are changing, and if the levels of extreme high waters are changing in a significantly different way to mean sea

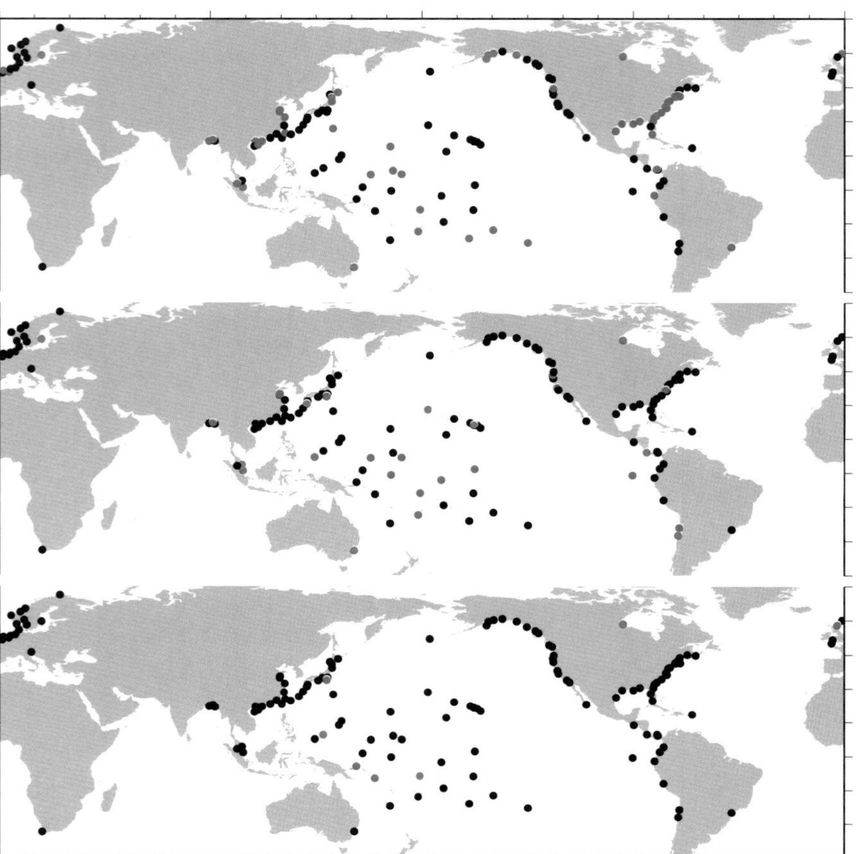

Figure 11.2 Distribution of tide-gauge stations selected for time-series analysis. Top: stations with observed trends in 99th percentile since 1975 significantly different from zero are shown in red (positive trend) or blue (negative trend) while others are shown in black. Middle: as before but with 99th percentile time series reduced to medians. Bottom: as before but with 99th percentile time series reduced to medians and with the tidal contributions to the percentiles removed. (From Woodworth and Blackman 2004. © American Meteorological Society.)

levels. The only study which has attempted a quasi-global investigation of this topic is that of Woodworth and Blackman (2004), who studied data from 141 stations, and concluded that there is indeed evidence for an increase in extreme high water levels worldwide since 1975, as reported frequently in the press. However, in most cases, the secular changes and the interannual variability in the extremes were found to be similar to those in mean sea level (i.e. compare top and middle maps of Figure 11.2). Zhang et al. (1997a, 2000) obtained a similar conclusion with regard to the rate of rise in the level of extremes and mean sea level along the US east coast, and considered that there had been no discernible long-term secular trend in storm-surge activity or severity during the past century.

A number of other studies of sea-level extremes at particular locations are available, although as they are for different epochs and use different methods, it is difficult to arrive at general conclusions. One of the longest data sets studied is that from Liverpool, UK (1768 onwards), from which Woodworth and Blackman (2002) observed annual maximum surge at high water to have been larger in the late-18th, late-19th, and late-20th centuries than for most of the 20th century, qualitatively consistent with knowledge of the regional variation in storminess.

(We use here the terms "surge" and "nontidal variation" almost interchangeably as the meteorologically-induced storm surge is responsible for much of the nontidal variability at mid and high latitudes). The Liverpool record is one of the few time series of extremes which is of comparable length to the corresponding mean-sea-level record (Woodworth 1999). Vassie, reported in Pugh and Maul (1999), concluded that there was no discernible trend over the last century in the statistics of nontidal sea-level variability around the UK above the considerable natural sea-level variability on decadal timescales. Other European studies include those of Bouligand and Pirazzoli (1999) and Pirazzoli (2000), who detected evidence for a slight decrease in the main factors contributing to surge development on the French Atlantic coast in the last half century. On the other hand, Pirazzoli et al. (2006) concluded that the medium-term (recent decades) coastal flood risk from high waters had increased on the English side of the Channel and less so on the French coast. Aráujo et al. (2002) computed trends in nontidal variability at six Channel sites. At Brest and Newlyn (120 and 84 years of data respectively) small but significant trends in extreme sea level were found. However, the authors cautioned that part of the observed trends may be due to changes in technology from float to bubbler pressure gauges.

In the Mediterranean, Raicich (2003) found no evidence for a trend in weak and moderate surges at Trieste over the period 1939–2001, while the frequency of both strong positive and negative surges decreased (see also Trigo and Davies 2002). Pirazzoli and Tomasin (2002) and Lionello (2005) determined that although Venice had experienced floods in the historic past, the frequency and intensity of floods had increased in the recent past, with positive trends in extreme surges due to changes in regional climate largely responsible for the observed changes. Ullmann et al. (2007) concluded that maximum annual sea levels had risen twice as fast as mean sea level during the 20th century (4 rather than 2 mm/year) in the Camargue (Rhone Delta) region of southern France, largely due to changes in the wind field in recent decades.

Bromirski et al. (2003) undertook a study of nontidal residuals at San Francisco since 1858, concluding that winter residuals had exhibited a significant increasing trend since about 1950. In one of the few studies from South America, D'Onofrio et al. (1999, 2008) observed a trend of extreme levels at Buenos Aires since 1905 similar to that in mean sea level. Church et al. (2006) provided examples of changes in extremes at a number of Australian locations over extended periods, and showed that high waters of the two longest records (Fort Denison in Sydney, New South Wales, and Fremantle, Western Australia) have risen during the 20th century faster than the median sea level.

Differences between observed sea levels and those expected from the tide can arise from a wide range of ocean processes in addition to storm surges (Pugh 1987). Firing and Merrifield (2004) studied extreme sea levels due to ocean eddy activity rather than storm surges. They found long-term increases in the number and height of extreme daily mean sea-level values at Honolulu, if levels were measured relative to a fixed datum (the highest ever value being due to an anti-cyclonic eddy system in 2003) but no evidence for an increase relative to the underlying upward trend in mean sea level.

Altogether, we conclude that there is little evidence for extreme sea levels changing over an extended period by amounts significantly different to mean sea level at most locations. This is an important finding with regard to coastal impact studies implying that the major uncertainties in the future projection of extremes are likely equivalent to those in mean sea level, at least over the next few decades. However, further into the future, changes in atmospheric storminess may also start to become important at some locations (see section 11.3 and Lowe and Gregory 2005).

Observational Considerations

Two technical aspects of the study of past extremes must be mentioned. The first is that research into extremes is far more difficult than into mean sea-level changes, owing to problems of access to raw sea-level data (e.g. hourly values). Most countries now make their data freely available for research, although the multiplicity of data formats and sampling frequencies means that a major effort in data processing is often required before scientific analysis can begin. Gradually, these difficulties are being addressed through international programs such as the Global Sea Level Observing System (GLOSS) and the Global Climate Observing System (GCOS). However, some countries continue to restrict access to raw data for reasons of cost recovery or national security. The result is a lack of century-timescale time series for analysis, especially in the Southern Hemisphere.

The second technical aspect is related to the clear identification of an extreme sea level. As high a recording frequency as possible is clearly required for the proper identification of an extreme (e.g. 6 or 15 min rather than the conventional hourly sampling) but in practice one often does not record the true extreme value due to the regular sampling. In addition, unusually high levels are sometimes outside the instrumental recording range (e.g. as for acoustic gauges during Hurricane Katrina, New Orleans, in 2005). There are also concerns about degradation of data quality at the limits of the range. In some circumstances, simple human errors can enter the historical data set and distort an extreme analysis (e.g. an error in handwritten tabulations of observed high waters). The study of the 99th or 99.9th percentile water-level values, instead of the high waters themselves (or 100th percentiles), can guard against errors in the recording of extremes to some extent (Woodworth and Blackman 2004; von Storch and Reichardt 1997).

A particular problem in defining an extreme arises when one wishes to investigate data from before the era of the automatic tide gauge. For study of the very longest records, extending back to the 18th or early 19th centuries, in which high waters were recorded rather than the full tidal curve, one is forced to use parameters such as annual maximum high water, annual maximum surge at high water, or surge at annual maximum high water, rather than annual maximum surge, which is clearly of greatest interest for climate research. Woodworth and Blackman (2002) discussed the relative merits of each parameter.

An important point with regard to extremes relates to the role of the tide and the fact that, in most regions with predominantly semi-diurnal tide, there will be perigean (quasi 4.5-year) variations in high- and low-water levels which have a similar period to El Niño variability. This may have contributed to confusion as to the reasons for flooding in some parts of the world. Locations where anecdotal evidence for sea-level rise, based on evidence of extreme sea levels and associated flooding, is particularly strong include low-lying Pacific islands such as Tuvalu (Hunter 2002, 2004; Woodworth and Blackman 2004). However, it is difficult to separate reasons for extremes and flooding (e.g. tide, surge, El Niño, global sea-level rise) simply from anecdotal evidence and short observational records. Of course, perigean spring tides in combination with storm surges are well known to often result in flooding and erosion in other parts of the world (e.g. northeast US coast; Wood 1978). At some locations (e.g. Hamburg), there is evidence that storm surges have been elevated by human-made water works, such as changes made to coastal defenses. This must be considered when interpreting historical records and simulations of past surge changes (von Storch and Woth 2008).

Modeling Considerations

The success of operational tide-surge modeling in many regions (Flather 2000), and the confidence acquired in the ability of models to describe observed surge events, has led to the use of barotropic models (two-dimensional depth-averaged models) for construction of long time series of historic water levels. The main requirements in such work are adequate spatial model resolution, accurate bathymetry, and reliable meteorological data sets of high spatial and temporal resolution. Examples of such work are by Flather et al. (1998), Langenberg et al. (1999), and Weisse and Plüß (2006), who undertook separate simulations of almost half a century of water levels for the North Sea area. Wakelin et al. (2003) and Tsimplis et al. (2005) used the Flather data set to investigate the correlation of the North Atlantic Oscillation (NAO) with mean-sea-level variability around the North Sea, while Woodworth et al. (2006) subsequently demonstrated a similar (if not identical) spatial dependence of correlations with the NAO of high and low waters and mean sea level. Langenberg et al. (1999) showed that positive trends in high percentiles (high waters) of winter sea levels during 1955–93 were largest in the German Bight, with only small trends along the Dutch and English coasts, while subtraction of winter mean values considerably reduced the larger trends (i.e. the range of short-term variations remained almost unchanged with time). Woodworth et al. (2006) confirmed the spatial pattern of trends in winter means and high and low waters around the North Sea, and the relationships between them, and pointed to the importance of the epoch studied (i.e. to periods of different NAO phase).

Bernier and Thompson (2006) modeled 40 years of tides and surges in the northwest Atlantic (using an approach similar to that of Flather et al. 1998) and observed a slight reduction in extreme sea levels between 1960 and 1999 due to a reduction in extreme storm surges, including a reduced contribution from the inverse barometer response of air-pressure minima. They combined their 40 years of modeled surges with data retrieved from short tide-gauge records to reconstruct total sea levels and map the spatial dependence of extreme-sea-level return periods (Bernier et al. 2007). They introduced seasonal dependence into their return-period reconstructions, and downscaled their results to the urban/ecosystem level using Digital Elevation Models to provide maps of return periods of coastal inundation under current conditions and under projected scenarios of sea-level rise and changing meteorology for the next century. These maps have the distinct advantage of displaying the current flooding risks and the plausible impacts of climate change in terms of inundated landmarks as opposed to elevation above arbitrary datums such as chart datum and mean sea level. Several groups are employing global barotropic models in order to investigate time series and return periods of storm surges worldwide; in such cases the requirements given above remain, notably the need for high-resolution meteorological data and bathymetry. The adequacy for using a barotropic model in such studies, compared with a full three-dimensional model, has been demonstrated by Kauker and Langenberg (2000).

Several further technical points can be mentioned here. One is that the Waves and Storms in the North Atlantic (WASA) Group (1998) contains an important digression on the problems of homogeneity in observational data sets and in the meteorological fields used to drive numerical surge and wave models. Using the particular examples of the Greenland and the North Sea regions, the authors found only the data sets from the latter to have acceptable homogeneity for their purposes. Such data sets must always be used with regard to the historical spatial variations in data coverage. A second point is that there are high-frequency, meteorologically driven, resonance-like processes (sometimes inappropriately called "meteorological tsunamis") which are not well resolved by the spatial and temporal resolution of certain models or tide gauges. These have been observed in the Irish and North Seas, Adriatic, and northern Pacific (e.g. Lennon 1963; Vilibić et al. 2004), and inevitably result in an underrecording of the true extreme. A third technical point relates to the estimation of return periods from either modeling or observational information, or both in combination. Different methods of estimating return periods may give different results. Methods may depend on the quality of the data, the local tidal regime, and assumptions concerning the duration of an exceedance event. Therefore, results obtained using different methods are sometimes difficult to compare, especially if no proper error estimates are presented.

At any location, ongoing observations of extreme events can sometimes result in a major revision of existing estimates of 100- or 1000-year return-period levels. For example, Hurricane Katrina levels at New Orleans were by far the highest recorded, and the 100-year return levels shown in updated Flood Insurance Rate Maps exceed previous estimates by 1–3 m (FEMA 2006).

11.2.2 Effects of River Flow

Sea-level extremes and subsequent coastal flooding can also be exacerbated by intense rainfall in river catchments, the changes in the frequency of which may be related to climate change, enhanced by channeling of rivers and loss of floodplains to industry and housing. Extreme sea levels in many coastal areas may result from hydrometerological extremes of a different nature. For instance, van den Brink et al. (2005) showed that extreme sea levels in the coastal area of the Netherlands may result from the extreme sea storm-surge levels, waves, and river discharges and that it is difficult to quantify which one is a major contributor to a particular extreme event. Singh (2001) studied the effect of rainfall and subsequent runoff on the sea-level variability in the Bay of Bengal, finding that at least 50% of the trend in mean tidal level off the coast of Bangladesh could be related to Monsoon effects, including runoff of Monsoon rainfall from the land through the Meghna, Ganges, and Brahmaputra river systems.

The need for investment in the science, forecasting, and mitigation of combined river and coastal (tide + surge + wave) extremes is now widely recognized. Extremes can sometimes occur due to a combination of factors within their normal range of variability (e.g. large but not unusual tide, surge, or river flow combining to produce a notable overall event) as much as due to long-term change or fluctuation in one or more parameters individually. Svensson and Jones (2004) looked at these joint probability events at several sites around the UK coastline, noting that the relative importance of river flow and surge to flooding can be highly site-dependent. The estimation of change in risk from modifications in extremes will need statistically rigorous appreciation of the interdependence of all relevant meteorological and ocean parameters.

11.2.3 Past Changes in Wave Characteristics

Wave heights in the North Atlantic are known to have increased during the past half century, with some of the observed variability in wave height related to fluctuations in the NAO (e.g. Carter and Draper 1988; Bacon and Carter 1993; Kushnir et al. 1997; Günther et al. 1998; Gulev and Hasse 1999; Wang and Swail 2001, 2002, 2006a; Woolf et al. 2002; Tsimplis et al. 2005; Weisse and Günther 2007). Numerical wave-modeling exercises have confirmed this general picture, although at many locations studied trends are only weakly positive (Vikebø et al. 2003). The WASA Group (1998, and references therein) conducted one of the largest studies of northeast Atlantic climate and, while also concurring that the storm and wave climate in most of the region had become more severe in recent decades (see also Alexander et al. 2005), it pointed out that there had been significant variations on timescales of decades (some of it related to the NAO) and that the then present (in 1998) intensity of the storm and wave climate was comparable to that at the beginning of the 20th century. One notes that northeast Atlantic storminess quantified using simple counts, estimated winds, and depth of the storm centers during the 1990s was also only slightly in excess of that in the 1880s and 1890s. Furthermore, some of the most recent years had the lowest storm activity on record (e.g. Alexandersson et al. 2000 and update in figure 3.41 of Trenberth et al. 2007; Schmidt 2001; Weisse et al. 2005; Weisse and Günther

2007). Lozano and Swail (2002) discussed the relationship between North Atlantic wave heights and storm tracks during the last four decades. They found the largest waves to be associated with latitude shifts of storm tracks, in turn related to the expansion and contraction of the polar front and the NAO. On the other hand, Wolf and Woolf (2006) concluded that it is the strength of the prevailing westerly winds (again related to the NAO) which is the most effective parameter for increasing mean and maximum monthly wave heights, rather than the frequency, intensity, track, and speed of storms.

Similar evidence exists for increasing eastern North Pacific wave heights during the past 20–30 years obtained from buoy records and hindcast wave models (Allan and Komar 2000; Wang and Swail 2001, 2006a; Gower 2002). Sasaki et al. (2005, 2006) concluded that a recent increase in summertime extreme wave heights in the western North Pacific was due to an increase in total duration of intense tropical cyclones. Quasi-decadal variability of autumn extreme wave heights in the same area were considered attributable to changes in storm tracks of intense tropical cyclones around the south of Japan.

There are fewer studies of changes in wave characteristics in the Southern Hemisphere, and most of these are concerned with the relationships of mean significant wave height (SWH) to the El Niño Southern Oscillation (ENSO) and other climate variability, rather than with longterm trends in SWH or changes in extreme wave heights. For example, Laing (2000) constructed a wave climatology for New Zealand waters from 1985 based on satellite altimeter data and studied relationships of empirical orthogonal function (EOF) patterns of SWH spatial variability to ENSO. In the subtropical north of the region, spatial patterns for high wave heights were found to differ considerably from those of SWH. Goodwin (2005) constructed a wave climatology for southeastern Australia and studied the relationship between mean wave direction and climate indices, notably ENSO. Hemer et al. (2007) collected Australian wave data spanning several decades from wave models, satellite altimetry, and a network of 30 wave-rider buoys. They determined long-term means, annual cycles, and interannual variability of the regional wave climate, with Southern Ocean wind anomalies found to be a dominant mechanism for much of the variability. Correlations between monthly mean SWH and ENSO were found to be significant along Australia's eastern margin. Studies of extreme wave heights remain to be performed.

The only recent study of changes in wave height throughout the global ocean based on wave measurements has been that of Gulev and Grigorieva (2004). Although in principle it was worldwide in scope, most of the information in that study was also from the Northern Hemisphere. The authors processed data from voluntary observing ships, with such data being regarded as relatively little affected by changes in observational practice. Trends in wave height for 1958–2002 were statistically significant and positive over most of the mid-latitudinal North Atlantic and North Pacific, in addition to the western subtropical South Atlantic, eastern equatorial Indian Ocean, and the East and South China seas. The largest positive trends were found in the northeast Atlantic and northeast Pacific, with negative values in the western tropical Pacific, eastern Indian Ocean, Tasman Sea, and southern Indian Ocean. Gulev and Grigorieva (2006) attempted to separate the contributions to change in wave height due to wind waves and swell, concluding that changes in the northeast Atlantic over the second half of the 20th century were primarily due to swell rather than wind waves, and indirectly to the number of cyclones, whereas in the North Pacific spatial patterns of change due to wind wave and swell were similar (Figure 11.3).

Gulev and Grigorieva (2004) also provided estimates for linear trends in wave heights on century (1902–2002) timescales. Trends in SWH were found to be significantly positive throughout the North Pacific, with a maximum of 8–10 cm/decade (up to 0.5% per year). These inferences are supported by the buoy records mentioned above, for both annual and winter means, and by hindcast wave modeling (e.g. Graham and Diaz 2001; Wang and Swail 2001). Trends in SWH in the North Atlantic were slightly negative (with marginal significance), with a decrease of 5.2 cm/decade (0.25% per year) in the western Atlantic storm-formation region. In the shorter 1950–2002 period, changes in the North Atlantic region were also found to be positive.

A more suitable database for studies of changes in waves on a global basis has been acquired through the European Centre for Medium-Range Weather Forecasts (ECMWF) ERA-40 wave climatology hindcast reanalysis project, and in particular the Royal Netherlands Meteorological Institute (KNMI)/ERA-40 wave atlas and related work based on the 45-year period 1957–2002 (Sterl and Caires 2005). The model results have been extensively validated against buoy and altimeter data. The model limitations are that the spatial resolution is 1.5° × 1.5° and the temporal resolution 6 h, so that tropical cyclones are not resolved and the details of mid-latitude storms are not well-resolved. The model tends to overestimate low waves and underestimate high waves which may be due to limitations of resolution. Section 5.5 of the atlas (available at http://www.knmi.nl/waveatlas) deals with trends and utilized the methods of Wang and Swail (2001). Statistically significant differences between the return values of the 100-year return-period SWHs estimated from three different decades (1958–67, 1972–81, and 1986–95) occur only in a small number of regions including the North Atlantic (an increase in the region around 20°W, 51–56°N and a decrease around 28°W, 42°N when comparing estimates for the third decade relative to the second) and the North Pacific (an increase around 150–180°E, 40°N when comparing estimates for the second decade to the first). Spatial patterns of trends in the high percentiles of wave height were similar to those of the means but magnitudes of trends were larger. As regards the North Atlantic during periods when the NAO index is positive the storms tend to move from North America in the direction of the Norwegian Sea. When the NAO index is negative the storms move in the direction of the Mediterranean Sea (figure 3 of Rogers 1997), and the wave conditions are milder (see Wang and Swail 2002). The North Atlantic part of the KNMI/ERA-40 reanalysis database provides findings comparable in all but the most extreme events to those from the regional AES40 hindcast (a North Atlantic wind and wave climatology developed at Oceanweather with support from the Climate Research Branch of Environment Canada), with

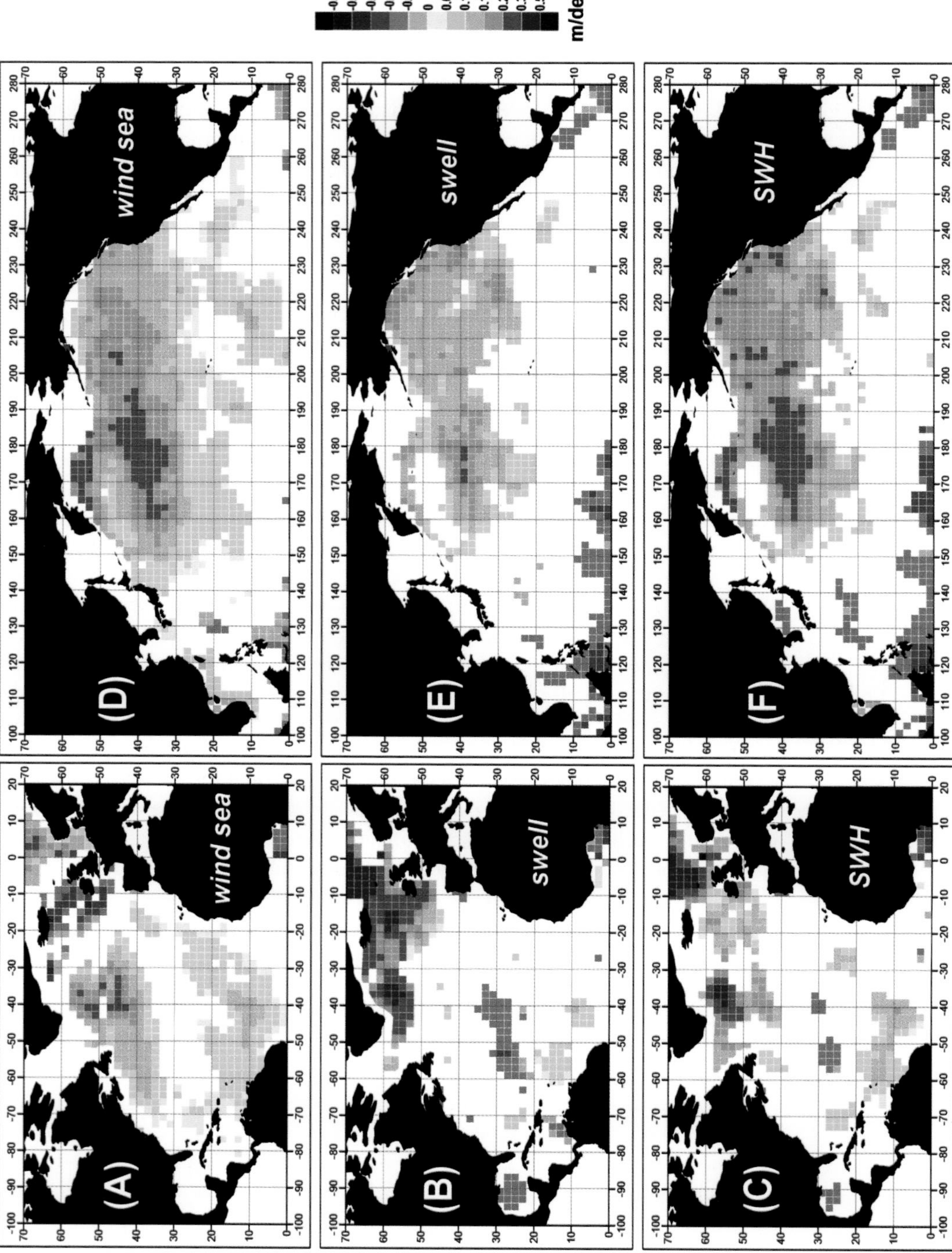

Figure 11.3 Linear trends (m/decade) in winter (January–March) wind sea height (A, D), swell height (B, E), and SWH (C, F) for the North Atlantic (A–C) and North Pacific (D–F) for 1958–2002. Only trends significant at 95% level and according to the Hayashi criterion are shown. (From Gulev and Grigorieva 2006. © American Meteorological Society.)

spatial patterns of trends in the region similar to those reported previously by the WASA Group (Günther et al. 1998).

There are two main limitations of the above body of research into past changes in wave characteristics. The first is that most studies are concerned with changes in annual or monthly mean SWH, rather than extreme (or high percentile) wave heights. The second is that relatively few studies are concerned with changes in wave direction, which are as important as changes in wave height when considering possible coastal impacts, through modification of longshore sediment transport and coastal evolution. (For example, Tsimplis et al. (2005) suggest a 20° change of direction per unit NAO index change in the one southern-England location at which they looked.)

The combined effects of extremes in tides, surges, and waves have maximum impact at the coast. For example, coral islands have elevations of typically 1–2 m above mean sea level and are often used in case studies of impacts of climate change (linked to coral mortality and increases in wave energy and erosion; see Sheppard et al. 2005) and sea-level rise. One of the first island floods of which the world took notice occurred in April 1987 in the Maldives when they were exposed to major ocean swell propagating from the Southern Ocean (Harangozo 1992). The floods were used by the Government of the Maldives and international environmental nongovernmental advocacy groups to alert the United Nations and other intergovernmental bodies to the dangers faced by reef islands from a major sea-level rise. The potential future modification of wind waves and swell as a consequence of climate change is clearly an issue which needs attention alongside sea-level rise. In many cases waves cause erosion as a consequence of which a sea-level extreme results in a coastal flood. However, waves are almost never recorded alongside sea levels at tide-gauge stations (see Vassie et al. 2004) and the literature contains only a limited number of examples of tides, surges, and waves in combination (e.g. Wolf and Flather 2005). More comprehensive studies of combined tide, surge, and wave extremes, and their interactions (see Mastenbroek et al. 1993), using observations and modeling of recent decades and simulations of future conditions, are clearly required around the global coastline.

11.3 Mid-Latitude and Tropical Storms: Changes in the Atmospheric Drivers of Extreme Sea Level

11.3.1 An Introduction to Storms

Both mid-latitude and tropical storms are associated with extremes of sea level. Storm surges are generated by low atmospheric pressure and intense winds over the ocean. The latter also cause high wave conditions.

Synoptic-scale extratropical cyclones, also known as depressions or storms, are baroclinic systems that form in mid-latitudes and are an important part of mid-latitude weather and climate. The main energy source for these cyclones is the temperature contrast between the cold polar and warm subtropical atmosphere. The storms tend to be organized in regions and these are known as storm tracks. The major Northern Hemisphere storm tracks extend across the North Atlantic and North Pacific. There is a smaller track in the Mediterranean region which is closely associated with North Atlantic mid-latitudinal track and influenced by the local diabatic heating over the Mediterranean Sea. These storms, being largely responsible for the extreme hydrometerological events in southern Europe, will contribute to extreme sea levels at the Mediterranean and Black Seas coasts.

Tropical cyclones are synoptic-scale low-pressure systems with a warm core (atmospheric temperature anomaly) relative to the surrounding environment in the middle to upper troposphere, and having a well-defined cyclonic circulation near the surface. The cyclones, which form in the tropics or subtropics, derive their energy from moisture evaporated from the ocean surface locally or converged in toward the storm from the surrounding environment. As moisture condenses in the cyclone to form precipitation and cloud, latent heat is released to help fuel the storm. Tropical cyclones can produce devastating storm surges and extreme-sea-level events. The question examined here is whether there have been any long-term trends in past tropical cyclone activity and what types of change might have been expected on the basis of current model simulations.

From the perspective of extreme sea levels, changes in many different aspects of cyclone characteristics are important. In addition to variations in cyclone number, track, intensity, size, and frequency, changes in the occurrence of extreme deepening (so-called meteorological bombs) and changes in cyclone propagation velocity are also important. Gulev et al. (2001) demonstrated on the basis of storm tracking of National Centers for Environmental Prediction (NCEP)/National Center for Atmospheric Research (NCAR) reanalysis data that some of these characteristics experienced significant changes in the last several decades, in particular a growing number of rapidly intensifying extratropical cyclones in parts of the Northern Hemisphere was noted.

11.3.2 Observed Changes in Atmospheric Storms and Their Drivers

The natural variability of the Northern Hemisphere storm tracks is related to the phase of many large-scale modes of variability, for example the NAO (Rogers 1990, 1997; Ueno 1993; Hurrell and van Loon 1997; Hurrell et al. 2003) and the ENSO (Zhang et al. 1997b; Sickmöller et al. 2000; Graham and Diaz 2001; Chang and Fu 2002; Bengtsson et al. 2006), and any future change in the frequency of one phase over the other may impact on the nature of future storms. Some studies have concentrated on the last half century, looking at changes in maps of atmospheric pressure at mean sea level (e.g. see section 3.5 of the Intergovernmental Panel on Climate Change (IPCC) Fourth Assessment Report (AR4); Trenberth et al. 2007) or using methods such as the tracking of individual storms in meteorological reanalysis products (Bengtsson et al. 2006). Gillett et al. (2003, 2005) noted some increases in Northern Hemisphere mid-latitude pressure gradients since the 1970s that could not be explained by natural variability alone. These changes might be related to increases in storm-track intensity. Trenberth et al. (2007) also highlighted the results of Wang et al. (2006), who noted a northward shift of the storm track in the North Atlantic region by around 200 km. The overall IPCC conclusion was that, in the Northern Hemisphere, it is likely that there has been an increase and poleward shift in the wintertime storm track during the second half of the 20th century, but uncertainties in the magnitude of changes remain. It also reported some evidence of decreases in extratropical cyclone number (and increases in their depth) in the Southern Hemisphere over the last two decades but noted even greater uncertainty than for the Northern Hemisphere results. The Northern Hemisphere results are consistent with observed trends in SWH since 1950 observed by Gulev and Grigorieva (2004).

These studies have typically focused on relatively short time periods, typically a few decades, but some modes of climate system variability take place on longer timescales. Therefore, studies of much longer time series can usefully place the more recent changes into context. For example, Bärring and von Storch (2004) and Jónsson and Hanna (2007) found no evidence for long-term changes in storminess in two Scandinavian and one Icelandic time series respectively, each approximately 200 years long, although considerable multidecadal variability was evident.

Turning to tropical storms, evidence is now emerging of significant sea-surface warming trends in both the Atlantic and northwest Pacific tropical cyclogenesis regions, with anthropogenic forcing likely playing a detectable role in the warming (Santer et al. 2006). The Indian Ocean/western tropical Pacific warm pool region also shows a particularly pronounced long-term warming, with a larger trend during the past half century than occurred in the tropical north Atlantic (e.g. Knutson et al. 2006). The emergence of such increases in sea-surface temperatures (SSTs) in the regions where tropical cyclones form and intensify raises the question of whether tropical cyclone characteristics may have changed significantly during the 20th century.

Several recent studies have considered a number of hurricane intensity-related indices, but provide conflicting information on long-term trends. One such index is Emanuel's (2005, 2007) power dissipation index (PDI) which is the accumulated sum of the cube of maximum surface wind speeds (one value per storm every 6 h) observed in tropical cyclones. Emanuel found a clear increase in this index in the Atlantic since the mid-1970s and an increase since 1950 as well, although data reliability decreases as one goes back earlier in the record, such as into the pre-satellite era (pre-1966). While Emanuel's (2007) Atlantic PDI is well correlated with tropical Atlantic SST on multiyear timescales, Swanson (2008) reports that Atlantic PDI is also correlated at least as well, if not better, with tropical Atlantic SST minus tropical mean SST. Landsea's (2005) PDI for US landfalling Atlantic tropical cyclones (1900–2004) shows no evidence of an upward trend. Mann and Emanuel (2006) present a time series of numbers of tropical cyclones in the Atlantic basin since the late 1800s which roughly tracks the low-frequency variations of SSTs in the Atlantic tropical cyclogenesis region, including the century-scale increases. The question remains as to how reliable the tropical cyclone count data are, particularly for the full basin for years prior to aircraft reconnaissance which began in the mid-1940s. For example, Vecchi and Knutson (2008) find that Atlantic tropical storm counts, after an estimated correction for missing storms in the pre-satellite years based on ship-track densities, does not exhibit a statistically significant increase since 1878. Landsea et al. (2009) discuss the impact of duration thresholds on Atlantic tropical cyclone counts.

Emanuel (2007) reports that multibasin PDI series since about 1950 for the northwest Pacific and Atlantic, or for the northeast Pacific, northwest Pacific, and Atlantic basins combined (wind.mit.edu/~emanuel/antho2.htm) show substantially increasing trends. The increase amounts to roughly a doubling over the past 30 years, due to both increased intensity and increased storm duration. These PDI series correlate to some degree with low-frequency variations and trends in tropical SSTs.

Webster et al. (2005) report a large increase in the number of category 4 and 5 tropical cyclones (but not overall tropical cyclone numbers) over the past three decades, with increases occurring in all six basins during that time. A follow-on study by Hoyos et al. (2006) establishes a stronger statistical link between the increasing numbers of category 4 and 5 storms and the SSTs in the various basins. However, Bogen et al. (2007) test whether the positive trend in North Atlantic SSTs since 1970 explains increased hurricane intensities and find only a "weak association" with PDI. Chan (2006) extends the analysis of Webster et al. for the northwest Pacific basin back to earlier years and argues that the "trend" in that basin is part of a large interdecadal variation. Chan uses unadjusted data from the earlier part of the record, in contrast to the adjustments for this period proposed by Emanuel (2005) for this basin. The Emanuel and Webster et al. studies continue to be the subject of much debate in the hurricane research community (e.g. Knaff and Sampson 2006; Landsea et al. 2006; Klotzbach 2006), particularly with regard to homogeneity of the tropical cyclone data over time and the required adjustments. Kossin et al. (2007) have attempted to present a more homogeneous (in time) and globally consistent analysis of hurricane variability and trends. Their analysis extends only over the period July 1983 to December 2005, a shorter period than analyzed by Emanuel (2005) and Webster et al. (2005). They use satellite-derived measures of intensity to infer past hurricane activity in a more homogeneous manner. Their results support the existence of an increase in activity in the Atlantic basin (1983–2005), but in other basins their results find trends which are contrary to those obtained using the existing "Best Track" data sets for tropical cyclones over that period. The implications of the various findings for the results obtained in earlier studies using longer records (e.g. Emanuel 2005) have not yet been thoroughly investigated. Emanuel's (2007) analysis indicates that the Kossin et al. (2007) results in the northwest Pacific basin, when combined with the longer record of Emanuel (2007), still yield an increase in PDI in that basin since the 1950s.

The literature on tropical cyclones continues to expand enormously. Recent studies include those of Wu et al. (2008) who examine how the annual frequency, lifetime, and intensity of tropical cyclones contribute to the changes in annual accumulated PDI. They find a significant upward trend between 1975 and 2004 in the North Atlantic only, together with some indication that the the trend was due to decreased wind shear and warming ocean. Briggs (2008) tests the number and rate of development of tropical cyclones in the North Atlantic and concludes that any increase in cumulative yearly storm intensity is due to the increasing number of storms rather than increase in intensity of individual storms.

11.3.3 Future Changes in Mid-Latitude Storms

The analyses of changes in mid-latitude cyclones in climate models have used a variety of analysis techniques for analyzing storms and their activity in the storm-track regions. These methods include band-pass filter statistics (Blackmon 1976) to look at variability on synoptic timescales (e.g. Hall et al. 1994; Christoph et al. 1997) and the number of gales (e.g. Carnell et al. 1996; Weisse et al. 2005; Fischer-Bruns et al. 2005). In addition, they have been applied to studies of eddy kinetic energy (e.g. Yin 2005), Eady parameter (Lindzen and Farrell 1980) to look at changes in the baroclinicity (e.g. Lunkeit et al. 1998), cyclone densities without tracking (e.g. Lambert 1995), and cyclone densities with tracking (e.g. König et al. 1993; Carnell et al. 1996). A variety of different methods have been used to identify the location and path of cyclones and these have been applied to 1000 hPa geopotential height, mean-sea-level pressure and 850 hPa relative vorticity at 24-, 12-, or 6-h intervals (e.g. König et al. 1993; Murray and Simmonds 1991; Carnell et al. 1996; Knippertz et al. 2000; Hoskins and Hodges 2002; Wang et al. 2006). The use of these different techniques makes it hard to compare the results of each study (Cubasch et al. 2001) or to assess if there is any underlying climate change signal. The majority of studies on future changes in storms (e.g. Carnell and Senior 1998; Knippertz et al. 2000; Fyfe 2003; Leckebusch and Ulbrich 2004) use data from models with low horizontal resolution and this can lead to errors with the simulation of mid-latitude cyclones. For example, the storms are too weak (Lambert et al. 2002) and the north Atlantic storm track is often located too far south (Lambert et al. 2002; Lambert and Fyfe 2006). These errors can often be greater than the climate change signal and so contribute to the uncertainty of the results. Improvements to the physics and dynamics of the models along with increased horizontal resolution have helped to improve the simulation of storms. Some groups have analyzed storms in higher-resolution models, but the simulation of storms does not necessarily improve as resolution is increased (Pope and Stratton 2002) and it is often possible to only run short experiments with these models and so the sample of data is small. The use of data at intervals of 6 h or less is becoming more common and this helps to improve the tracking of cyclones.

There is a common assumption that global warming may lead to increased storminess. However, the evidence is somewhat contradictory. The IPCC Third

Assessment Report (TAR) report found that climate change simulations of the impact of global warming on mid-latitude storm frequency and intensity were inconclusive (Cubasch et al. 2001). Sinclair and Watterson (1999) used the Commonwealth Scientific and Industrial Research Organisation (CSIRO) general circulation model (GCM) of the atmosphere to examine changes in the frequency and strength of mid-latitude storms under a doubled-carbon dioxide scenario. The polar regions were found to warm more than the tropics, reducing the equator-to-pole temperature difference. Reducing this gradient results in fewer and weaker storms, but there is some evidence of increased winter cyclone activity near the downstream end of the principal storm tracks. The results of modeling studies carried out since the TAR show that large uncertainties remain (Geng and Sugi 2003; Leckebusch et al. 2006; Lambert and Fyfe 2006; Bengtsson et al. 2006). Some modeling studies are tending to agree on there being fewer storms in winter in both hemispheres but there is no agreement on local changes in frequency.

Results are available from atmosphere-only models (HadAM3P and JMA; see list of abbreviations and acronyms at the front of the book for a list of full model names), and models which have both an atmosphere and an ocean (ECHAM5-OM, HadCM2, CSIRO2, ECHAM4/OPYC3, and ECHAM5/MPI-OM1). The ECHAM5-OM (Bengtsson et al. 2006), JMA (Geng and Sugi 2003), HadAM3P (Leckebusch et al. 2006), and CSIRO2 (McInnes et al., unpublished work) models show more storms around the UK in the future, but the ECHAM4/OPYC3 and ECHAM5/MPI-OM1 models show reduced storminess around the UK (Leckebusch et al. 2006) and HadCM2 shows no change (Carnell and Senior 1998). There was little change in the location of the cyclones in an average of GCM models prepared for the AR4 (Lambert and Fyfe 2006), although this may be because of the low horizontal resolution of the cyclone frequency data. The Northern Hemisphere storm tracks in McInnes et al. (unpublished work) became more zonal in structure producing an increase in storms over Europe and the eastern Pacific due to enhanced baroclinicity from warmer SSTs on the eastern edge of the ocean gyres. A similar relationship between changes in storm tracks and SSTs was also found in the CSIRO Mark 3 model. Storm-track predictability was investigated by Compo and Sardeshmukh (2004). They found that a predictable SST-forced storm-track signal exists in winter, but its strength and pattern can change substantially from winter to winter. The wide disparity in SST response in coupled ocean–atmosphere GCMs reported by Liu et al. (2005) suggests that the lack of agreement between GCM simulations on the specific geographic changes in storm tracks may continue to be a source of uncertainty in future projections of storm-track changes.

Yin (2005) found a consistent poleward shift and intensification of storm tracks in an ensemble of 21st-century climate simulations using 15 coupled climate models using eddy kinetic energy as an indicator of storminess. Lambert and Fyfe (2006), using the same ensemble of climate models, report a reduction of cyclone frequency and an increase in intensity using cyclone density. These seemingly conflicting results suggest that the increase in storm tracks reported in Yin (2005) occurs due to a change in the intensity of the systems which more than compensates the reduction in numbers. This finding is consistent with the Southern Hemisphere results of McInnes et al. (unpublished work), using both measures of storminess, and Lim and Simmonds (2007) who show that the tendency for fewer but more intense cyclones extends from the surface to 500 hPa in the CSIRO Mark 2 model. Yin (2005) refers to recent results observed in reanalysis-based studies showing that there has been a poleward shift in the mean latitude of extratropical storms and cyclones have been fewer but more intense in the latter half of the 20th century. However, it is difficult to attribute this to the effects of increasing atmospheric concentrations of greenhouse gases as yet.

Chang et al. (2002) provided a useful review of storm-track dynamics and found storm tracks exhibit notable variation in intensity on decadal timescales. Fischer-Bruns et al. (2005) studied storm variability on even longer timescales in a 500-year historical simulation and noted that anomalous temperature regimes, such as the late Maunder minimum, are not associated with systematic changes in storm conditions. It remains to establish a causal relationship between temporal variability of storm-track eddies (cyclones and anticyclones) and that of the background flow.

In summary, the AR4 (Meehl et al. 2007, section 10.3.6.4) concluded that the majority of future climate models show a northward shift in storm-track position in the Northern Hemisphere in a warming climate. Many of the models used in the IPCC Assessment also showed some indication of an intensification of their storm tracks over the northeast Atlantic near the UK (Lowe et al. 2009). Although the focus here has been on the Northern Hemisphere, many of the studies mentioned above also simulate Southern Hemisphere storms. The AR4 concluded that there is a tendency for models to show a poleward shift in their Southern Hemisphere storm tracks in a warming climate.

11.3.4 Future Changes in Tropical Storms

Two techniques for diagnosing tropical storms have been commonly used in global climate model experiments. The first technique locates and tracks individual cyclones, as centers of maximum relative vorticity which have a warm core (e.g. Haarsma et al. 1993; Bengtsson et al. 1995; Vitart et al. 1997). The second technique provides an estimate of tropical storm activity using a genesis parameter, which is calculated from seasonal means of the large-scale fields and so avoids the problems of simulating individual cyclones. The parameter needs to be chosen with care as it may not be appropriate to use a parameter that has been tuned for present-day conditions for global warming experiments (see Royer et al. 1998; McDonald et al. 2005 and Chauvin et al. 2006). Tropical storm-like features have been analyzed in studies with numerical models that range from low-resolution climate GCMs, to regional climate models, to very high-resolution hurricane prediction models.

The horizontal scale of tropical storms is much smaller than the horizontal grid-scale of most GCMs and, because of this, there has been some debate over the utility of GCMs in studying tropical storm behavior (e.g. Lighthill et al. 1994; Henderson-Sellers et al. 1998). However, some analysis has been carried out using models with higher horizontal resolution (e.g. Bengtsson et al. 1996; Sugi et al. 2002; McDonald et al. 2005) but the grid-scale is still larger than the scale of tropical cyclones and so these models are unable to simulate the intense core of tropical cyclones. Experiments have been run at 20-km model resolution (Oouchi et al. 2006), which improves the simulation of tropical storms, but models of this resolution are too expensive to be run at present by most climate modeling centers. Confidence in predictions would be increased by demonstrating the skill in the simulation of present-day mean winds and variability (Walsh 2004). In addition, objectively derived, resolution-dependent criteria for the detection of tropical cyclones in model simulations are required (Walsh et al. 2007).

Coupled atmosphere–ocean GCM experiments, run with increasing greenhouse gas concentrations, project enhanced SSTs and atmospheric moisture in the tropical region (Cubasch et al. 2001) that may change the intensity and location of tropical storms (Henderson-Sellers et al. 1998). However, factors other than SSTs are also likely to be important (see for instance, Bengtsson et al. 1995, 1996, 2006).

There is currently large uncertainty in the future changes in tropical cyclone frequency predicted by climate models forced with future greenhouse gases. The changes in frequency of storms simulated by models are often smaller than those due to natural variability. The IPCC TAR concluded that the results of GCM experiments are inconclusive (Giorgi et al. 2001). The JMA T106 (a JMA GCM with T106 spatial resolution ($1.1° \times 1.1°$); Sugi et al. 2002), HadAM3 (McDonald et al. 2005), and ECHAM5-OM (Bengtsson et al. 2006) models all have fewer tropical storms in the future, whereas the NCAR Community Climate Model version 2 (CCM2) has slightly more storm days (Tsutsui 2002), although not all of these changes are greater than natural variability. There are regional variations in the sign of the changes and these vary between models. For example there are more storms in the North Atlantic region in the future in the JMA model (Sugi et al. 2002) but fewer in the HadAM3 model (McDonald et al. 2005) or the coupled model of Bengtsson et al. (2007). Yoshimura and Sugi (2005) found that the decrease in tropical storm frequency in their model was due to the direct effect of increased carbon dioxide on the atmosphere with the SST changes having a relatively small impact. The frequency of tropical storms also decreased in the higher-resolution 20-km JMA future simulations (Oouchi et al. 2006). While there were more storms in the North Atlantic with climate warming in this model, tropical cyclone frequency decreased overall. Nevertheless, the number of intense tropical cyclones increased markedly. However, one notes that a relatively short coupled model simulation period was used to investigate the climate-change signal.

Several high-resolution modeling studies of the relation between greenhouse warming and hurricane intensity have been conducted using the Geophysical Fluid Dynamics Laboratory (GFDL) hurricane model in idealized mode (e.g. Figure 11.4, taken from Knutson and Tuleya 2004). Under warmer, high-carbon dioxide conditions, simulated hurricanes are more intense (and have higher precipitation rates) than under present-day conditions. The simulated sensitivity is

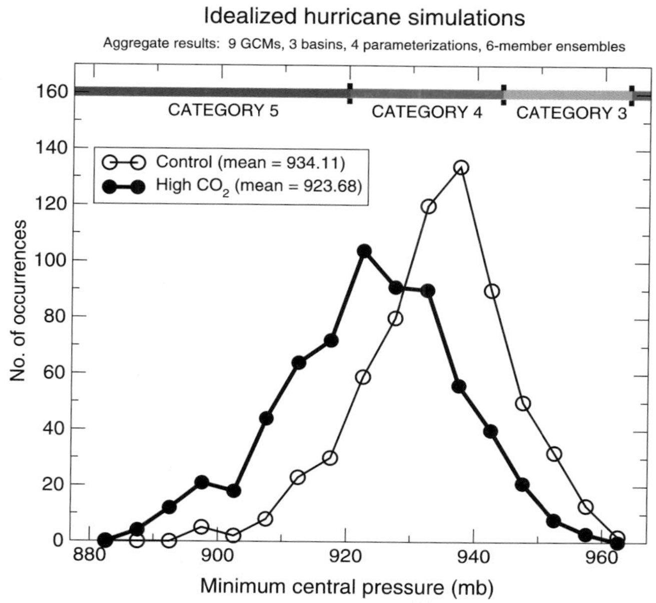

Figure 11.4 Histograms showing simulated hurricane intensity results (mb) from a series of idealized hurricane simulations. The histograms are formed from the minimum central pressures, averaged over the final 24 h from each 5-day model experiment. The thin line with open circles shows results for the control-CO_2 cases and the dark line with solid circles shows the high-CO_2 cases. High-CO_2 experiments represent conditions in which atmospheric CO_2 concentrations approximately double over a period of 80 years, relative to the control values. (From Knutson and Tuleya 2004. © American Meteorological Society.)

roughly a half category on the Saffir–Simpson scale (14% in terms of pressure fall, 6% in terms of maximum surface winds) for the warming associated with an 80-year buildup of carbon dioxide at 1%/year compounded. The GFDL results are broadly robust to the use of different climate models to define the high-carbon dioxide conditions, and to details of the treatment of moist convection in the hurricane model. Earlier studies with a similar model (Knutson et al. 2001) indicate the results are robust to inclusion of ocean coupling. The SST changes due to increased carbon dioxide in these experiments ranged from about +0.8 to +2.4°C (average 1.75°C), which is substantially greater than the approximately 0.5°C warming experienced in the tropical Atlantic and other tropical basins during the 20th century. The sensitivity of storm intensity to SST warming obtained in the Knutson and Tuleya study is similar, although slightly smaller, than that obtained using theories of hurricane potential intensity (about 5%/°C according to Emanuel 2005). Some additional modeling evidence for more intense hurricanes at the high end of the intensity distribution has been reported from other simulations using relatively high-resolution global or regional models (Oouchi et al. 2006; Bengtsson et al. 2007; Walsh et al. 2004; Stowasser et al. 2007).

In comparing the GFDL simulations with the historical intensity trends reported by Emanuel (2005) it is necessary to make assumptions about lapse-rate trends during the past three decades, for which the observed lapse-rate trends are uncertain. Assuming historical lapse-rate trends (per degree of SST warming) are close to those from +1%/year carbon dioxide experiments, the results imply that any trends in tropical cyclone intensity in the various basins should be too small to be detectable at this time, in contrast to the observational findings of Emanuel

(2005) and Webster et al. (2005). The Knutson and Tuleya (2004) simulations focused on intensity alone, and did not address the effects of changes in either storm frequency or duration, which were additional important factors in the historical tropical cyclone increases reported by Emanuel (2005). Efforts are continuing to reconcile the model-based and observation-based sensitivities.

11.4 Future Extreme Water Levels

Currently, predictions of extreme water level a few days ahead are produced for a number of regions using output from numerical weather-prediction models to drive storm-surge models (Flather 2000). Here we report on potential changes in extreme sea level over the coming decades. Changes in the number, path, and strength of atmospheric cyclonic storms, as discussed in section 11.3, may alter the formation and evolution of storm surges. Such changes could result in either increases or decreases in the surge climatology, with results being highly location-dependent. Individual surge events depend on the driving meteorology, so individual surges cannot be predicted more than a few days ahead at most. For times that are 50 or 100 years into the future, the best we can do is to predict, for a given set of assumptions on future greenhouse gas emissions, how many events of a given size or type will occur, on average, in a given length of time.

Extreme water levels will also increase as local mean sea levels rise in the future, so predictions of the regional change in mean sea level will also be necessary for predicting changes in extreme water levels a few decades ahead. The limitations of climate models in simulating future changes in mean sea level are discussed in Chapter 13.

11.4.1 Tools for the Simulation of Future Extreme Water Levels

Two main approaches to simulate future changes in extreme sea level have been employed: the statistical method and the dynamical method.

In the statistical approach, relationships between large-scale driving meteorology and local storm-surge heights are developed from observations or atmospheric and storm-surge model simulations of the recent past or present day. These relationships must be capable of explaining a significant fraction of the surge-height variability. Next, a projection is made of future large-scale meteorology using a global climate model and the future storm-surge characteristics are estimated from these using statistical relationships (von Storch and Reichardt 1997; Langenberg et al. 1999; Grossman et al. 2007). This category of future extreme-water-level estimation also includes combining relationships between extreme sea level and large-scale indices, such as the NAO (as described by Wakelin et al. 2003; Tsimplis et al. 2005; Woodworth et al. 2006), with climate model projections of these indices. However, in the case of the NAO the current generation of climate models appear unable to adequately replicate the observed variability of this index (Osborne 2004; Kuzmina et al. 2005). While removing the complexity of the need to run a storm-surge model, the statistical techniques assume that the relationship between the large-scale variables and the surges, which hold in the present-day climate, remain unchanged in a future perturbed climate. The validity of this assumption at particular locations is uncertain, although the general approach has been tested using climate models. For instance, Mearns et al. (1999) noted the inability of a statistical approach to reproduce some of the simulated changes in a regional model study of atmospheric parameters over eastern Nebraska. Busuioc et al. (2006) noted some skill in a statistical method when applied to simulate the future climate over Romania. However, even for the best predictor case there were notable differences in the magnitude of response compared to the dynamical method.

In the alternative, dynamical approach, physically based models of shallow-water dynamics are used to simulate storm-surge levels in past/present-day and future periods. The storm-surge models are usually barotropic in formulation (e.g. the Proudman Oceangraphic Laboratory (POL) CS3/CSX model, the Tidal Residual and Intertidal Mudflat Model (TRIMGEO; Casulli and Cattani 1994), or the barotropic versions of Princeton Ocean Model (POM) and Global Coastal Ocean Model, depth-average version (GCOM2D)) although some baroclinic models (e.g. POL Coastal-Ocean Modelling System (POLCOMS), a three-dimensional model for shelf regions)) are now being used in studies of storm surges. Kauker and Langenberg (2000) found that for investigating storm surges in the North Sea the barotropic models are adequate. The barotropic models solve the equation of continuity and a depth-averaged version of Newton's second law of motion in a rotating fluid, given the tidal and meteorological (wind stress and air pressure) forcings across the model domain. Parameterizations of the surface stress and bottom friction, and effects of horizontal viscosity, are required (e.g. Dyke 2001). Tidal and meteorological (often simple inverse barometer) boundary conditions are applied at the open boundary.

In many studies using the dynamic approach, the driving winds and pressure are taken directly from atmospheric climate models for both past/present and future periods. For example, Flather and Smith (1998) used the ECHAM3 climate model to drive a storm-surge model for two 5-year time slices, which represented present-day and future climate conditions. Global atmospheric climate models, or the atmospheric component of coupled ocean–atmosphere climate models, may not have adequate horizontal resolution to credibly simulate the local-scale meteorology; the horizontal resolution of the atmosphere in HadCM3, for example, is $2.5° \times 3.75°$. Some studies have used a one-way nested high-resolution regional atmospheric model to derive fine-scale winds and atmospheric surface pressure from the large-scale climate simulated by the coarser global climate models (e.g. Jones et al. 1995). These drivers are then used as the input to the dynamic storm-surge models, as in Lowe et al. (2001), Lowe and Gregory (2005), Woth (2005), Woth et al. (2006), and Unnikrishnan et al. (2006). The details of these studies are elaborated in the case studies section below. McInnes et al. (2003,

2005b) use an alternative approach to generate the meteorological drivers, typically using observed storm characteristics for the 20th century and perturbing these using the large-scale predictions from global climate models. In the tropics, synthetic 20th-century tropical cyclones are constructed based on sampling the range of storm characteristics from observations noting that cyclone intensity was represented using extreme-value statistical theory applied to observed cyclone intensities, whereas in mid-latitudes extreme-value statistical theory is used to extrapolate the modeled surges from actual storms in the observational record to longer return periods. These dynamic methods do not rely on a universal relationship between past storm behavior and extreme water levels.

Once a population of extreme future water-level events has been generated, there are numerous ways in which they can be presented and used. The simplest is to use percentiles (e.g. Woth et al. 2006 looked at changes in and exceedence of the 99.5th percentile of Northern Hemisphere winter extreme water levels; the water level will exceed this for approximately 12 h per winter season). This method has the significant advantage of not requiring a fit to a particular parametric distribution.

The alternative method to analyzing the extreme water levels is to fit them to an extreme-value parametric distribution (for example, Reiss and Thomas 1997; Coles 2001). This has the advantage that it enables results to be extrapolated beyond the relatively short periods of some climate model experiments. For instance it allows an estimate of the 50-year return-period event from 30-year time-slice experiments. The parametric method also enables the changes in the extreme water levels to be characterized and described by a small number of parameters (e.g. a location and shape parameter). Two commonly used parametric approaches are the generalized-extreme-value method and the peak-over-threshold method. The latter approach uses more of the original data. An obvious problem with these methods is that with relatively small samples the uncertainty in the fit might be large. With the generalized-extreme-value method it is quite easy to increase the sample size by including more than one maximum from each year (e.g. Smith 1986) but one has to ensure that the events can still be considered extreme. Furthermore, generalized-extreme-value or peak-over-threshold methods sometimes give a poor fit to the observed distribution of extremes, for example when surges induced by relatively rare but intense tropical cyclones are present. One should also be cautious when extrapolating results for long return periods, given that using a short sample to estimate the parameter values will not include all the effects of long period variability, which is known to affect the climate (Lowe et al. 2009). For example, there have been significant increases during the late 20th century in the number of severe storms over the UK since the 1950s (Alexander et al. 2005), which appear to be related to changes in the NAO to a more positive phase, although since the mid-1990s there is evidence of a reduction in storminess in this region as discussed in section 11.2.3. It remains uncertain whether the multidecadal NAO variability is related to climate change (Tsimplis et al. 2005).

The statistical analysis methods considered so far typically treat points in space independently so that information on the joint probability of the extreme events at different locations is discarded. Work by Dixon and Tawn (1992) and more recent work by the same group attempts to retain information on the spatial structure of the extreme events. In future, it might also be useful to combine aspects of the dynamic and statistical approaches. For instance, the dynamic method could be used to estimate changes in sea-level extremes on a scale of a few kilometers and the statistical method used to downscale these to individual point locations.

11.4.2 Case Studies

Extreme sea levels are affected by local meteorology and small-scale geographical features, such as the shape of the coastline, making results specific to particular sites. It is thus not possible to generalize the results from one region to another or the entire globe. Here we present case studies for three regions: the European coastline, the Bay of Bengal, and the Australian coast.

European Shelf Region

A number of studies have looked at storm surges in the shelf seas around Western Europe, especially in the southern North Sea, where surges are driven by mid-latitude storms and are strongly affected by the region's topography. It is also a region where significant damage and loss of life have resulted from coastal flooding (McRobie et al. 2005).

von Storch and Reichardt (1997) found that the change in storm-surge height (relative to mean sea level) resulting from changes in climate associated with a doubling of atmospheric carbon dioxide fell within the limits of natural variability. This was also found by Langenberg et al. (1999), who used different techniques for analyzing the surge heights along the coastline of the North Sea, including both the statistical and dynamic methods.

Flather and Smith (1998) used the ECHAM3 climate model to drive a storm-surge model for present-day and future climate time slices. They found that the surge extremes were different for future and present-day simulations but the differences were mostly within the range of natural variability, as estimated from a longer storm-surge simulation driven by surface forcing from a meteorological reanalysis of the period 1955–94 (Flather et al. 1998). This work was updated in the Regional Storm, Wave and Surge Scenarios for the 2100 century (STOWASUS) study (dmiweb.dmi.dk/pub/STOWASUS-2100), which used the ECHAM4 model to project changes in the driving meteorology and used longer time slices. The study did find significant changes in storm-surge height in the future climate for some locations. Lowe et al. (2001) looked at the projected 21st-century changes in extreme water levels using 20- and 30-year time slices from the Hadley Centre global ocean–atmosphere coupled climate model (HadCM2) and downscaled these to 50 km using the atmospheric regional model HadRM2. Again, significant changes were projected at some locations. Debernard et al. (2002) ran a storm-surge model for two 20-year time slices, representing the present day and a period centered on 2040, using the 50-km high-resolution limited-area regional climate model (HIRLAM) to provide driving meteorology. Significant changes in surge extremes were found in the southern and western North Sea in the autumn season but these did not greatly affect the annual results. Lowe and Gregory (2005) used 30-year time slices, the HadCM3/HadRM3 climate models and the Special Report on Emissions Scenarios (SRES) A2 and B2 emission scenarios to look at changes in extreme water levels during the 21st century. In this study, the changes in surge height were combined with both mean sea-level rise and vertical land movements, with a sizeable increase resulting at the southern end of the North Sea. The studies by Woth et al. (2006) and Woth (2005) took results from the Prediction of Regional Scenarios and Uncertainties for Defining European Climate Change Risks and Effects (PRUDENCE) project (prudence.dmi.dk) and used global and regional climate models (HIRLAM, Rossby Centre Regional Atmosphere-Ocean model (RCAO), Climate Version of the Local Model (CLM; a non hydrostatic regional climate model developed from the LM by the CLM Community; clm.gkss.de), and the Hamburg regional climate model (REMO)) to drive a storm-surge model for 30-year time slices of the present day and of future (approximately the end of 21st century) climates. Projected changes in extreme (high percentile in this case) storm-surge height along the UK coastline tended to be within the estimate of natural variability but significant changes were seen along part of the coast of continental Europe. However, there were no significant differences between the changes in extreme sea levels derived from different combinations of global and regional models and different emissions scenarios (see Figure 11.5).

Svensson and Jones (2006) have combined results from the Lowe and Gregory study with simulated river discharge to look at the future joint probability of combined surge and river flooding for selected UK sites. This type of work is currently being repeated for the River Thames as part of the Thames Estuary in 2100 (TE2100) project of the UK Environment Agency (Lowe et al. 2009), with river flow being calculated using the most sophisticated grid-to-grid river routing models available, and with a focus on quantifying uncertainty.

Bay of Bengal

If the severity of a storm surge is measured in terms of the number of lives lost during the 20th century, then the most severe events have occurred in the Bay of Bengal, where extreme water levels are driven by tropical cyclones (e.g. Cyclone Aila, Figure 11.6). Several hundred thousand people (perhaps 300 000, although estimates vary) are thought to have been killed by the 1970 cyclone alone (Murty et al. 1986; Murty and Flather 1994). Cyclones Sidr (2007) and Nargis (2008) have provided more recent examples of major loss of life due to both the storm itself and the accompanying surge.

In this section, we concentrate primarily on the northern part of the Bay; the storm-surge problem in the Bay of Bengal as a whole has been discussed in detail by Murty et al. (1986). This region contains one of many megadeltas in Southeast

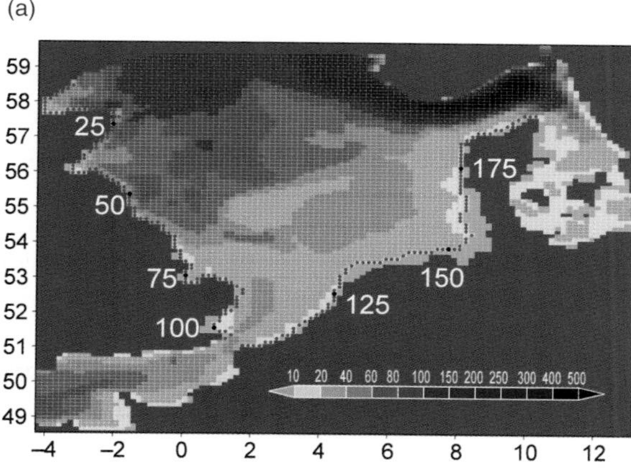

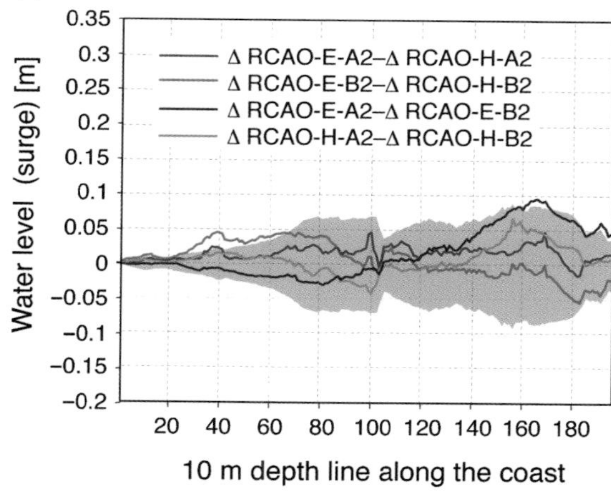

Figure 11.5 (a) Model domain of the tide-surge model TRIMGEO: the bathymetry (m) and the 196 near-coastal grid cells (red points) located along the 10-m depth line along the North Sea coast beginning with 1 in Scotland and ending with 196 in Denmark. (b) Differences (Δ) in the height of the simulated 99.5th percentile of future (2071–2100) winter water level/surge relative to today's climate (1961–90) calculated for a range of climate model and emission scenario combinations: H denotes the HadAM3H climate model, E denotes the ECHAM4/OPYC3 model, while A2 and B2 denote SRES emission scenarios. Each calculation is performed with the RCAO regional atmosphere–ocean model (Döscher et al. 2002). The plot shows the differences between combinations of Δ values as a function of grid cell number. For methods see Woth (2005).

Figure 11.6 Temporary shelters became home to thousands of people on the embankments surrounding the island of Padma Pakur, southern Bangladesh following Cyclone Aila and its storm surge in May 2009. Over 300 people were killed during this one storm. On the left side is the Bay of Bengal, on the right the flooded island, which used to contain villages, cultivated land and fishing farms. (Photo credit: Espen Rasmussen/Panos Pictures.)

and East Asia undergoing rapid change due to sea-level rise and human activities (Parry et al. 2007). Das (1972) and Das et al. (1974) used a linear shallow-water model for the northern Bay of Bengal to simulate the severe cyclone of 1970. A simple superposition of their model surge with the astronomical tide overestimated observed sea-level elevation at the time of landfall, suggesting that interactions between the surge and tide may be important. More recently, Dube et al. (1985) used a nonlinear surge model with improved representation of the coastal boundary to simulate storm surges using winds and pressure for three observed cyclonic storms. The simulated water levels compared well with tide-gauge observations at Chittagong port.

Flather and Khandker (1993) used a depth-averaged (two-dimensional) surge model to investigate the effect of a rise in mean sea level on the height of storm surges in the northern Bay of Bengal. Their model included a simple one-dimensional representation of the many channels of the delta which feed into the bay and which complicate the study of surges in this region. Using low-level winds and atmospheric surface pressure from an observed cyclone in May 1985, the authors predicted that a 2-m increase in mean sea level would cause lower surges in some parts of the bay (measured relative to the sum of mean sea level and tide level) but higher surges in other parts. Flather (1994) made use of an improved version of the Flather and Khandker scheme to produce surge hindcasts for two cyclones that occurred during 1970 and 1991.

As-Selek and Yasuada (1995) employed a model of the Northern Bay of Bengal with a resolution more than 10 times greater than that of Flather and Khandker to demonstrate the importance of the Swatch of No Ground (an underwater trench) to the bay's circulation and to investigate the effect of a range of mean sea level rise scenarios. The results supported those from the lower-resolution model of Flather and Khandker, again suggesting that mean sea-level rise could, in some parts of the bay, reduce the height of water levels below that expected by simply adding together the mean-sea-level rise and storm surges predicted for present sea level. The authors also examined the effect of changes in the bathymetry, which are brought about by the large sediment transports through the delta, predicting that small changes in bathymetry will not increase the height of storm-surge peaks but will alter the horizontal span of the surges.

Unnikrishnan et al. (2006) and Mitchell et al. (2006) have used output from the HadCM2 and HadRM2 climate models to drive barotropic surge models, although the methodologies used were somewhat different. Both noted changes in the height of storm surges in some parts of the bay. This work has been repeated and combined with fluvial flooding estimates using an improved experimental design as part of the international Climate and Sea Level in parts of the Indian Subcontinent (CLASIC) project. For example, Figure 11.7 shows the simulated change in the height of the 99.9th percentile of annual surge residuals in the northern Bay of Bengal (measured relative to the tide) between present day and the 2080s for a future in which human-made emissions into the atmosphere follow the mid-range SRES B2 scenario (Nakicenovic and Swart 2000), with

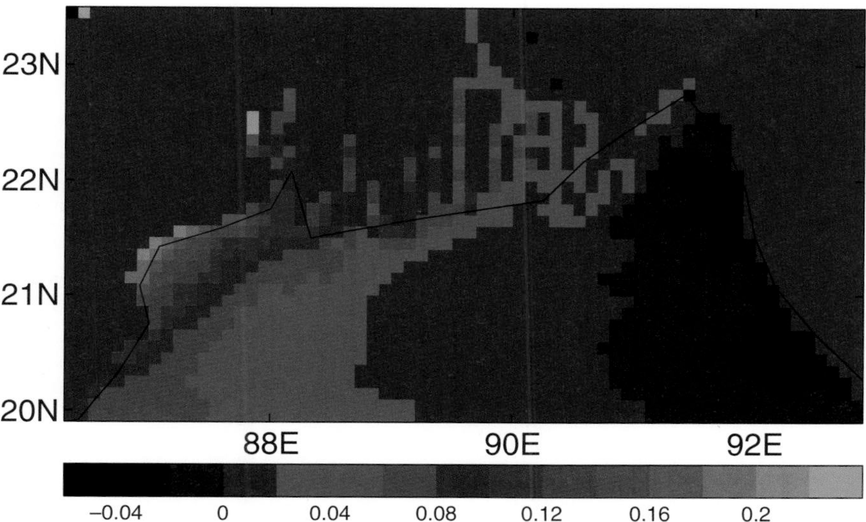

Figure 11.7 CLASIC project findings for simulated change in the height (m) of the 99.9th percentile of annual surge residuals in the northern Bay of Bengal (measured relative to the tide) between the present day and the 2080s assuming the SRES B2 emission scenario. The thin black line indicates the coastline.

simulations made using the HadRM3 regional climate model and the POL storm-surge model.

Recent studies on the occurrence of cyclones in the Bay of Bengal have not shown any trends during the last century. However, studies of intensification of cyclones have shown that the frequency of intense cyclones in the bay during November has been increasing (SMRC 2000). Unnikrishnan et al. (2006) made a similar conclusion for the future while analyzing the HadRM2 data for the northern Indian Ocean, in which the frequency of intense cyclones in the bay was found to be larger in an increased carbon dioxide model run (IS92a scenario) than in a control run. A corresponding increase in large surges was found in storm-surge simulations for the bay, when the surge model was forced with winds from the increased carbon dioxide run.

Australian Region

The effects of climate-driven changes in extreme sea levels around the coastline of Australia have been extensively studied. The northerly Cairns region is affected by surge events driven by tropical cyclones. McInnes et al. (2003) found that including an estimate of climate change to 2050 (with both changes in cyclone characteristics and a mean sea-level rise being considered) led to an increase in the area inundated by the most severe 5% of storms. McInnes et al. (2005b, 2006, 2009) examined change in extreme water levels to the south of Australia in the Bass Strait, where surges are driven by mid-latitude storms. McInnes et al. (2005b, 2006) included climate driven changes in wind speeds from 13 climate models, some of which predicted an increase and some a decrease. The lower projections

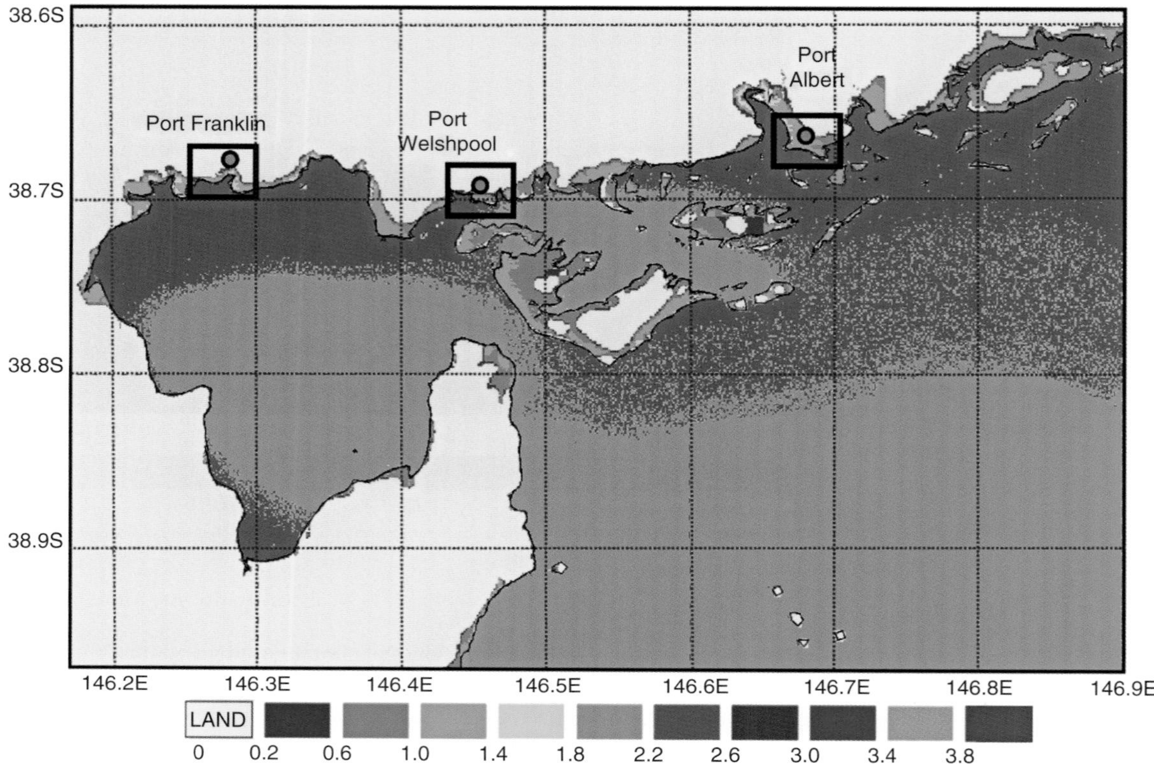

Figure 11.8 The coastal areas around Corner Inlet (south coast of Victoria, Australia) likely to experience inundation during a projected future one-in-100-year event (as presently estimated) for epoch 2070 following a high mean-sea-level rise scenario (McInnes et al. 2006). Levels are shown in meters relative to present-day mean sea level over the sea, and relative to the land surface over the land. The coastline is shown by a thin black line and flooded areas of land can be readily identified in blue.

of the climate models yielded little change in extreme water levels. However, the higher climate model projections gave rise to a significant increase in extreme-level events by 2070, driven mainly by the increase in mean sea level rather than by stronger winds.

Inundation levels due to the projected future one-in-100-year event around coastal townships within Corner Inlet along the southeastern coastline of Victoria were found to increase by between 15 and 30% (relative to land level) under a 2070 worst-case scenario (Figure 11.8). However, approximately 100 km along the coast to the northeast, around the Gippsland Lakes, inundation levels were found to increase by 166% under the same scenario (McInnes et al. 2006). The authors note that land subsidence could exacerbate the problem considerably.

11.4.3 Uncertainty in Storm-Surge Projections

Lowe and Gregory (2005) produced a simple comparison of some of the major sources of uncertainty in projected changes in the extreme water levels in the North Sea region using the IPCC TAR's estimate of global mean-sea-level rise (Church et al. 2001), the range of spatial sea level changes simulated in coupled climate models (Gregory et al. 2001), three storm-surge simulations from different driving models, and simulations using the same model set up but for different SRES emissions scenarios. They found that all of these contributions are important, but the relative importance can vary over quite small spatial scales. Natural variability is also important.

Woth et al. (2006) used a number of regional climate models to examine the uncertainty from the downscaling step from coarse global climate models to the regional-scale meteorological models. The spread in the results made with the alternative regional climate models suggests that this downscaling step does introduce some uncertainty but this is relatively small. Woth (2005) compared projected changes in extreme storm surges using results from two different global climate models, but found these were not statistically separable. The authors noted that these two models had similar physical formulations, and thus it is unlikely that they sampled a large fraction of the range of plausible global model response space.

McInnes et al. (2005b) studied the uncertainty in surge projections using a range of projections in future climate from several different climate models using the pattern scaling technique of Whetton et al. (2005) (see McInnes et al. 2005a). This approach incorporates an estimate of the range of uncertainty due to future emissions, climate sensitivity, and spatial variation, although the probability distribution of different climate model projections was not estimated. The changes in wind speed per degree of warming were estimated from 13 global and regional climate models then combined with estimates of temperature rise for nonintervention and stabilization emission scenarios to give a range of potential changes to meteorological conditions in the future. These, in turn, were combined with current meteorological conditions and then used to drive their storm-surge model, producing an estimate of the uncertainty range in extreme-water-level conditions.

11.4.4 Contribution of Waves to Future Coastal Extremes

Wave setup can contribute to changes in mean sea level at the coast (Longuet-Higgins and Stewart 1962). During a storm event, the wave setup may be of the order of a meter at certain locations, although at most locations it is usually much less. The actual value depends on the direction of wave approach and details of the nearshore bathymetry. Waves can cause overtopping of sea defenses with consequent failure or coastal erosion which may increase the future exposure of the coastline to further attack. Changes in the direction of storm waves or swell may change longshore drift and associated sediment transport. As with the prediction of future changes in storm surges, future changes in wave characteristics can be determined using both dynamic modeling and statistical techniques. The same advantages and disadvantages of each approach, as discussed in section 11.4.1, apply here.

Wang et al. (2004) employed an observed relationship between the observed wave conditions and the NAO (see section 11.2.3), and assumed the relationship will continue to hold in the future. Projected changes in waves were then linked to a projection that the occurrence of the positive phase of the NAO will be more frequent under global warming. A study by Terray et al. (2004) supports this view and suggests that the main change in wintertime weather patterns will result in more positive NAO patterns. On the other hand, Jones et al. (2003) showed that the influence of the NAO has varied significantly over the last 150–200 years and has been particularly strong recently. It is notable that the availability of satellite measurements of waves over the last quarter century coincides with a particular phase of the NAO. Thus, it is still not clear how important the NAO will be in the future and how much NAO changes will be linked with anthropogenic climate change (Tsimplis et al. 2005).

Caires et al. (2006) derived return value estimates of SWH up to the end of the 21st century using projections of the sea-level pressure under three different forcing scenarios from the Canadian coupled climate model. The methodology employed used regression methods to link the climate model pressure simulations to wave height. Under all forcing scenarios, significant changes are to be expected in different regions of the globe with the larger and more significant changes occurring under the more severe emission scenarios. Under all future scenarios considered, significant positive trends are to be expected in the North Pacific. Similar patterns were seen in a study using different climate models (Wang and Swail 2006b).

The WASA and STOWASUS projects both included an element to estimate future climate-driven changes in waves (see Kaas and Andersen 2000). The simulated time slices in the WASA experiment were rather short to isolate a clear signal but the experimental design was improved for STOWASUS. Using a dynamical wave model driven with atmospheric winds from a set of global high-resolution simulations a future increase in high waves was found in the northeastern part of the North Atlantic but decreases occurred further southwest. Andrade et al. (2007) studied possible changes in wave conditions around Portugal by the end of the 21st century with the use of the HadCM3 climate model coupled with a spectral wave model. Results suggest that mean wave heights will be essentially unchanged although intra-annual distributions may change, in addition to wave direction, with possible modifications in longshore transports.

Wolf and Woolf (2006) used a dynamic wave model approach to show how different climate change effects (e.g. increase in wind speed or change in wind direction) are likely to alter wave conditions in the waters around the UK. This type of sensitivity study is a valuable step towards being able to understand the predicted changes in waves resulting from a range of policy relevant future emissions scenarios. Changes in sea-ice extent have important implications for waves, in changing the fetch over which the wind is blowing. Recently Arctic sea ice has been reducing (e.g. Rigor and Wallace 2004).

11.4.5 Tsunamis

There are many processes other than storm surges which can result in a high coastal sea level. These include seiches (e.g. Pugh 1987, chapter 6), "meteorological tsunamis" (a combination of storm surge and resonance, e.g. Monserrat et al. 2006), and tsunamis themselves. The latter can be localized to a particular ocean area such as the Mediterranean (e.g. Tinti et al. 2006), or global in scale as demonstrated by the Sumatra 2004 event (Titov et al. 2005; Woodworth et al. 2005).

The very rare and unpredictable nature of earthquakes and their associated tsunamis makes it very difficult to include them in probability estimates of extreme events as they are less likely (than storm surges, for example) to have been included in the instrumental record. This introduces further grounds for caution in application of published extreme-level curves derived from periods without tsunamis.

Attempts at inclusion of tsunami risk in extreme-level estimates are best performed with the use of numerical modeling, combined with geological and other insight into the likely frequency and magnitude of tsunami events, and with due attention to their often very localized coastal impact (Horsburgh et al. 2008). There is no reason to expect more, or larger, tsunamis due to projected century-scale climate change. However, with the increase in coastal populations, the impacts of any such events are likely to be greater.

11.5 Future Research Needs

Extreme sea levels are closely linked to serious impacts in the coastal zone, and so their continued monitoring and improved prediction are important.

11.5.1 Extending the Evidence Base for Change

The limitations of historical data sets of extreme sea levels referred to in section 11.2.1 need to be addressed urgently. This applies to extremes for which the probability of occurrence is consistent with the present-day distribution of levels (e.g. Figure 11.9a). In other cases, extremes occur as a consequence of individual catastrophic events such as the Katrina storm surge (or a number of major surges in the Bay of Bengal discussed above, or even tsunamis), which lead to existing probability distribution curves needing to be revised (Figure 11.9b). In each case, an underlying mean sea level rise can add to the surge contribution.

Figure 11.9 (a) The Ponte della Paglia in Venice, an example of regular flooding which occurs from a spectrum of large surge events coupled with a mean-sea-level rise. (b) New Orleans after the Katrina hurricane and storm surge, an example of an individual catastrophic "outlier" event. (Photographs copyright Sarah Quill and UK Met Office.)

As regards sea-level data sets, historical information which exists still in paper form (e.g. tide-gauge charts) needs to be converted into computer-accessible formats, so that studies of extremes can take place alongside those of mean sea-level change (a process called "data archaeology"). Such information is known to exist in several European countries as a consequence of measurements in their former colonies. India, Brazil, and several other countries are also known to hold such information. In addition, some historic data from tide gauges lack vertical datum information which it would be good to determine if possible. For the present day, monitoring at sites in the GLOSS and GCOS networks needs to be undertaken to modern standards (i.e. at sufficiently high frequency) to establish where and at what rate extreme sea level is changing. In a complementary fashion, spatial information on sea-level extremes in the open ocean should also become possible if the present altimetric record can be extended into the future. However, if altimeters are to be used to study extremes, higher spatial and temporal sampling than provided by an individual present-day, nadir-pointing instrument is required. More routine observations are also necessary to adequately monitor changes in the wave climate, including wave direction. In addition to their direct use, wave observations provide a proxy for storminess. Further observational studies of changes in coastal morphodynamics (by direct surveying or by aerial or satellite remote sensing) would also be extremely useful.

Barotropic modeling provides understanding of historical storm surges over a region, complementing the time-series information from gauges, while the necessary construction of high-resolution meteorological data sets needed for such work can also be applied to investigate the temporal variability in wave heights. The development of such historical regional and global meteorological data sets (as free as possible from time-dependent biases) is a priority. A prerequisite for reconstructing past developments with models, possibly augmented by assimilating homogeneous time series into the models, requires the availability of homogeneous space-time detailed analyses of wind and air pressure for the past decades. Efforts such as regional reanalysis, making use of a large variety of local observations as well as dynamical downscaling efforts, constrained by large-scale analyses, are needed globally, not only for Europe (e.g. Feser et al. 2001) but also for tropical regions affected by storms. Long runs of regional and global barotropic models, forced by historical meteorological information, also need to be employed to derive extreme-level curves for the entire global coastline, enabling coastal planners to estimate changes in risk to potential mean-sea-level change at any location.

In order to build a case on which decisions regarding future emissions reductions can be made there is a need, at each location where large sea level (and wave) variations have been observed, to identify as far as possible that part of historical change in both mean levels and extremes that have resulted from human activity and that which is due to natural drivers of change (see Woodworth and Blackman 2004; Hunter 2002, 2004; Woodworth 2005; Sheppard et al. 2005). Such investigations are necessarily multidisciplinary and might usefully be similar to the formal detection and attribution methodology already applied to temperature, rainfall, and runoff.

11.5.2 Increasing the Mechanistic Understanding of the Drivers of Change

There are significant gaps in our understanding of global mean sea level rise, the spatial pattern of these time mean changes, and changes in atmospheric storminess as the climate warms. These drivers of extreme sea level change have been discussed in this and previous chapters.

Advances are likely to require more observations in order to construct and test detailed models of key elements of the climate system. Work is also needed to understand the importance of large-scale planetary changes on the local scale. For instance, there are still gaps in our knowledge of how individual local feedbacks combine to produce the large-scale pattern of temperature rise, which in turn affect the rate and pattern of sea-level rise and atmospheric storminess changes.

11.5.3 Improving Prediction

Improved understanding of changes in the meteorological drivers of extreme water level should be combined with improved models of regional sea level and waves (for instance at higher spatial resolution and with coupling to river-routing models) in order to produce better predictions of future extreme sea level at a fine spatial scale. The improvements in meteorological insight will benefit the projection of changes in waves also. There is also the need to improve experimental design with longer simulated time slices or larger sets of initial condition ensembles, and better awareness of available statistical methods. Because results cannot be generalized from region to region, these models need to be applied to a wider range of regions, perhaps with the technology to make these simulations being transferred to the nations likely to be affected. This is especially important for developing countries where the spending on climate change modeling and climate change adaptation is limited.

A rigorous comparison of the dynamical and statistical methods in a wide range of regions is desirable, in order to allow recommendations to be made on the appropriate techniques for future work in each area. The dynamical methods will require the best possible information on bathymetry and its temporal evolution near the coast.

In parallel with improved model predictions, we also need to better quantify the uncertainty in the extreme sea level predictions. Initial results have been described here but further developments are required. One such development is the use of perturbed parameter ensembles of global climate models (e.g. Murphy et al. 2004) to produce estimated changes in driving meteorology. Another development is the wider use of observational constraints to validate models and assess the likelihood of different model configurations.

Finally, a key reason for modeling extreme sea-level changes, waves, and river flows is to provide advice on the impacts in the coastal zone, such as inundation and coastal erosion. Often it is the joint probability of occurrence of extreme events in these quantities rather than the occurrence of an individual extreme that results in the most serious damage. Such studies require that investigators of extremes participate in a more widespread communication with scientists and engineers in possession of models of inundation, coastal erosion, and other impacts (different types of model employed for different applications). While this is starting to happen (e.g. in the UK Tyndall Coastal Simulator Project; http://

vrlab.env.uea.ac.uk/wiki/index.php), improvements are required. These impacts models will need detailed information on natural and human-made coastal defenses, which do not exist in many countries.

11.6 Conclusions

A general conclusion based on tide-gauge observations complemented by numerical modeling is that at most locations there is so far little evidence for extreme sea levels changing by amounts significantly different to changes in mean sea level. However, that conclusion is limited to those parts of the world where adequate historical data exist. In the last few years, there has been great progress in collecting and interpreting observations and in the making of predictions of extreme sea level. There is an urgent need for sustained observational data sets and for a range of improved numerical modeling and statistical techniques. Together these will lead to the improved understanding of past changes and projections of future ones.

One concludes that improvement of the quantification of uncertainty for regions of interest is a research priority. Since one is interested in the extreme water levels and waves, the uncertainties in meteorological drivers and mean-sea-level changes are both required. Local vertical land-movement information is also required. Ultimately, projections of the probability distributions of extreme water-level changes, not just their ranges, are needed in order to undertake quantitative risk assessments of the effects of inundation events on the coastal natural environment and on strategic coastal infrastructure.

Acknowledgments

We thank Dr X.L. Wang (Environment Canada), Dr. Roger Flather, and Professor David Pugh (POL) for useful comments. We also thank the reviewers for valuable comments.

References

Alexander L.V., Tett S.F.B. and Jónsson T. (2005) Recent observed changes in severe storms over the United Kingdom and Iceland. *Geophysical Research Letters* **32**, L13704.

Alexandersson H., Tuomenvirta H., Schmith T. and Iden K. (2000) Trends of storms in NW Europe derived from an updated pressure data set. *Climate Research* **14**, 71–3.

Allan J. and Komar P. (2000) Are ocean wave heights increasing in the eastern North Pacific? *EOS Transactions of the American Geophysical Union* **47**, 561–7.

Andrade C., Pires H.O., Taborda R. and Freitas M.C. (2007) Projecting future changes in wave climate and coastal response in Portugal by the end of the 21st century. *Journal of Coastal Research, Special Issue* **50**, 253–7.

Aráujo I., Pugh D. and Collins M. (2002) *Trends in Components of Sea Level around the English Channel*. Proceedings of Littoral 2002, Porto, Portugal.

As-Salek J.A. and Yasuda T. (1995) Comparative study of the storm surge models proposed for Bangladesh: last developments and research needs. *Journal of Wind Engineering and Industrial Aerodynamics* **54–5**, 595–610.

Bacon S. and Carter D.J.T. (1993) A connection between mean wave height and atmospheric pressure gradient in the North Atlantic. *International Journal of Climatology* **13**, 423–36.

Bärring L. and von Storch H. (2004) Scandinavian storminess since about 1800. *Geophysical Research Letters* **31**, L20202.

Bengtsson L., Botzet M. and Esch M. (1995) Hurricane type vortices in a general circulation model. *Tellus A* **47**, 175–96.

Bengtsson L., Botzet M. and Esch M. (1996) Will greenhouse gas-induced warming over the next 50 years lead to a higher frequency and greater intensity of hurricanes? *Tellus A* **48**, 57–73.

Bengtsson L., Hodges K. and Roeckner E. (2006) Storm tracks and climate change. *Journal of Climate* **19**, 3518–43.

Bengtsson L., Hodges K.I., Esch M., Keenlyside N., Kornblueh L., Luo J.-J. and Yamagata T. (2007) How may tropical cyclones change in a warmer climate? *Tellus A* **59**, 539–61.

Bernier N.B. and Thompson K.R. (2006) Predicting the frequency of storm surges and extreme sea levels in the Northwest Atlantic. *Journal of Geophysical Research* **111**, C10009.

Bernier N.B., Thompson K.R., Ou J. and Ritchie H. (2007) Mapping the return periods of extreme sea levels: allowing for short sea level records, seasonality, and climate change. *Global and Planetary Change* **57**, 139–50.

Blackmon M.L. (1976) A climatological spectral study of the 500mb geopotential height of the northern hemisphere. *Journal of the Atmospheric Sciences* **33**, 1607–23.

Bogen K.T., Jones E.D. and Fischer L.E. (2007) Hurricane destructive power predictions based on historical storm and sea surface temperature data. *Risk Analysis* **27**, 1497–517.

Bouligand R. and Pirazzoli P.A. (1999) Les surcotes et les décotes marines à Brest, étude statistique et évolution. *Oceanologica Acta* **22**, 153–66.

Briggs W.A. (2008) On the changes in the number and intensity of north Atlantic tropical cyclones. *Journal of Climate* **21**, 1387–402.

Bromirski P.D., Flick R.E. and Cayan D.R. (2003) Storminess variability along the California coast: 1858–2000. *Journal of Climate* **16**, 982–93.

Busuioc A., Giorgi F., Bi X. and Ionita M. (2006) Comparison of regional climate model and statistical downscaling simulations of different winter precipitation change scenarios over Romania. *Theoretical and Applied Climatology* **86**, 101–23.

Caires S., Swail V. and Wang X. (2006) Projection and analysis of extreme wave climate. *Journal of Climate* **19**, 5581–605.

Carnell R.E. and Senior C.A. (1998) Changes in mid-latitude variability due to increasing greenhouse gases and sulphate aerosols. *Climate Dynamics* **14**, 369–83.

Carnell R.E., Senior C.A. and Mitchell J.F.B. (1996) An assessment of measures of storminess: simulated changes in northern hemisphere winter due to increasing CO_2. *Climate Dynamics* **12**, 467–76.

Carter D.J.T. and Draper L. (1988) Has the north-east Atlantic become rougher? *Nature* **332**, 494.

Casulli V. and Cattani E. (1994) Stability, accuracy and efficiency of a semi-implicit method for three dimensional shallow water flow. *Computers and Mathematics with Applications* **27**, 99–112.

Chan J.C.L. (2006) Comment on "Changes in tropical cyclone number, duration, and intensity in a warming environment". *Science* **311**, 1713.

Chang E.K.M. and Fu Y. (2002) Interdecadal variations in Northern Hemisphere winter storm track intensity. *Journal of Climate* **15**, 642–58.

Chang E.K.M., Lee S. and Swanson K.L. (2002) Storm track dynamics. *Journal of Climate* **15**, 2163–83.

Chauvin F., Royer J-F. and Déqué M. (2006) Response of hurricane-type vortices to global warming as simulated by ARPEGE-Climat at high resolution. *Climate Dynamics* **27**, 377–99.

Christoph M., Ulbrich U. and Speth P. (1997) Midwinter suppression of northern hemisphere storm track activity in the real atmosphere and in GCM experiments. *Journal of the Atmospheric Sciences* **54**, 1589–99.

Church J.A., Gregory J.M, Huybrechts P., Kuhn M., Lambeck K., Nhuan M.T. et al. (2001) Changes in sea level. In: *Climate Change 2001: The Scientific Basis.* Contribution of Working Group 1 to the Third Assessment Report of the Intergovernmental Panel on Climate Change (Houghton J.T., Ding Y., Griggs D.J., Noguer M., van der Linden P.J., Dai X. et al., eds), pp. 639–94. Cambridge University Press, Cambridge.

Church J.A., Hunter J.R., McInnes K.L. and White N.J. (2006) Sea-level rise around the Australian coastline and the changing frequency of extreme sea-level events. *Australian Meteorological Magazine* **55**, 253–60.

Coles S. (2001) *An Introduction to Statistical Modeling of Extreme Values.* Springer, London.

Compo G.P. and Sardeshmukh P.D. (2004) Storm track predictability on seasonal and decadal scales. *Journal of Climate* **17**, 3701–20.

Cubasch U., Meehl G.A., Boer G.J., Stouffer R.J., Dix M., Noda A. et al. (2001) Projections of future climate change. In: *Climate Change 2001: The Scientific Basis.* Contribution of working group I to the third assessment report of the Intergovernmental Panel on Climate Change (Houghton J.T., Ding Y., Griggs D.J., Noguer M., van der Linden P.J., Dai X. et al., eds), pp. 526–82. Cambridge University Press, Cambridge.

Das P.K. (1972) Prediction model for storm surges in the Bay of Bengal. *Nature* **239**, 211–13.

Das P.K., Sinha M.C. and Balasubramanyam V. (1974) Storm surges in the Bay Bengal. *Quarterly Journal of the Royal Meteorological Society* **100**, 437–49.

Debernard J., Saetra O. and Røed L.P. (2002) Future wind, wave and storm surge climate in the northern North Atlantic. *Climate Research* **23**, 39–49.

Dixon M.J. and Tawn J.A. (1992) Trends in U.K. extreme sea levels: a spatial approach. *Geophysical Journal International* **111**, 607–16.

D'Onofrio E.E., Fiore M.M.E. and Romero S.I. (1999) Return periods of extreme water levels estimated for some vulnerable areas of Buenos Aires. *Continental Shelf Research* **19**, 1681–93.

D'Onofrio E.E., Fiore M.M.E. and Pousa J.L. (2008) Changes in the regime of storm surges in Buenos Aires, Argentina. *Journal of Coastal Research* **24**, 260–5.

Döscher R., Willén U, Jones C., Rutgersson A., Meier H.E.M., Hansson U. and Graham L.P. (2002) The development of the regional coupled ocean-atmosphere model RCAO. *Boreal Environment Research* **7**, 183–92.

Dube S.K., Sinha P.C. and Roy G.D. (1985) The numerical simulation of storm surges along the Bangladesh coast. *Dynamics of Atmospheres and Oceans* **9**, 121–33.

Dyke P.P.G. (2001) *Coastal and Shelf Sea Modelling*. Kluwer Academic Publishers, Dordrecht.

Emanuel K.A. (2005) Increasing destructiveness of tropical cyclones over the past 30 years. *Nature* **436**, 686–8.

Emanuel K.A. (2007) Environmental factors affecting tropical cyclone power dissipation. *Journal of Climate* **20**, 5497–509.

FEMA (2006) *Reconstruction Guidance using Hurricane Katrina Surge Inundation and Advisory Base Flood Elevation Maps*. Federal Emergency Management Agency. http://www.fema.gov/hazard/flood/recoverydata/katrina/katrina_about.shtm.

Feser F., Weisse R. and von Storch H. (2001) Multi-decadal atmospheric modelling for Europe yields multi-purpose data. *EOS Transactions of the American Geophysical Union* **82**, 305–10.

Firing Y.L. and Merrifield M.A. (2004) Extreme sea level events at Hawaii: the influence of mesoscale eddies. *Geophysical Research Letters* **31**, L24306.

Fischer-Bruns I., von Storch H., Gonzáles-Rouco J.F. and Zorita E. (2005) Modelling the variability of midlatitude storm activity on decadal to century time scales. *Climate Dynamics* **25**, 461–76.

Flather R.A. (1994) A storm surge prediction model for the northern Bay of Bengal with application to the cyclone disaster in April 1991. *Journal of Physical Oceanography* **24**, 172–90.

Flather R.A. (2000) Existing operational oceanography. *Coastal Engineering* **41**, 13–40.

Flather R.A. and Khandker H. (1993) The storm surge problem and possible effects of sea level changes on coastal flooding in the Bay of Bengal. In: *Climate and Sea Level Change: Observations, Projections and Implications* (Warrick R.A., Barrow E.M. and Wigley T.M., eds), pp. 229–45. Cambridge University Press, Cambridge.

Flather R.A. and Smith J.A. (1998) First estimates of changes in extreme storm surge elevations due to the doubling of CO2. *The Global Atmosphere and Ocean System* **6**, 193–208.

Flather R.A., Smith J.A., Richards J.D., Bell C. and Blackman D.L. (1998) Direct estimates of extreme storm surge elevations from a 40-year numerical model simulation and from observations. *The Global Atmosphere and Ocean System* **6**, 165–76.

Fyfe J.C. (2003) Extratropical Southern Hemisphere cyclones: harbingers of climate change? *Journal of Climate* **16**, 2802–5.

Geng Q. and Sugi M. (2003) Possible change of extratropical cyclone activity due to enhanced greenhouse gases and sulfate aerosols – study with a high-resolution AGCM. *Journal of Climate* **16**, 2262–74.

Gillett N.P., Graf H.F. and Osborn T.J. (2003) Climate change and the North Atlantic Oscillation. In: *North Atlantic Oscillation: Climatic Significance and Environmental Impact* (Hurrell J.W., Kushnir Y., Ottersen G. and Visbeck M., eds), pp. 193–209. Geophysical Monograph 134. American Geophysical Union, Washington DC.

Gillett N.P., Allan R.J. and Ansell T.J. (2005) Detection of external influence on sea level pressure with a multi-model ensemble. *Geophysical Research Letters* **32**(19), L19714.

Giorgi F., Hewitson B., Christensen J., Hulme M., von Storch H., Whetton P. et al. (2001) Regional climate information – evaluation and projections. In: *Climate Change 2001: The Scientific Basis*. Contribution of Working Group I to the Third Assessment Report of the Intergovernmental Panel on Climate Change (Houghton J.T., Ding Y., Griggs D.J., Noguer M., van der Linden P.J., Dai X. et al., eds), pp. 583–638. Cambridge University Press, Cambridge.

Goodwin I.D. (2005) A mid-shelf, mean wave direction climatology for south-eastern Australia and its relationship to the El Niño – Southern Oscillation since 1878 A.D. *International Journal of Climatology* **25**(13), 1715–29.

Gower J.F.R. (2002) Temperature, wind and wave climatologies, and trends from marine meteorological buoys in the Northeast Pacific. *Journal of Climate* **15**, 3709–18.

Graham N. and Diaz H. (2001) Evidence for intensification of North Pacific winter cyclones since 1948. *Bulletin of the American Meteorological Society* **82**, 1869–83.

Gregory J.M., Church J.A., Boer G.J., Dixon K.W., Flato G.M., Jackett D.R. et al. (2001) Comparison of results from several AOGCMs for global and regional sea-level change 1900–2100. *Climate Dynamics* **18**, 225–40.

Grossmann I., Woth K. and von Storch H. (2007) Localization of global climate change: Storm surge scenarios for Hamburg in 2030 and 2085. *Die Küste* **71**, 169–82.

Gulev S.K. and Hasse L. (1999) Changes of wind waves in the North Atlantic over the last 30 years. *International Journal of Climatology* **19**, 1091–1117.

Gulev S.K. and Grigorieva V. (2004) Last century changes in ocean wind wave height from global visual wave data. *Geophysical Research Letters* **31**, L24302.

Gulev S.K. and Grigorieva V. (2006) Variability of the winter wind waves and swell in the North Atlantic and North Pacific as revealed by the Voluntary Observing ship data. *Journal of Climate* **19**, 5667–85.

Gulev S., Zolina K. and Grigoriev S. (2001) Extratropical cyclone variability in the Northern Hemisphere winter from the NCEP/NCAR-Reanalysis data. *Climate Dynamics* **17**, 795–809.

Günther H., Rosenthal W., Stawarz M., Carretero J.C., Gomez M., Lozano I. et al. (1998) The wave climate of the Northeast Atlantic over the period 1955–94: the WASA wave hindcast. *The Global Atmosphere Ocean System* **6**, 121–63.

Haarsma R.J., Mitchell J.F.B. and Senior C.A. (1993) Tropical disturbances in a GCM. *Climate Dynamics* **8**, 247–57.

Hall N.M.J., Hoskins B.J., Valdes P.J. and Senior C.A. (1994) Storm tracks in a high-resolution GCM with doubled carbon dioxide. *Quarterly Journal of the Royal Meteorological Society* **120**, 1209–30.

Harangozo S.A. (1992) Flooding in the Maldives and its implications for the global sea level rise debate. In: *Sea Level Changes: Determination and Effects* (Woodworth P.L., Pugh D.T., De Ronde J.G., Warrick R.G. and Hannah J., eds), pp. 95–9. American Geophysical Union, Washington DC.

Hemer M.A., Church J.A. and Hunter J.R. (2007) Waves and climate change on the Australian coast. *Journal of Coastal Research* **SI 50**, 432–37.

Henderson-Sellers A., Zhang H., Berz G., Emanuel K., Gray W., Landsea C. et al. (1998) Tropical cyclones and global climate change: a post-IPCC assessment. *Bulletin of the American Meteorological Society* **79**, 19–38.

Horsburgh K.J., Wilson C., Baptie B.J., Cooper A., Cresswell D., Musson R.M.W. et al. (2008) Impact of a Lisbon-type tsunami on the UK coastline, and the implications for tsunami propagation over broad continental shelves. *Journal of Geophysical Research* **113**, C04007.

Hoskins B.J. and Hodges K.I. (2002) New perspectives on the northern hemisphere winter storm tracks. *Journal of the Atmospheric Sciences* **59**, 1041–61.

Hoyos C.D., Agudelo P.A., Webster P.J. and Curry J.A. (2006) Deconvolution of the factors contributing to the increase in global hurricane intensity. *Science* **312**, 94–7.

Hunter J.R. (2002) *A Note on Relative Sea Level Change at Funafuti, Tuvalu.* University of Tasmania internal report. http://staff.acecrc.org.au/~johunter/tuvalu.pdf.

Hunter J.R. (2004) Comments on "Tuvalu not experiencing increased sea level rise". *Energy and Environment* **15**, 925–30.

Hurrell J.W. and van Loon H. (1997) Decadal variations in climate associated with the north Atlantic oscillation. *Climate Change* **36**, 301–26.

Hurrell J.W., Kushnir Y., Ottersen G. and Visbeck M. (2003) An overview of the North Atlantic Oscillation. In: *The North Atlantic Oscillation: Climatic Significance and Environmental Impact* (Hurrell J.W., Kushnir Y., Ottersen G. and Visbeck M.), pp. 1–35, Geophysical Monograph Series 134. American Geophysical Union, Washington DC.

Jones P.D., Osborn T.J. and Briffa K.R. (2003) Pressure-based measurements of the North Atlantic Oscillation (NAO): a comparison and an assessment of changes in the strength of the NAO and in its influence on surface climate parameters. In: *The North Atlantic Oscillation: Climatic Significance and Environmental Impact* (Hurrell J.W., Kushnir Y., Ottersen G. and Visbeck M., eds), pp. 51–62, Geophysical Monograph 134. American Geophysical Union, Washington DC.

Jones R.G., Murphy J.M. and Noguer M. (1995) Simulation of a climate change over Europe using a nested regional climate model. I: Assessment of control climate including sensitivity to location of lateral boundary conditions. *Quarterly Journal of the Royal Meteorological Society* **121**, 1413–99.

Jónsson T. and Hanna E. (2007) A new day-to-day pressure variability index as a proxy of Icelandic storminess and complement to the North Atlantic Oscillation index 1823–2005. *Meteorologische Zeitschrift* **16**, 25–36.

Kaas E. and Andersen U. (2000) Scenarios for extra-tropical storm and wave activity: methodologies and results. In: *ECLAT-2 KNMI Workshop Report No. 3* (Beersma J., Agnew M., Viner D. and Hulme M., eds). KNMI, The Netherlands.

Kauker F. and Langenberg H. (2000) Two models for the climate change related development of sea levels in the North Sea: a comparison. *Climate Research* **15**, 61–7.

Klotzbach P.J. (2006) Trends in global tropical cyclone activity over the past twenty years (1986–2005). *Geophysical Research Letters* **33**, L10805.

Knaff J.A. and Sampson C.R. (2006) *Reanalysis of West Pacific Tropical Cyclone Intensity 1966–1987*. Proceedings of 27th AMS Conference on Hurricanes and Tropical Meteorology, #5B.5. http://ams.confex.com/ams/pdfpapers/108298.pdf.

Knippertz P., Ulbrich U. and Speth P. (2000) Changing cyclones and surface wind speeds over the North Atlantic and Europe in a transient GHG experiment. *Climate Research* **15**, 109–22.

Knutson T.R. and Tuleya R.E. (2004) Impact of CO_2-induced warming on simulated hurricane intensity and precipitation: sensitivity to the choice of climate model and convective parameterization. *Journal of Climate* **17**, 3477–95.

Knutson T.R., Tuleya R.E., Shen W. and Ginis I. (2001) Impact of CO2-induced warming on hurricane intensities as simulated in a hurricane model with ocean coupling. *Journal of Climate* **14**, 2458–68.

Knutson T.R., Delworth T.L., Dixon K.W., Held I.M., Lu J., Ramaswamy V. et al. (2006) Assessment of twentieth-century regional surface temperature trends using the GFDL CM2 coupled models. *Journal of Climate* **9**, 1624–51.

König W., Sausen R. and Sielman F. (1993) Objective identification of cyclones in GCM simulations. *Journal of Climate* **6**, 2217–31.

Kossin J.P., Knapp K.R., Vimont D.J., Murnane R.J. and Harper B.A. (2007) A globally consistent reanalysis of hurricane variability and trends. *Geophysical Research Letters* **34**, L04815.

Kushnir Y., Cardone V.J., Greenwood J.G. and Cane M.A. (1997) The recent increase in North Atlantic wave heights. *Journal of Climate* **10**, 2107–13.

Kuzmina S.I., Bengtsson L., Johannessen O.M., Drange H., Bobylev L.P. and Miles M.W. (2005) The North Atlantic Oscillation and greenhouse-gas forcing. *Geophysical Research Letters* **32**, L04703.

Laing A.K. (2000) New Zealand wave climate from satellite observations. *New Zealand Journal of Marine and Freshwater Research* **34**, 727–44.

Lambert S.J. (1995) The effect of enhanced greenhouse warming on winter cyclone frequencies and strengths. *Journal of Climate* **8**, 1447–52.

Lambert S.J. and Fyfe J.C. (2006) Changes in winter cyclone frequencies and strengths simulated in enhanced greenhouse warming experiments: results from the models participating in the IPCC diagnostic exercise. *Climate Dynamics* **26**, 713–28.

Lambert S.J., Sheng J. and Boyle J. (2002) Winter cyclone frequencies in thirteen models participating in the Atmospheric Model Intercomparison Project (AMIP1). *Climate Dynamics* **19**, 1–16.

Landsea C.W. (2005) Hurricanes and global warming. *Nature* **438**, E11–12.

Landsea C.W., Harper B.A., Hoarau K. and Knaff J.A. (2006) Can we detect trends in extreme tropical cyclones? *Science* **313**, 452–4.

Landsea C.W., Vecchi G.A., Bengtsson L. and Knutson T.R. (2009) Impact of duration thresholds on Atlantic tropical cyclone counts. *Journal of Climate* doi:10.1175/2009JCLI3034.1.

Langenberg H., Pfizenmayer A., von Storch H. and Sündermann J. (1999) Storm related sea level variations along the North Sea coast: natural variability and anthropogenic change. *Continental Shelf Research* **19**, 821–42.

Leckebusch G. and Ulbrich U. (2004) On the relationship between cyclones and extreme windstorm events over Europe under climate change. *Global and Planetary Change* **44**, 181–93.

Leckebusch G.C., Koffi B., Ulbrich U., Pinto J.G., Spangehl T. and Zacharias S. (2006) Analysis of frequency and intensity of European winter storm events from a multi-model perspective, at synoptic and regional scales. *Climate Research* **31**, 59–74.

Lennon G.W. (1963) The identification of weather conditions associated with the generation of major storm surges along the west coast of the British Isles. *Quarterly Journal of the Royal Meteorological Society* **89**, 381–94.

Lighthill J., Holland G.J., Gray W.M., Landsea C., Craig G., Evans J. et al. (1994) Global climate change and tropical cyclones. *Bulletin of the American Meteorological Society* **75**, 2147–57.

Lim E.-P. and Simmonds I. (2007) Southern hemisphere winter extratropical cyclone characteristics and vertical organization observed with the ERA-40 data in 1979–2001. *Journal of Climate* **20**, 2675–90.

Lindzen R.S. and Farrell B. (1980) Simple approximate result for the maximum growth rate of baroclinic instabilities. *Journal of the Atmospheric Sciences* **37**, 1648–54.

Lionello P. (2005) Extreme surges in the Gulf of Venice, present and future climate. In: *Venice and its Lagoon, State of Knowledge* (Fletcher C. and Spencer T., eds), pp. 59–65. Cambridge University Press, Cambridge.

Liu Z., Vavrus S., He F., Wen N. and Zhong Y. (2005) Rethinking tropical ocean response to global warming: the enhanced equatorial warming. *Journal of Climate* **18**, 4684–4700.

Longuet-Higgins M.S. and Stewart R.W. (1962) Radiation stress and mass transport in gravity waves, with applications to "surf beats". *Journal of Fluid Mechanics* **13**, 481–504.

Lowe J.A. and Gregory J.M. (2005) The effects of climate change on storm surges around the United Kingdom. *Philosophical Transactions of the Royal Society A* **363**, 1313–28.

Lowe J.A., Gregory J.M. and Flather R.A. (2001) Changes in the occurrence of storm surges around the United Kingdom under a future climate scenario using a dynamic storm surge model driven by the Hadley Centre climate models. *Climate Dynamics* **18**, 179–88.

Lowe J.A., Howard T., Pardaens A., Tinker J., Holt J., Wakelin S. et al. (2009) *UK Climate Projections Science Report: Marine and Coastal Projections*. Met Office Hadley Centre, Exeter.

Lozano I. and Swail V. (2002) The link between wave height variability in the North Atlantic and the storm track activity in the last four decades. *Atmosphere-Ocean* **40**, 377–88.

Lunkeit F., Fraedrich K. and Bauer S.E. (1998) Storm tracks in a warmer climate: sensitivity studies with a simplified global circulation model. *Climate Dynamics* **14**, 813–26.

Mann M. and Emanuel K. (2006) Atlantic hurricane trends linked to climate change. *EOS Transactions of the American Geophysical Union* **87**, 24.

Mastenbroek C., Burgers G. and Janssen P.A.E.M. (1993) The dynamical coupling of a wave model and storm surge model through the atmospheric boundary layer. *Journal of Physical Oceanography* **23**, 1856–66.

McDonald R.E., Bleaken D.G., Cresswell D.R, Pope V.D. and Senior C.A. (2005) Tropical storms: representation and diagnosis in climate models and the impacts of climate change. *Climate Dynamics* **25**, 19–36.

McInnes K.L., Walsh K.J.E., Hubbert G.D. and Beer T. (2003) Impact of sea-level rise and storm surges on a coastal community. *Natural Hazards* **30**(2), 187–207.

McInnes K.L., Abbs D.J. and Bathols J.A. (2005a) *Climate Change in Eastern Victoria. Stage 1 Report: The Effect of Climate Change on Coastal Wind and Weather Patterns: a Project Undertaken for the Gippsland Coastal Board, Aspendale, Victoria*. CSIRO Marine and Atmospheric Research. http://www.cmar.csiro.au/e-print/open/mcinnes_2005a.pdf.

McInnes K.L., Macadam I., Hubbert G.D., Abbs D.J. and Bathols J.A. (2005b) *Climate change in Eastern Victoria. Stage 2 Report: The Effect of Climate Change on Storm Surges: a Project Undertaken for the Gippsland Coastal Board. Aspendale, Victoria:* CSIRO Marine and Atmospheric Research. http://www.cmar.csiro.au/e-print/open/mcinnes_2005b.pdf.

McInnes K. L., Macadam I. and Hubbert G.D. (2006) *Climate change in Eastern Victoria. Stage 3 report: the effect of climate change on extreme sea levels in Corner Inlet and the Gippsland Lakes*. Report to Gippsland Coastal Board. July 2006. www.cmar.csiro.au/e-print/open/mcinneskl_2006a.pdf.

McInnes K.L., Macadam I., Hubbert G.D. and O'Grady J.G. (2009) A modelling approach for estimating the frequency of sea level extremes and the impact of climate change in southeast Australia. *Natural Hazards* **51**(1), 115–37.

McRobie A., Spencer T. and Gerritsen H. (2005) The big flood: North Sea storm surge. *Philosophical Transactions of the Royal Society A* **363**, 1263–70.

Mearns L.O., Mavromatis T., Tsvetsinskaya E., Hays C. and Easterling W. (1999) Comparative responses of EPIC and CERES crop models to high and low resolution climate change scenarios. *Journal of Geophysical Research* **104**(D6), 6623–46.

Meehl G.A., Stocker T.F., Collins W.D., Friedlingstein P., Gaye A.T., Gregory J.M. et al. (2007) Global climate projections. In: *Climate Change 2007: The Physical Science Basis*. Contribution of Working Group 1 to the Fourth Assessment Report of the Intergovernmental Panel on Climate Change (Solomon S., Qin D., Manning M., Marquis M., Averyt K., Tignor M.M.B. et al., eds), pp. 747–845. Cambridge University Press, Cambridge.

Mitchell J.F.B., Lowe J.A., Wood R.A. and Vellinga M. (2006) Extreme events due to human induced climate change. *Philosophical Transactions of the Royal Society A* **364**, 2117–33.

Monserrat S., Vilibić I. and Rabinovich A.B. (2006) Meteotsunamis: atmospherically induced destructive ocean waves in the tsunami frequency band. *Natural Hazards and Earth System Sciences* **6**, 1035–51.

Murphy J.M., Sexton D.M.H., Barnett D.N., Jones G.S., Webb M.J., Collins M. and Stainforth D.A. (2004) Quantification of modelling uncertainties in a large ensemble of climate change simulations. *Nature* **430**, 768–72.

Murray R.J. and Simmonds I. (1991) A numerical scheme for tracking cyclone centres from digital data. Part 1: development and operation of the scheme. *Australian Meteorological Magazine* **39**, 155–66.

Murty T.S. and Flather R.A. (1994) Impact of storm surges in the Bay of Bengal. *Journal of Coastal Research, Special Issue* **12**, 149–61.

Murty T.S., Flather R.A. and Henry R.F. (1986) The storm surge problem in the Bay of Bengal. *Progress in Oceanography* **16**, 195–233.

Nakicenovic N. and Swart R. (2000) *Special Report on Emissions Scenarios*. Contribution to the Intergovernmental Panel on Climate Change. Cambridge University Press, Cambridge.

Oouchi K., Yoshimura J., Yoshimura H., Mizuta R., Kusunoki S. and Noda A. (2006) Tropical cyclone climatology in a global-warming climate as simulated in a 20km-mesh global atmospheric model: frequency and wind intensity analysis. *Journal of the Meteorological Society of Japan* **84**, 259–76.

Osborne T.J. (2004) Simulating the winter North Atlantic Oscillation: the roles of internal variability and greenhouse gas forcing. *Climate Dynamics* **22**, 605–23.

Parry M.L., Canziani O.F., Palutikof J.P., van der Linden P.J. and Hanson C.E. (eds) (2007) Cross-chapter case study. In: *Climate Change 2007: Impacts, Adaptation and Vulnerability*. Contribution of Working Group II to the Fourth Assessment Report of the Intergovernmental Panel on Climate Change (Parry M.L., Canziani O.F., Palutikof J.P., van der Linden P.J. and Hanson C.E., eds), pp. 843–68. Cambridge University Press, Cambridge.

Pirazzoli P.A. (2000) Surges, atmospheric pressure and wind change and flooding probability on the Atlantic coast of France. *Oceanologica Acta* **23**, 643–61.

Pirazzoli P.A. and Tomasin A. (2002) Recent evolution of surge-related events in the northern Adriatic area. *Journal of Coastal Research* **18**, 537–54.

Pirazzoli P.A., Costa S., Dornbusch U. and Tomasin A. (2006) Recent evolution of surge-related events and assessment of coastal flooding risk on the eastern coast of the English Channel. *Ocean Dynamics* **56**, 498–512.

Pope V.D. and Stratton R.A. (2002) The processes governing horizontal resolution sensitivity in a climate model. *Climate Dynamics* **19**, 211–36.

Pugh D.T. (1987) *Tides, Surges and Mean Sea-Level: a Handbook for Engineers and Scientists*. Wiley, Chichester.

Pugh D.T. and Maul G.A. (1999) Coastal sea level prediction for climate change. In: *Coastal Ocean Prediction* (Mooers C., ed.), pp. 377–404. Coastal and Estuarine Studies vol. 56. American Geophysical Union, Washington DC.

Raicich F. (2003) Recent evolution of sea-level extremes at Trieste (Northern Adriatic). *Continental Shelf Research* **23**(3), 225–35.

Reiss R.-D. and Thomas M. (1997) *Statistical Analysis of Extreme Values*. Birkhäuser, Basel.

Rigor I.G. and Wallace J.M. (2004) Variations in the age of Arctic sea-ice and summer sea-ice extent. *Geophysical Research Letters* **31**, L09401.

Rogers J.C. (1990) Patterns of low-frequency monthly sea level pressure variability (1899–1986) and associated wave cyclone frequencies. *Journal of Climate* **3**, 1364–79.

Rogers J.C. (1997) North Atlantic storm track variability and its association to the North Atlantic Oscillation and climate variability of northern Europe. *Journal of Climate* **10**, 1635–47.

Royer J.-F., Chauvin F., Timbal B., Araspin P. and Grimal D. (1998) A GCM study of impact of greenhouse gas increase on the frequency of occurrence of tropical cyclones. *Climate Dynamics* **38**, 307–43.

Santer B.D., Wigley T.M.L., Gleckler P.J., Bonfils C., Wehner M.F., AchutaRao K. et al. (2006) Forced and unforced ocean temperature changes in Atlantic and Pacific tropical cyclogenesis regions. *Proceedings of the National Academy of Sciences USA* **103**, 13905–10.

Sasaki W., Iwasaki S.I., Matsuura T. and Iizuka S. (2005) Recent increase in summertime extreme wave heights in the western North Pacific. *Geophysical Research Letters* **32**, L15607.

Sasaki W., Iwasaki S.I., Matsuura T. and Iizuka S. (2006) Quasi-decadal variability of fall extreme wave heights in the western North Pacific. *Geophysical Research Letters* **33**, L09605.

Schmidt H. (2001) Die Entwicklung der Sturmhäufigkeit in der Deutschen Bucht zwischen 1879 und 2000. In: *Klimastatusbericht 2001*, pp. 199–205. Deutscher Wetterdienst, Offenbach/Main.

Sheppard C., Dixon D.J., Gourlay M., Sheppard A. and Payet R. (2005) Coral mortality increases wave energy reaching shores protected by reef flats: examples from the Seychelles. *Estuarine, Coastal and Shelf Science* **64**, 223–34.

Sickmöller M., Blender R. and Fraedrich K. (2000) Observed winter cyclone tracks in the northern hemisphere in re-analysed ECMWF data. *Quarterly Journal of the Royal Meteorological Society* **126**, 591–620.

Sinclair M.R. and Watterson I.G. (1999) Objective assessment of extratropical weather systems in simulated climates. *Journal of Climate* **12**, 3467–85.

Singh O.P. (2001) Cause-effect relationships between sea surface temperature, precipitation and sea level along the Bangladesh coast. *Theoretical and Applied Climatology* **68**, 233–43.

Smith R.L. (1986) Extreme value theory based on the r-largest annual events. *Journal of Hydrology* **86**, 27–43.

SMRC (2000) Variability of sea surface temperature and intensity of tropical cyclones. In: *The Vulnerability Assessment of the SAARC Coastal Region Due to Sea level Rise: Bangladesh Case*, pp. 67–89, SMRC-No.3 Publication. SAARC Meteorological Research Center, Dhaka, Bangladesh.

Sterl A. and Caires S. (2005) Climatology, variability and extrema of ocean waves: the web-based KNMI/ERA-40 wave atlas. *International Journal of Climatology* **25**, 963–77.

Stowasser M., Wang Y. and Hamilton K. (2007) Tropical cyclone changes in the western North Pacific in a global warming scenario. *Journal of Climate* **20**, 2378–96.

Sugi M., Noda A. and Sato N. (2002) Influence of the global warming on tropical cyclone climatology: an experiment with the JMA global model. *Journal of the Meteorological Society of Japan* **80**, 249–72.

Svensson C. and Jones D. (2004) Dependence between sea surge, river flow and precipitation in south and west Britain. *Hydrology and Earth System Sciences* **8**(5), 973–92.

Svensson C. and Jones D. (2006) Climate change impacts on the dependence between sea surge, precipitation and river flow around Britain. In: *Proceedings from the 40th Defra Flood and Coastal Management Conference 2005*, University of York, York 5–7 July 2005, pp. 6A.3.1–6A.3.10.

Swanson K. (2008) Nonlocality of Atlantic tropical cyclone intensities. *Geochemistry, Geophysics and Geosystems* **9**, Q04V01.

Terray L., Demory M-E., Déqué M., de Coetlogon G. and Maisonnave E. (2004) Simulation of late-twenty-first-century changes in wintertime atmospheric circulation over Europe due to anthropogenic causes. *Journal of Climate* **17**, 4630–5.

Tinti S., Armigliato A., Pagnoni G. and Zaniboni F. (2006) Scenarios of giant tsunamis of tectonic origin in the Mediterranean. *ISET Journal of Earthquake Technology* **42**, 171–88.

Titov V., Rabinovich A.B., Mofjeld H.O., Thomson R.E. and González F.I. (2005) The global reach of the 26 December 2004 Sumatra tsunami. *Science* **309**, 2045–8.

Trenberth K.E., Jones P.D., Ambenje P., Bojariu R., Easterling D., Tank A.K. et al. (2007) Observations: Surface and Atmospheric Climate Change. In: *Climate Change 2007: The Physical Science Basis, IPCC AR4 WG1 Final Report* (Solomon S., Qin D., Manning M., Chen Z., Marquis M.C., Avery K.B. et al., eds), pp. 235–6. Cambridge University Press, Cambridge.

Trigo I.F. and Davies T.D. (2002) Meteorological conditions associated with sea surges at Venice: a 40 year climatology. *International Journal of Climatology* **22**, 787–803.

Tsimplis M.N., Woolf D.K., Osborn T.J., Wakelin S., Wolf J., Flather R. et al. (2005) Towards a vulnerability assessment of the UK and northern European coasts: the role of regional climate variability. *Proceedings of the Royal Society of London* **363**, 1329–58.

Tsutsui J. (2002) Implications of anthropogenic climate change for tropical cyclone activity: a case study with the NCAR CCM2. *Journal of the Meteorological Society of Japan* **80**, 45–65.

Ueno K. (1993) Interannual variability of surface cyclone tracks, atmospheric circulation patterns and precipitation patterns, in winter. *Journal of the Meteorological Society of Japan* **71**, 655–71.

Ullmann A., Pirazzoli P.A. and Tomasin A. (2007) Sea surges in Camargue: trends over the 20th century. *Continental Shelf Research* **27**, 922–34.

Unnikrishnan A.S., Kumar K.R., Fernandes S.E., Michael G.S. and Patwardhan S.K. (2006) Sea level changes along the Indian coast: observations and projections. *Current Science* **90**(3), 362–8.

van den Brink H.W., Können G.P., Opsteegh J.D., van Oldenborgh G.J. and Burgers G. (2005) Estimating return periods of extreme events from ECMWF seasonal forecast ensembles. *International Journal of Climatology* **25**, 1345–54.

Vassie J.M., Woodworth P.L. and Holt M.W. (2004) An example of North Atlantic deep ocean swell impacting Ascension and St. Helena islands in the central South Atlantic. *Journal of Atmospheric and Oceanic Technology* **21**(7), 1095–1103.

Vecchi G.A. and Knutson T.R. (2008) On estimates of historical North Atlantic tropical cyclone activity. *Journal of Climate* **21**, 3580–3600.

Vikebø F., Furevik T., Furnes G., Kvamstø N.G. and Reistad M. (2003) Wave height variations in the North Sea and on the Norwegian continental shelf, 1881–1999. *Continental Shelf Research* **23**, 251–63.

Vilibić I., Domijan N., Orlić M., Leder N. and Pasarić M. (2004) Resonant coupling of a traveling air pressure disturbance with the east Adriatic coastal waters. *Journal of Geophysical Research* **109**, C10001.

Vitart F., Anderson J.L. and Stern W.F. (1997) Simulation of interannual variability of tropical storm frequency in an ensemble of GCM integrations. *Journal of Climate* **10**, 745–60.

von Storch H. and Reichardt H. (1997) A scenario of storm surge statistics for the German Bight at the expected time of doubled atmospheric Carbon Dioxide concentration. *Journal of Climate* **10**, 2653–62.

von Storch H. and Woth K. (2008) Storm surges: perspectives and options. *Sustainability Science* **3**, 33–43.

Wakelin S.L., Woodworth P.L., Flather R.A. and Williams J.A. (2003) Sea-level dependence on the NAO over the NW European Continental Shelf. *Geophysical Research Letters* **30**(7), 1403.

Walsh K. (2004) Tropical cyclones and climate change: unresolved issues. *Climate Research* **27**, 77–83.

Walsh K.J.E., Nguyen K.-C. and McGregor J.L. (2004) Fine-resolution regional climate model simulations of the impact of climate change on tropical cyclones near Australia. *Climate Dynamics* **22**, 47–56.

Walsh K.J.E., Fiorino M., Landsea C.W. and McInnes K.L. (2007) Objectively-determined resolution-dependent threshold criteria for the detection of tropical cyclones in climate models and reanalyses. *Journal of Climate* **20**, 2307–14.

Wang X.L. and Swail V.R. (2001) Changes of extreme wave heights in northern hemisphere oceans and related atmospheric circulation regimes. *Journal of Climate* **14**, 2204–21.

Wang X.L. and Swail V.R. (2002) Trends of Atlantic wave extremes as simulated in a 40-yr wave hindcast using kinematically reanalyzed wind fields. *Journal of Climate* **15**, 1020–35.

Wang X.L. and Swail V.R (2006a) Historical and possible future changes of wave heights in northern hemisphere oceans. In: *Atmosphere Ocean Interactions*, vol. 2 (Perrie W., ed.), Advances in Fluid Mechanics Series, vol. 39. Wessex Institute of Technology Press, Southampton.

Wang X.L. and Swail V. (2006b) Climate change signal and uncertainty in the projections of ocean wave heights. *Climate Dynamics* **26**, 109–26.

Wang X.L., Zwiers F.W. and Swail V.R. (2004) North Atlantic Ocean wave climate change scenarios for the twenty-first century. *Journal of Climate* **17**, 2368–83.

Wang X.L., Swail V.R. and Zwiers F.W. (2006) Climatology and changes of extra tropical cyclone activity: comparison of ERA40 with NCEP-NCAR reanalysis for 1958-2001. *Journal of Climate* **19**, 3145–66.

WASA Group (1998) Changing waves and storms in the Northeast Atlantic? *Bulletin of the American Meteorological Society* **79**, 741–60.

Webster P.J., Holland G.J., Curry J.A. and Chang H.-R. (2005) Changes in tropical cyclone number, duration, and intensity in a warming environment. *Science* **309**, 1844–6.

Weisse R. and Plüß A. (2006) Storm related sea level variations along the North Sea coast as simulated by a high-resolution model 1958–2002. *Ocean Dynamics* **56**(1), 16–25.

Weisse R. and Günther H. (2007) Wave climate and long-term changes for the Southern North Sea obtained from a high-resolution hindcast 1958–2002. *Ocean Dynamics* **57**(3), 161–72.

Weisse R., von Storch H. and Feser F. (2005) Northeast Atlantic and North Sea storminess as simulated by a regional climate model during 1958–2001 and comparison with observations. *Journal of Climate* **18**, 465–79.

Whetton P.H., McInnes K.L., Jones R.N., Hennessy K.J., Suppiah R., Page C.M. et al. (2005) *Australian Climate Change Projections for Impact Assessment and Policy Application: a Review*. CSIRO technical report. http://www.cmar.csiro.au/e-print/open/whettonph_2005a.pdf. CSIRO.

Wolf J. and Flather R.A. (2005) Modelling waves and surges during the 1953 storm. *Philosophical Transactions of the Royal Society* **363**, 1359–75.

Wolf J. and Woolf D.K. (2006) Waves and climate change in the north-east Atlantic. *Geophysical Research Letters* **33**, L06604.

Wood F.J. (1978) *The Strategic Role of Perigean Spring Tides in Nautical History and North American Coastal Flooding, 1635–1976*. US Dept. of Commerce, National Oceanic and Atmospheric Administration, Rockville, MD.

Woodworth P.L. (1999) High waters at Liverpool since 1768: the UK's longest sea level record. *Geophysical Research Letters* **26**(11), 1589–92.

Woodworth P.L. (2005) Have there been large recent sea level changes in the Maldive Islands? *Global and Planetary Change* **49**, 1–18.

Woodworth P.L. and Blackman D.L. (2002) Changes in extreme high waters at Liverpool since 1768. *International Journal of Climatology* **22**, 697–714.

Woodworth P.L. and Blackman D.L. (2004) Evidence for systematic changes in extreme high waters since the mid-1970s. *Journal of Climate* **17**, 1190–7.

Woodworth P.L., Blackman D.L., Foden P., Holgate S., Horsburgh K., Knight P.J. et al. (2005) Evidence for the Indonesian tsunami in British tidal records. *Weather* **60**(9), 263–7.

Woodworth P.L., Flather R.A., Williams J.A., Wakelin S.L. and Jevrejeva S. (2006) The dependence of extreme UK and NW European sea levels on the NAO. *Continental Shelf Research* **27**, 935–46.

Woolf D.K., Challenor P.G. and Cotton P.D. (2002) Variability and predictability of the North Atlantic wave climate. *Journal of Geophysical Research* **107**(C10), 3145.

Woth K. (2005) North Sea storm surge statistics based on projections in a warmer climate: How important are the driving GCM and the chosen emission scenario? *Geophysical Research Letters* **32**, L22708.

Woth K., Weisse R. and von Storch H. (2006) Climate change and North Sea storm surge extremes: An ensemble study of storm surge extremes expected in a changed climate projected by four different Regional Climate Models. *Ocean Dynamics* **56**(1), 3–15.

Wu L.G., Wang B. and Braun S.A. (2008) Implications of tropical cyclone power dissipation index. *International Journal of Climatology* **28**, 727–31.

Yin J.H. (2005) A consistent poleward shift of the storm tracks in simulations of 21st century climate. *Geophysical Research Letters* **32**, L18701.

Yoshimura J. and Sugi M. (2005) Tropical cyclone climatology in a high-resolution AGCM – impacts of SST warming and CO2 increase. *SOLA* **1**, 133–6.

Zhang K., Douglas B.C. and Leatherman S.P. (1997a) East coast storm surges provide unique climate record. *EOS Transactions of the American Geophysical Union* **78**(37), 389.

Zhang K., Douglas B.C. and Leatherman S.P. (2000) Twentieth-century storm activity along the U.S. east coast. *Journal of Climate* **13**, 1748–61.

Zhang Y., Wallace J.M. and Battisti D.S. (1997b) ENSO-like interdecadal variability: 1900–1993. *Journal of Climate* **10**, 1004–20.

12 Observing Systems Needed to Address Sea-Level Rise and Variability

W. Stanley Wilson, Waleed Abdalati, Douglas Alsdorf, Jérôme Benveniste, Hans Bonekamp, J. Graham Cogley, Mark R. Drinkwater, Lee-Lueng Fu, Richard Gross, Bruce J. Haines, D.E. Harrison, Gregory C. Johnson, Michael Johnson, John L. LaBrecque, Eric J. Lindstrom, Mark A. Merrifield, Laury Miller, Erricos C. Pavlis, Stephen Piotrowicz, Dean Roemmich, Detlef Stammer, Robert H. Thomas, Eric Thouvenot, and Philip L. Woodworth

12.1 Introduction

The workshop, *Understanding Sea-Level Rise and Variability*, was conducted in part to support implementation of the Global Earth Observation System of Systems[1] (GEOSS). As such, it developed international scientific consensus for those observational requirements needed to address sea-level rise and its variability, which are especially relevant to GEOSS activities focused on climate and hazards. This chapter is organized according to those consensus requirements. It provides a descriptive summary of the observing systems, both exist ing systems to be sustained and new systems to be developed, as well as the terrestrial reference frame in which they are used to collect observations that are needed to address sea-level rise and variability.

An overarching observational requirement is the need for an open data policy, together with timely, unrestricted access for all. Using the Argo and Jason policies as a guide, this access would include real-time, high-frequency sea-level data from the Global Sea Level Observing System (GLOSS) tide gauges and co-located geo-

[1] http://earthobservations.org/

Understanding Sea-Level Rise and Variability, 1st edition. Edited by John A. Church, Philip L. Woodworth, Thorkild Aarup & W. Stanley Wilson.

detic stations, as well as data from satellite missions and other *in situ* observing systems. Further requirements include appropriate data archaeology: retrieving and making accessible historical, paper-based sea-level (and other) records, especially those extending over long periods and in the Southern Hemisphere. Moreover, satellite observations need to be as continuous as possible, with overlap between successive missions and coincident with the collection of appropriate *in situ* observations. In general, ongoing satellite and *in situ* observing systems should adhere to the Global Climate Observing System (GCOS) climate-monitoring principles[2].

12.2 Sustained, Systematic Observing Systems (Existing Capabilities)

12.2.1 Sea Level

The collection of direct observations of sea-level change requires the continuation and, where needed, the completion of existing observing capabilities. This involves:

- sustaining the Jason class of high-accuracy satellite altimeters for the foreseeable future through the implementation of a Jason-3 and follow-on high-accuracy missions with equivalent performance; these serve as a reference for, and are supplemented by, complementary altimeter missions.
- completing the GLOSS Core Network (GCN) of approximately 300 gauges, each with high-frequency sampling and real-time data availability. Gauges should be linked to absolute positioning where possible (either directly at the gauge or by leveling to nearby absolute networks) to enable an assessment of the coastal signatures of the open-ocean patterns of sea-level variability and the incidence of extreme events, as well as the calibration of satellite altimeters.

It is critical to extend the Jason series of satellite altimeters for the foreseeable future to resolve the spatial and temporal variability, as well as any acceleration, in the rate of global sea-level rise. This series[3] was initiated by the National Aeronautics and Space Administration (NASA) and Centre National d'Etudes Spatiales (CNES) with TOPEX/Poseidon (1992–2005) and is currently being extended with Jason-1 (2001–present; Fu and Cazenave 2001) and Jason-2 (2008–present). These missions have demonstrated technical feasibility and scientific utility of high-accuracy satellite altimetry. They have also served, and continue to serve, as a reference for complementary higher-inclination altimeter missions, thereby extending observations of sea level to higher latitudes and finer spatial scales (see Benveniste and Ménard 2006).

---

[2] http://www.wmo.ch/pages/prog/gcos/index.php?name=monitoringprinciples

[3] http://sealevel.jpl.nasa.gov and http://www.aviso.oceanobs.com/msl/

In June 2008, NASA and CNES launched the Ocean Surface Topography Mission (OSTM; also known as Jason-2) satellite, with the US National Oceanic and Atmospheric Administration (NOAA) and the European Organisation for the Exploitation of Meteorological Satellites (EUMETSAT) as operational agencies providing ground support and routine data processing for this mission. With OSTM/Jason-2 in place to extend the Jason series, planning is underway by CNES, the European Space Agency (ESA), EUMETSAT, and NOAA to define and implement a continuing operational system for the longer term – including Jason-3 and further follow-on missions – that would feature sufficient lead time to ensure an overlap for cross-calibration between successive missions. In late January 2010, NOAA and EUMETSAT succeeded in securing the funding required to implement Jason-3.

The sea-surface topography observed by altimeters, combined with the temperature and salinity fields observed by Argo floats profiling in the upper ocean, are fundamental variables required to prepare global analyses and forecasts, characterize seasonal and decadal variability in the oceans, and advance our understanding of the role of the oceans in climate. As such, they are helping establish the basis for *operational oceanography*[4].

GLOSS (Woodworth et al. 2003) is conducted under the auspices of the Joint Technical Commission for Oceanography and Marine Meteorology (JCOMM) of the World Meteorological Organization (WMO) and the Intergovernmental Oceanographic Commission (IOC). GLOSS[5] includes data assembly and archiving centers in the UK and the USA, and a group of experts that provides scientific and technical oversight. Nearly 90 countries are represented in the GCN. The scientific utility of the GCN has been demonstrated through contributions to World Ocean Circulation Experiment (WOCE), Climate Variability and Predictability project (CLIVAR), Global Ocean Data Assimilation Experiment (GODAE), and other international research programs.

There are currently four main data streams within GLOSS:

1. delayed-mode higher-frequency data to the University of Hawai'i Sea Level Center or Permanent Service for Mean Sea Level (PSMSL) as GLOSS archiving centers;
2. real-time or fast higher-frequency data to the University of Hawai'i Sea Level Center;
3. mean-sea-level data to PSMSL;
4. Global Positioning System (GPS) data to the Tide Gauge (TIGA) data center of the International GNSS System (IGS) in Potsdam.

With regard to the requirements of high-frequency sampling (on the order of minutes) and real-time data availability (within an hour), the current status of the 290 GCN stations is that 57% of the stations meet the standards; 17% of the stations provide high-frequency data, but not in real-time; 15% have provided

---

[4] http://www.myocean.eu.org/index.php/products-services/catalogue

[5] http://www.gloss-sealevel.org

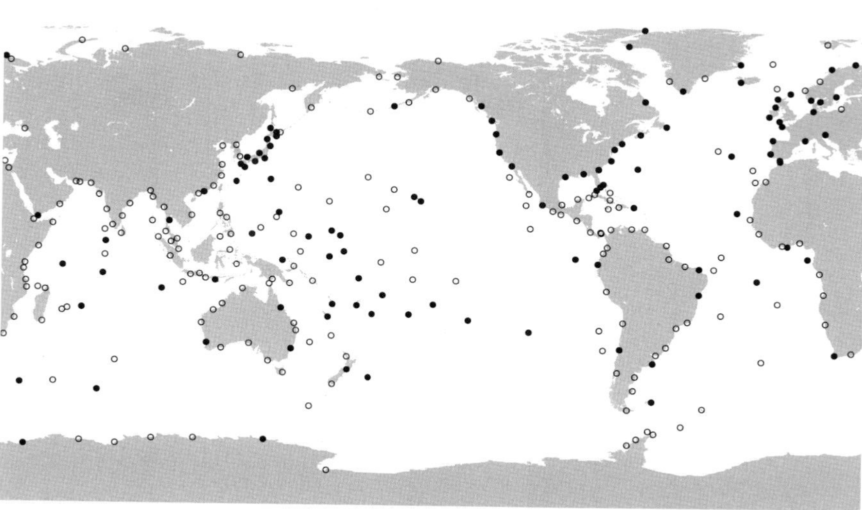

Figure 12.1 Of the 290 tide gauges (circles and black dots) identified as elements of the GCN, only approximately 110 (black dots) are *complete*; that is, each provides high-frequency data (sampling at least hourly) in real-time or fast-delivery mode (monthly), as well as has a nearby capability to monitor vertical motion of the tide gauge itself. (Figure courtesy of Mark Merrifield, University of Hawai'i Sea Level Center.)

high-frequency data, but not over the past 6 years; and no high-frequency data have been contributed to the GLOSS Real-time Data Center at the University of Hawai'i for 11% of the stations.

Approximately 132 GCN tide gauges have GPS or Doppler Orbitography and Radiopositioning Integrated by Satellite (DORIS) stations near the tide gauge to provide information on vertical land movements; and about 192 meet the high-frequency/real-time requirements. Notably, the African tide-gauge network has been improved considerably via the Ocean Data and Information Network for Africa (ODINAfrica) project and the Indian Ocean Tsunami Warning System.

The PSMSL considers that approximately 60% of the GCN was operational as of August 2009, based on the reporting of monthly mean-sea-level values in delayed mode.

To complete the GCN (i.e. the approximately 100 GCN stations that are not high frequency/real time), a number of challenges remain (see Figure 12.1). National restrictions on the distribution of high-frequency data hinder access to approximately 30 stations, for which diplomatic solutions will be needed. National and commercial restrictions also apply to the continuous GPS data for about 25% of the 132 GCN tide gauges that have GPS or DORIS stations nearby. Approximately 70 stations have little or no infrastructure in place and typically limited or no technical capability to install or maintain a station. Major funding and a supporting agency dedicated to the installation/maintenance of the equipment and the training of local operators are needed to bring these stations into operation. For most of the remaining stations, equipment upgrades (modern water-level sensors, data-collection platforms, satellite transmitters) are needed to achieve the desired sampling and transmission benchmarks. These improvements are likely to be achieved through partnerships with tsunami warning and other operational initiatives that require real-time water-level data. GLOSS is committed to multiple-use platforms and coordination efforts are underway to improve GCN network coverage in tsunami-vulnerable areas in the Caribbean, the Mediterranean, the Indian Ocean, and the Pacific Rim; however, funds are needed for the successful completion of these projects.

12.2.2 Ocean Volume

Measuring properties of the ocean that determine its density, and hence estimate their corresponding effect on ocean volume, requires:

- sustaining the Argo array of profiling floats to obtain broad-scale, upper-ocean observations of the temperature and salinity fields.

With more than 40 countries participating, the Argo[6] program uses autonomous instruments to collect high-quality temperature and salinity profiles from the upper 2000 m of the ice-free oceans, as well as to observe currents at intermediate depths (Gould et al. 2004). The initial deployment objective to have a global array distributed on a $3° \times 3°$ grid (i.e. one float for every $3° \times 3°$ box) requires approximately 3000 units. The instruments are battery-powered floats that drift at depth (generally 1000 m) for approximately 9 days after which a pump transfers a fluid between a reservoir inside the instrument and an external bladder, enabling the instruments to sink to 2000 m and then to ascend to the surface recording temperature and salinity. Upon surfacing, a satellite determines their positions and telemeters their profile data to shore. Then arriving at a receiving station, the data are subjected to real-time quality control and then distributed via the Global Telecommunications System to various operational centers, as well as the two Argo global Data Assembly Centers. A fundamental principle of Argo is free and open data access to anyone – ideally a principle for all observing systems – either via the Global Telecommunications System or a Data Assembly Center[7]. About 90% of the Argo data pass the real-time quality control procedures and are available in less than 24 h. More than half of the remaining data are placed onto the Global Telecommunications System within 48 h of receipt after a visual inspection of those data and application of any needed quality-control flags.

The origin of the profiling float is directly traceable to the development by John Swallow in the 1950s of neutrally buoyant floats tracked acoustically by ship to observe subsurface currents. In the 1990s, similar floats – but tracked by satellite when they surfaced every few weeks – were deployed during WOCE. It was then straightforward to extend their capability by adding sensors enabling the collection of observations of the broad-scale, time-varying temperature and salinity fields of the upper ocean.

---

[6] http://www.argo.net

[7] Argo data are available from either http://www.coriolis.eu.org/cdc/argo_rfc.htm or http://www.usgodae.org/argo/argo.html

While the initial deployment objective of Argo was attained in November 2007, some gaps remain (e.g. marginal seas, south of 60 °S). These gaps will be addressed as technology and deployment platforms become available. The main challenge is to sustain the array, both financially and logistically, long enough to fully evaluate its capabilities and to optimize the design (see Figure 12.2). Sustaining the array will require funding to be transitioned from short-term, research-based programs to sustained operations, including support for dedicated platforms to deploy floats in remote regions.

As noted in Chapter 6, it is critical that Argo-derived profiles of temperature and salinity and temperature profiles from expendable bathythermographs (XBTs) are cross-calibrated on a continuing basis with standard ship-borne systems measuring accurate temperature and salinity profiles to ensure data quality. Without close attention to data quality, it is possible for small systematic offsets or drifts to creep into the data and accumulate into sizable sources of error.

Satellite-based observations of surface temperature and salinity offer the potential to provide complementary data to the profiles collected by Argo profiling floats. While a number of microwave systems have flown in space to observe all-weather sea surface temperature[8], the first two missions to demonstrate the feasibility of observing surface salinity will be the Soil Moisture and Ocean Salinity[9] (launched in November 2009) and the Aquarius[10] satellites (to be launched in 2010).

12.2.3 Ocean, Terrestrial Water, and Ice Mass

Sea-level change is driven by the exchange of water mass between the oceans, land, and cryosphere, as well as the changing volume of the oceans from temperature and salinity changes, tectonic motions, and glacial isostatic adjustment (GIA). Measurement of these large-scale mass changes requires:

- continuation of observations of the time-varying gravity field by the Gravity Recovery and Climate Experiment (GRACE) and an appropriate follow-on mission.

Terrestrial geophysicists have long used gravity and topography to map density variations within the Earth's crust and the location of mass concentrations. Precision topographic changes (land, ocean, and ice) and time-variable gravity from satellites are now being applied similarly to map the transport of water mass, the effects of thermal expansion, and estimates of global isostatic adjustment.

Gravity measurements alone can provide significant insight into mass flux, but maximum benefit from these measurements will come from combined precision

---

[8] http://www.remss.com/sst/microwave_oi_sst_data_description.html
[9] http://www.esa.int/smos
[10] http://aquarius.nasa.gov/

Figure 12.2 Completion of the global Argo array. Demonstration deployment of the 3000th Argo float by the Hon Steve Maharey, (then) New Zealand Minister of Research, Science and Technology, assisted by Simon Wadsworth, Captain (left) and John Hunt, Second Mate (right), both of the RV *Kaharoa*. With initial floats deployed in 2000, Argo reached its initial implementation goal – 3000 active floats covering the ice-free global oceans – in October 2007. During that month, over 100 floats were deployed by various Argo scientists around the world, so it was impossible to tell exactly which float was the 3000th. Float deployers were invited to submit candidate photographs of the Argo 3000th float, and this photograph is one of those submitted. The RV *Kaharoa* has played a key role in the global deployment and maintenance of the Argo array, a challenging issue and significant expense. Transiting research vessels and commercial ships are used for float deployment wherever possible. However, in remote ocean regions, particularly in the South Pacific and Indian Oceans, opportunistic traffic is not sufficient. Through a collaboration of the US and New Zealand Argo programs a series of dedicated deployment cruises has been carried out, including ten voyages by RV *Kaharoa* since 2004 and over 750 floats deployed. This vessel is cost-effective due to its small size (28 m length, crew of five) but large capability. Without this or a comparable program, a global Argo array would not be possible. Photograph taken on October 2, 2007 at Queen's Wharf, Wellington, by Alan Blacklock and provided through the courtesy of the National Institute of Water and Atmospheric Research.

topography and gravity. Inferring ice-sheet mass change from measurements of changing ice volume is complicated by changing surface density due to snow accumulation and by GIA of the lithosphere. Gravity field changes reflect mass changes, and therefore are not influenced by snow–ice conversion. However, the gravity field is strongly affected by GIA because the density of the crust and lithosphere far exceeds that of water, snow, and ice. On the other hand, altimetry estimates of ice-volume changes – whether laser or radar – are susceptible to snow accumulation and firn compaction (varying near-surface density due to snow accumulation; its compaction introduces significant uncertainty when converting measured surface-elevation change to changes in ice mass), but are only weakly affected by GIA. The combined analysis of gravity and precision topography, given independent estimates of compaction, has been suggested as a means of estimating GIA through remote sensing.

The combined analysis of precision ocean topography and time-variable gravity is now providing both the thermosteric signal and the variation in total water mass responsible for sea-level change. Currently, seasonal changes in water-mass flux from hemisphere to hemisphere are observable, and interannual signals in mass exchange between the oceans, cryosphere, and land have been detected, although the 5-year period of the GRACE[11] mission is not yet sufficient to discern trends from periodic variability (see Willis et. al 2008).

The geoid is currently estimated to be accurate to better than 7 mm at a 300-km spatial resolution (Tapley et al. 2005), with a significant improvement in accuracy at longer wavelengths. The determination of the geoid at these longer wavelengths is now several times more accurate than that of the surface topography determined by satellite altimetry and the terrestrial reference frame used to define that topography. Improvements in the global geodetic networks have the potential to provide more accurate tracking of the satellite altimeters and a better reference frame.

Launched by ESA on March 17, 2009 with a nominal lifetime of 20 months, the Gravity Field and Steady-State Ocean Circulation Explorer (GOCE)[12] satellite will provide the most accurate snapshot of the gravity field obtained to date. While both GRACE and GOCE currently have projected lifetimes of a few more years, several space agencies have begun studying follow-on missions. NASA has begun planning for a GRACE-II mission as recommended by the National Research Council's *Decadal Survey* (Committee on Earth Science and Applications from Space 2007); this is in the long-term set of missions with a projected launch date in the 2016–20 time frame[13]. More recently, NASA has been considering a GRACE

---

[11] http://www.csr.utexas.edu/grace

[12] http://www.esa.int/goce

[13] The satellite missions recommended for NASA in the National Research Council's *Decadal Survey* are grouped into three broad categories: *near-term* for launch in 2010–13, *mid-term* in 2013–16, and *long-term* in 2016–20. The two missions in the *near-term* category that concern this chapter are ICESat-II and DesDynI; NASA has a planned launch for the former in 2015; as of June 2009, it does not yet have a similar date for the latter. Given that the planned launch for ICESat-II is already 2 years later than the stated dates for the *near-term* category, one could assume a corresponding slip in each of these categories, unless there is a significant infusion of new funding.

follow-on, a carbon copy of GRACE, to be available in the nearer term. Meanwhile in Europe, a recent workshop on *The Future of Satellite Gravimetry* (Koop and Rummel 2008)[14], building on the results of the earlier *Earth Gravity Field from Space* (Beutler et al. 2003) workshop[15], has been active in the development of future mission concepts.

At least two concepts are among those being considered in the formulation of future missions, such as GRACE-II. First, the inclusion of more accurate and sensitive accelerometers as developed for GOCE can improve the determination of non-conservative forces such as atmospheric drag. Second, the use of laser interferometry, as developed for the joint ESA/NASA Laser Interferometer Space Antenna mission, for inter-satellite ranging has the potential for more precise inter-satellite ranging than microwave techniques. There is little doubt that international collaboration in the utilization of such concepts in the definition and development of future missions will lead to improvements in the determination of the gravity field.

12.2.4 Ice Sheet and Glacier Volume

Key elements of an observing system designed to quantify ice-sheet and glacier contributions to sea-level change include:

- continued surveys by satellite radar and laser altimeters, to provide elevation changes over broad areas on a routine basis;
- continued time series of interferometric synthetic aperture radar (InSAR) measurements for glacier velocities, needed to understand changes in ice thickness and ice discharge;
- continued time series of satellite measurements of the gravity field from GRACE and similar missions, to enable accurate conversion of volume changes to large-scale mass changes;
- repeated aircraft surveys using scanning laser altimeters and radar depth sounders to map in detail glacier drainage basins that other sources of observation, for example InSAR or GRACE, show to be undergoing significant change.

Satellite radar altimetry (SRALT) provides the longest continuous time series from 1991 to the present day, but data interpretation is influenced by the effects of time-variable penetration into the snow, as well as elevation changes on scales smaller than the radar beam width, especially as associated with rugged topography. On the other hand, satellite laser altimeters lack the long time series available from radar, and they can be attenuated by cloud cover, but they are not influenced by penetration below the surface; thus radar and laser altimetry data are comple-

[14] http://www.dgfi.badw.de/typo3_mt/fileadmin/Dokumente/Final_report_satellite_gravity_workshop_12-13.04.2007_complete.pdf

[15] http://www.issi.unibe.ch/workshops/SatGrav/Programme.htm

mentary. As noted earlier, altimetry – whether radar or laser – is susceptible to snow accumulation or firn compaction. However, if elevation measurements are well-calibrated, data from different periods and different altimeters – both radar and laser – should in principle be comparable, offering the prospect to link the record of change from future laser altimeters back to the European Remote Sensing satellites 1 and 2 (ERS-1/-2) and ENVISAT radar data, together with the current Ice, Cloud, and Land Elevation Satellite (ICESat) (Zwally et al. 2002) and aircraft surveys[16] that began in 1993.

Cross-calibration of radar altimeters requires a substantial period (approximately a year) of overlap to account for changes in characteristics such as beam width and radar frequency. In the absence of such overlap, there is a high level of uncertainty (approximately decimeter) that requires many years of altimeter observations (roughly a decade) so that those uncertainties do not dominate the signal when comparing data from the different missions. Moreover, it is not yet clear to what extent SRALT data are affected by time-variable radar penetration of snow and firn resulting, for instance, from increased summer melting in a changing climate. Successfully launched on April 8, 2010, CryoSat-2[17] with its synthetic aperture radar interferometric altimeter (SIRAL) has a narrower effective beam width, approximately 250 m comparable to the laser's approximately 70 m, thereby approaching a similar capability to resolve topographic effects. Although it will still be influenced by time-variable radar penetration, CryoSat-2 offers significantly improved performance as its much denser orbit crossovers in time and space have been optimized to provide rates of change in ice-sheet topography.

Based on what we have learned from more than 10 years of aircraft laser surveys over Greenland, seasonal and interannual elevation variability is so large that identification of longer-term trends will require decades of measurements. In the case of a satellite laser system, while temporal overlap is not necessarily required if elevation measurements are well calibrated, each survey should be continuous over several years to quantify seasonal and interannual variability.

Aircraft surveys have the distinct advantage of covering individual glaciers in great spatial detail, and of measuring glacier thickness with ice-penetrating radar along the same flight lines. This information is essential to help understand observed changes in glacier behavior, and to provide boundary conditions for any model attempting to predict future glacier responses to climate change.

But aircraft surveys are unable to provide long, detailed time series over entire ice sheets or resolve seasonal/interannual variability of ice-sheet mass balance. For this, a combination of satellite radar and laser surveys is needed, complemented with periodic aircraft missions to provide near-coastal details that are "missed" by the satellite orbit pattern, particularly in regions of persistent cloud cover. While it was recommended by the *Decadal Survey* in the near-term set of missions for the 2010–13 timeframe, ICESat-II is now scheduled for launch in approximately 2015.

---

[16] http://icesat.gsfc.nasa.gov/
[17] http://www.esa.int/cryosat

Although the glaciers outside Greenland and Antarctica contain enough ice to raise sea level by only 15–37 cm (section 7.5.1), they are currently losing ice at rates comparable to the far larger polar ice sheets. Consequently, it is important to continue *in situ* mass-balance measurements on glaciers for which time series of such measurements already exist. In addition, many of the techniques used for monitoring larger ice sheets and bigger polar glaciers (e.g. GRACE, InSAR, and aircraft laser altimetry) are increasingly being applied at lower latitudes, and these should be continued and extended.

A capability for *in situ* verification of remote observations of ice sheets and glaciers of all sizes needs to be sustained for the foreseeable future. This requires surface observations that include calibrating and validating new sensors, reconciling conflicting results from existing sensors, and, for smaller glaciers in particular, resolving changes on spatial scales too small to be seen by some orbiting sensors. A hierarchical network of well-supported stations, ranging from those as part of specific process studies to routine seasonal and annual measurements of accumulation and ablation by traditional methods (snow-pit and stake), could provide this capability. Such a network, the Global Terrestrial Network for Glaciers[18] (GTN-G) already exists in principle, but it does not yet extend to the ice sheets. It is obliged to rely on the ability and willingness of field workers to submit measurements and suffers, as far as permanency is concerned, from the short-term character of funding for field work at the national level.

12.2.5 Reference Frame

An accurate and stable reference frame is essential for all the techniques discussed in this chapter. Maintaining and improving it requires:

- strengthened and sustained support for the International Terrestrial Reference Frame (ITRF), integrating the geodetic components – satellite laser ranging (SLR), very-long-baseline interferometry (VLBI), DORIS, and Global Navigation Satellite System (GNSS; GPS together with the Global Orbiting Navigation Satellite System (GLONASS) and Galileo once launched and in operation) – to make them more robust and stable;
- inclusion of observations of the time-invariant gravity field from GOCE, and other stand-alone missions to determine the precise geoid.

Understanding sea-level change requires the comparison of a multitude of global measurements – tide gauges, satellite altimetry, and crustal deformation – taken over several decades with an accuracy in an aggregate sense on the order 0.1 mm/year or better. Further, they must be made relative to a global reference frame of equivalent stability and accuracy, the ITRF[19] being the internationally

[18] http://www.fao.org/gtos/gt-netGLA.html
[19] http://itrf.ensg.ign.fr/

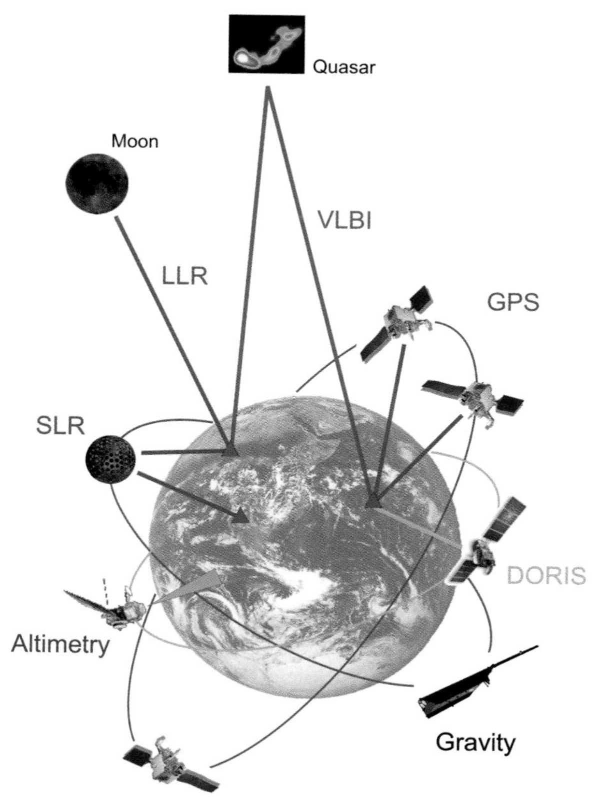

Figure 12.3 Satellites contributing to better definition of the ITRF. The five levels of infrastructure contributing to the GGOS: (1) terrestrial geodetic infrastructure; (2) low-earth-orbiting (LEO) satellite missions (e.g. altimetry and gravity); (3) navigation (GPS and DORIS) and Lageos-type satellite laser ranging (SLR) satellites; (4) planetary missions and geodetic infrastructure on Moon (LLR, lunar laser ranging) and planets; and (5) extragalactic objects (VLBI and quasars). Ground networks and navigation satellites are crucial for maintaining the reference frame and they allow, for example, the monitoring of volcanoes, earthquakes, and tectonically active regions. The LEO satellites monitor sea level, ice sheets, water storage on land, high-resolution surface motion, and variations in the Earth's gravity field mainly originating from regional and global mass transport in the hydrological cycle. (see Rothacher et al. 2009). (Figure prepared by Markus Rothacher of the Institute of Geodesy and Photogrammetry, Swiss Institute of Technology (ETH), Zürich.)

accepted standard. The ITRF is determined by the International Earth Rotation and Reference Systems Service using data from what is collectively known as the Global Geodetic Observing System[20] (GGOS), a consortium with contributions from over 200 international agencies and institutions (see Figure 12.3) The goal of the GGOS is to achieve an ITRF accurate to approximately 1 mm and stable to approximately 0.1 mm/year.

The ITRF is likely accurate to 1 or 2 ppb and this translates to a vertical positioning uncertainty of approximately 0.6–1.2 cm at the Earth's surface. However, concerning the associated impact on estimating sea-level rise, these uncertainties can be significantly reduced with long-term averaging since a significant portion of the geocenter variation is at seasonal scales. While the absolute scale may be erroneous at the 1 ppb level (0.6 cm), the difference between the ITRF2000 and ITRF2005 scale rates – a value of 0.08 ppb/year which translates to about 0.5 mm/year – may be a better proxy for the error in the scale rate. However, when using data with different reference frames, scale offsets between the two may cause an apparent error in the scale rate.

[20] http://www.ggos.org/

These important uncertainties stem largely from uncertainties in the determination of the geocenter and the scale parameters used to fit the modeled reference frame to the observational data. For example, Beckley et al. (2007) report regional-to-hemispherical sea-level changes of 1.5 mm/year due in large part to changes in the geocenter determination when comparing recomputed sea-surface measurements referenced to the ITRF2000 and ITRF2005 respectively. Furthermore, Altamimi et al. (2007) report 1–2 mm/year relative drifts during recent years in the VLBI and SLR scale parameters. While these studies demonstrate that global and regional averages of sea level are susceptible to changes in the scale and geocenter parameters for a given reference frame, regional estimates of sea-level change can also be affected by positioning errors in the Earth's center of mass, local circulation, regional geoid changes, and land-elevation changes such as GIA and tide-gauge measurements. Therefore, the aim is to improve the accuracy of the ITRF[21] by a factor of 10 in order to meet the challenge of global sea-level measurement. Furthermore, the existing global geodetic networks that provide the observations necessary for the determination of the ITRF – which have developed since the 1970s as independent capabilities – are themselves in a state of decay with aging infrastructure and a poorly distributed global infrastructure.

Given this situation, survival of the GGOS has risen as a critical objective, given that it underpins the basic metrology for all of the other missions. Numerous geodetic observatories have been closed in the past 15 years, with the Southern Hemisphere significantly underrepresented in space geodetic observing sites. At the same time, robust national plans for the expansion and deployment of the new GNSS by the USA, European Union, Russia, China, and Japan represents both an opportunity and a challenge. These GNSS satellites will bring much improved geodetic measurement capability, but will also require a retooling of the networks at a time of diminishing financial resources, as well as improved coordination between these evolving systems.

There is significant agreement within the space geodetic community as to what needs to be done: add SLR retro-reflectors to all GNSS satellites; implement and/or renew a globally well-distributed set of geodetic observatories with co-located VLBI, SLR, DORIS, and GNSS instrumentation with continuous monitoring, including the associated SLR; integrate the analysis of resulting data to eliminate systematic processing errors; develop an analytical model relating reference-frame accuracy to the design of the global network, and explore innovative spaceborne collocation and GNSS-enhancing mission concepts, such as the proposed Geodetic Reference Antenna in Space (GRASP; Bar-Sever et al. 2009).

The origin of the ITRF is the Earth's center of mass, the geocenter. Unmodeled motions of the geocenter with respect to the Earth's surface – that is, errors in the ITRF – can induce significant errors in the estimate of regional sea-level change. Furthermore, the determination of precision orbits for the GNSS and altimetry satellites are dependent on an accurate gravity field. Therefore, improving the

[21] The new ITRF2008 is currently under preparation and is expected to be released in 2010.

gravity-field measurements has a direct effect upon the accuracy of the reference frame and of sea-level change.

The gravity field constantly varies in both time and space reflecting the transport of mass within the Earth system due to climatic and tectonic forces. Space geodesy can resolve the Earth's gravity field with an accuracy sufficient to resolve these variations in gravity and the related changes in density and mass transport. These daily-to-decadal temporal changes are for the most part very small perturbations on the order of a few parts per billion of the Earth's gravity field. Therefore, there is considerable utility in the mean gravity field taken as an average over time. For example, the mean gravity field or geoid is required to estimate the dynamic ocean topography for ocean circulation, geopotential heights for orthometric land topography, crustal structure, and perhaps most importantly the geocenter. There are currently millimeter-scale (1.8 mm/year) annual drifts in the estimate of the location of the geocenter with respect to the Earth's surface, as expressed in the difference between the ITRF2000 and ITRF2005 realizations, with an accompanying 5.8-mm offset. The geocenter drift directly affects the estimated rate of global sea-level change by approximately 0.3 mm/year, with larger impact on regional changes.

The global mean gravity field will continue to improve in accuracy and spatial resolution as gravity satellite measurements continue, with differing technologies required to measure components of the gravity field. These include:

- 40 000-km (center of mass) to 5000-km spatial resolution: SLR ranging to Medium Earth Orbit (MEO) geodetic satellites (e.g. Lageos I and II);
- 5000- to 100-km resolution: low-earth-orbit (LEO) satellite gravity (e.g. GRACE, launched in March 2002, and GOCE, launched in March 2009);
- better than 500-km resolution: ocean altimetry, airborne, and land gravity surveys become increasingly important sources of information.

The GRACE satellite has provided a dramatic improvement in the mean Earth's gravity field with a near-uniform accuracy of better than 7 mm at 300-km spatial resolution. Unfortunately the measurement accuracy of GRACE degrades significantly at shorter spatial scales (e.g. Tapley et al. 2005). GOCE will extend the GRACE accuracy in the determination of the geoid to approximately 1 cm at better than 100-km spatial resolution. At the lower or global end of the spatial-resolution scale, improved knowledge of the center of mass will come from an improved global geodetic observing network and a small number of new geodetic satellites in MEO or high LEO orbits such as the proposed Italian LARES[22] satellite.

Continuation and improvement of GRACE gravity measurements will be of particular importance to the future measurement and understanding of sea-level change. The *Decadal Survey* has recommended the launch of a GRACE follow-on mission in the 2016–20 time frame. While NASA is studying the methodology for such a mission, ESA, CNES, and the German and Italian Space Agencies, as well as China and Japan, have all expressed interest in future high-resolution gravity missions.

---

[22] http://www.diaa.uniroma1.it/docenti/a.paolozzi/lares.htm

12.3 Development of Improved Observing Systems (New Capabilities)

12.3.1 Ocean Volume

Measuring properties over the full extent of the ocean, hence enabling a more refined estimate of the effect of changes in its density field on ocean volume, will require:

- extending the Argo-type capability to enable the collection of similar observations under both the sea ice and the floating ice shelves into which most Antarctic glaciers flow;
- designing and implementing an effort to obtain observations for the deep ocean complementary to those from Argo for the upper ocean.

Two approaches are being pursued to obtain profiles in ice-covered regions. One approach at the Alfred Wegener Institute in Germany uses Argo floats with temperature-sensing algorithms to determine whether an instrument may encounter ice at the surface and be unable to transmit data (see Figure 12.4). In such instances an instrument will terminate its ascent prior to encountering ice, store the data for later transmission, return to its normal profiling cycle, and continue that cycle until a free-ocean surface is encountered and all of the data can be transmitted. Positions of the instruments are determined using underwater sound sources. This approach has been used in Antarctic regions, specifically the Weddell Sea. For example, see Klatt et al. (2007).

The second approach pursued at the Japan Agency for Marine-Earth Science and Technology involves tethering a profiling float to an ice platform using a cable

Figure 12.4 (*Opposite*) Two ice-resilient Argo floats in the Antarctic. These floats are equipped with ice-detection software, so they can store profiles collected while operating under the ice; once the ice has melted during the austral summer the floats can then surface and transmit all of their stored data. (a) The deployment of Argo float 2900127 on September 11, 2007 at coordinates 64.2 °S, 127.9 °E during a cruise of the Australian icebreaker, *Aurora Australis* (in the background). In early spring before the seasonal ice has melted, a hole is drilled in the 1.5-m-thick ice through which the float is lowered. Twenty-four floats provided by Steve Riser of the University of Washington were deployed in this manner during that cruise. (b) Argo float 6900588 drifting amidst sea-ice floes just after its deployment on December 23, 2008 at 65.5 °S, 3.0 °E during a cruise of the RV *Polarstern*. The Alfred Wegener Institute has been deploying such floats since 2002–3, with six out of 10 floats of this first batch (2008–9) operating in their seventh year, having surpassed 225 profiles. Utilization of ships such as the *Aurora Australis* and *Polarstern* is essential to facilitate extension of the Argo array into ice-covered areas of the Southern Ocean. (Photographs taken by and provided through the courtesy of (a) Guy Williams, then with the Australian Antarctic Division, now at the Institute of Low Temperature Science, Hokkaido University, and (b) Olaf Boebel of the Alfred Wegener Institute.)

(a)

(b)

with a terminal weight at the bottom[23]. The profiling instrument moves up and down the tether collecting observations while the surface platform collects meteorological observations and contains the telecommunications equipment. Data from the profiling instrument are transmitted acoustically to the ice platform for telemetry via satellite. Operations in this manner mean those data do not have to be stored on the instrument, but rather are transmitted in real time for dissemination using the Argo data-distribution systems (Global Telecommunications System and Data Assembly Centers). The ice-tethered instrument is employed in Arctic operations.

There is a significant lack of year-round, routine oceanographic profile data from the region around the Antarctic continent and from the Arctic Ocean. The technology exists to obtain these observations. With supply and research vessels in the Antarctic providing numerous deployment opportunities, the major limitation in sustaining a profiling float array capable of operating in seasonal ice are the financial and logistical resources to acquire and deploy the sound sources. A limited number of acoustic sound sources would be required to cover the oceanographic regions subject to large seasonal ice cover. For the Arctic, the availability of resources to acquire and deploy the equipment represents a significant limitation.

The technology for deep-moored observations and for repeat hydrographic/tracer observations is in hand. However, at present, the ocean at depths below those traversed by Argo floats is sampled only by a handful of moorings and infrequent repeat hydrographic and tracer cruises. Given the amount of variability being observed by Argo profiling floats in the upper 2000 m of the ocean, the deeper variability observed in some repeat hydrographic data, and the amount of heat that some numerical ocean models and observations indicate can get into the sub-Argo-depth ocean, it is critical that the sub-Argo ocean be sampled well enough – both in space and time – to determine long-term variability and trends. Much better coverage in space and time is likely needed, but presently there is no proven technology to make observations sufficient to meet this need.

Appropriate observing platforms need to be developed to carry at least temperature and salinity sensors, in order to understand and project the role of this part of the ocean on global sea-level variability and change. A thorough review of what is known about the variability and trends in the sub-Argo ocean, supplemented with analysis of ocean model results, would be necessary to produce a preliminary observing system strategy for the number and sampling density thought to be required. At present, the most likely candidate technology to address these additional sampling requirements is some sort of autonomous underwater vehicle. It is thought to be possible that both freely drifting and "glider" technologies may be feasible to meet the need. A global system to deploy, and possibly recover, such autonomous underwater vehicles would also be required.

There is ongoing development of ocean gliders, with field testing of one with the needed capability imminent. With sufficient developmental support, full-

[23] http://www.jamstec.go.jp/arctic/scosa/pops/pops.htm

water column autonomous underwater vehicles might be sufficiently robust for widespread use within a few years. Several countries have the marine engineering experience to participate in system development if they choose to. There is ongoing development and pilot project work with ocean gliders, but none as yet has the needed capability. There is also ongoing development work on deeper-profiling Argo-class floats, but development of full-ocean depth profilers is not taking place at the time of writing, although groups with the necessary experience are in place to undertake the work required.

12.3.2 Ice-Sheet and Glacier Volume

Refined estimates of ice-sheet contributions to sea-level change will include:

- developing and implementing a suitable follow-on capability based on experience gained with radar and laser satellite altimeters.

High-latitude radar altimetry has been providing information on ice-sheet changes since the launch of ERS-1 in 1991. ERS-2 launched in 1995 and ENVISAT launched in 2002 – with their 82° inclined orbits – have each sequentially contributed to a continuous SRALT time series. ESA's CryoSat-2 features a radar altimeter using synthetic aperture radar and interferometry, effectively making measurements at a known angle to small (≈70 m) footprints, thus overcoming the spatial sampling limitation associated with broad-beam altimeters such as those on ERS and ENVISAT missions. Following CryoSat-2, continuity of high-latitude radar altimetry will be extended in 2012–13 by the Sentinel-3 series of synthetic aperture radar altimeters.

Laser altimetry from aircraft and satellites continues to provide accurate information on ice-sheet surface topography, and is instrumental in assessing ice-sheet contributions to sea-level rise. Airborne laser altimetry measurements have provided detailed topographic information over sinuous outlet glaciers and ice-sheet margins in Greenland and parts of Antarctica. These measurements, begun in earnest in Greenland in 1993, accurately capture changes in these steep and rough regions, where ice losses have been quite significant. Aircraft measurements have also provided the basis for an assessment of changes in glaciers in Alaska, the Canadian Archipelago, Svalbard, and Patagonia. These small ice masses are contributing significantly to sea level, and cannot be systematically observed by any other means.

The ICESat mission has been providing 33-day snapshots of ice-surface topography three times per year from 2003 to 2007, and twice per year since 2007, allowing ice-sheet-wide detailed observations of ice-sheet elevation changes. Premature degradation of the first two of ICESat's three lasers has cut the spatial and temporal sampling of the ice sheets each to about one-third of what was originally planned, yet despite these limitations it is providing repeat measurements of ice-sheet surface elevation changes in Greenland and Antarctica. The

Figure 12.5 NASA's P-3B aircraft on its final approach into Thule Air Base, Greenland, in the summer of 2007. The aircraft is carrying the NASA Airborne Topographic Mapper lidar, a scanning laser system for mapping the surface topography of the ice sheet. Under the wings can be seen the antennas for the University of Kansas Coherent Radar Depth Sounder, a radar capable of penetrating the ice sheet to measure the elevation of the land surface below. By combining elevation data for the surface and base of the ice, taking into account the aircraft position using GPS, researchers can determine ice thickness at any given location. This P-3B flies routes that take it directly under the path of ICESat and other satellites, allowing the satellite and plane to measure the same features. Each has its strength: the satellite provides regular, continental-scale coverage of Greenland and hard-to-reach regions like Antarctica, while the aircraft can make more detailed surveys of areas where scientists expect to see rapid change. (Photograph courtesy of Bill Krabill, NASA WFC.)

follow-on ICESat-II mission is needed to achieve this full level of sampling; it is projected to be launched in approximately 2015.

Assuming the elevation measurements are well calibrated, the ICESat-II and CryoSat-2 observations will be compared to the current ICESat data to assess the change history dating back to 2003 and to the aircraft data for the change history in Greenland dating back to 1993 (see Figure 12.5).

In order to address the problem of "missing" data, next-generation ice altimeters should have an extensive swath to increase areal coverage, as well as fine spatial resolution for application in smaller, rougher glacier regions. Efforts to develop such a capability are currently underway, and have been recommended by the *Decadal Survey* for the long-term set of missions in the 2016–20 time frame; this is the Lidar Surface Topography (LIST) mission with a 5-km-wide swath, 1000-beam, full-waveform, imaging laser altimeter that will provide increased spatial coverage of the ice sheets. While also susceptible to snow accumulation and firn compaction, these laser missions (ICESat-II and LIST) avoid the issues of penetration into the snowpack and the broader spatial resolution associated with radar altimeters.

Remote sensing of density profiles on glaciers, down at least as far as snow that is 1 year old, is not possible at present. It would, however, solve many of the problems that afflict radar and laser altimetry as tools for the monitoring of glacier mass changes. Research should be encouraged on the modeling of optical- and microwave-band radiative interactions with snow covers, to explore the prospects for inferring density remotely.

12.3.3 Ice Velocity

Observing ice velocity is essential to characterize the dynamics of ice sheets, the flow of ice to the ocean, and how those characteristics may change with time. This will require:

- development and implementation of an improved InSAR mission to measure detailed flow rates in glaciers and ice sheets.

Ice-sheet velocity measurements indicate that the flow rates of outlet glaciers, and ice streams can vary dramatically on a range of timescales. These variations have the potential to cause rapid adjustments in ice-sheet mass balance, calving flux, and in turn sea level. If the thickness can be estimated for the fastest flowing glaciers, it is possible to compute the flux of ice into the oceans. We do not know how much faster glaciers can flow, and there is significant residual uncertainty in the magnitude of their response to a warming climate. Consequently, it is essential to develop a comprehensive understanding of their time-varying flow characteristics and factors governing their flow, as well as thickness estimates for the fastest-flowing glaciers. Such information will enable a better understanding of ice-sheet stability and predict their future contributions to sea-level rise.

Over the last decade, the InSAR technique has matured – using C-band imaging synthetic aperture radar on the first ERS satellite (ERS-1), ERS-2, ENVISAT, and the Canadian RADARSAT – and is revolutionizing the study of ice sheets. InSAR is demonstrating the remarkable capability to track the motion of ice sheet and glacier surfaces by cross-correlating the reflection patterns of ice surfaces over time acquired from two nearly identical observation locations in space. If the ice surface remains stable, this permits a detection of the displacement of the icy surface in the radar-looking direction with a typical precision of fractions of the radar wavelength; that is, centimeters.

To characterize the three-dimensional flow vector, and to cancel out the influence of topography, three unique radar looking directions are required. Since the polar orbits of previously operating synthetic aperture radar satellites have not offered this possibility – with the recent Terra SAR-X and Cosmo-Skymed missions representing an opportunity to address this issue – the present InSAR strategy has been limited and has operated on long-repeat passes to detect ice motion in two dimensions (projected on the surface). Ice velocity maps of Greenland and Antarctica are now being produced even though the missions that acquired the data were not designed with this objective in mind; for example, see Rignot and Kanagaratnam (2006) and Rignot et al. (2008), respectively.

With the launch of RADARSAT-2[24] in 2007, the Canadian Space Agency is planning a three-satellite constellation for RADARSAT-C (to be launched in 2010 or later). In continuity with RADARSAT-1, these satellites will be capable

[24] http://www.radarsat2.info/

of left- and right-looking with a 24-day repeat cycle. Meanwhile the Phased Array L-band Synthetic Aperture Radar (PALSAR) launched in 2006 aboard the Japanese Space Exploration Agency's Advanced Land Observing Satellite[25] has been delivering L-band acquisitions over Antarctic ice sheets with a 44-day repeat. The Advanced Synthetic Aperture Radar on ESA's ENVISAT[26], with a 35-day revisit cycle, was tuned in 2007 (via orbit control) to provide shorter baselines and better data for InSAR ice-sheet mapping; it will be followed by the C-band synthetic aperture radar on ESA's Sentinel-1A and -1B[27], with the first to be launched in the 2012/13 time frame and the second some time later. All these can contribute to ice-sheet dynamics, but their respective mission planning needs to be done with this in mind. They also offer the potential to look for differential land motion in the area surrounding tide gauges, thereby complementing the use of GPS as noted in the first section of this chapter.

The lack of a dedicated mission, plus limitations on the use of the InSAR imaging mode together with inherent limitations of existing missions, hinders further progress with the existing missions. To develop realistic ice-sheet models, dedicated InSAR – optimized for ice-sheet use – and the capability to deliver free data frequently to the science community are absolutely critical.

The *Decadal Survey* has recommended that NASA fly the Deformation, Ecosystem Structure and Dynamics of Ice (DESDynI[28]) satellite with a payload including both an L-band InSAR and a multi-beam laser altimeter in an 8–16-day repeat orbit. While this mission concept is optimized for solid Earth and natural hazard applications, it would make valuable – though not optimal – contributions to ice-sheet studies. The *Decadal Survey* has proposed DESDynI in the near-term set of missions in the 2010–13 time frame.

12.3.4 Sea and Terrestrial Water Levels; Ice-Sheet and Glacier Topography

The sampling capability of satellite altimetry has always been a compromise between the spatial and temporal requirements; hence, to broaden its utility will require:

- developing and implementing a mission with an advanced wide-swath (≈500 km) radar altimeter to observe the:
  - sea level associated with the oceanic mesoscale field, coastal variability, and marine geoid/bathymetry;
  - surface water levels on land and their changes in space and time;
  - surface topography of glaciers and ice sheets.

---

[25] http://www.jaxa.jp/projects/sat/alos/index_e.html
[26] http://envisat.esa.int/instruments/asar/
[27] http://www.esa.int/esaLP/SEMZHM0DU8E_LPgmes_0.html
[28] http://desdyni.jpl.nasa.gov

Satellite altimetry has revealed a wealth of information about the height of the sea surface and of water levels of rivers and lakes on land. Despite its revolutionary accomplishments, high spatial resolution can only be achieved in the along-track direction, leading to an asymmetry in the radar's mapping capability. For example, the zonal currents of the ocean tend to be better determined than the meridional currents, and only a one-dimensional measurement can be made of the water levels of rivers and lakes. Both the ocean surface topography and terrestrial surface water hydrology communities recognize the need for a capability for high-resolution spatial mapping of water surface heights. For example, our understanding of oceanic circulation at the energetic mesoscale is poor, and we have even less knowledge of the global dynamics of terrestrial surface waters.

A new technology has been demonstrated by the Shuttle Radar Topography Mission[29] for mapping the Earth's land topography using the technique of radar interferometry. This same technique, employing a synthetic aperture radar system to achieve spatially uniform high resolution for mapping the ocean surface topography and the water levels of rivers, lakes, and wetlands, has been proposed with a resolution on the order of tens of meters. However, after averaging, the radar interferometer can achieve centimetric precision over the ocean at 1-km resolution; this is smaller than the typical eddy scales in the ocean by an order of magnitude, offering a revolutionary new view of ocean dynamics. For land hydrology applications, a 10-cm precision can be achieved at 100-m resolution: this is sufficient for measuring the world's most important rivers, lakes, and wetlands.

These measurements will allow fundamental advancements in the understanding of (1) the dynamics of ocean currents, eddies, fronts, and internal tides that are key to improving the modeling of ocean circulation and its energy budgets, (2) the dynamics of terrestrial surface waters that are key to the understanding of global water cycles, flow hydraulics, aquatic ecosystem dynamics, flood hazards, water resources, and management, and (3) the dynamics of coastal ocean and its interaction with river discharges in estuaries. Additionally, the measurements will significantly improve the knowledge of marine gravity and ocean bathymetry at small scales. And while laser systems represent the preferred option for the cyospheric community in the future, these radar-based measurements will contribute to observing the surface topography of glaciers and ice sheets.

Recognizing the revolutionary impacts of the proposed new measurement from space, the ocean surface topography community and the land-surface hydrology community have joined forces in the development of the Surface Water Ocean Topography[30] (SWOT) mission to make this proposal a reality. The *Decadal Survey* has proposed SWOT for the mid-term set of missions in the 2013–16 time frame.

---

[29] www2.jpl.nasa.gov/srtm/

[30] http://swot.jpl.nasa.gov/ and http://www.legos.obs-mip.fr/recherches/missions/water/

12.4 Summary

Improving our understanding of sea-level rise and variability, as well as reducing the associated uncertainties, critically depends on the availability of adequate observations, as well as an open data policy with timely, unrestricted access for all.

The existing systems that should be sustained include those observing sea level: the Jason series of satellite altimeters, as well as the complementary GLOSS network of approximately 300 gauges (each with high-frequency sampling, real-time reporting, and geodetic positioning). Together, these satellite and *in situ* observations enable the measurement of both absolute and relative changes in sea level. To estimate the change in sea level resulting from changes in the volume of the oceans due to thermal expansion and salinity changes, the Argo array of profiling floats needs to be sustained to observe the upper-ocean in ice-free areas. In order to estimate the change in sea level resulting from changes in mass of the oceans due to melting ice caps and glaciers and changes in terrestrial water storage, observations of changes in these characteristics and the processes that control them must be maintained.

For terrestrial water storage, observations of the time-varying gravity field from the GRACE mission, as well as deployment of an appropriate follow-on, are necessary to provide continuity in these data. For ice-sheet and glacier contributions to sea level, a combination of altimetry, gravity, and InSAR observations is essential. Altimetry will provide detailed information on the spatial character of change, along with important insights into the mechanisms that drive that change (melt, flow, or accumulation) based on the observed topographic signatures. Gravity measurements will continue to provide large-scale direct assessments of actual mass change and are important for assessing large-scale ice-sheet contributions to sea level (see Figure 12.6). InSAR will provide detailed information on the spatial and temporal variability of flow, which is the key to detecting and assessing ice-sheet stability. Detailed elevation change measurements and flow observations are essential for the development of models capable of reliably predicting future ice-sheet contributions to sea-level rise. Finally, detailed aircraft measurements of ice topography, elevation change, and ice thickness are needed to complement the larger-scale observations of ice behavior in support of model development. The full benefits of these capabilities will only be realized if they are deployed simultaneously. As such, every effort must be made to execute these observations concurrently or with as much overlap as possible.

A key component of altimetry measurements of sea-level, ice-topography, and ice-elevation changes is the ITRF. The success of these measurements requires that the ITRF be made more robust and stable. Furthermore, observations of the time-invariant gravity field from the GOCE satellite and other stand-alone missions are needed to determine the precise geoid.

New and improved observing systems which need to be developed include those directed at changes in the ocean volume, specifically extending the Argo-

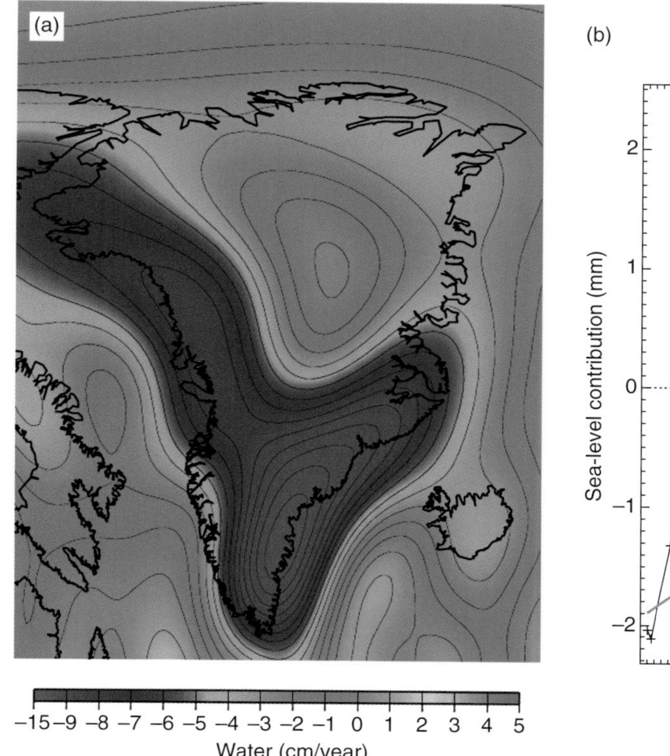

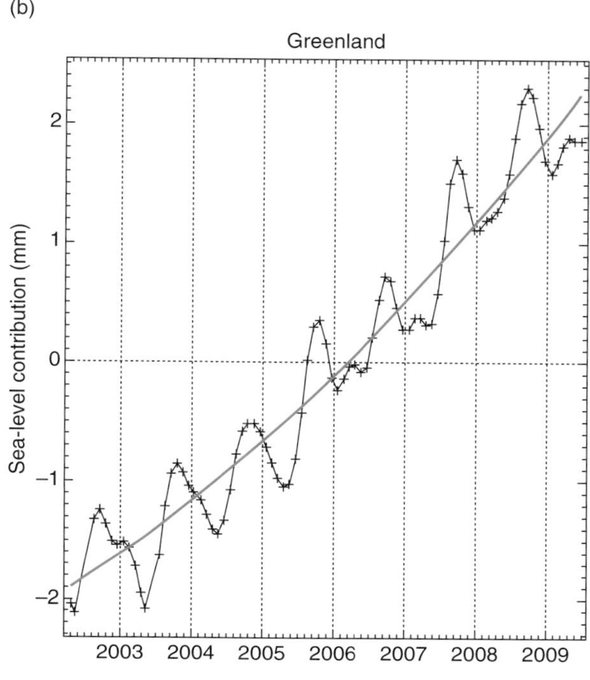

Figure 12.6 Estimates of the rate of mass loss for the Greenland Ice Sheet based on gravity observations collected by the NASA/DLR GRACE satellite between April 2002 and June 2009. (a) Spatial distribution of the rate of mass loss, expressed in cm/year of equivalent water thickness (note the break in the color scale between −9 and −15 cm/year; the solid contour lines occur at 1 cm/year intervals); most of the mass loss occurs along the coast, especially in the southeast. (b) Time series of the rate of mass loss for the entire Greenland Ice Sheet expressed in terms of its contribution to global sea level; the orange line – the best-fitting quadratic – shows an increase in the rate of mass loss during this time period. (Figures prepared by and provided through the courtesy of John Wahr, University of Colorado.)

type capability to enable the collection of similar observations under the sea ice, as well as the design and implementation of an effort to obtain observations for the deep ocean. Based on experience gained with radar and laser satellite altimeters, the development of a suitable follow-on capability is needed to improve observations of ice-sheet and glacier topography. Access to InSAR data and ongoing InSAR missions are needed to observe flow rates in glaciers and ice sheets. Finally, the development of an advanced wide-swath altimeter is needed to observe: sea level associated with the oceanic mesoscale field, coastal variability, and marine geoid/bathymetry; surface-water levels on land and their changes in space and time; and surface topography of glaciers and ice sheets at high spatial resolution.

References

Altamimi Z., Collilieux X., LeGrand J., Garyt B. and Boucher C. (2007) ITRF2005: a new release of the International Terrestrial Reference Frame based on time series of station positions and Earth Orientation Parameters. *Journal of Geophysical Research* V **112**, B09401.

Bar-Sever Y., Haines B. and Wu S. (2009) The Geodetic Reference Antenna in Space (GRASP) Mission Concept. *Geophysical Research Abstracts* **11**, EGU2009-1645.

Beckley B.D., Lemoine F.G., Luthcke S.B., Ray R.D. and Zelensky N.P (2007) A reassessment of global rise and regional mean sea level trends from TOPEX and Jason-1 altimetry based on revised reference frame and orbits, *Geophysical Research Letters* **34**, L14608.

Benveniste J. and Ménard Y. (eds) (2006) *Proceedings of the 15 Years of Progress in Radar Altimetry Symposium, Venice, Italy*, 13–18 March 2006. European Space Agency Special Publication SP-614.

Beutler G., Drinkwater M.R., Rummel R. and von Steiger R. (eds) (2003) *Earth Gravity Field from Space – from Sensors to Earth Sciences*. Space Sciences Series of ISSI, vol. 18. Kluwer Academic Publishers, Dordrecht.

Committee on Earth Science and Applications from Space. (2007) *Earth Science and Applications from Space: National Imperatives for the Next Decade and Beyond [Decadal Survey]*. Space Studies Board, National Academies Press, Washington DC. http://www.nap.edu/catalog/11820.html.

Fu L-L. and Cazenave A. (eds) (2001) *Satellite Altimetry & Earth Sciences, A Handbook of Techniques & Applications*. International Geophysical Series, vol. 69. Academic Press, London.

Gould J. and the Argo Science Team (2004) Argo profiling floats bring new era of in situ ocean observations. *EoS Transactions of the American Geophysical Union* **85**(19), 179, 190–1.

Klatt O., Boebel O. and Fahrbach E. (2007) A profiling float's sense of ice. *Journal of Atmospheric and Oceanic Technology* **24**(7), 1301–8.

Koop R. and Rummel R. (eds) (2008) *Report from the Workshop on The Future of Satellite Gravimetry*, ESTEC, Noordwijk, The Netherlands, 12–13 April 2007. Technische Universität München, Institute for Advanced Study Publication.

Rignot E. and Kanagaratnam P. (2006) Changes in the velocity structure of the Greenland Ice Sheet. *Science* **311**(5763), 986–90.

Rignot E., Bamber J.L., Van den Broecke M.R., Davis C., Li Y., Van de Berg W.J. and Van Meijgaard E. (2008) Recent Antarctic ice mass loss from radar interferometry and regional climate modelling. *Nature Geoscience* **1**, 106–10.

Rothacher M., Beutler G., Bosch W., Donnellan A., Gross R., Hinderer J. et al. (2009) The future Global Geodetic Observing System. In: *Global Geodetic Observing System: Meeting the Requirements of a Global Society on a Changing Planet in 2020* (Plag H.-P. and Pearlman M., eds), pp. 237–72. Springer, Berlin.

Tapley B., Ries J., Bettadpur S., Chambers D., Cheng M., Condi F. et al. (2005) GGM02 – an improved Earth gravity field model from GRACE. *Journal of Geodesy* **79**, 467–78.

Willis J.K., Chambers D.T. and Nerem R.S. (2008) Assessing the globally averaged sea level budget on seasonal to interannual time scales. *Journal of Geophysical Research* **113**, C06015.

Woodworth P.L., Aarup T., Merrifield M., Mitchum G.T. and Le Provost C. (2003) Measuring progress of the Global Sea Level Observing System. *EOS Transactions of the American Geophysical Union* **84**(50), 565.

Zwally H.J., Schutz R., Abdalati W., Abshire J., Bentley C., Bufton J. et al. (2002) ICESat's laser measurements of polar ice, atmosphere, ocean, and land. *Journal of Geodynamics* **34**, 405–45.

13 Sea-Level Rise and Variability: Synthesis and Outlook for the Future

John A. Church, Thorkild Aarup, Philip L. Woodworth, W. Stanley Wilson, Robert J. Nicholls, Ralph Rayner, Kurt Lambeck, Gary T. Mitchum, Konrad Steffen, Anny Cazenave, Geoff Blewitt, Jerry X. Mitrovica, and Jason A. Lowe

Coastal zones have changed profoundly during the 20th century with growing populations and economies (Figure 1.2). Increasing urbanization was a major driver of this change. Today, many of the world's megacities are situated at the coast. At the same time, sea level has been rising and is projected to continue to rise further. However, coastal developments have generally occurred with little regard to the consequences of rising sea levels, even in developed regions such as Europe (Tol et al. 2008). An improved understanding of sea-level rise and variability is required to reduce the uncertainties associated with projections for sea-level rise, and hence contribute to more effective coastal planning and management.

The preceding chapters have provided an overview of our understanding of sea-level change. These chapters benefitted from the discussion of position papers at a workshop on sea-level rise and variability held under the auspices of the World Climate Research Programme (WCRP) at the Intergovernmental Oceanographic Commission of the United Nations Educational, Scientific and Cultural Organization (UNESCO) in Paris in 2006. The workshop was attended by 163 scientists from 29 countries representing a wide range of expertise and supported by 34 organizations. The workshop prompted and underpinned new research initiatives. This chapter provides a synthesis of the findings and recommendations from this community discussion, including a summary

Understanding Sea-Level Rise and Variability, 1st edition. Edited by John A. Church, Philip L. Woodworth, Thorkild Aarup & W. Stanley Wilson.

of the contributions to 20th century sea-level rise and a survey of the outlook for the future. It ends with a discussion of the implications for society.

13.1 Historical Sea-Level Change

Sea level has changed and continues to change on all timescales. The last era when there were essentially no permanent grounded ice sheets anywhere on the Earth occurred more than 35 million years ago when the atmospheric carbon dioxide concentration was 1250 ± 250 ppm. At that time, the Earth was warmer as a result of larger greenhouse gas concentration in the atmosphere and sea level was about 70 m above present-day values (Alley et al. 2005). About 32 million years ago, carbon dioxide contributions dropped to 500 ± 150 ppm, corresponding with the formation of the Antarctic Ice Sheet and with sea level falling to about 30 m higher than today (Alley et al. 2005). In comparison, the pre-industrial concentration was about 280 ppm and the 2009 concentration was about 387 ppm (and increasing at about 2 ppm/year).

Over the glacial cycles of the last 500 000 years, sea level has oscillated by more than 100 m as the great ice sheets, particularly those of northern Europe and North America, waxed and waned (Chapter 4; Figure 13.1; Rohling et al. 2009).

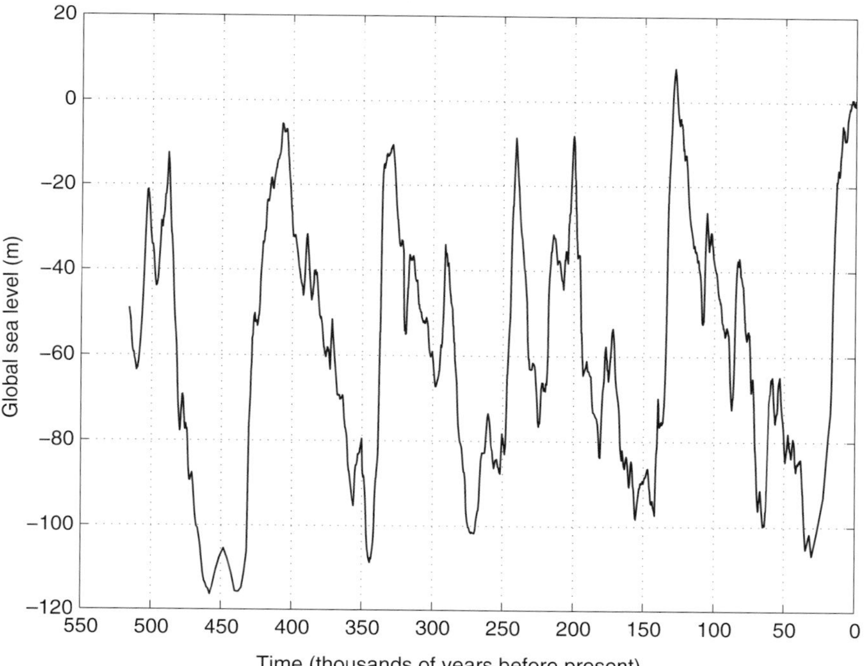

Figure 13.1 Sea level over the last 500 000 years. This sea-level estimate is from Rohling et al. (2009) and is based on carbonate $\delta^{18}O$ measurements in the central Red Sea rather than a direct measurement of sea-level change.

These changes in sea level and the related global average temperature changes were a direct response to changes in the solar radiation reaching the Earth's surface as a result of variations in Earth's orbit around the sun and feedbacks associated with the related changes in the Earth's albedo and greenhouse gas concentrations that amplified the initial solar radiation changes.

The climatic conditions most similar to those expected in the latter part of the 21st century occurred during the last interglacial, about 125 000 years ago. At that time, some paleodata (Rohling et al. 2008) suggest rates of sea-level rise perhaps as high as 1.6 ± 0.8 m/century and sea level about 4–6 m above present-day values (Overpeck et al. 2006; Chapter 4), with global average temperatures about 3–5°C higher than today (Otto-Bliesner et al. 2006). Much of this higher sea level is thought to have come from a smaller Greenland Ice Sheet, but with additional contributions from the Antarctic Ice Sheet (Otto-Bliesner et al. 2006). These conditions serve as a useful analog for the 21st century.

Over the following hundred thousand years, sea level fell to about 130 m below today's values as the northern European and American ice sheets formed (Figures 4.2 and 13.1). From 20 000 years ago to about 7000 years ago these ice sheets collapsed, and sea level rose rapidly at average rates of 1 m/century for many millennia, with peak rates during the deglaciation potentially exceeding several meters per century (Figures 4.2 and 4.3; Fairbanks 1989; Lambeck et al. 2002; Alley et al. 2005). However, these peak rates of sea-level rise during the last deglaciation are not a particularly good analogue to 21st-century conditions because the ice-sheet distribution was very different. From about 6000 to 2000 years ago, sea level rose more slowly, about 2.5 m over 4000 years, and then more slowly again over the last 2000 years up to the 18th century (Figure 4.14). Over this latter period, models of the Antarctic Ice Sheet (Chapter 7) suggest a continuing slow dynamic response to changes in climate since the Last Glacial Maximum (LGM), contributing only about 0.2 mm/year to global averaged sea-level rise, slightly larger than but consistent with the indication of very low rates of sea-level change from paleodata.

Coastal sediment cores and other paleo sea-level data, the few long (pre-1900) tide-gauge records, reconstructions of 20th-century sea levels, and satellite altimetry data all indicate that the rate of sea-level rise has increased by about an order of magnitude: from at most a few tenths of a millimeter per year over previous millennia to about 1.7 mm/year during the 20th century (Figures 5.1 and 5.3), and to over 3 mm/year since 1993 (Figure 5.5). The various estimates of sea level from the late 19th through to the early 21st century are discussed in Chapter 5 and the tide-gauge estimates are summarized in Figure 13.2. The main feature common to all of these estimates is a higher rate of rise during the 20th century compared with the late 19th century. There are also indications of an acceleration during the first half of the 20th century and a deceleration during the 1960s, possibly associated with the volcanic eruptions since 1960, and a substantially faster rate of rise since the late 1980s/early 1990s (about a third or more faster than the rate during any previous 20-year period since 1870; Church et al. 2008). Further

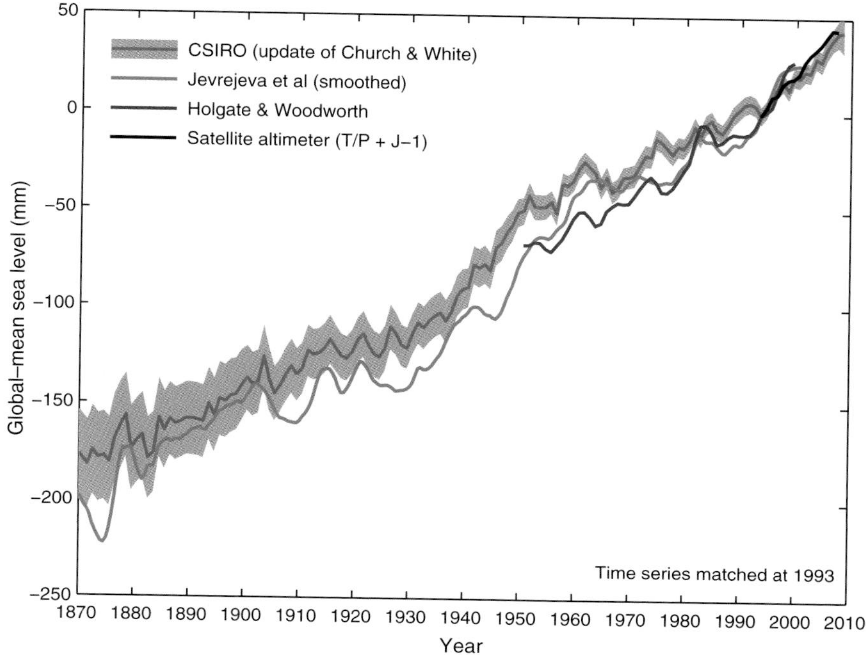

Figure 13.2 Global mean sea level from 1870 to 2008 with 1 standard deviation error estimates updated from Church and White (2006; red), a low-passed filtered sea level from Jevrejeva et al. (2006; green) and from 1950 to 2000 from Holgate and Woodworth (2004; blue). The TOPEX/Poseidon/Jason-1 and -2 global mean sea level (based on standard processing as in Church and White 2006) from 1993 to 2008 is in black. All series have been set to a common value at the start of the altimeter record in 1993.

investigation, especially given the sparse historical database, is required to better quantify the timing and the magnitude of all these changes.

13.2 Why is Sea Level Rising?

A major inadequacy of sea-level science over recent decades, including that in the four IPCC Assessments completed to date, has been the failure to quantitatively explain the observed 20th-century sea-level rise; the observed sea-level rise has been larger than the sum of estimated contributions. Munk (2002) highlighted this discrepancy when he argued that observations of changes in the Earth's rotational parameters (an increase in the length of the day) constrained the possible magnitude of the ice-sheet contributions. The sum of the possible ice-sheet contribution and the estimated contribution from ocean thermal expansion was too small to explain the observed sea-level rise. He called this conundrum "the sea-level enigma".

There has been significant progress in resolving this "enigma" over recent years. Firstly, it is now clear that the observed Earth rotational parameters are not as strong a constraint on ice-sheet contributions as argued by Munk (Chapter 10; Mitrovica et al. 2006). Secondly, revised estimates of ocean thermal expansion

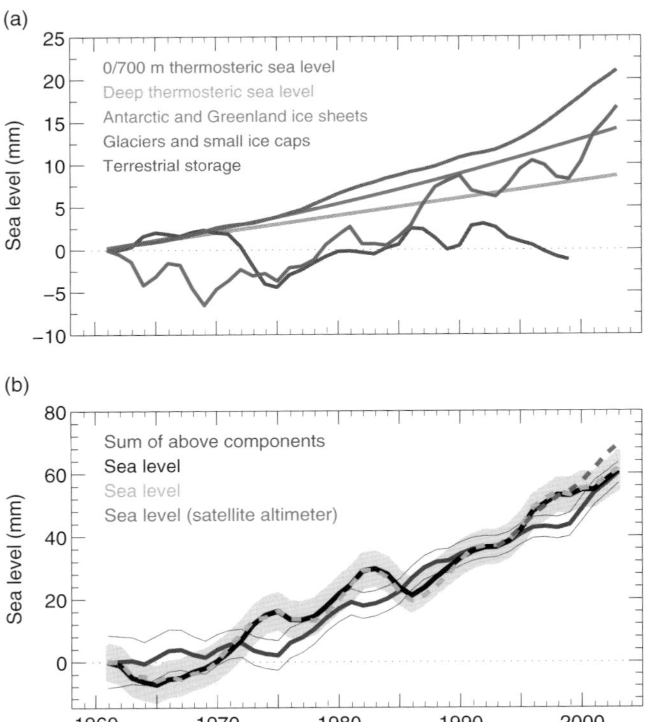

Figure 13.3 Total observed sea-level rise and its components. (a) The components are thermal expansion in the upper 700m (red), thermal expansion in the deep ocean (orange), the ice sheets of Antarctica and Greenland (cyan), glaciers and ice caps (dark blue), and terrestrial storage (green). (b) The estimated sea levels are indicated by the black line from Domingues et al. (2008), the yellow dotted line from Jevrejeva et al. (2006), and the red dotted line from satellite altimeter observations. The sum of the contributions is shown by the blue line. Estimates of 1 standard deviation error for the sea level are indicated by the grey shading. For the sum of components, the estimates of 1 standard deviation error for upper-ocean thermal expansion are shown by the thin blue lines. All time series were smoothed with a 3-year running average and are relative to 1961 (from Domingues et al. 2008).

and the melting of glaciers and ice caps[1], at least since 1961, are somewhat larger than earlier estimates (Chapters 6 and 7; Domingues et al. 2008; Levitus et al. 2009; Ishii and Kimoto 2009). By combining these revised estimates for upper-ocean thermal expansion and glacier and ice-cap contributions with reasonable but more poorly known estimates of contribution from deep-ocean thermal expansion and the Greenland and Antarctic Ice-Sheet contributions, Domingues et al. (2008) produced an approximate closure of the sea-level budget (Figure 13.3; Table 13.1). On these decadal timescales the dominant contributions are the melting of glaciers and ice caps and upper-ocean thermal expansion, with smaller but significant contributions from deep-ocean thermal expansion and the ice sheets. Over 11 years of the relatively short satellite altimeter era, 1993–2003, the IPCC Fourth Assessment Report (AR4) managed to explain the observed sea-level rise using both satellite and *in situ* data.

[1] Clarification of terminology: two general categories of ice are considered as contributors to global sea-level rise: (1) the ice sheets of Antarctica and Greenland, all of which drain into the surrounding ocean via a number of ice streams or outlet glaciers. Ice sheets are sufficiently thick to cover most of the bedrock topography. (2) Glaciers and ice caps, mostly outside Antarctica and Greenland. A glacier is a mass of ice on the land flowing downhill under gravity and an ice cap is a mass of ice that typically covers a highland area.

Table 13.1 Contributions to sea-level rise for the period 1961 to 2003, from Domingues et al. (2008).

Contribution	Amount of rise
Ocean thermal expansion for the upper 700 m	0.5 ± 0.1 mm/year
Ocean thermal expansion below 700 m	0.2 ± 0.1 mm/year
Glaciers and ice caps	0.5 ± 0.2 mm/year
Greenland Ice Sheet	0.1 ± 0.1 mm/year
Antarctic Ice Sheet	0.2 ± 0.4 mm/year
Sum of contributions	1.5 ± 0.4 mm/year
Observed sea-level rise	1.6 ± 0.2 mm/year

The use of satellite altimetry to measure changes in ocean volume, satellite gravity to measure changes in ocean mass, and Argo profiling floats to measure changes in upper-ocean temperatures and thermal expansion is also leading to improved understanding of the sea-level budget since 2003 (Chapter 6; Willis et al. 2008; Cazenave et al. 2009; Leuliette and Miller 2009). Both Cazenave et al. (2009) and Leuliette and Miller (2009) have closed the sea-level budget within error bars over the short record. However, they had rather different contributions for slightly different periods. Cazenave et al. estimated the mass and thermal-expansion increases were about 2.2 and 0.4 mm/year, respectively, whereas the equivalent estimates of Leuliette and Miller were both about 0.8 mm/year (see Chapter 6 for more detailed discussion). A longer record of these three complementary measurements is likely to lead to significant further progress in understanding 21st-century sea-level rise.

One thing is clear from all of the recent analyses: observations indicate an increasing glacier and ice-cap contribution and also increasing ice-sheet contributions as a result of the flow of ice into the ocean from both Greenland and Antarctica. Of particular concern is the rapid dynamic thinning of the margins of the Greenland and Antarctic Ice Sheets (Chapter 7; Pritchard et al. 2009). However, the record is still short, some discrepancies remain, and physically based quantitative estimates for the 21st century are lacking.

An additional contribution to changing sea level comes from the storage of water on land: in lakes, dams, rivers, wetlands, soil moisture, snow cover, permafrost, and aquifers. These respond to both climate variations and to anthropogenic activities through, for example, the building of dams and the mining of water from aquifers (Chapter 8). Although these terms are important to understanding 20th-century sea-level rise, and hence in improving our projections for the future, their contribution to sea-level change through the 21st century and beyond is likely to be substantially smaller than other changes (continued contribution at the 20th-century rate would alter the projections by centimeters).

This improved understanding of the contributions to sea-level rise is important as it is likely to lead to better observational constraints on the climate models used for projections of sea-level rise in the IPCC Fifth Assessment Report, which has an anticipated completion date of 2013–14.

13.3 The Regional Distribution of Sea-Level Rise

The regional distribution of sea-level rise is important as a scientific question, and also because it is the regional or local sea-level change (and local land motion) that impacts society and the environment. Satellite altimeter data show significant regional variations in the rate of sea-level rise (Figure 13.4), with some regions having experienced about five times the global-averaged rate of rise since 1993. However, this regional variation in the relatively short altimeter record is largely a result of climate variability, particularly in the equatorial Pacific Ocean (Chapters 5 and 6). The pattern is associated with the movement of water within the oceans in response to varying wind patterns associated with climate phenomena like the El Niño Southern Oscillation and is largely reflected in regional patterns of ocean thermal expansion (Figure 6.3). During the 21st century, climate variability will continue and coastal communities will be impacted by the combination of the pattern of long-term sea-level rise and the natural variability in sea level.

Changes in the mass of the ice sheets (and glaciers and ice caps) also influence the regional distribution of sea-level rise through corresponding changes in the Earth's gravitational field and the elastic movement of the Earth's crust (Figure 10.5). That is, the contribution from the ice sheets results in a lower relative sea level near decaying ice sheets and a larger than the globally averaged rise (up to

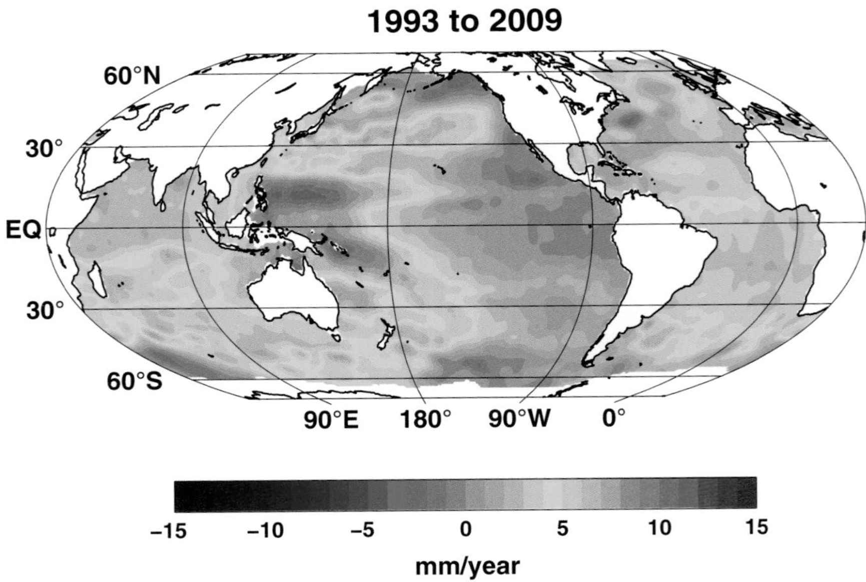

Figure 13.4 The spatial distribution of the rates of sea-level rise, plotted about the global averaged rate of rise for the period January 1993 to December 2009, as measured from satellite altimeter data (available at http://www.cmar.csiro.au/sealevel/).

about 20%) far from the decaying ice sheets. As a result, ice-sheet contributions to future sea-level rise are likely to have a disproportionate impact in some far-field and potentially vulnerable regions.

13.4 Projections of Sea-Level Rise for the 21st Century and Beyond

The IPCC Third Assessment Report (TAR; IPCC 2001) projections of sea-level rise were expressed in the form of globally averaged levels for 2100 compared with 1990 levels (the lines and shaded regions in Figure 13.5), while the AR4 (IPCC 2007) were expressed for the 2090–2100 decade (shown as the bars plotted at 2095 in Figure 13.5) compared with 1980–2000 averages (approximately equal to the 1990 values).

The average of the TAR model projections for the full range of greenhouse gas scenarios is about 30–50 cm (dark shading in Figure 13.5). The range of all model projections over all scenarios is about 20–70 cm (light shading). The full range of projections, including an allowance for uncertainty in estimates of contributions from land-based ice, were for a sea-level rise of 9–88 cm (outer black lines).

The AR4 model projections are composed of two parts. The first part consists of the estimated sea-level rise (with a 90% confidence range) from ocean thermal expansion, glaciers and ice caps, and modeled ice-sheet contributions and is for a sea-level rise of 18–59 cm in 2095 (the magenta bar). This contribution is similar to, but slightly smaller than, the equivalent range from the TAR (the light shaded

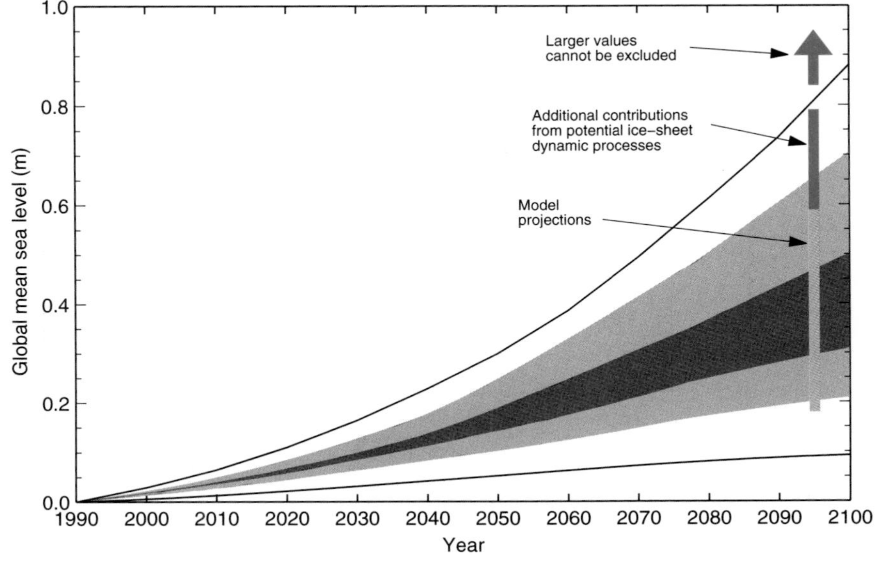

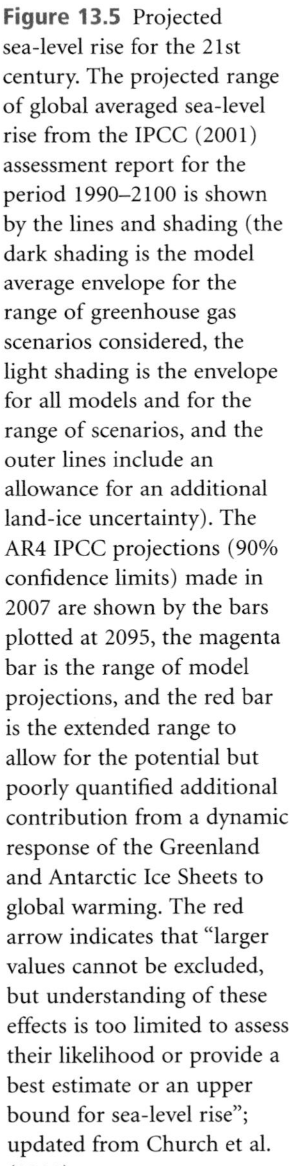
Figure 13.5 Projected sea-level rise for the 21st century. The projected range of global averaged sea-level rise from the IPCC (2001) assessment report for the period 1990–2100 is shown by the lines and shading (the dark shading is the model average envelope for the range of greenhouse gas scenarios considered, the light shading is the envelope for all models and for the range of scenarios, and the outer lines include an allowance for an additional land-ice uncertainty). The AR4 IPCC projections (90% confidence limits) made in 2007 are shown by the bars plotted at 2095, the magenta bar is the range of model projections, and the red bar is the extended range to allow for the potential but poorly quantified additional contribution from a dynamic response of the Greenland and Antarctic Ice Sheets to global warming. The red arrow indicates that "larger values cannot be excluded, but understanding of these effects is too limited to assess their likelihood or provide a best estimate or an upper bound for sea-level rise"; updated from Church et al. (2008).

region). The second part consists of a possible rapid dynamic response of the Greenland and West Antarctic Ice Sheets, which could result in an accelerating contribution to sea-level rise. This additional contribution was not included in the AR4 range of projections noted above because adequate models for quantitative estimates were not available. Recognizing this deficiency, an *ad hoc* allowance for a dynamic response of the ice sheets was made resulting in an additional allowance of 10–20 cm of sea-level rise (the red bar). However, there is currently insufficient understanding of this dynamic response, and IPCC (2007) also clearly stated that a larger contribution cannot be excluded.

When compared in this way, the TAR and AR4 projections of sea-level rise for the 21st century are similar, especially at the upper end of the projected range. However, the wide range of the current projections is a significant hindrance in planning adaptation measures to sea-level rise. It was this wide range that was one of the main motivations that led to the WCRP sea-level workshop underpinning this book.

Another of the main motivations concerned where the observations of sea-level rise lay relative to the IPCC projections. Rahmstorf et al. (2007) demonstrated that sea levels, observed with satellite altimeters from 1993 to 2006 and estimated from coastal sea-level measurements from 1990 to 2001, are tracking close to the upper bound of the TAR projections of 2001, or equivalently as shown above, the upper bound of the AR4 projections of 2007, after the allowance for land-ice uncertainties is included. Recent altimeter measurements indicate sea level is continuing to rise at a rate near the upper bound of the projections since 1993. Updated sea-level reconstructions following the methods of Church and White (2006) also shows the rate of sea-level rise near the upper bound of the projections from 1993. However, the reconstruction also indicates a sea-level fall of about 6 mm between 1991 and 1993 (Figure 13.2), possibly resulting from the explosive volcanic eruption of Mt Pinatubo in the Philippines in 1991. Model simulations indicate that the recovery from the eruption might have resulted in a rate of sea-level rise about 0.6 mm/year higher than what otherwise might have been expected (Chapter 6). The rapid dynamic thinning of the margins of the Greenland and Antarctic Ice Sheets discussed above is likely to be a significant contribution to the observed rapid rate of sea-level rise over the last decade or so. These observations do not necessarily indicate that sea level will continue to track the upper edge of the projections; it may diverge above or below these values.

Recognizing that sea level is currently rising near the upper bound of the IPCC projections, a number of authors (Rahmstorf 2007; Horton et al. 2008; Grinsted et al. 2010) developed relatively simple parameterizations of sea-level rise, based on the relationship between observed historical global sea-level and atmospheric surface temperature records. The processes leading to sea-level rise are not explicitly considered but are represented by a few statistically determined parameters. These models have generally produced higher projections than in the IPCC AR4. While these models are an attempt to overcome the limited understanding of potential future ice-sheet contributions, most of them have been "trained" with observations of 20th-century sea-level rise and some paleo sea-level data, during a period when ocean thermal expansion and glacier melt were the largest contributors to sea-level rise. The Rahmstorf (2007) model has been criticized on statistical grounds and its inability to adequately reproduce observed (Holgate et al. 2007) and modelled (von Storch et al. 2008) sea-level rise.

Sea levels will continue to rise long after 2100. Glaciers and ice caps (outside the polar regions) only contain a limited amount of ice (less than 40 cm of equivalent sea-level rise if they were all to melt), so in the longer term their rate of contribution to sea-level rise will diminish. However, ocean thermal expansion will continue for centuries, even after greenhouse gas concentrations in the atmosphere have been stabilized. The eventual sea-level rise would be dependent on the concentration of greenhouse gases and atmospheric temperatures. Estimates vary but a millennial climate model simulation suggests the order of 0.5 m/°C of global warming (Meehl et al. 2007). The Antarctic and Greenland Ice Sheets are the biggest concern for longer-term sea-level rise. The area and mass of melt from the Greenland Ice Sheet (which contains enough water to raise sea level by about 7 m) is increasing. Model simulations indicate that surface melting of the Greenland Ice Sheet will increase more rapidly than snowfall, leading to a threshold stabilization temperature above which there is an ongoing decay of the Greenland Ice Sheet over millennia. This threshold is estimated as a global-averaged temperature rise of just 3.1 ± 0.8°C (Gregory and Huybrechts 2006) above pre-industrial temperatures. With unmitigated emissions of greenhouse gases, the world is likely to pass this threshold during the 21st century.

In addition, both the Greenland and Antarctic Ice Sheets are showing signs of a dynamic response, potentially leading to a more rapid rate of rise than can occur from surface melting alone (Chapter 7). In an attempt to put bounds on the magnitude of the response, Pfeffer et al. (2008) used kinematic constraints on the potential cryospheric contributions. They estimated that sea-level rise greater than 2 m by 2100 was physically untenable and that a more plausible estimate was about 80 cm, consistent with the upper end of the IPCC estimates and the present rate of rise. This value still requires an acceleration of the ice-sheet contributions. Recent analysis of space-based gravity data (from 2002 to 2009) from the Gravity Recovery and Climate Experiment (GRACE) satellite mission does indicate an accelerating contribution from both Greenland and Antarctica (Velicogna 2009), consistent with the range discussed by Pfeffer et al. (2008). Improved understanding of the processes responsible for ice-sheet changes are urgently required to improve estimates of the rate and timing of 21st-century and longer-term sea-level projections.

Sea-level rise during the 21st century and beyond is not expected to be spatially uniform (as shown in Figures 6.7–6.9). However, there is as yet little agreement in climate models of this regional distribution, indicating systematic uncertainty in the model representation of ocean heat uptake, and transport processes in particular (Chapter 6). This contrasts with quantities relating to surface climate, especially the distribution of projected surface air temperature change, in which current models show reasonable agreement. As discussed above, in addition to this regional distribution of sea-level rise resulting from changes in the coupled atmosphere–ocean climate system, contributions from ice stored on land will also result in changes in the Earth's gravitational field, the shape of the Earth, and hence in the regional distribution of relative sea-level rise (Mitrovica et al. 2009; Bamber et al. 2009).

13.5 Changes in Extreme Events

Rising sea levels have been and will continue to be felt most acutely through extreme events (periods of above average sea level). These include the many cyclones and associated storm surges that have resulted in major loss of life over many years in low-lying nations such as Bangladesh and the 1953 and 1962 storm surges in northwest Europe (Figure 11.1). Among the most recent examples are Hurricane Katrina in New Orleans and Cyclone Nargis in Myanmar.

Analysis of 20th-century sea-level extreme events indicates that coastal flooding events of a given height are now happening more frequently than at the start of the 20th century (Chapter 11). This is primarily a response to changes in mean sea level rather than a change in the frequency or intensity of storm events. Indeed, a change in the frequency of flooding when a given level (such as the height of a storm surge barrier or dyke) is exceeded, can be dramatic. Building on the analysis at a number of locations such as San Francisco (Chapter 1; Bromirski et al. 2003) and Sydney (Chapter 11), it is likely that by 2100 the present-day "one in 100 years" flood will be experienced several times per year at many locations. Also, the most severe sea-level events will be higher and thus have a greater impact during the 21st century.

A warmer climate means that the atmosphere can hold more water vapor and projections are for an increase in precipitation during storms. There is also an expectation of more severe winds in storms, leading to larger storm surges and surface waves, but not necessarily an increase in the frequency of storm events (Chapter 11). Each of these phenomena will impact coastal regions.

13.6 Sea Level and Society

Variations in sea level have always had an impact on society since *Homo sapiens* evolved in eastern Africa about 200 000 years ago. Initially, the oceans restricted human migration. However, lower sea levels more than 50 000 years ago allowed migration through Southeast Asia and Indonesia to New Guinea and Australia, and during the LGM to North America.

It was only well after the end of the last glacial cycle when climate and sea level stabilized that the precursors of our modern society, with coastal cities, trade, and shipping, began to develop. Indeed, much of the development of our coastal civilization and the occupation of mid-ocean islands occurred when the rate of global averaged sea-level change was only a few tenths of a millimeter per year, and in many regions relative sea level was falling slightly as a result of ongoing glacial isostatic adjustment (GIA) (Chapters 4 and 10). This coastal development accelerated over the last century, particularly over recent decades (e.g. Figures 1.1 and 1.2). With coastal development continuing at a rapid pace, society is becoming increasingly vulnerable to sea-level rise and variability, as Hurricane Katrina demonstrated in New Orleans in 2005.

Climate-change mitigation will be essential to avoid the most severe impacts of sea-level rise, as might occur from ongoing ocean thermal expansion or a collapse of the Greenland or West Antarctic Ice Sheets. While sea-level rise over the next few decades is not sensitive to future greenhouse gas emissions (Church et al. 2001), sea-level projections for 2100 for the highest greenhouse-gas emission scenario (A1FI) considered in the IPCC AR4 (IPCC 2007) are about 50% larger than for the lowest-emission scenario (B1). On the longer term, ocean thermal expansion is roughly proportional to the amount of global warming (see above) and Gregory et al. (2004) indicate that the Greenland Ice Sheet is likely to be eliminated by anthropogenic climate change unless much more substantial emission reductions are made than those considered by the IPCC in either the 2001 or 2007 report.

Even with successful mitigation, adaptation to rising sea levels will be essential (Nicholls et al. 2007b). It is critically important to recognize that during the 20th century global averaged sea level moved outside the range of sea level over previous centuries when much coastal development occurred (shown schematically in Figure 13.6). During the 21st century, sea level will move substantially further outside the range experienced by our society to date. Where coasts are subsiding due to natural and human-induced processes, such as in many densely populated deltas and associated cities, this effect will be exacerbated (Ericson et al. 2006;

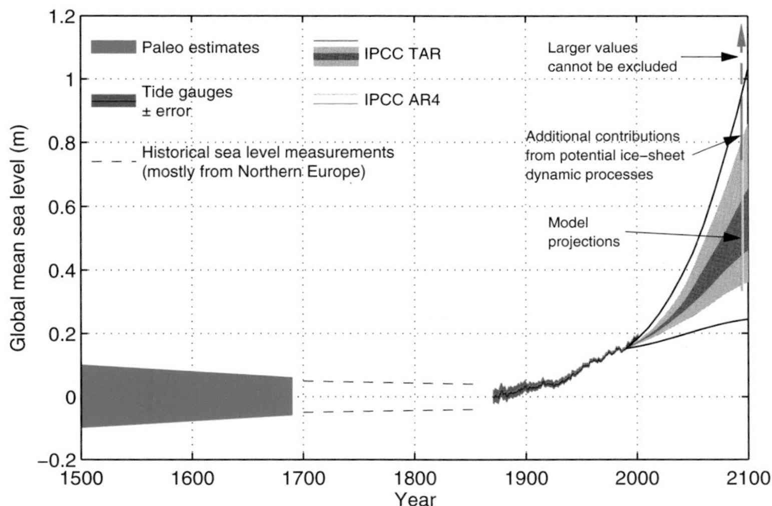

Figure 13.6 Sea levels from 1500 to 2100. The blue band indicates the range of paleo sea-level estimates from Chapters 4 and 5, the dashed lines from 1700 to 1860 indicate the range of sea levels inferred from a limited number of (mostly European) long sea-level records, the black line from 1870 to 2006 is an estimate of global averaged sea level updated from Church and White (2006), and the curves from 1990 to 2100 are the projections from Figure 13.5 (updated from Church et al. 2008).

Syvitski et al. 2009). As shown in Chapter 2, rising sea levels will result in a number of impacts including (1) more frequent coastal inundation/submergence (Figure 1.12), (2) ecosystem change, such as salt-marsh and mangrove loss, (3) increased erosion of beaches (70% of which have been retreating over the past century with less than 10% prograding; Bird 1993) and soft cliffs (Figure 1.3), and (4) salinization of surface and ground-waters. Low-lying islands and deltaic regions are especially vulnerable. Indicative estimates suggest that about 200 million people, and infrastructure worth several trillion dollars, are threatened by coastal floods today; the actual exposure may be larger (Chapters 2 and 3; Nicholls et al. 2007a). This exposure continues to grow at a rapid rate, primarily due to socioeconomic trends, and in the absence of adaptation, risks are growing as sea levels rise. To effectively manage these increasing risks, appropriate information on how and why sea level is changing and will change during the 21st century and beyond is essential.

Appropriate adaptation can significantly reduce the impact of sea-level rise. Planned adaptation will range from retreat from rising sea levels, through planning and zoning of vulnerable coastal regions (Figure 13.7a), accommodation through modification of coastal infrastructure, and the construction of facilities like the cyclone centers used so effectively in Bangladesh (Figure 13.7b), to protection of highly valued coastal regions through highly sophisticated barriers like the Thames Barrage protecting London and the Maeslantkering storm-surge barrier protecting Rotterdam (Figures 13.7c and d). Planned adaptation is more cost-effective and less disruptive than forced adaptation in response to the impacts of extreme events. For example, the estimated cost of strengthening the levees protecting New Orleans, while large, was substantially less than the cost of the damage caused by Hurricane Katrina.

Science has an important role to play in assisting societies to respond to sea-level rise. Improved understanding and narrowing of the uncertainties of projected rise at both the global and regional/local level and its impacts are critical elements in assisting society. The broad range of current projections of global averaged sea-level rise for the 21st century is primarily the result of model uncertainty, and there is currently inadequate understanding of the factors controlling the global-averaged sea-level rise and its regional distribution. Improving monitoring, understanding, and modeling of the global oceans, of glaciers and ice caps, and of the Greenland and Antarctic Ice Sheets, and detecting early signs of any growing ice-sheet contributions, are critical to informing decisions about the required level of greenhouse gas mitigation and for adaptation planning. Quantifying how the Greenland and Antarctic Ice Sheets will contribute to sea-level rise during the 21st century and beyond is currently the largest single uncertainty.

Today, planning for and early warning of extreme events, through improved storm-surge modeling and its operational application, are important aspects of coastal zone management in some regions. This approach goes hand-in-hand with the building and operation of storm surge barriers and cyclone centers (Figure 13.7b). Coastal planning as well as warning systems need to be improved and applied in regions where they do not currently exist and where substantial

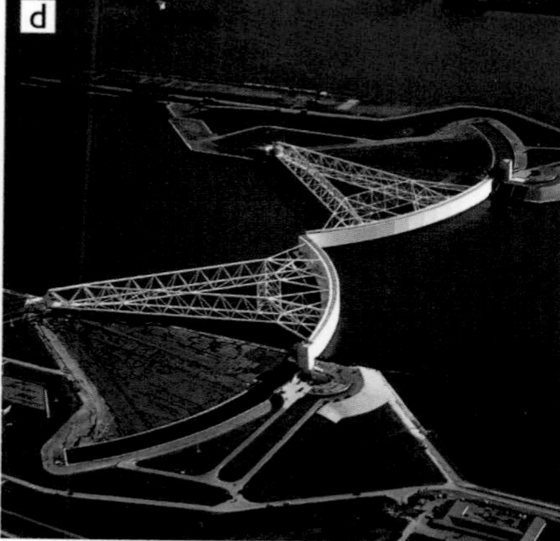

Figure 13.7 Examples of adaptation to sea-level variability and rise. (a) Retreat. Managed realignment at Wallasea, Essex, UK, on the estuary of the River Crouch where a defense line was deliberately breached in 2006 – a planned retreat of the shoreline often reduces protection costs and also allows the development of intertidal habitat as intended here (www.abpmer.net/wallasea/). This is likely to become a widespread response to sea-level rise across Europe. (b) Accommodate. Khajura Cyclone Center, Kalapara, Patuakhali, Bangladesh, on June 3, 2007. The Cyclone Center is also used as a school building. (c) Protect. The Thames Barrage protecting the City of London from storm surges. (d) Protect. The Maeslantkering storm-surge barrier for protecting the City of Rotterdam from storm surges. (a, © Department of Environment Food and Rural Affairs (DEFRA), London; b, © Shehab Uddin/Drik/Red Cross; c, © UK Environment Agency; d, photo credit: Rijkswaterstaat, Dutch Ministry of Transport, Public Works and Water Management.)

loss of life and damage to infrastructure and the environment has occurred or is likely to occur in the future. As a minimum, the coastal planning effort and the warning systems will require significant improvements in bathymetric, near-shore topographic and forecast meteorological information (including surface waves) for storm-surge modeling and detailed inundation mapping.

The understanding of sea-level rise and variability has progressed considerably over the last decade, largely as a result of dramatically improved *in situ* and satellite observational systems and improved models of the climate system. These observing systems need to be completed, improved and sustained, as described in the plans of the Global Climate Observing System, if we are to continue to reduce uncertainties. Another critical component is the development and maintenance of an accurate International Terrestrial Reference Frame (ITRF) based on an ongoing Global Geodetic Observing System (GGOS; Chapter 9). The 2006 Paris WCRP sea-level workshop which led to this book identified the research and observational needs and they were documented in the summary statement from the workshop (see http://wcrp.wmo.int/AP_SeaLevel.html). These needs are documented in each of the chapters of this book with the observational priorities brought together in Chapter 12.

Ensuring that nations have access to the necessary information for adaptation planning is dependent on continued progress in the implementation of these observing systems and improvement of models of the climate system. This requires implementation of needed local observing systems by individual nations, international cooperation, and exchange of data, including an open-data policy with timely and unrestricted access for all. Finally, the scientific information must be translated into practical adaptation plans and this requires the development and strengthening of partnerships between science, different levels of governments, business, and the public.

References

Alley R.B., Clark P.U., Huybrechts P. and Joughin I. (2005) Ice-sheets and sea-level changes. *Science* **310**, 456–60.

Bamber J.L., Riva R.E.M., Vermeersen B.L.A. and LeBrocq A.M. (2009) Reassessment of the potential sea-level rise from a collapse of the West Antarctic Ice Sheet. *Science* **334**, 901–3.

Bird, E.C.F. (1993) *Submerging Coasts: the Effects of a Rising Sea Level on Coastal Environments*. John Wiley and Sons, Chichester.

Bromirski P.D., Flick R.E. and Cayan D.R. (2003) Storminess variability along the California coast: 1858–2000. *Journal of Climate* **16**, 982–93.

Cazenave A., Dominh K., Guinehut S., Berthier E., Llovel W., Ramillien G., Ablain M. and Larnicol G. (2009) Sea level budget over 2003–2008: a reevaluation from GRACE space gravimetry, satellite altimetry and Argo. *Global Planetary Change* **65**, 83–8.

Church J.A. and White N.J. (2006) A 20th century acceleration in global sea-level rise. *Geophysical Research Letters* **33**, L01602.

Church J.A., Gregory J.M, Huybrechts P., Kuhn M., Lambeck K., Nhuan M.T., Qin D. and Woodworth P.L. (2001) Changes in Sea Level. In: *Climate Change 2001: The Scientific Basis*. Contribution of Working Group 1 to the Third Assessment Report of the Intergovernmental Panel on Climate Change (Houghton J.T., Ding Y., Griggs D.J., Noguer M., van der Linden P.J., Dai X. et al., eds), pp. 639–94. Cambridge University Press, Cambridge.

Church J.A., White N.J., Aarup T., Wilson W.S., Woodworth P.L., Domingues C.M. et al. (2008) Understanding global sea levels: past, present and future. *Sustainability Science* **3**, 9–22.

Domingues C.M., Church J.A., White N.J., Gleckler P.J., Wijffels S.E., Barker P.M. and Dunn J.R. (2008) Improved estimates of upper-ocean warming and multi-decadal sea-level rise. *Nature* **453**, 1090–3.

Ericson J.P., Vörösmarty C.J., Dingman S.L., Ward L.G. and Meybeck M. (2006) Effective sea-level rise and deltas: causes of change and human dimension implications. *Global and Planetary Change* **50**, 63–82.

Fairbanks R.G. (1989) A 17,000-year glacio-eustatic sea level record: influence of glacial melting rates on the Younger Dryas event and deep-ocean circulation. *Nature* **342**, 637–42.

Gregory J.M. and Huybrechts P. (2006) Ice-sheet contributions to future sea-level change. *Philosophical Transactions of the Royal Society of London A* **364**, 1709–31.

Gregory J.M., Huybrechts P. and Raper S.C.B. (2004) Threatened loss of the Greenland ice-sheet. *Nature* **428**, 616.

Grinsted A., Moore J.C. and Jevrejeva S. (2010) Reconstructing sea level from paleo and projected temperatures 200 to 2100. *Climate Dynamics* **34**, 461–72.

Holgate S.J. and Woodworth P.L. (2004) Evidence for enhanced coastal sea level rise during the 1990s. *Geophysical Research Letters* **31**, L07305.

Holgate S., Jevrejeva S., Woodworth P. and Brewer S. (2007) Comment on "A semi-empirical approach to projecting future sea-level rise". *Science* **317**, 1866.

Horton R., Herweijer C., Rosenzweig C., Lu J., Gornitz V. and Ruane A.C. (2008) Sea level rise projections for current generation CGCMs based on the semi-empirical method. *Geophysical Research Letters* **35**, L02725.

IPCC (2001) *Climate Change 2001: The Scientific Basis*. Contribution of Working Group I to the Third Assessment Report of the Intergovernmental Panel on Climate Change (Houghton J.T., Ding Y., Griggs D.J., Noguer M., van der Linden P.J., Dai X. et al., eds). Cambridge University Press, Cambridge.

IPCC (2007) *Climate change 2007: The Physical Science Basis*. Contribution of Working Group I to the Fourth Assessment report of the Intergouvernmental Panel on Climate Change (Solomon S., Qin D., Manning M., Chen Z., Marquis M., Averyt K.B. et al., eds). Cambridge University Press, Cambridge.

Ishii M. and Kimoto M. (2009) Reevaluation of historical ocean heat content variations with time-varying XBT and MBT depth bias. *Journal of Oceanography* **65**, 287–99.

Jevrejeva S., Grinsted A., Moore J.C. and Holgate S. (2006) Nonlinear trends and multiyear cycles in sea level records. *Journal of Geophysical Research* **111**, C09012.

Lambeck K., Yokoyama Y. and Purcell A. (2002) Into and out of the last glacial maximum sea level change during oxygen isotope stages 3-2. *Quaternary Science Reviews* **21**, 343–60.

Leuliette E. and Miller L. (2009) Closing the sea level budget with altimetry, Argo and GRACE, *Geophysical Research Letters* **36**, L04608.

Levitus S., Antonov J.I., Boyer T.P., Locarnini R.A., Garcia H.E. and Mishonov A.V. (2009) Global ocean heat content 1955–2007 in light of recently revealed instrumentation problems. *Geophysical Research Letters* **36**, L07608.

Meehl G.A., Stocker T.F., Collins W.D., Friedlingstein P., Gaye A.T., Gregory J.M. et al. (2007) Global climate projections. In: *Climate Change 2007: The Physical Science Basis*. Contribution of Working Group I to the Fourth Assessment Report of the Intergovernmental Panel on Climate Change (Solomon S., Qin D., Manning M., Marquis M., Averyt K., Tignor M.M.B. et al., eds), pp. 747–845. Cambridge University Press, Cambridge.

Mitrovica J.X., Wahr J., Matsuyama I., Paulson A. and Tamisiea M.E. (2006) Reanalysis of ancient eclipse, astronomic and geodetic data: A possible route to resolving the enigma of global sea-level rise. *Earth and Planetary Science Letters* **243**, 390–9.

Mitrovica J.X., Gomez N. and Clark P.U. (2009) The sea-level fingerprint of West Antarctic collapse. *Science* **323**, 753.

Munk W. (2002) Twentieth century sea level: An enigma. *Proceedings of the National Academy of Sciences USA* **99**, 6550–5.

Nicholls R.J., Hanson S., Herweijer C., Patmore N., Hallegatte S., Corfee-Morlot J. et al. (2007a) *Ranking Port Cities with High Exposure and Vulnerability to Climate Extremes – Exposure Estimates*. Environmental Working Paper No. 1. Organisation for Economic Co-operation and Development (OECD), Paris.

Nicholls R.J., Wong P.P., Burkett V.R., Codignotto J.O., Hay J.E., McLean R.F., Ragoonaden S. and Woodroffe C.D. (2007b) Coastal systems and low-lying areas. In: *Climate Change 2007: Impacts, Adaptation and Vulnerability*. Contribution of Working Group II to the Fourth Assessment Report of the Intergovernmental Panel on Climate Change (Parry M.L., Canziani O.F., Palutikof J.P., van der Linden P.J. and Hanson C.E., eds). Cambridge University Press, Cambridge.

Otto-Bliesner B.L., Marshall S.J., Overpeck J.T., Miller G.H., Hu A. and CAPE Last Interglacial Project Members (2006) Simulating Arctic climate warmth and icefield retreat in the last interglaciation. *Science* **311**, 1751–3.

Overpeck J.T., Otto-Bliesner B.L., Miller G.H., Muchs D.R., Alley R.B. and Kiehl J.T. (2006) Paleoclimatic evidence for future ice-sheet instability and rapid sea-level rise. *Science* **311**, 1747–50.

Pfeffer W.T., Harper J.T. and O'Neel S. (2008) Kinematic constraints on glacier contributions to 21st-century sea-level rise. *Science* **321**, 1340–3.

Pritchard H.D., Arthern R.J., Vaughan D.G. and Edwards L.A. (2009) Extensive dynamic thinning on the margins of the Greenland and Antarctic Ice Sheets. *Nature* **461**, 971–5.

Rahmstorf S. (2007) A semi-empirical approach to future sea-level rise. *Science* **315**, 368–70.

Rahmstorf S., Cazenave A., Church J.A., Hansen J.E., Keeling R.F., Parker D.E. and Somerville R.C.J. (2007) Recent climate observations compared to projections. *Science* **316**, 709.

Rohling E.J., Grant K., Hemleben C.H., Siddall M., Hoogakker B.A.A., Bolshaw M. and Kucera M. (2008) High rates of sea-level rise during the last interglacial period. *Nature Geoscience* **1**, 38–42.

Rohling E.J., Grant K., Bolshaw M., Roberts A.P., Siddall M., Hemleben C.H. and Kucera M. (2009) Antarctic temperature and global sea level closely coupled over the past five glacial cycles. *Nature Geoscience* **2**, 500–4.

Syvitski J.P.M., Kettner A.J., Overeem I., Hutton E.W.H., Hannon M.T., Brakenridge G.R. et al. (2009) Sinking deltas due to human activities. *Nature Geoscience* **2**(10), 681–6.

Tol R.S.J., Klein R.J.T. and Nicholls R. J. (2008) Towards successful adaptation to sea-level rise along Europe's coasts. *Journal of Coastal Research* **24**(2), 432–50.

Velicogna I. (2009) Increasing rates of ice mass loss from the Greenland and Antarctic ice sheets revealed by GRACE. *Geophysical Research Letters* **36**, L19503.

von Storch H., Zorita E. and Gonzáles-Rouco J.F. (2008) Relationship between global mean sea-level and global mean temperature and heat-flux in a climate simulation of the past millennium. *Ocean Dynamics* **58**, 227–36.

Willis J.K., Chambers D.T. and Nerem R.S. (2008) Assessing the globally averaged sea level budget on seasonal to interannual time scales. *Journal of Geophysical Research* **113**, C06015.

Index

Note: page numbers in *italics* refer to figures; those in **bold** refer to tables; n indicates information in a note.

Understanding Sea-Level Rise and Variability, 1st edition. Edited by John A. Church, Philip L. Woodworth, Thorkild Aarup & W. Stanley Wilson.